MW00814829

Studies in Systems, Decision and Control

Volume 161

Series editor

Janusz Kacprzyk, Polish Academy of Sciences, Warsaw, Poland
e-mail: kacprzyk@ibspan.waw.pl

The series "Studies in Systems, Decision and Control" (SSDC) covers both new developments and advances, as well as the state of the art, in the various areas of broadly perceived systems, decision making and control-quickly, up to date and with a high quality. The intent is to cover the theory, applications, and perspectives on the state of the art and future developments relevant to systems, decision making, control, complex processes and related areas, as embedded in the fields of engineering, computer science, physics, economics, social and life sciences, as well as the paradigms and methodologies behind them. The series contains monographs, textbooks, lecture notes and edited volumes in systems, decision making and control spanning the areas of Cyber-Physical Systems, Autonomous Systems, Sensor Networks, Control Systems, Energy Systems, Automotive Systems, Biological Systems, Vehicular Networking and Connected Vehicles, Aerospace Systems, Automation, Manufacturing, Smart Grids, Nonlinear Systems, Power Systems, Robotics, Social Systems, Economic Systems and other. Of particular value to both the contributors and the readership are the short publication timeframe and the world-wide distribution and exposure which enable both a wide and rapid dissemination of research output.

More information about this series at http://www.springer.com/series/13304

Dawid Taler

Numerical Modelling and Experimental Testing of Heat Exchangers

 Springer

Dawid Taler
Faculty of Environmental Engineering
Cracow University of Technology
Cracow
Poland

ISSN 2198-4182   ISSN 2198-4190   (electronic)
Studies in Systems, Decision and Control
ISBN 978-3-319-91127-4   ISBN 978-3-319-91128-1   (eBook)
https://doi.org/10.1007/978-3-319-91128-1

Library of Congress Control Number: 2018940626

Printed on acid-free paper

This Springer imprint is published by the registered company Springer International Publishing AG part of Springer Nature
The registered company address is: Gewerbestrasse 11, 6330 Cham, Switzerland

Symbols

a	Thermal diffusivity, $a = \lambda / (c\rho)$, m^2/s
$\mathbf{a}$	Vector of sought parameters
$\tilde{\mathbf{a}}$	Vector of coefficients found using the least squares method
A	Surface area, m^2
A_{123}	Surface area of a triangle with vertices in points 1, 2 and 3, m^2
a, b	Side length, m
$\tilde{a}_i$	i-th coefficient determined by means of the least squares method
A_m	Tube mean surface area, $A_m = (A_{wrg} + A_{zrg}) / 2$, m^2
$A_{\min}$	Surface area of the narrowest cross section with the maximum flow velocity, m^2
$A_{m\dot{z}}$	Tube surface area in between the fins, m^2
A_{owal}	Surface area of the oval opening in the fin, m^2
a_w, a_z	Semiminor axis of the oval or elliptical tube inner and outer surface, respectively, m
A_{wr}	Surface area of the fluid flow cross section in a tube or channel, m^2
A_{wrg}, A_{zrg}	Surface area of the bare tube inner and outer surface, respectively, m^2
$A_{\dot{z}}$	Surface area of the fin opening, m^2
b	Half of the flat slot width, m
Bi	Biot number, $\mathrm{Bi} = \alpha r_w/\lambda$
b_w, b_z	Semimajor axis of the oval or elliptical tube inner and outer surface, respectively, m
c	Specific heat, J/(kg K)
$C = mc$	Thermal capacity, J/K
$\mathbf{C}_{\tilde{\mathbf{a}}}, \mathbf{C}_{\tilde{\eta}}$	Covariance matrices of coefficients and functions determined by means of the least squares method

$\dot{C}_c$, $\dot{C}_h$	Heat capacity rate of the cold and hot fluid, respectively, W/K
$\dot{C}_{min}$	Minimum heat capacity rate of the fluid in the heat exchanger, W/K
c_p, c_v	Specific heat at constant pressure and constant volume, respectively, J/(kg K)
$\mathbf{C_x}$, $\mathbf{C_y}$, $\mathbf{C_z}$	Covariance matrices
D	Diameter, m
d_1, d_2	Coordinates of fixing points of thermocouples in the heat flux meter, m
d_h	Equivalent hydraulic diameter, m
$d_{h,a}$	Air-side hydraulic diameter, m
$d_{h,w}$	Water-side hydraulic diameter, m
d_w	Inner diameter of the tube, m
d_z	Outer diameter of the tube, m
e	Specific energy of the medium, J/kg
$e(t)$	Difference between set and measured values
E	Energy, J
Eu	Euler number, Eu $= \Delta p / \left(\rho w_\infty^2 \right)$
f	Frequency, Hz
f	Fanning friction factor, $f = \xi / 4$
F	Correction factor taking account of the reduction in the heat exchanger heat flow rate compared to a counter-current exchanger
$\mathbf{F}$	Force vector
f_i	Temperature measured in time instant t_i, K or °C
F_{ij}	Angle factor of the radiative heat transfer
Fr	Froude number, Fr $= w_\infty^2 / (g L)$
G	Gravitational acceleration, $g = 9.81$ m/s^2
$\mathbf{G}$	Gravitational acceleration vector
$\mathbf{G_y}$	Weight matrix in the least squares method, $\mathbf{G_y} = \mathbf{C_y^{-1}}$
H	Height, m
H_{ch}	Radiator active height, m
i	Specific enthalpy, J/kg
$\mathbf{i_r}$, $\mathbf{i_\theta}$, $\mathbf{i_z}$	Axis versors in a cylindrical system of coordinates
$\mathbf{i_r}$, $\mathbf{i_\theta}$, $\mathbf{i_z}$	Axis versors in a spherical system of coordinates
I	Radiation intensity, W/m^2
$\mathbf{i}$, $\mathbf{j}$, $\mathbf{k}$	Axis versors in a Cartesian system of coordinates
$I_0(x)$, $K_0(x)$	Modified first- and second-kind zero-order Bessel functions, respectively
j	Colburn parameter, $j = $ Nu/(RePr$^{1/3}$)
$\mathbf{J}$	Jacobian matrix
$J_0(x)$, $J_1(x)$	First-kind zero- and first-order Bessel functions, respectively

k	Overall heat transfer coefficient, $W/(m^2 \, K)$
k_d	Measuring instrument accuracy class
K_d	PID controller derivative gain
K_i	PID controller integral gain
K_p	PID controller proportional gain
l	Prandtl mixing length, m
L	Fin height, m
L_c	Equivalent fin height, $L_c = L + \delta_f/2$, m
L_{ch}	Radiator active length, m
L_H	Hydraulic entrance length, m
L_t	Shaft work, J
$\dot{L}_t$	Rate of doing work (power), W
L_T	Thermal entrance length, m
L_x	Tube length, m
$\mathbf{M}$	Angular momentum vector
m	Mass, kg
m	Fin parameter, $m = \sqrt{2\alpha/(\lambda\delta_f)}$, 1/m
$\dot{m}$	Mass flow rate, kg/s
m_{cv}	Substance mass in the control volume, kg
$\dot{m}_g, \dot{m}_l$	Gas and liquid mass flow rate, respectively, kg/s
m_f	Fin mass, kg
n	Fan revolution number, rpm
$\mathbf{n}$	Unit vector normal to the surface and directed outside the area
N	Number of control volumes (finite volumes) on the tube length
N	Number of tubes
N_g	Number of heat transfer units for the gas
N_i	Number of heat transfer units for the liquid
NTU	Number of heat transfer units, $NTU = kA/\dot{C}_{\min}$
n_f	Number of fins on a single tube with length L_x
p	Pressure, Pa
P	Perimeter, m
P	Probability, confidence level
p_1, p_2	Transverse and longitudinal pitch of the exchanger tubes, respectively
P_l, P_t	Longitudinal and transverse pitch of the exchanger tubes, respectively
Pr	Prandtl number, $Pr = c_p \, \mu/\lambda$
Pr_t	Turbulent Prandtl number, $Pr_t = \varepsilon_\tau/\varepsilon_q$
Pe	Peclet number, $Pe = Re \, Pr = w \, d_w/a$
Q	Heat, J
$\dot{q}$	Heat flux, W/m^2
$\dot{q}^*$	Dimensionless heat flux, $\dot{q}^* = \dot{q}/\dot{q}_w$

$\dot{Q}$	Heat flow rate, W
$\dot{Q}_c$	Heat flow rate calculated for the cold fluid, W
$\dot{Q}_h$	Heat flow rate calculated for the hot fluid, W
$\dot{Q}_m$	Mean heat flow rate in the exchanger, $\dot{Q}_m = (\dot{Q}_c + \dot{Q}_h)/2$, W
$\mathbf{\dot{q}_r}$	Radiative heat flux vector
$\dot{q}_v$	Heat flow rate generated inside a body per volume, W/m^3
$\dot{q}_w$	Heat flux at the inner surface, W/m^2
$\dot{q}_z$	Heat flux at the outer surface, W/m^2
r	Radius or radial coordinate, m
r	Number of freedom degrees
r	Correlation coefficient
r	Rank of matrices
$\mathbf{r}$	Point radius vector
R	Dimensionless radius, $R = r/r_w$
r^2	Determination coefficient
Re	Reynolds number, Re $= w\,d/\nu$
Re$_{\delta_2}$	Reynolds number based on the momentum thickness, Re$_{\delta_2} = w_{x\infty}\delta_2/\nu$
r_{in}, r_o	Radius of a circular fin base and tip, respectively
$r_w,$	Inner radius of the tube, m
r_w^+	Dimensionless radius, $r_w^+ = r_w\,u_\tau/\nu$
r_z	Outer radius of the tube, m
$S_{\min}$	Minimum value of the sum of squares
s	Specific entropy, J/(kg K)
s	Curvilinear coordinate, m
s	Fin pitch, m
s_1, s_2	Tube pitch perpendicular and parallel to the flow direction, tube pitch, m
s_t	Root-mean-square deviation
s_x	Root-mean-square error in physical quantities measured directly
t	Time, s
T	Temperature, K or °C
T_b	Fin base temperature, K or °C
T_f	Temperature of the fluid, K or °C
T_0	Initial temperature distribution, K or °C
$T_{l,i}, T_{l,i+1}$	Liquid temperature at the i-th finite volume inlet and outlet, K or °C, respectively
$\bar{T}_l$	Liquid mean temperature on the length of the i-th finite volume, K or °C
T_m	Mean temperature, K or °C

T_{mg}	Gas mean temperature on the thickness of a single tube row, K or °C
T_{wm}	Water temperature in the car cooler downstream the first pass, K or °C
T'_g, T'_l	Gas and liquid temperature at the heat exchanger inlet, respectively, K or °C
T'_{am}, T''_{am}	Air mean temperature at the heat exchanger inlet and outlet, respectively, K or °C
$t_{n-1,\,\alpha}$	**t**-Student's distribution quantile
T'_w, T''_w	Water temperature at the heat exchanger inlet and outlet, respectively, K or °C
$T''_{w,\,set}$	Water temperature set at the heat exchanger outlet, K or °C
$T''_{w,\,meas}$	Water temperature measured at the heat exchanger outlet, K or °C
T'''_{um}	Air mean temperature in the car cooler downstream the first pass, K or °C
u	Internal specific energy, J/kg
U	Velocity component in the direction of axis x, m/s
U	Ratio between the radii of the fin tip and base, $u = r_o/r_{in}$
$u(\tau)$	Controller output signal
u^+, v^+	Dimensionless velocity components in the direction of axis x and axis y, respectively
U	Perimeter, m
U_m	Mean perimeter, $U_m = (U_w + U_z)\,/\,2$, m
u_τ	Friction velocity, m/s
v	Specific volume, m³/kg
v	Velocity component in the direction of axis y, m/s
V	Volume, m³
$\dot{V}$	Volume flow rate, m³/s
w	Velocity, velocity component in the direction of axis z, m/s
w_0	Mean air flow velocity in a rectangular channel upstream the heat exchanger, m/s
$\mathbf{w_1}$	Velocity vector at the control area inlet
$\mathbf{w_2}$	Velocity vector at the control area outlet
w_i	i-th weight coefficient
w_m	Mean air flow velocity in a channel with a circular cross section, m/s
w_{max}	Maximum fluid flow velocity in the channel cross section or in the heat exchanger, m/s
w_x, w_y, w_z	Velocity vector components in a Cartesian system of coordinates, m/s

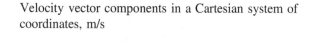

$\bar{w}_x,\ \bar{w}_y,\ \bar{w}_z$	Time-averaged velocity vector components in a Cartesian system of coordinates, m/s
$w'_x,\ w'_y,\ w'_z$	Random fluctuations in velocity components in a Cartesian system of coordinates, m/s
$w^*_{a,i},\ w^*_{w,i}$	Dimensionless weight coefficients in the weighted least squares method
w_c	Velocity vector of the centre of gravity
$x,\ y,\ z$	Cartesian coordinates, m
$\mathbf{x}$	Vector of parameters determined using the Levenberg–Marquardt method
$\mathbf{x}$	Vector of coordinates of measuring points in the least squares method
$x^+,\ y^+$	Dimensionless coordinates
$x_1,\ldots,x_n$	Physical quantities measured directly
$x_a,\ y_a,\ z_a$	Coordinates of the midpoint of side 1-2, m
$x_b,\ y_b,\ z_b$	Coordinates of the midpoint of side 2-3, m
$x_c,\ y_c,\ z_c$	Coordinates of the midpoint of side 3-1, m
$x_o,\ y_o$	Coordinates of the triangle centre of gravity, m
$(x_1, y_1),\ (x_2, y_2),\ (x_3, y_3)$	Coordinates of the triangle vertices, m
$\bar{x}_n$	Arithmetic mean of a series of measurement results
X^*	Dimensionless coordinate, $X^* = \frac{1}{\text{Pe}}\frac{x}{b}$
y_i	Measured quantity
z	Physical quantity measured indirectly
z	Elevation, m

Greek Symbols

α	Heat transfer coefficient, W/(m^2 K)
α	Absorptivity
α_m	Mean heat transfer coefficient on a set segment, W/(m^2 K)
α_{zr}	Weighted heat transfer coefficient for a finned surface, W/(m^2 K)
β	Linear or volume thermal expansion coefficient, 1/K
Γ	Boundary of the domain
δ_2	Momentum thickness being a measure of the reduction in the momentum flux due to the boundary layer, m
δ_f	Fin thickness, m
δ_r	Tube wall thickness, m
Δp	Pressure drop on the duct set length, Pa
Δt	Time step, s
ΔT	Temperature difference, K
ΔT_m	Logarithmic mean temperature difference between the hot and the cold fluid, K

$\Delta x, \Delta y, \Delta z$	Spatial step in the direction of axis x, y and z, respectively
Δx_i	Uncertainty of a physical quantity measured directly
Δz_i^c	Total uncertainty of an indirect measurement of physical quantity z_i
ε	Heat exchanger effectiveness
ε	Emissivity
ε_i	Random measurement error
ε_w	Wall emissivity
ε_q	Turbulent thermal diffusivity (eddy diffusivity for heat transfer), m^2/s
ε_τ	Momentum turbulent diffusivity (eddy diffusivity for momentum transfer), m^2/s
ζ	Pressure loss coefficient (coefficient of local resistance)
η	Flat slot dimensionless coordinate, $\eta = y/b$
$\eta(\mathbf{x}, \mathbf{a})$	Measurement data approximation function
$\tilde{\eta}$	Approximation function determined by means of the least squares method
η_f	Fin efficiency
θ	Dimensionless temperature
θ	Surplus of the fin temperature to the temperature of the fin base, K
κ	Specific-heat ratio, $\kappa = c_p/c_v$
λ	Thermal conductivity, W/(m K)
λ	Wavelength, m
λ_x, λ_y	Thermal conductivity in the direction of x-axis and y-axis, respectively, W/(m K)
μ	Dynamic viscosity, Pa s
μ_n	n-th root of a characteristic equation
ν	Kinematic viscosity, m^2/s
ξ	Darcy–Weisbach friction factor
ρ	Density, kg/m^3
σ	Stefan–Boltzmann constant, $\sigma = 5.67 \times 10^{-8}$ W/(m^2 K^4)
σ	Mean standard deviation
σ	Normal stress
$\boldsymbol{\sigma}$	Stress tensor
$\sigma_{a,i}$	Standard deviation of the air mean temperatures measured downstream the heat exchanger, K
σ_i^2	Variance characterizing deviation of measurement results y_i from expected value $E(y_i)$
$\sigma_{w,i}$	Standard deviation of the water temperatures measured at the heat exchanger outlet, K
τ	Shear stress, Pa
τ	Time constant, s
τ_d	Differentiation time in the PID controller equation

τ_g, τ_l	Gas and liquid medium time constant, respectively, s
τ_i	Integration time in the PID controller equation
τ	Turbulent shear stress, Pa
φ	Angular coordinate, rad
φ	Angle of the channel inclination to the horizontal plane, rad
$\dot{\phi}$	Dissipated energy rate related to a unit of volume, W/m^3
Ψ	Waterwalls heat efficiency
ω	Angular velocity, rad/s
Ω	Domain under analysis

Subscripts

i	Node number
m	Mean value
n	Old time step, $t = t_n$
$n+1$	New time step, $t = t_{n+1} = t_n + \Delta t$
p	Pressure
w	Wall or water
∞	Free jet
$\partial z / \partial x_i$	Sensitivity coefficient of quantity z measured indirectly with respect to quantity x_i measured directly

Superscripts

$'$	Random fluctuation
+	Dimensionless quantity
C	Calculated quantity
CFD	Quantity determined by means of CFD (Computational Fluid Dynamics) modelling
exp	Quantity determined experimentally
m	Measured quantity

Overscores

$-$	Mean value
$\sim$	Coefficient or function values determined by means of the least squares method

Chapter 1
Introduction

This monograph presents numerical modeling and experimental testing of heat exchangers with a particular focus on finned-tube cross-flow devices. Such exchangers are widely used in many industries, e.g., in power plants, oil refineries, chemical plants, as well as in gas and heating engineering and air-conditioning. Their characteristic feature is that liquid flows inside the tubes and gas flows in the direction perpendicular to the tube axis.

In thermal power plants, finned exchangers are used for hot water cooling in what is referred to as dry cooling systems. In closed-loop cooling systems of the turbine condensers, the cooling water is first heated in the turbine condenser and then cooled in an air-cooled heat exchanger located in a cooling stack or a forced-draught cooling tower. The heat exchangers can be made of circular or elliptical tubes with individual round or rectangular fins or continuous plate fins (lamellae).

Finned exchangers with finned circular tubes are used as economizers, evaporators or steam superheaters in waste-heat boilers downstream gas turbines, where the flue gas maximum temperature at the boiler inlet is at the level of 600 °C. Finned water heaters are also used in large steam boilers as economizers in gas- or oil-fired facilities.

Finned cross-flow heat exchangers are commonly used in chemical and petrochemical plants and in gas transmission installations to cool liquids or gases using air.

Finned-tube flow water heaters are popular in heating engineering to prepare hot tap water. Finned-tube air heaters, where the air is heated by hot water flowing through tubes, are another example application. They are used to enable a fast rise in the temperature of air in rooms used periodically with insufficient thermal insulation of the walls.

Finned evaporators and condensers are common in air-conditioning installations and heat pumps.

This monograph presents the entire range of issues related to the single-phase convective heat transfer in heat exchangers, including design and operation calculations, as well as experimental testing of the devices. Much attention is devoted

© Springer International Publishing AG, part of Springer Nature 2019
D. Taler, *Numerical Modelling and Experimental Testing of Heat Exchangers,*
Studies in Systems, Decision and Control 161,
https://doi.org/10.1007/978-3-319-91128-1_1

to the experimental determination of correlations for the gas- and the liquid-side Nusselt number that enable calculation of mean heat transfer coefficients in the whole heat exchanger. A detailed discussion of the procedures for the determination of the uncertainty of physical quantities or parameters determined indirectly, including parameters found using the least squares method based on a high number of measuring series, is presented.

In Chap. 2, the mass, momentum, and energy conservation equations are derived using three different formulations. The equations are first derived treating the substance-filled control area as an object with a concentrated mass, momentum and energy. Then the mass, momentum, and energy conservation equations are presented in the integral and differential formulation. The equations are discussed in detail, assuming that the flow and transfer of heat in the duct is transient and one-dimensional.

Chapter 3 presents an analysis of the laminar fluid flow in tubes and flat slots. The heat transfer process is analyzed for the laminar fluid flow in channels at constant heat flux or temperature set on the inner surface of the duct. Formulae are first determined for the temperature distribution of the fluid flowing in the tube, assuming that the flow is laminar and fully developed at constant values of the heat flux or temperature on the tube inner surface. The Nusselt number values are found, too. An analysis is also conducted of the heat transfer on the tube inlet section for constant heat flux or temperature values set on the tube inner surface and assuming the fluid plug flow. Then, using identical boundary conditions on the tube or the flat slot surface, formulae are derived for the fluid temperature distribution assuming that the flow is laminar and fully developed hydraulically. Due to the rather slow convergence of series in the expressions used for the calculation of the Nusselt number close to the channel inlet and derived based on the distributions of temperature found using the variables separation method, formulae are derived which are valid for very small distances from the inlet. Formulae are given for the calculation of the Nusselt numbers on the inlet section that, if appropriately high accuracy is kept, can be used in engineering practice. The work also presents formulae for the calculation of the local and mean Nusselt numbers in the tube inlet section at constant values of the heat flux or temperature at the tube inner surface, assuming that the fluid flow is still developing both hydraulically and thermally.

Chapter 4 is devoted to issues related to the turbulent fluid flow, including the Reynolds equations, the fluid turbulent viscosity and diffusivity, the mixing model, and the universal velocity profiles put forward by Prandtl, von Kármán, Deissler, Reichardt and van Driest.

Chapter 5 offers a brief description of three basic analogies between the heat and the momentum transfer: the Reynolds and Chilton-Colburn analogy, the Prandtl analogy and the von Kármán analogy.

In Chap. 6, the analytical method and the finite element method are used to determine the velocity distribution in the tube cross-section and the friction factor coefficient for the turbulent fluid flow in a tube. A survey of the formulae applied to calculate the friction factor for the turbulent fluid flow in tubes with a smooth and a rough surface is performed. Analytical formulae are derived for the fluid temperature distribution and the Nusselt number in the tube at a set heat flux on the tube inner surface. The fluid temperature and Nusselt number are also determined numerically. A two-point boundary value problem for two ordinary differential equations is solved using the finite difference method, finding the distributions of temperature and the heat flux as functions of the radius. After determining the tube inner surface temperature and mean-mass fluid temperature for given Reynolds and Prandtl numbers, the Nusselt number is found. The chapter includes examples of velocity and temperature distributions in the tube cross-section for a few selected values of the Reynolds number. The Nusselt number values are determined as a function of the Reynolds and the Prandtl numbers. Several $Nu = f(Re, Pr)$ functions are then applied to approximate the results presented in a table format. Three groups of correlations are defined, corresponding to the following ranges of the Reynolds and the Prandtl number:

- $0.1 \leq Pr \leq 1000, 3000 \leq Re \leq 10^6$,
- $0.1 \leq Pr \leq 1000, 2300 \leq Re \leq 10^6$,
- $0.0001 \leq Pr \leq 0.1, 3000 \leq Re \leq 10^6$.

The second correlation group covers the transition range, where the flow changes from laminar to turbulent. The third group concerns liquid metals, which have excited an increased interest in recent years due to the development of fast neutron reactors. The reactors are cooled using liquid metals.

Chapter 7 focuses on the basics of the heat exchanger calculations. In the beginning, the chapter presents simplified equations of the mass, momentum and energy conservation used in the process. Then, determination of the temperature distribution in the wall separating the two fluids is discussed. Much attention is given to determination of the temperature field in walls with a complex cross-section shape using the control volume based finite element method (CVFEM). The differential equation describing time-dependent temperature changes in the control area node are derived using three different methods, the last two of which can be applied not only to finite triangular elements but rectangular or polygonal ones as well. The overall heat transfer coefficient formulae are derived for both bare and finned tubes with different shapes of the cross-section. In the case of tubes with individual fins or continuous ones (lamellae), the heat transfer through the outer surface is calculated introducing the weighted heat transfer coefficient, which can be found provided that the fin efficiency is known. The chapter presents exact analytical relations needed to calculate the efficiency of straight and round fins. In the case of rectangular or hexagonal fins fixed on elliptical or oval tubes, the fin temperature distribution and efficiency can be determined using the CVFEM mentioned above. A detailed description is given of the process of determining the efficiency of a rectangular fin fixed on an oval tube using the CVFEM. The results obtained from the analysis of a fin divided into a small number of finite elements are compared to those produced by the classical finite element method (FEM) based on the Galerkin method, with the fin division into a large number of finite elements.

Chapter 8 presents basic methods of the heat exchanger calculation applied in the engineering practice, i.e., the logarithmic mean temperature difference method (LMTD) and the ε–NTU (Effectiveness-Number of Transfer Units) method.

Chapter 9 includes mathematical models of selected heat exchangers: the tube-in-tube co- and counter-current exchangers, the single-row cross-flow tube exchanger, and the plate cross-flow exchanger.

Chapter 10 deals with modeling tube cross-flow heat exchangers. First, differential equations that describe the water, wall, and air temperature distributions are derived. Two methods of the tube cross-flow heat exchanger modeling are presented next. In the first, the gas mean temperature on the thickness of a single tube row is calculated as the arithmetic mean of the gas temperatures up- and downstream the tube row, whereas in the second—a mean integral temperature across the single row thickness is used. An analytical relationship gives the gas temperature distribution in the direction of the flow. Mathematical models are developed of a two-pass radiator with two rows of tubes. The heat flow rate between the fluids obtained using the two numerical models are compared to the analytical model results. Better agreement with the analytical mathematical model results can be obtained using the numerical model based on integral averaging of the air temperature on the thickness of a single tube row.

The second part of the monograph (Chaps. 11–18) presents a discussion of issues related to the heat exchanger experimental testing.

Chapter 11 deals with the indirect measurement uncertainty assessment.

It presents an extensive discussion of the calculation of the uncertainty of parameters found using the least squares method. The least squares method linear and nonlinear problems are analyzed. Considerable attention is given to the linear and nonlinear weighted least squares method, including the nonlinear least squares method with equality constraints. The chapter presents determination of the confidence interval limits for known and unknown mean standard deviations of physical quantities which are measured directly.

Numerous examples illustrate the indirect determination of parameters. The methods of assessment of the uncertainty of indirect measurements presented in Chap. 11 are used further on in the monograph to assess the uncertainty of unknown coefficients that appear in correlations for the Nusselt number on the air and the water side. The correlations are determined using the nonlinear least squares method based on measured mass flow rates and temperatures of the fluids measured at the exchanger inlet and outlet.

Chapter 12 includes necessary information needed to perform correct measurements of the heat flow rate transferred from the hot to the cold fluid and measurements of the heat transfer coefficient. Particular attention is devoted to the measurement of the fluid mean velocity and mass averaged temperature in the channel cross-section. The measurement of these two quantities is of particular importance in the measurement of the mean velocity and temperature of the gas.

Chapter 13 deals with the experimental determination of the local and the mean heat transfer coefficient on the inner surface of a single tube and with finding ways of determining correlations for the Nusselt number calculation depending on the Reynolds and the Prandtl number.

First, methods of measuring the local and the mean heat transfer coefficient on the tube inner surface at a constant heat flux set on the tube outer surface are discussed. Measuring the tube inner surface temperature in different cross-sections on the tube length and determining the fluid mass average temperature from the energy conservation equation, it is easy to find the variation of the heat transfer coefficient on the tube length, and—after that—the coefficient mean value. This method of the heat transfer coefficient determination is widespread in laboratory testing due to the easy access to the tube outer surface. Experimental testing results are usually generalized by establishing the relationship between the Nusselt and the Reynolds and the Prandtl numbers. It is demonstrated how, in the general case, dimensionless numbers can be found writing the boundary conditions or the momentum and energy conservation equations in a dimensionless form. Dimensional analysis is an alternative way of finding dimensionless numbers. Based on the matrix of dimensions and the Buckingham theorem, it is possible to determine dimensionless numbers that are taken into account in the Nusselt number formula. An example of determining dimensionless formulae for the pressure drop and the Nusselt number for the forced fluid flow in the tube is presented.

Chapter 14 presents the Wilson method, which is used to determine mean heat transfer coefficients in heat exchangers. The method is based on linear regression and finds wide application in the heat exchanger testing. In order to find the formula for the Nusselt number for one of two fluids, this fluid velocity and temperature should vary, whereas on the side of the other medium the heat transfer coefficient should remain constant during the tests. The condition that the heat transfer coefficient is constant on one side of the wall is difficult to meet in the case of the liquid. The variations in thermal and flow conditions on one side have an impact on changes in the wall temperature, which in turn involves changes in the liquid viscosity and the heat transfer coefficient on the other side, despite the constancy of the liquid flow velocity. The Wilson method application is illustrated by determining the air and the water side Nusselt number in a car radiator. The calculation results confirm that the correlations on the water side can be found with higher accuracy compared to the air side because during tests it is easier to keep the heat transfer coefficient constant on the air side.

Determination of correlations for the Nusselt number on the air side, assuming that the Nusselt number correlation on the water side is known, is the subject matter of Chap. 15. When different literature correlations for the Nusselt number are used to calculate the heat transfer coefficient on the water side, then air side correlations are obtained, which differ from each other considerably. In the air side correlation, two unknown coefficients are assumed, which are determined based on experimental testing using the least squares method. The calculations performed for a car radiator indicate that in order to be determined correctly, the correlations for the Nusselt number on the water and the air side should be found in parallel based on experimental testing results.

In Chap. 16 it is proposed that the unknown coefficients appearing in the correlation for the Nusselt number on the water and the air side should be found simultaneously using the weighted least squares method. The correlations are determined for a car radiator based on experimental testing assuming a different number of the unknown coefficients: three, four or five. The differences between the

Nusselt number correlations are smaller on the air side. Like in Chap. 15, 95% confidence intervals are determined for the coefficients.

Chapter 17 is focused on determination of correlations for the heat transfer coefficient on the air side using CFD simulations. Air-side thermal and flow processes are modeled. A method is presented for determining the mean heat transfer coefficient on the air side based on the results of a computer simulation. It is shown that CFD simulations enable determination of the correlation for the air side Nusselt number.

Chapter 18 presents the application of the car radiator mathematical model with the empirical correlations for the air and the water side Nusselt number for automatic control of the temperature of water at the radiator outlet by changing the number of fan revolutions. A comparison is made between the operation of the control system based on the radiator mathematical model and the digital PID controller.

The final remarks and directions of further research on tube cross-flow heat exchangers are included in Chap. 19.

Then come appendices to selected chapters.

The sources referred to and cited herein are included in the list of references at the end of the book.

Part I
Heat Transfer Theory

Chapter 2
Mass, Momentum and Energy Conservation Equations

Heat can be transferred by means of the following three fundamental processes:

- conduction,
- convection,
- radiation.

In practice, these three kinds of the heat transfer co-occur. Heat conduction plays the most significant role in solid bodies, whereas in gases and liquids convection and radiation prevail. Radiation differs from conduction and convection in that it can also occur in a vacuum. Like thermodynamics, the heat transfer phenomenon is based on mass, momentum, and energy conservation equations. The first law of thermodynamics is the law of energy conservation or, simply speaking, the energy balance equation.

The heat transfer phenomenon differs from thermodynamics. Classical thermodynamics deals with systems in an equilibrium state, which makes it possible for example to calculate a change in the body internal energy during the transition from one equilibrium state to another. According to the methodology adopted in thermodynamics, it is possible for example to calculate the increment in internal energy ΔU of a steel ball with the mass of 0.5 kg after it is heated from temperature $T_1 = 20\,°C$ to temperature $T_2 = 700\,°C$. Assuming that the sphere specific heat is $c = 480$ J/(kgK), the internal energy increment ΔU is:

$$\Delta U = mc(T_2 - T_1) = 0.5 \cdot 480 \cdot (700 - 20) = 163,200\,\text{J}$$

However, from the point of view of classical thermodynamics, nothing can be said about the duration of the process, or about time-dependent changes in the temperatures of the body individual points, e.g., the temperature differences between the sphere surface and center. Nor can the arising thermal stresses be defined. Heat transfer is based on mass, momentum and energy balance equations [9, 10, 16, 21, 24, 45, 47, 79, 109, 131, 132, 137, 140, 167, 187, 201, 218, 225, 226, 327, 345, 346, 361]. In the case of impeller pumps, fans, and water, steam or

© Springer International Publishing AG, part of Springer Nature 2019   9
D. Taler, *Numerical Modelling and Experimental Testing of Heat Exchangers*,
Studies in Systems, Decision and Control 161,
https://doi.org/10.1007/978-3-319-91128-1_2

gas turbines, the angular momentum conservation equation should also be taken into account [80, 110, 221, 369]. An analysis will be conducted of the control volume presented in Fig. 2.1.

The control volume boundaries are constant over time. The boundaries can be real or conventional. The system is open: mass flow rate $\dot{m}_1$ enters the control volume at one end, and mass flow rate $\dot{m}_2$ leaves it at the other end. Two time points: t and $t + \Delta t$ which are close to each other will be analyzed.

At first, the conservation equations will be derived assuming that mass, momentum, and energy are concentrated in the control volume center C.

2.1 Mass Conservation Equation

In the time interval from t to $t + \Delta t$, $\dot{m}_1 \Delta t$ and $\dot{m}_2 \Delta t$ flow into and out of control volume cv, respectively (Fig. 2.1). Mass Δm_{cv} is accumulated in the control volume in time Δt. The mass balance equation then takes the following form:

$$\dot{m}_1 \Delta t = \Delta m_{cv} + \dot{m}_2 \Delta t \tag{2.1}$$

Dividing both sides of Eq. (2.1) by Δt, the following is obtained:

$$\dot{m}_1 - \dot{m}_2 = \frac{\Delta m_{cv}}{\Delta t} \tag{2.2}$$

If $\Delta t \rightarrow 0$, Eq. (2.2) gives:

$$\frac{dm_{cv}}{dt} = \dot{m}_1 - \dot{m}_2. \tag{2.3}$$

In steady-state conditions $dm_{cv}/dt = 0$ and $\dot{m}_1 = \dot{m}_2$.

Fig. 2.1 Diagram of control volume cv

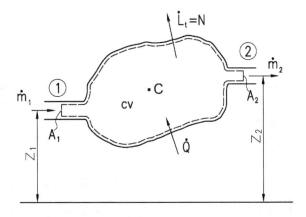

In steady-state problems the mass flow rate is constant:

$$\dot{m} = \rho \dot{V} = \rho w A = \text{const}. \tag{2.4}$$

The mass flow rates through the inlet or outlet, and—similarly—through the machine or device, are identical. However, it should be noted that in steady-state problems the volume flow rate $\dot{V}$ is not constant, i.e., the volume flow rate at the inlet $\dot{V}_1$ is usually different from $\dot{V}_2$ because ρ_1 differs from ρ_2.

Example 2.1 Saturated steam with pressure p_s = 3 MPa, temperature T_s = 233.8 °C and specific volume v_s = 0.0666 m³/kg is superheated at constant pressure to temperature T = 600 °C. The specific volume of superheated steam with pressure p_s = 3 MPa and temperature T = 600 °C is v_s = 0.1323 m³/kg. What should be the superheated steam pipeline diameter to keep the steam flow velocity constant?

Solution
The steam flow is a steady-state one so the steam mass flow rate is constant:

$$\dot{m} = \rho w A = \text{const}.$$

Assuming that the saturated steam pipeline inner diameter is d_s, the superheated steam pipeline diameter is found from the following equation:

$$\frac{\pi d_s^2}{4} \frac{1}{v_s} w = \frac{\pi d^2}{4} \frac{1}{v} w,$$

which gives:

$$d = d_s \sqrt{\frac{v}{v_s}} = d_s \sqrt{\frac{0.1323}{0.0666}} = 1.41 d_s.$$

It can be seen that the superheated steam pipeline diameter should be 41% bigger than the diameter of the saturated steam pipeline.

2.2   Momentum Conservation Equation

The fundamental principle of dynamics is Newton's second law.

$$\sum \mathbf{F} = \frac{d}{dt}(m\mathbf{w}), \tag{2.5}$$

where $\sum \mathbf{F}$ is the sum of forces acting on a material particle with mass m, and moving with velocity $\mathbf{w}$. Other body forces are also taken into consideration, such as the gravitational force, magnetomotive and electromotive forces, as well as surface forces acting on the body, e.g., pressure, or forces tangential to the surface.

In the case of fluid flow, the account is also taken of the momentum flow rates flowing into and out of the control volume. The momentum balance equation for the control volume presented in Fig. 2.1 has the following form:

$$\sum \mathbf{F} + \dot{m}_1 \mathbf{w}_1 = \dot{m}_2 \mathbf{w}_2 + \frac{d}{dt}(m_{cv}\mathbf{w}_c), \tag{2.6}$$

where m_{cv} is the control volume, and $\mathbf{w}_c$—velocity of the control volume center of gravity.

Example 2.2 Calculate the driving force in a jet engine propelling a jet aircraft flying horizontally at the velocity of 160 m/s. In relation to the engine, exhaust gases flow out with the velocity of 1000 m/s. Assume that the exhaust gas mass flow rate is 9 kg/s.

Solution
Assuming that the intake and the exhaust pressure values are equal to atmospheric pressure and omitting the fuel mass and momentum, force F acting on the exhaust gases and opposite to the flight direction (Fig. 2.2) can be calculated from the following formula:

$$-F + \dot{m}(-w_1) = \dot{m}(-w_2), \tag{2.7}$$

which gives:

$$F = \dot{m}(w_2 - w_1),$$

where w_1 and w_2 are the exhaust gas velocities at the engine inlet and outlet (measured using sensors installed on the aircraft).

An identical result is obtained if the air and flue gas velocities are measured by instruments located on the Earth's surface. In this case, air velocity w_1 equals zero, and the exhaust gas measured velocity is: $(w_2 - w_1)$.

Introducing data into formula (2.7), the result is as follows:

$$F = 9(1000 - 160) = 7560\,\text{N}.$$

The direction of the force acting on the aircraft is the same as that of the aircraft velocity vector.

Fig. 2.2 Jet engine diagram

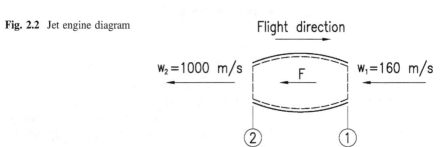

2.3 Angular Momentum Conservation Equation

In the case of turbomachinery such as pumps, fans, and water, steam or gas turbines, the angular momentum balance equation is a convenient tool. In fluid-flow machines, the quantity that has to be determined is not force but the shaft net torque.

Multiplying Eq. (2.5) by $\mathbf{r}$ (Fig. 2.3), the following is obtained:

$$\sum \mathbf{r} \times \mathbf{F} = \frac{d}{dt}(\mathbf{r} \times m\mathbf{w}), \qquad (2.8)$$

where $\mathbf{r}$ is the radius connecting the origin of coordinates to the material point. Considering that $\mathbf{M} = \mathbf{r} \times \mathbf{F}$, Eq. (2.8) gives:

$$\sum \mathbf{M} = \frac{d}{dt}(\mathbf{r} \times m\mathbf{w}). \qquad (2.9)$$

In a Cartesian system of coordinates, the momentum value is calculated using the following formula:

$$M_z = \sum r F_t = \frac{d}{dt}(r m w_t). \qquad (2.10)$$

In the case of fluid flow, the account is taken of the angular momentum at the inlet and outlet of the control volume rotating around axis z (Fig. 2.4).

The angular momentum conservation equation for the control volume is of the following form:

$$\sum \mathbf{M} + \dot{m}_1 \mathbf{r}_1 \times \mathbf{w}_1 = \dot{m}_2 \mathbf{r}_2 \times \mathbf{w}_2 + \frac{d}{dt}(m_{cv}\mathbf{r}_c \times \mathbf{w}_c), \qquad (2.11)$$

where $\mathbf{r}_c$ and $\mathbf{w}_c$ are, respectively, the radius and the velocity of the center of gravity of control volume cv. For a steady-state one-dimensional flow, the angular momentum balance equation for the control volume presented in Fig. 2.4 is written as:

Fig. 2.3 Diagram illustrating Newton's second law for the angular momentum; single force $\mathbf{F}$ acts on a material particle with mass m

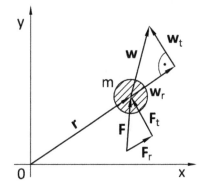

Fig. 2.4 Control volume
diagram illustrating the
angular momentum
conservation principle

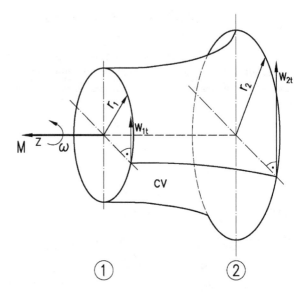

$$M_z = \dot{m}(r_2 w_{2t} - r_1 w_{1t}). \qquad (2.12)$$

Multiplying both sides of Eq. (2.12) by angular velocity ω, the result is:

$$N = M_z\,\omega = \dot{m}(\omega\,r_2 w_{2t} - \omega\,r_1 w_{1t}) = \dot{m}(u_2 w_{2t} - u_1 w_{1t}), \qquad (2.13)$$

where $u_1 = \omega_1 r_1$ and $u_2 = \omega_2 r_2$ denote peripheral speeds of the rotor with radii r_1 and r_2, respectively. For the impeller of a pump or a compressor rotating with angular velocity ω, the power transferred to the liquid jet by the impeller is described by Eq. (2.13), which is referred to as the Euler pump equation. For turbines, the corresponding Euler equation is of the following form:

$$N = M_z\,\omega = \dot{m}(u_1 w_{1t} - u_2 w_{2t}). \qquad (2.14)$$

In a turbine, power N is transferred to the rotor from the liquid, steam or gas stream.

The angular momentum conservation Eq. (2.11) finds application in turbomachinery calculations.

Example 2.3 Calculate the power of the turbine single stage presented in Fig. 2.5. The fluid mass flow rate $\dot{m} = 20$ kg/s flows axially into nozzles, where—upon expansion—it reaches velocity $w_1 = 500$ m/s. The peripheral speed of the turbine blades is $u_1 = u_2 = u = 1050$ m/s. The (absolute) velocity vector w_1 deviates from the vertical by 45°. At the inlet from the blade ring, the fluid flow direction is axial.

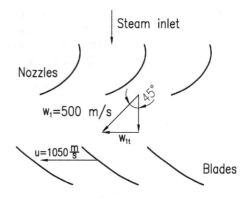

Fig. 2.5 Flat model of the
fluid flow though a turbine
stage

Solution

The rotor shaft power can be calculated using the Euler turbine Eq. (2.14).

$$N = \dot{m}(u_1 w_{1t} - u_2 w_{2t}).$$

It can be seen from Fig. 2.5 that:

$$w_{1t} = w_1 \sin 45° = 500 \cdot 0.7071 = 353.55 \, \text{m/s},$$

$$w_{2t} = 0.$$

Introducing data into the Euler formula, the result is as follows:

$$N = 20(1050 \cdot 353.55 - 0) = 7.42455 \times 10^6 \text{W} = 7.42 \, \text{MW}.$$

2.4   Energy Conservation Equation

Energy E of a material particle with mass m is:

$$E = E_w + E_k + E_p = mu + \frac{mw^2}{2} + mgz = m\left(u + \frac{w^2}{2} + gz\right) = me, \qquad (2.15)$$

where:

$$e = u + \frac{w^2}{2} + gz \qquad (2.16)$$

is the energy of 1 kg of the medium. The energy of 1 m^3 of the medium is:

$$\frac{E}{V} = \rho\left(u + \frac{w^2}{2} + gz\right). \qquad (2.17)$$

In formulae (2.15)–(2.17), u denotes internal energy related to 1 kg of the medium, w—the material point velocity, and z—the point elevation compared to the adopted reference level (Fig. 2.1). The energy balance equation for the control volume presented in Fig. 2.1 has the following form:

$$\dot{m}\left(u_1 + \tfrac{w_1^2}{2} + gz_1\right) + \dot{L}_{p1} + \dot{Q} + \dot{Q}_v =$$
$$= \dot{m}\left(u_2 + \tfrac{w_2^2}{2} + gz_2\right) + \dot{L}_{p2} + \dot{L}_t + \frac{d}{dt}\left[m\left(u + \tfrac{w_c^2}{2} + gz_c\right)\right], \tag{2.18}$$

where $\dot{L}_{p1}$ is the power needed to overcome pressure forces in cross-section 1, and $\dot{L}_{p2}$ is the power transferred to the environment in cross-section 2. Subscript c denotes the analyzed system center of gravity. Power $\dot{L}_t$ is the internal power of the engine or working machine.

Power $\dot{L}_t$ is positive if the control volume does work on the environment. Heat flow rate $\dot{Q}$ is positive if heat is supplied to the control volume. Heat flow rate $\dot{Q}_v$ is generated by internal heat sources. The power needed to overcome pressure forces can be calculated as follows (Fig. 2.6).

The work done by a piston with the front surface area A at the piston shift by Δx in time Δt can be found using the following formula:

$$L_p = F \cdot \Delta x = pA \cdot \Delta x = p \cdot \Delta V. \tag{2.19}$$

The power needed to overcome pressure forces is:

$$\dot{L}_p = p\frac{\Delta V}{\Delta t}. \tag{2.20}$$

If $\Delta t \to 0$, then $\Delta V/\Delta t \to dV/dt = \dot{V}$ and formula (2.20) takes the following form:

$$\dot{L}_p = p\dot{V} = \dot{m}\frac{p}{\rho} = \dot{m}pv. \tag{2.21}$$

The following nomenclature is adopted in formulae (2.19)–(2.21): F—force acting on the piston and needed to balance pressure forces, $\dot{V}$—volume flow rate of the medium, $\dot{m}$—mass flow rate of the medium, v—specific volume of the medium.

Fig. 2.6 Diagram illustrating determination of power needed to pump the medium

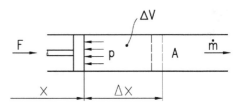

Formula (2.21) enables calculation of power $\dot{L}_{p1}$ and $\dot{L}_{p2}$.

Taking account of Eq. (2.21), Eq. (2.18) can be transformed into:

$$\dot{m}_1\left(u_1 + p_1 v_1 + \tfrac{w_1^2}{2} + g z_1\right) + \dot{Q} + \dot{Q}_v - \dot{L}_t =$$
$$= \dot{m}_2\left(u_2 + p_2 v_2 + \tfrac{w_2^2}{2} + g z_2\right) + \tfrac{\mathrm{d}}{\mathrm{d}t}\left[m\left(u + \tfrac{w_c^2}{2} + g z_c\right)\right]. \tag{2.22}$$

Introducing specific enthalpy i, defined as:

$$i = u + pv,$$

the balance Eq. (2.22) takes the following form:

$$\dot{m}_1\left(i_1 + \tfrac{w_1^2}{2} + g z_1\right) + \dot{Q} + \dot{Q}_v - \dot{L}_t = \dot{m}_2\left(i_2 + \tfrac{w_2^2}{2} + g z_2\right) +$$
$$+ \tfrac{\mathrm{d}}{\mathrm{d}t}\left[m\left(u + \tfrac{w_c^2}{2} + g z_c\right)\right]. \tag{2.23}$$

It can be noticed that Eq. (2.23) does not take account of the resistance resulting from friction on the stationary channel wall even if tangential stresses arising on the channel surface are high.

In steady-state problems, the time derivative equals zero, and the energy balance equation is reduced to the following form:

$$\dot{m}_1\left(i_1 + \frac{w_1^2}{2} + g z_1\right) + \dot{Q} + \dot{Q}_v - \dot{L}_t = \dot{m}\left(i_2 + \frac{w_2^2}{2} + g z_2\right). \tag{2.24}$$

Equation (2.24) takes account of the fact that in the steady state the following equality occurs: $\dot{m}_1 = \dot{m}_2 = \dot{m}$

2.5   Averaging of Velocity and Temperature

In the balance equations presented in Sects. 2.1, 2.2 and 2.3, velocity and temperature in the duct cross-section are assumed as constant.

In real conditions, however, this is not the case—the two quantities vary. The velocity values are usually the highest in the duct axis, whereas on the duct wall velocity equals zero. Attention should be paid to the manner in which velocity and temperature are averaged in the duct cross-section. Mean velocity w_m is found to satisfy the mass conservation equation:

$$\dot{m} = \int \rho\, w\, \mathrm{d}A = w_m \int \rho\, \mathrm{d}A, \tag{2.25}$$

which gives:

$$w_m = \frac{\dot{m}}{\int \rho \, dA} = \frac{\int \rho w dA}{\int \rho \, dA}. \tag{2.26}$$

Similarly, mean temperature T_m satisfies the energy continuity equation:

$$\dot{E} = \int \rho w c_p T dA = T_m \int \rho w c_p dA, \tag{2.27}$$

which gives:

$$T_m = \frac{\int \rho w c_p T dA}{\int \rho w c_p dA}. \tag{2.28}$$

If a single velocity, temperature or enthalpy value is given for the entire cross-section, it should be treated as the mean value for the entire cross-section of the duct.

2.6 Basic Equations of Fluid Mechanics and Heat Transfer in the Integral Form

The equations presented in Sects. 2.1, 2.2., 2.3 and 2.4 are derived assuming that the mass of the substance contained in control volume cv with surface area cs is concentrated in the center of gravity C. In the general case of a continuous medium, mass is distributed across the entire control area. Similarly, the other quantities—momentum, angular momentum, and temperature—are not constant within the control area (control volume), but they are a function of location. The boundaries of control volume cv are constant over time.

The balance equations are of the following form:

– mass conservation equation (Figs. 2.7 and 2.8):

$$0 = \frac{\partial}{\partial t} \int_{cv} \rho \, dV + \int_{cs} \rho \mathbf{w} \cdot dA, \tag{2.29}$$

– momentum conservation equation:

$$\sum \mathbf{F} = \frac{\partial}{\partial t} \int_{cv} \rho \mathbf{w} dV + \int_{cs} \rho \mathbf{w} \mathbf{w} \cdot dA, \tag{2.30}$$

Fig. 2.7 Control volume diagram

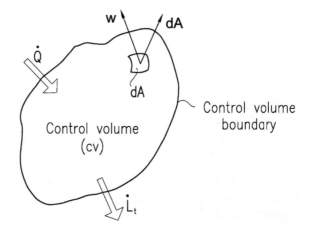

Fig. 2.8 Diagram illustrating calculation of surface integrals during the fluid flow through a channel

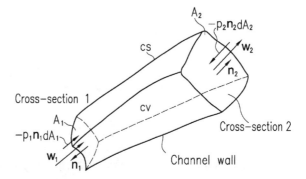

– angular momentum balance equation:

$$\sum \mathbf{M} = \frac{\partial}{\partial t} \int_{cv} \rho \, (\mathbf{r} \times \mathbf{w}) dV + \int_{cs} \rho \, (\mathbf{r} \times \mathbf{w}) \mathbf{w} \cdot d\mathbf{A}, \qquad (2.31)$$

– energy balance equation:

$$\dot{Q} + \dot{Q}_v - \dot{L}_t + \dot{L}_\mu = \frac{\partial}{\partial t} \int_{cv} \rho \, e \, dV + \int_{cs} \left(e + \frac{p}{\rho} \right) \rho \mathbf{w} \cdot d\mathbf{A}, \qquad (2.32)$$

where the power of heat sources is defined as:

$$\dot{Q}_v = \int_{cv} \dot{q}_v dV. \tag{2.33}$$

Power $\dot{L}_\mu$ used to overcome friction forces reduces power $\dot{L}_t$ transferred from the control volume to the environment.

The balance Eqs. (2.29)–(2.32) include surface integrals, the calculation of which will be illustrated in the example of the fluid flow through a stationary channel.

The channel cross-sections 1 and 2 are flat, so normal vectors $\mathbf{n}_1$ and $\mathbf{n}_2$ are constant in the respective cross-sections. The surface integral in Eq. (2.29) is calculated in the following way:

$$\int_{cs} \rho \mathbf{w} \cdot d\mathbf{A} = \int \rho_1 \mathbf{w}_1 \cdot d\mathbf{A}_1 + \int \rho_2 \mathbf{w}_2 \cdot d\mathbf{A}_2$$

$$= \int \rho_1 \mathbf{w}_1 \cdot \mathbf{n}_1 dA_1 + \int \rho_2 \mathbf{w}_2 \cdot \mathbf{n}_2 dA_2, \tag{2.34}$$

If it is assumed that the planes of cross-sections 1 and 2 are perpendicular to velocity vectors $\mathbf{w}_1$ and $\mathbf{w}_2$, respectively, then:

$$\begin{aligned} \mathbf{w}_1 \cdot \mathbf{n}_1 &= w_1 \cdot 1 \cdot \cos 180° = -w_1, \\ \mathbf{w}_2 \cdot \mathbf{n}_2 &= w_2 \cdot 1 \cdot \cos 0° = w_2. \end{aligned} \tag{2.35}$$

Considering Eq. (2.35), the surface integral expressed by Eq. (2.34) takes the following form:

$$\int_{cs} \rho \mathbf{w} \cdot d\mathbf{A} = \int \rho_1 w_1 dA_1 + \int \rho_2 w_2 dA_2. \tag{2.36}$$

Substitution of (2.36) in (2.29) results in:

$$\int \rho_1 w_1 dA_1 = \int \rho_2 w_2 dA_2 + \frac{\partial}{\partial t} \int \rho dV, \tag{2.37}$$

which, considering that:

$$\begin{aligned} \int \rho_1 w_1 dA_1 &= \dot{m}_1, \\ \int \rho_2 w_2 dA_2 &= \dot{m}_2, \end{aligned} \tag{2.38}$$

$$\int \rho \, dV = m_c, \tag{2.39}$$

gives:

$$\dot{m}_1 = \dot{m}_2 + \frac{dm_{cv}}{dt}.$$ (2.40)

Equations (2.40) and (2.3) are identical.

2.7   Basic Equations of Fluid Mechanics and Heat Transfer in the Differential Form

Equations (2.29)–(2.32) will be transformed using the Gauss-Ostrogradsky (divergence) theorem. If $\mathbf{F}$ is a vector field with first-order continuous partial derivatives and cs—a closed, smooth, internally oriented surface limiting control volume cv, then:

$$\int_{cs} \mathbf{F} \cdot \mathbf{n} dA = \int_{cv} \nabla \cdot \mathbf{F} dV.$$ (2.41)

After the surface integrals in Eqs. (2.29)–(2.32) are transformed using the divergence theorem (2.41), the equations can be written in the following form of:

– the continuity equation:

$$\int_{cv} \left[\frac{\partial \rho}{\partial t} + \nabla \cdot (\rho \mathbf{w}) \right] dV = 0,$$ (2.42)

– the momentum conservation equation:

$$\sum \mathbf{F} = \int_{cv} \left[\frac{\partial(\rho \mathbf{w})}{\partial t} + \nabla \cdot \rho \mathbf{w}\mathbf{w} \right] dV,$$ (2.43)

– the angular momentum conservation equation:

$$\sum \mathbf{M} = \int_{cv} \left[\frac{\partial(\mathbf{r} \times \mathbf{w})}{\partial t} + \nabla \cdot \rho(\mathbf{r} \times \mathbf{w})\mathbf{w} \right] dV,$$ (2.44)

– the energy conservation equation:

$$\dot{Q} + \dot{Q}_v - \dot{L}_t + \dot{L}_\mu = \int_{cv} \left[\frac{\partial(\rho e)}{\partial t} + \nabla \cdot \left(e + \frac{p}{\rho}\right)\rho\mathbf{w}\right] dV. \qquad (2.45)$$

Product $\mathbf{ww}$ in Eq. (2.43) and product $(\mathbf{r} \times \mathbf{w})\mathbf{w}$ in Eq. (2.44) are second-order tensors [Appendix A.2, formula (A.17)].

Transformation of Eqs. (2.42)–(2.45) can be facilitated using the concept of substantial derivative D/Dt defined as:

$$\frac{D}{Dt} = \frac{\partial}{\partial t} + \mathbf{w} \cdot \nabla. \qquad (2.46)$$

Derivative (2.46) represents the rate of changes in a quantity over time from the perspective of an observer moving with the substance. In order to explain the derivative physical significance, temperature T will be determined of a small fluid particle moving with the fluid with velocity $\mathbf{w}(w_x, w_y, w_z)$. The fluid particle is so small that temperature $T[x(t), y(t), z(t), t]$ is constant across its entire volume. The rate of changes in temperature DT/Dt is expressed as:

$$\frac{DT}{Dt} = \frac{\partial T}{\partial t} + \frac{\partial T}{\partial x}\frac{\partial x}{\partial t} + \frac{\partial T}{\partial y}\frac{\partial y}{\partial t} + \frac{\partial T}{\partial z}\frac{\partial z}{\partial t}, \qquad (2.47)$$

which, considering the definition:

$$w_x = \frac{\partial x}{\partial t}, w_y = \frac{\partial y}{\partial t}, w_z = \frac{\partial z}{\partial t} \qquad (2.48)$$

gives:

$$\frac{\partial T}{\partial t} \qquad (2.49)$$

The derivative $\frac{\partial T}{\partial t}$ is the local rate of temperature changes and $w_x \frac{\partial T}{\partial x} + w_y \frac{\partial T}{\partial y} + w_z \frac{\partial T}{\partial z}$ is the rate of changes in temperature being the effect of the fluid motion.

Substantial derivative DT/Dt defined by formula (2.49) can also be written in the following form:

$$\frac{DT}{Dt} = \frac{\partial T}{\partial t} + \mathbf{w} \cdot \nabla T, \qquad (2.50)$$

where:

$$\nabla T = \mathrm{grad} T = \mathbf{i}\frac{\partial T}{\partial x} + \mathbf{j}\frac{\partial T}{\partial y} + \mathbf{k}\frac{\partial T}{\partial z}. \tag{2.51}$$

If the fluid velocity $\mathbf{w} = 0$, substantial derivative $DT/Dt = \partial T/\partial t$. The changes in the fluid temperature are caused in this case by the inflow or outflow of heat to/ from the particle or by heat generated inside the particle. In steady-state problems, when $\partial T/\partial t = 0$, the change in the particle temperature is the effect of the particle displacement with velocity $\mathbf{w}$. The next three subsections discuss Eqs. (2.42), (2.43) and (2.45).

2.7.1 Continuity Equation

Continuity Eq. (2.42) is valid also for an infinitely small control volume. In such a case, integration in (2.42) is omitted, and the result is:

$$\frac{\partial \rho}{\partial t} + \nabla \cdot (\rho \mathbf{w}) = 0. \tag{2.52}$$

Transforming Eq. (2.52) to:

$$\frac{\partial \rho}{\partial t} + \mathbf{w} \cdot \nabla \rho + \rho \nabla \cdot \mathbf{w} = 0 \tag{2.53}$$

and considering (2.50), the following is obtained:

$$\frac{D\rho}{Dt} + \rho \nabla \cdot \mathbf{w} = 0. \tag{2.54}$$

Continuity Eq. (2.52) written in different systems of coordinates (Fig. 2.9) is presented in Table 2.1.

For incompressible fluids, Eqs. (2.55)–(2.57) are reduced since $\rho = $ const, so $\frac{\partial \rho}{\partial t} = 0$.

2.7.2 Momentum Balance Equation

The momentum balance Eq. (2.43) includes the sum of forces acting on the control volume. These are primarily pressure forces acting in the direction perpendicular to the surface, viscosity forces—acting in the tangential direction, and volume forces, such as the gravity force or the electric or magnetic field forces. Only three of the forces acting on the control volume will be analysed: the forces arising due to

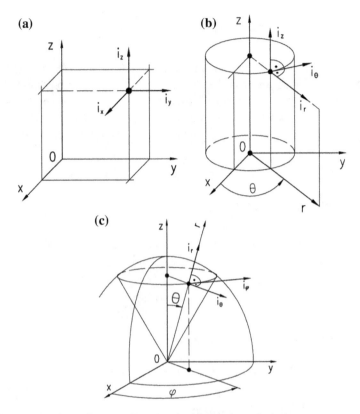

Fig. 2.9 Systems of coordinates: **a** Cartesian, **b** cylindrical, **c** spherical

Table 2.1 Continuity Eq. (2.52) in different systems of coordinates presented in Fig. 2.9

System of coordinates	Continuity equation
Cartesian (x, y, z)	$\frac{\partial \rho}{\partial t} + \frac{\partial}{\partial x}(\rho\, w_x) + \frac{\partial}{\partial y}(\rho\, w_y) + \frac{\partial}{\partial z}(\rho\, w_z) = 0$ (2.55)
Cylindrical (r, θ, z)	$\frac{\partial \rho}{\partial t} + \frac{1}{r}\frac{\partial}{\partial r}(\rho\, r w_r) + \frac{1}{r}\frac{\partial}{\partial \theta}(\rho\, w_\theta) + \frac{\partial}{\partial z}(\rho\, w_z) = 0$ (2.56)
Spherical (r, θ, φ)	$\frac{\partial \rho}{\partial t} + \frac{1}{r^2}\frac{\partial}{\partial r}(\rho r^2 w_r) + \frac{1}{r \sin\theta}\frac{\partial}{\partial \theta}(\rho w_\theta \sin\theta)$ (2.57) $+ \frac{1}{r \sin\theta}\frac{\partial}{\partial \varphi}(\rho w_\varphi) = 0$

pressure, the viscosity forces and the gravity force. The sum of the forces $\sum \mathbf{F}$ can be written as follows:

$$\sum \mathbf{F} = \int \boldsymbol{\sigma} \cdot \mathbf{n}\, dA + \int \rho \mathbf{g}\, dV, \qquad (2.58)$$

where stress tensor σ takes account of the pressure and the tangential stresses arising in the flow of a viscous fluid. The stress state in a selected point is characterized by the stress tensor, which in the Cartesian system of coordinates is given as (Fig. 2.10):

$$\boldsymbol{\sigma} = \begin{bmatrix} \sigma_{xx} & \tau_{xy} & \tau_{xz} \\ \tau_{yx} & \sigma_{yy} & \tau_{yz} \\ \tau_{zx} & \tau_{zy} & \sigma_{zz} \end{bmatrix}. \tag{2.59}$$

Of the stress state six shear components, only three are independent. Figure 2.10b presents shear stresses acting in plane xy. The condition of equality between momentums for point A or point C, assuming that the fluid particle is infinitely small, gives $\tau_{xy} = \tau_{yx}$. In the general case, the shear components satisfy the following equalities:

$$\tau_{xy} = \tau_{yx}, \quad \tau_{yz} = \tau_{zy}, \quad \tau_{zx} = \tau_{xz}. \tag{2.60}$$

Substitution of (2.58) in (2.30) results in:

$$\int_{cs} \boldsymbol{\sigma} \cdot \mathbf{n} \, dA + \int_{cv} \rho \mathbf{g} \, dV = \int_{cv} \frac{\partial(\rho \mathbf{w})}{\partial t} \, dV + \int_{cs} (\rho \mathbf{w}) \mathbf{w} \cdot \mathbf{n} \, dA, \tag{2.61}$$

where σ is the stress tensor.

After the divergence theorem (2.41) is applied to the surface integrals in Eq. (2.61), the following is obtained:

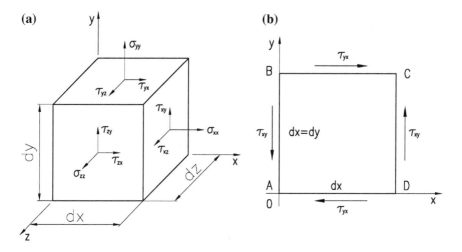

Fig. 2.10 Normal and shear stresses acting on the fluid element: **a** stresses on three walls of the fluid element, **b** shear stresses acting on the element in plane x-y

$$\int\limits_{cv} \left[\nabla \cdot \boldsymbol{\sigma} + \rho \, \mathbf{g} - \frac{\partial}{\partial t}(\rho \, \mathbf{w}) - \nabla \cdot \rho \mathbf{w} \, \mathbf{w} \right] dV = 0, \qquad (2.62)$$

where $\mathbf{w} \, \mathbf{w} = \mathbf{w} \otimes \mathbf{w} = \mathbf{T}$ is a dyadic (i.e., binary or two-argument) product. In the case of a dyadic product of two vectors, the symbol $\otimes$ can be omitted. Dyadic product $\mathbf{T}$ of two vectors $\mathbf{u}$ and $\mathbf{w}$ with components u_i and w_i, $i = 1,2,3$ is a tensor whose elements T_{ij} are defined by the following formula:

$$T_{ij} = u_i, w_j, \quad i = 1, 2, 3.$$

Equation (2.62) is satisfied for any control volume, even if it is infinitely small. Ignoring the integrals in (2.62), a differential form of the momentum balance equation is obtained:

$$\frac{\partial}{\partial t}(\rho \mathbf{w}) + \nabla \cdot \rho \mathbf{w} \mathbf{w} = \nabla \cdot \boldsymbol{\sigma} + \rho \, \mathbf{g}. \qquad (2.63)$$

It follows from the following identity:

$$\nabla \cdot \mathbf{v} \, \mathbf{w} = \mathbf{v} \cdot \nabla \mathbf{w} + \mathbf{w}(\nabla \cdot \mathbf{v}) \qquad (2.64)$$

that:

$$\nabla \cdot \rho \, \mathbf{w} \, \mathbf{w} = \rho \, \mathbf{w} \cdot \nabla \mathbf{w} + \mathbf{w}(\nabla \cdot \rho \, \mathbf{w}), \qquad (2.65)$$

where $\nabla \mathbf{w} = \mathbf{T}$ is the gradient of vector $\mathbf{w}$.

Gradient $\mathbf{T}$ of vector $\mathbf{w}$, with components w_i, $i = 1,2,3$, is a second-order tensor whose elements T_{ij} are derivatives of each component w_i, $i = 1,2,3$ with respect to each x_j, $j = 1,2,3$, i.e.:

$$T_{ij} = \frac{\partial w_i}{\partial x_j}, \quad i = 1, 2, 3; \quad j = 1, 2, 3.$$

Considering (2.65) in (2.64), the following is obtained after simple transformations:

$$\mathbf{w} \left[\frac{\partial \rho}{\partial t} + \nabla \cdot (\rho \, \mathbf{w}) \right] + \rho \left(\frac{\partial \mathbf{w}}{\partial t} + \mathbf{w} \cdot \nabla \mathbf{w} \right) = \nabla \cdot \boldsymbol{\sigma} + \rho \, \mathbf{g}. \qquad (2.66)$$

Taking account of the continuity Eq. (2.52) and the definition of the substantial derivative from velocity vector w, Eq. (2.66) is reduced to the following form:

$$\rho \frac{D\mathbf{w}}{Dt} = \rho \mathbf{g} + \nabla \cdot \boldsymbol{\sigma}. \qquad (2.67)$$

In a Cartesian system of coordinates, Eq. (2.67) is expressed as follows:

$$\rho \frac{Dw_x}{Dt} = \rho g_x + \frac{\partial \sigma_{xx}}{\partial x} + \frac{\partial \tau_{yx}}{\partial y} + \frac{\partial \tau_{zx}}{\partial z},$$
$$\rho \frac{Dw_y}{Dt} = \rho g_y + \frac{\partial \tau_{xy}}{\partial x} + \frac{\partial \sigma_{yy}}{\partial y} + \frac{\partial \tau_{zy}}{\partial z}, \tag{2.68}$$
$$\rho \frac{Dw_z}{Dt} = \rho g_z + \frac{\partial \tau_{xz}}{\partial x} + \frac{\partial \tau_{yz}}{\partial y} + \frac{\partial \sigma_{zz}}{\partial z}.$$

At the end of the 17th century, Isaac Newton found that shear stresses in a fluid are proportional to the rate of changes in strains, i.e., to the velocity gradient. Fluids that satisfy the law discovered by Newton are referred to as Newtonian fluids.

A complete system of equations expressing the Newtonian viscosity law and written in a Cartesian system of coordinates has the following form:

$$\tau_{xy} = \tau_{yx} = \mu \left(\frac{\partial w_x}{\partial y} + \frac{\partial w_y}{\partial x} \right), \tag{2.69}$$

$$\tau_{yz} = \tau_{zy} = \mu \left(\frac{\partial w_y}{\partial z} + \frac{\partial w_z}{\partial y} \right), \tag{2.70}$$

$$\tau_{zx} = \tau_{xz} = \mu \left(\frac{\partial w_z}{\partial x} + \frac{\partial w_x}{\partial z} \right), \tag{2.71}$$

$$\sigma_{xx} = -p + 2\mu \frac{\partial w_x}{\partial x} - \frac{2}{3} \mu \nabla \cdot \mathbf{w}, \tag{2.72}$$

$$\sigma_{yy} = -p + 2\mu \frac{\partial w_y}{\partial y} - \frac{2}{3} \mu \nabla \cdot \mathbf{w}, \tag{2.73}$$

$$\sigma_{zz} = -p + 2\mu \frac{\partial w_z}{\partial z} - \frac{2}{3} \mu \nabla \cdot \mathbf{w}. \tag{2.74}$$

The expressions for normal stresses (2.72)–(2.74) take account of compressive stresses $(-p)$ arising due to pressure and stresses caused by viscosity forces.

The coefficient: $-2/(3\mu)$ results from the hypothesis formulated by Stokes in 1845. The hypothesis is frequently applied, even though it has not been fully confirmed yet. If stress components (2.69)–(2.74) are substituted in the momentum balance Eqs. (2.68), the result is as follows:

$$\rho \frac{Dw_x}{Dt} = \rho g_x - \frac{\partial p}{\partial x} + \frac{\partial}{\partial x} \left[\mu \left(2 \frac{\partial w_x}{\partial x} - \frac{2}{3} \nabla \cdot \mathbf{w} \right) \right] +$$
$$+ \frac{\partial}{\partial y} \left[\mu \left(\frac{\partial w_y}{\partial x} + \frac{\partial w_x}{\partial y} \right) \right] + \frac{\partial}{\partial z} \left[\mu \left(\frac{\partial w_z}{\partial x} + \frac{\partial w_x}{\partial z} \right) \right], \tag{2.75}$$

$$\rho\frac{Dw_y}{Dt} = \rho\, g_y - \frac{\partial p}{\partial y} + \frac{\partial}{\partial x}\left[\mu\left(2\frac{\partial w_y}{\partial x} + \frac{\partial w_x}{\partial y}\right)\right] +$$
$$+ \frac{\partial}{\partial y}\left[\mu\left(\frac{\partial w_y}{\partial y} - \frac{2}{3}\nabla\cdot\mathbf{w}\right)\right] + \frac{\partial}{\partial z}\left[\mu\left(\frac{\partial w_z}{\partial y} + \frac{\partial w_y}{\partial z}\right)\right], \tag{2.76}$$

$$\rho\frac{Dw_z}{Dt} = \rho\, g_z - \frac{\partial p}{\partial z} + \frac{\partial}{\partial x}\left[\mu\left(2\frac{\partial w_z}{\partial x} + \frac{\partial w_x}{\partial z}\right)\right] +$$
$$+ \frac{\partial}{\partial y}\left[\mu\left(\frac{\partial w_z}{\partial y} + \frac{\partial w_y}{\partial z}\right)\right] + \frac{\partial}{\partial z}\left[\mu\left(\frac{\partial w_z}{\partial z} - \frac{2}{3}\nabla\cdot\mathbf{w}\right)\right], \tag{2.77}$$

Assuming that the fluid density and viscosity are constant, the Navier-Stokes Eqs. (2.75)–(2.77) take the following form:

$$\rho\frac{Dw_x}{Dt} = \rho\, g_x - \frac{\partial p}{\partial z} + \mu\left(\frac{\partial^2 w_x}{\partial x^2} + \frac{\partial^2 w_x}{\partial y^2} + \frac{\partial^2 w_x}{\partial z^2}\right), \tag{2.78}$$

$$\rho\frac{Dw_y}{Dt} = \rho\, g_y - \frac{\partial p}{\partial y} + \mu\left(\frac{\partial^2 w_y}{\partial x^2} + \frac{\partial^2 w_y}{\partial y^2} + \frac{\partial^2 w_y}{\partial z^2}\right), \tag{2.79}$$

$$\rho\frac{Dw_z}{Dt} = \rho\, g_z - \frac{\partial p}{\partial z} + \mu\left(\frac{\partial^2 w_z}{\partial x^2} + \frac{\partial^2 w_z}{\partial y^2} + \frac{\partial^2 w_z}{\partial z^2}\right). \tag{2.80}$$

Equations (2.78)–(2.80) can also be written in a vector form:

$$\rho\frac{D\mathbf{w}}{Dt} = \rho\mathbf{g} - \nabla p + \mu\nabla^2\mathbf{w}, \tag{2.81}$$

where $\nabla^2\mathbf{w}$ denotes the Laplacian from vector $\mathbf{w}$, the components of which can be calculated using the following formula:

$$\nabla^2\mathbf{w} = \nabla(\nabla\cdot\mathbf{w}) - \nabla\times(\nabla\times\mathbf{w}). \tag{2.82}$$

The momentum balance Eq. (2.66)–(2.68) can be derived directly from Newton's second law written for an elementary cuboid with the dimensions: dx × dy × dz (Fig. 2.1).

The value of function $f(x+\Delta x)$ can be calculated using the Taylor series, knowing the values of the function and derivatives at the point with coordinate x:

$$f(x+\Delta x) = f(x) + \frac{df}{dx}\bigg|_x \Delta x + \frac{1}{2!}\frac{d^2 f}{dx^2}(\Delta x)^2 + \cdots \tag{2.83}$$

Taking account of the first two terms on the right side only, and assuming that $\Delta x \to dx$, Eq. (2.83) gives:

$$f(x+dx) = f(x) + \frac{df}{dx}\bigg|_x dx. \tag{2.84}$$

From Newton's second law:

$$\frac{D(m_{cv}\mathbf{w})}{Dt} = m_{cv}\frac{D\mathbf{w}}{Dt} = \sum \mathbf{F} \tag{2.85}$$

written for the elementary cuboid presented in Fig. 2.10a, the following is obtained for direction x:

$$\rho\, dx\, dy\, dz\frac{Dw_x}{Dt} = g_x\rho\, dx\, dy\, dz + dy\, dz\left(-\sigma_{xx} - \tau_{yx} - \tau_{zx} + \right.$$
$$\left. + \sigma_{xx} + \frac{\partial\sigma_{xx}}{\partial x}dx + \tau_{yx} + \frac{\partial\tau_{yx}}{\partial x}dx + \tau_{zx}\frac{\partial\tau_{zx}}{\partial x}dx\right). \tag{2.86}$$

Dividing both sides of Eq. (2.86) by $dx\, dy\, dz$, the result is:

$$\rho\frac{Dw_x}{Dt} = \rho g_x + \frac{\partial\sigma_{xx}}{\partial x} + \frac{\partial\tau_{yx}}{\partial y} + \frac{\partial\tau_{zx,}}{\partial y} \tag{2.87}$$

which means that the equation is identical to Eq. (2.68).

The Navier-Stokes Eq. (2.81) written in a cylindrical and a spherical system of coordinates is presented in Appendix B.

2.7.3 Energy Conservation Equation

The mechanical energy balance equation will be derived to transform the energy conservation equation.

2.7.3.1 Mechanical Energy Balance Equation

In a flow system, mechanical energy is not conserved. However, it is possible to derive an equation defining changes in mechanical energy by multiplying the momentum balance Eq. (2.65) by $\mathbf{w}$:

$$\rho\mathbf{w}\cdot\frac{D\mathbf{w}}{Dt} = \rho\mathbf{g}\cdot\mathbf{w} + \mathbf{w}\cdot\nabla\cdot\boldsymbol{\sigma}. \tag{2.88}$$

Considering the substantial derivative definition (2.46), Eq. (2.88) takes the following form:

$$\rho\,\mathbf{w}\cdot\left(\frac{\partial\mathbf{w}}{\partial t}+\mathbf{w}\cdot\nabla\mathbf{w}\right)=\rho\,\mathbf{g}\cdot\mathbf{w}+\mathbf{w}\cdot\nabla\cdot\boldsymbol{\sigma}. \tag{2.89}$$

Equation (2.89) will be transformed using the following identities:

$$\nabla(\mathbf{v}.\mathbf{w})=(\nabla\mathbf{v})\cdot\mathbf{w}+(\nabla\mathbf{w})\cdot\mathbf{v}, \tag{2.90}$$

$$\mathbf{w}\cdot\nabla\cdot\boldsymbol{\sigma}=\nabla\cdot(\boldsymbol{\sigma}\cdot\mathbf{w})-\boldsymbol{\sigma}:\nabla\mathbf{w}. \tag{2.91}$$

It follows from equality (2.90) that:

$$\nabla(\mathbf{w}.\mathbf{w})=(\nabla\mathbf{w})\cdot\mathbf{w}+(\nabla\mathbf{w})\cdot\mathbf{w}, \tag{2.92}$$

which gives:

$$\nabla\mathbf{w}\cdot\mathbf{w}=\frac{1}{2}\nabla(w^2). \tag{2.93}$$

The first term on the left side of Eq. (2.89) will be transformed using the following equality:

$$\frac{\partial}{\partial t}\left(\frac{1}{2}\rho\,w^2\right)=\frac{1}{2}w^2\frac{\partial\rho}{\partial t}+\rho\,\mathbf{w}\cdot\frac{\partial\mathbf{w}}{\partial t}. \tag{2.94}$$

Equation (2.94) gives:

$$\rho\,\mathbf{w}\cdot\frac{\partial\mathbf{w}}{\partial t}=\frac{\partial}{\partial t}\left(\frac{1}{2}\rho\,w^2\right)-\frac{1}{2}w^2\frac{\partial\rho}{\partial t}. \tag{2.95}$$

Considering equalities (2.91), (2.93) and (2.95) in the equation of changes in mechanical energy (2.89), the following is obtained:

$$\frac{\partial}{\partial t}\left(\frac{1}{2}\rho\,w^2\right)-\frac{1}{2}w^2\frac{\partial\rho}{\partial t}+\rho\,\mathbf{w}\cdot\frac{1}{2}\nabla(w^2)=\rho\mathbf{g}\cdot\mathbf{w}+\nabla\cdot(\boldsymbol{\sigma}\cdot\mathbf{w})-\boldsymbol{\sigma}:\nabla\mathbf{w}. \tag{2.96}$$

The third term on the right side will be transformed using the following equality:

$$\mathbf{v}\cdot\nabla s=\nabla\cdot(s\cdot\mathbf{v})-s\nabla\cdot\mathbf{v}, \tag{2.97}$$

from which it follows that:

$$\frac{1}{2}\rho\,\mathbf{w}\,\nabla(w^2)=\nabla\cdot\left(\frac{1}{2}\rho\,w^2\mathbf{w}\right)-\frac{1}{2}w^2\nabla(\rho\,\mathbf{w}). \tag{2.98}$$

Substitution of (2.98) in (2.96) results in:

$$\frac{\partial}{\partial t}\left(\frac{1}{2}\rho w^2\right) = -\nabla \cdot \left(\frac{1}{2}\rho w^2 \mathbf{w}\right) + \frac{1}{2}w^2\left[\frac{\partial \rho}{\partial t} + \nabla \cdot (\rho \mathbf{w})\right] +$$
$$\rho \mathbf{g} \cdot \mathbf{w} + \nabla \cdot (\boldsymbol{\sigma} \cdot \mathbf{w}) - \boldsymbol{\sigma} : \nabla \mathbf{w}. \tag{2.99}$$

Taking account of the continuity Eq. (2.52), the equation defining changes in mechanical energy (2.99) takes the following form:

$$\frac{\partial}{\partial t}\left(\frac{1}{2}\rho w^2\right) = -\nabla \cdot \left(\frac{1}{2}\rho w^2 \mathbf{w}\right) + \rho \mathbf{g} \cdot \mathbf{w} + \nabla \cdot (\boldsymbol{\sigma} \cdot \mathbf{w}) - \boldsymbol{\sigma} : \nabla \mathbf{w}. \tag{2.100}$$

In Eq. (2.100) $\nabla \mathbf{w}$ denotes the vector gradient, which in Cartesian coordinates is defined as:

$$\nabla \mathbf{w} = \begin{bmatrix} \frac{\partial w_x}{\partial x} & \frac{\partial w_x}{\partial y} & \frac{\partial w_x}{\partial z} \\ \frac{\partial w_y}{\partial x} & \frac{\partial w_y}{\partial y} & \frac{\partial w_y}{\partial z} \\ \frac{\partial w_z}{\partial x} & \frac{\partial w_z}{\partial y} & \frac{\partial w_z}{\partial z} \end{bmatrix}, \text{ where } \mathbf{w} = \begin{bmatrix} w_x \\ w_y \\ w_z \end{bmatrix}. \tag{2.101}$$

The two dot product denotes the scalar product of two tensors. In Cartesian coordinates the product is expressed in the following way:

$$\boldsymbol{\sigma} : \nabla \mathbf{w} = \begin{bmatrix} \sigma_{xx} & \tau_{xy} & \tau_{xz} \\ \tau_{yx} & \sigma_{yy} & \tau_{yz} \\ \tau_{zx} & \tau_{zy} & \sigma_{zz} \end{bmatrix} : \begin{bmatrix} \frac{\partial w_x}{\partial x} & \frac{\partial w_x}{\partial y} & \frac{\partial w_x}{\partial z} \\ \frac{\partial w_y}{\partial x} & \frac{\partial w_y}{\partial y} & \frac{\partial w_y}{\partial z} \\ \frac{\partial w_z}{\partial x} & \frac{\partial w_z}{\partial y} & \frac{\partial w_z}{\partial z} \end{bmatrix}. \tag{2.102}$$

After multiplication, and considering that $\tau_{xy} = \tau_{yx}$, $\tau_{yz} = \tau_{zy}$ and $\tau_{xz} = \tau_{zx}$, the following is obtained:

$$\boldsymbol{\sigma} : \nabla \mathbf{w} = \sigma_{xx}\frac{\partial w_x}{\partial x} + \sigma_{yy}\frac{\partial w_y}{\partial y} + \sigma_{zz}\frac{\partial w_z}{\partial z} +$$
$$+ \tau_{xy}\left(\frac{\partial w_y}{\partial x} + \frac{\partial w_x}{\partial y}\right) + \tau_{yz}\left(\frac{\partial w_z}{\partial y} + \frac{\partial w_y}{\partial z}\right) + \tau_{zx}\left(\frac{\partial w_x}{\partial z} + \frac{\partial w_z}{\partial x}\right). \tag{2.103}$$

The scalar product $\nabla \cdot (\boldsymbol{\sigma} \cdot \mathbf{w})$ can be obtained by determining first:

$$\boldsymbol{\sigma} \cdot \mathbf{w} = \begin{bmatrix} \sigma_{xx} & \tau_{xy} & \tau_{xz} \\ \tau_{yx} & \sigma_{yy} & \tau_{yz} \\ \tau_{zx} & \tau_{zy} & \sigma_{zz} \end{bmatrix} \cdot \begin{bmatrix} w_x \\ w_y \\ w_z \end{bmatrix} = \begin{bmatrix} \sigma_{xx}w_x + \tau_{xy}w_y + \tau_{xz}w_z \\ \tau_{yx}w_x + \sigma_{yy}w_y + \tau_{yz}w_z \\ \tau_{zx}w_x + \tau_{zy}w_y + \sigma_{zz}w_z \end{bmatrix}. \tag{2.104}$$

Then, the product $\nabla \cdot (\boldsymbol{\sigma} \cdot \mathbf{w})$ is expressed as:

$$\nabla \cdot (\boldsymbol{\sigma} \cdot \mathbf{w}) = \frac{\partial}{\partial x}\left[\sigma_{xx}w_x + \tau_{xy}w_y + \tau_{xz}w_z\right] + \frac{\partial}{\partial y}\left[\tau_{yx}w_x + \sigma_{yy}w_y + \tau_{yz}w_z\right]$$
$$+ \frac{\partial}{\partial y}\left[\tau_{zx}w_z + \tau_{yz}w_z + \sigma_{zz}w_z\right]. \tag{2.105}$$

Equation (2.100) defines changes in kinetic energy. Moreover, the account will be taken of potential energy V_p related to a mass unit and defined as follows:

$$\mathbf{g} = -\nabla V_p. \tag{2.106}$$

The scalar product $\rho\mathbf{g}\cdot\mathbf{w} = \rho\mathbf{w}\cdot\mathbf{g}$ in Eq. (2.100) can be written in the following way:

$$-\rho(\mathbf{w}\cdot\nabla V_p) = -\nabla(\rho\mathbf{w}V_p) + V_p\nabla\cdot(\rho\mathbf{w}). \tag{2.107}$$

Considering (2.106) and continuity Eq. (2.52) in (2.107), the result is:

$$\rho\mathbf{w}\cdot\mathbf{g} = -\nabla\cdot(\rho\mathbf{w}V_p) - V_p\frac{\partial\rho}{\partial t}. \tag{2.108}$$

If potential energy V_p is independent of time t, which is the case for the Earth's gravitational field, Eq. (2.108) can be written as:

$$\rho\mathbf{w}\cdot\mathbf{g} = -\nabla\cdot(\rho\mathbf{w}V_p) - V_p\frac{\partial(\rho V_p)}{\partial t}. \tag{2.109}$$

Considering (2.109) in (2.100), the following is obtained:

$$\frac{\partial}{\partial t}\left(\frac{1}{2}\rho w^2 + \rho V_p\right) = -\nabla\cdot\left[\left(\frac{1}{2}\rho w^2 + \rho V_p\right)\mathbf{w}\right] + \nabla(\boldsymbol{\sigma}\cdot\mathbf{w}) - \boldsymbol{\sigma}:\nabla\mathbf{w}. \tag{2.110}$$

In the case of the Earth's gravitational field:

$$\mathbf{g} = -\mathbf{k}g, \quad V_p = gz, \tag{2.111}$$

where $g = 9.81\,\text{m/s}^2$ is gravitational acceleration, and z is the control volume elevation above the adopted reference level. The term on the left side of Eq. (2.110) denotes changes in energy over time related to a unit of volume. The first term on the right side denotes the flux of kinetic and potential energy related to a unit of volume. The last two terms on the right side include the work done against pressure forces and the work done by viscosity forces on the flowing fluid, as well as the irreversible increment in internal energy being the effect of the conversion of mechanical work done by the viscosity forces to internal energy. This irreversible increment in internal energy related to a unit of volume is referred to as viscous dissipation.

2.7.3.2 Energy Conservation Equation

The analysis presented in subsection 2.6.3.1 dealt with changes in mechanical energy only. This subsection presents balance equations resulting from the principle of total energy conservation, i.e. conservation of both mechanical and internal energy. In the energy balance Eq. (2.32), the power used to overcome viscosity forces $\dot{L}_\mu$ and the power needed to overcome pressure-related forces $\dot{L}_p$, defined as:

$$\dot{L}_p = -\int_{cs} p\,\mathbf{w}\cdot d\mathbf{A} = -\int_{cv} \nabla\cdot[p\,\mathbf{w}]dV, \tag{2.112}$$

will be treated jointly to give:

$$\dot{L}_p + \dot{L}_\mu = -\int_{cv} \nabla\cdot[p\,\mathbf{w}]dV + \dot{L}_\mu = \int_{cv} \nabla\cdot(\boldsymbol{\sigma}\cdot\mathbf{w})dV. \tag{2.113}$$

The surface integral in expression (2.112) is transformed into a volume integral using the divergence theorem (2.41). Considering (2.113) in Eq. (2.32), and assuming that $\dot{L}_t = 0$, the following is obtained:

$$\frac{\partial}{\partial t}\int_{cv} \rho\,e\,dV = -\int_{cs} \rho\,e\,\mathbf{w}\cdot d\mathbf{A} + \int_{cv} \nabla\cdot(\boldsymbol{\sigma}\cdot\mathbf{w})dV + \dot{Q} + \dot{Q}_v, \tag{2.114}$$

where the product $\boldsymbol{\sigma}\cdot\mathbf{w}$ is defined by formula (2.104). If the potential of (2.111) is taken into account, the energy of 1 kg of the medium defined by formula (2.16) can be written as:

$$e = u + \frac{w^2}{2} + V_p. \tag{2.115}$$

The calculation of the power needed to overcome pressure-related forces needs an additional explanation. The force acting on an elementary surface is $-p\,\mathbf{n}\,dA$, and the power consumed to pump a fluid into or out of the control volume through that surface is:

$$d\dot{L}_p = -p\,\mathbf{n}\,dA\cdot\mathbf{w} = -p\mathbf{w}\cdot d\mathbf{A}, \tag{2.116}$$

where $d\mathbf{A} = \mathbf{n}\,dA$.

The power needed to overcome the pressure forces is therefore expressed as:

$$\dot{L}_p = -\int_{cs} p\mathbf{w}\cdot d\mathbf{A}. \tag{2.117}$$

If the fluid is pumped into the control volume, the power transferred from the environment to the control volume is (Fig. 2.8):

$$d\dot{L}_{p,1} = -p_1 \mathbf{w}_1 \cdot \mathbf{n}_1 dA_1. \tag{2.118}$$

Considering that the scalar product $\mathbf{w}_1 \cdot \mathbf{n}_1$ is negative, power $d\dot{L}_{p,1}$ is positive. If the fluid flows out of the control volume, power $d\dot{L}_{p,2}$ is defined as (Fig. 2.8):

$$d\dot{L}_{p,2} = -p_2 \mathbf{w}_2 \cdot \mathbf{n}_2 dA_2. \tag{2.119}$$

Considering that product $\mathbf{w}_2 \cdot \mathbf{n}_2$ is positive, power $d\dot{L}_{p,2}$ is negative. The obtained results comply with the principles adopted for the formulation of the energy balance Eq. (2.18).

The first term on the right side of Eq. (2.114) will be transformed using the divergence theorem.

$$\int_{cs} \rho e \mathbf{w} \cdot d\mathbf{A} = \int_{cv} \nabla \cdot (\rho e \mathbf{w}) dV. \tag{2.120}$$

Heat flow rate $\dot{Q}$ is positive if heat is supplied to the control volume. If two kinds of the heat transfer are taken into consideration: conduction and radiation, heat flow rate $\dot{Q}$ transferred through the control volume surface is expressed as:

$$\dot{Q} = -\int_{cs} \dot{\mathbf{q}} \cdot d\mathbf{A} - \int_{cs} \dot{\mathbf{q}}_r \cdot d\mathbf{A}, \tag{2.121}$$

where the heat flux vector $\dot{\mathbf{q}}$ is defined by the Fourier law:

$$\dot{\mathbf{q}} = -\lambda \nabla T, \tag{2.122}$$

and the radiation heat flux vector $\dot{\mathbf{q}}_r$ satisfies the radiation energy conservation equation:

$$\nabla \cdot \dot{\mathbf{q}}_r = \chi(4\pi I_b - G), \tag{2.123}$$

where: χ—absorptivity, $I_b = 4\sigma T^4$—black body radiation intensity, $\sigma = 5.67 \times 10^{-8}$ W/(m^2 · K^4)—Stefan-Boltzmann constant, G—incident radiation (directional integral of radiation intensity).

After the divergence theorem (2.41) is applied to the surface integrals in Eq. (2.117), the following is obtained:

$$\dot{Q} = \int_{cv} (-\nabla \cdot \dot{\mathbf{q}} - \nabla \cdot \dot{\mathbf{q}}_r) dV. \tag{2.124}$$

Heat flow rate $\dot{Q}_v$ can be generated inside the control volume due to the effect of internal volume heat sources with unit power $\dot{q}_v$ [W/m^3].

The symbol $\dot{q}_v$ stands for the thermal power (heat flow rate) generated in a unit of the body volume. The heat flux inside a body can be generated due to nuclear reactions or due to electric heating by means of resistance or induction. Heat flow rate $\dot{Q}_v$ is defined as follows:

$$\dot{Q} = \int_{cv} \dot{q}_v dV. \tag{2.125}$$

Considering (2.116), (2.124) and (2.125) in (2.114), the following is obtained:

$$\frac{\partial}{\partial t} \int_{cv} \rho e \, dV = - \int_{cv} \nabla \cdot (\rho e \mathbf{w}) dV + \int_{cv} \nabla \cdot (\boldsymbol{\sigma} \cdot \mathbf{w}) dV + \\ + \int_{cv} (-\nabla \cdot \dot{\mathbf{q}} - \nabla \cdot \dot{\mathbf{q}}_r) dV + \int_{cv} \dot{q}_v \, dV. \tag{2.126}$$

Equation (2.126) is satisfied for every control volume, even a tiny one with the dimensions: dx, dy, dz, as presented in Fig. (2.10a). In such a case, volume integration in Eq. (2.126) can be omitted, which gives a differential form of the energy balance equation:

$$\frac{\partial(\rho e)}{\partial t} = -\nabla \cdot (\rho e \mathbf{w}) + \nabla \cdot (\boldsymbol{\sigma} \cdot \mathbf{w}) - \nabla \cdot \mathbf{q} - \nabla \cdot \dot{\mathbf{q}}_r + \dot{q}_v. \tag{2.127}$$

Taking account of (2.115), the equation can be written in a slightly different from:

$$\frac{\partial[\rho(u + w^2/2 + V_p)]}{\partial t} = -\nabla \cdot \left[\rho\left(u + \frac{w^2}{2} + V_p\right)\mathbf{w}\right] + \nabla \cdot (\boldsymbol{\sigma} \cdot \mathbf{w}) - \\ -\nabla \cdot \dot{\mathbf{q}} - \nabla \cdot \dot{\mathbf{q}}_r + \dot{q}_r. \tag{2.128}$$

Equation (2.128) will be reduced by subtracting from it the mechanical energy balance Eq. (2.110):

$$\frac{\partial(\rho u)}{\partial t} = -\nabla \cdot (\rho u \mathbf{w}) - \nabla \cdot \dot{\mathbf{q}} - \nabla \cdot \dot{\mathbf{q}}_r + \dot{q}_v + \boldsymbol{\sigma} : \nabla \mathbf{w}. \tag{2.129}$$

The last term of Eq. (2.129) will be discussed in detail first. Considering relations (2.69–2.74) in expression (2.103), the following is obtained:

$$\boldsymbol{\sigma} : \nabla \mathbf{w} = -p\left(\frac{\partial w_x}{\partial x} + \frac{\partial w_y}{\partial y} + \frac{\partial w_z}{\partial z}\right) - \frac{2}{3}\nabla \cdot \mathbf{w}\left(\frac{\partial w_x}{\partial x} + \frac{\partial w_y}{\partial y} + \frac{\partial w_z}{\partial z}\right) \\ + 2\mu\left[\left(\frac{\partial w_x}{\partial x}\right)^2 + \left(\frac{\partial w_y}{\partial y}\right)^2 + \left(\frac{\partial w_z}{\partial z}\right)^2\right] + \mu\left[\left(\frac{\partial w_x}{\partial y} + \frac{\partial w_y}{\partial x}\right)^2 \\ + \left(\frac{\partial w_y}{\partial z} + \frac{\partial w_z}{\partial y}\right)^2 + \left(\frac{\partial w_z}{\partial x} + \frac{\partial w_x}{\partial z}\right)^2\right]. \tag{2.130}$$

Noting that:

$$\nabla \cdot \mathbf{w} = \frac{\partial w_x}{\partial x} + \frac{\partial w_y}{\partial y} + \frac{\partial w_z}{\partial z} \tag{2.131}$$

and introducing:

$$\dot{\phi} = -\frac{2}{3}\mu(\nabla \cdot \mathbf{w})^2 + 2\mu\left[\left(\frac{\partial w_x}{\partial x}\right)^2 + \left(\frac{\partial w_y}{\partial y}\right)^2 + \left(\frac{\partial w_z}{\partial z}\right)^2\right]$$
$$+ \mu\left[\left(\frac{\partial w_x}{\partial y} + \frac{\partial w_y}{\partial x}\right)^2 + \left(\frac{\partial w_y}{\partial z} + \frac{\partial w_z}{\partial y}\right)^2 + \left(\frac{\partial w_z}{\partial x} + \frac{\partial w_x}{\partial z}\right)^2\right], \tag{2.132}$$

expression (2.130) can be written as:

$$\boldsymbol{\sigma} : \nabla \mathbf{w} = -p\nabla \cdot \mathbf{w} + \dot{\phi}, \tag{2.133}$$

where $\dot{\phi}$ is the dissipated energy flux related to a unit of volume. Substitution of (2.133) in the internal energy balance Eq. (2.129) results in:

$$\frac{\partial(\rho u)}{\partial t} + \nabla \cdot (\rho u \mathbf{w}) = -p\nabla \cdot \mathbf{w} - \nabla \cdot \dot{\mathbf{q}} - \nabla \cdot \dot{\mathbf{q}}_r + \dot{\mathbf{q}}_v + \dot{\phi}. \tag{2.134}$$

The left side of Eq. (2.130) can be transformed as follows:

$$\frac{\partial(\rho u)}{\partial t} + \nabla \cdot (\rho u \mathbf{w})$$
$$= \rho\frac{\partial u}{\partial t} + u\frac{\partial \rho}{\partial t} + \frac{\partial}{\partial x}(u\rho w_x) + \frac{\partial}{\partial y}(u\rho w_y) + \frac{\partial}{\partial z}(u\rho w_z)$$
$$= u\left(\frac{\partial \rho}{\partial t} + \frac{\partial(\rho w_x)}{\partial x} + \frac{\partial(\rho w_y)}{\partial y} + \frac{\partial(\rho w_z)}{\partial z}\right) + \rho\left(\frac{\partial u}{\partial t} + w_x\frac{\partial u}{\partial x}\right) \tag{2.135}$$
$$+ w_y\frac{\partial u}{\partial y} + w_z\frac{\partial u}{\partial z}\right) = u\left[\frac{\partial \rho}{\partial t} + \nabla \cdot (\rho \mathbf{w})\right] + \rho\frac{Du}{dt},$$

which, considering the continuity Eq. (2.52), gives:

$$\frac{\partial(\rho u)}{\partial t} + \nabla \cdot (\rho u \mathbf{w}) = \rho\frac{Du}{dt}. \tag{2.136}$$

The first term on the right side will be transformed using the continuity equation written in the form of (2.54), from which it follows that:

$$\nabla \cdot \mathbf{w} = -\frac{1}{\rho}\frac{D\rho}{Dt}. \tag{2.137}$$

Therefore, the first term on the right side can be transformed in the following way:

$$-p\nabla \cdot \mathbf{w} = \frac{p}{\rho}\frac{D\rho}{Dt} = \frac{p}{\rho}\frac{D\rho}{Dt} + \rho\frac{D}{Dt}\left(\frac{p}{\rho}\right) - \rho\frac{D}{Dt}\left(\frac{p}{\rho}\right) =$$
$$\frac{p}{\rho}\frac{D\rho}{Dt} + \rho\left(\rho\frac{Dp}{Dt} - p\frac{D\rho}{Dt}\right)\frac{1}{\rho^2} - \rho\frac{D}{Dt}(p\,v) \qquad (2.138)$$
$$= \frac{Dp}{Dt} - \rho\frac{D(p\,v)}{Dt}.$$

Substitution of (2.136) and (2.138) in the energy balance Eq. (2.134) results in:

$$\rho\frac{D(u+p\,v)}{Dt} = -\nabla \cdot \dot{\mathbf{q}} - \nabla \cdot \dot{\mathbf{q}}_r + \dot{q}_v + \frac{Dp}{Dt} + \dot{\phi}. \qquad (2.139)$$

Considering the definition of specific enthalpy i:

$$i = u + p\,v \qquad (2.140)$$

and the Fourier law:

$$\dot{\mathbf{q}} = -\lambda\,\nabla\,T, \qquad (2.141)$$

Equation (2.139) can be written as:

$$\rho\frac{Di}{Dt} = \nabla \cdot (\lambda\,\nabla\,T) - \nabla \cdot \dot{\mathbf{q}}_r + \dot{q}_v + \frac{Dp}{Dt} + \dot{\phi}. \qquad (2.142)$$

This is the form of the energy balance equation which is most often applied in practice. If the fluid flow isobaric, then $Dp/Dt = 0$. In many practical applications, e.g., those considering the flow of the water/steam mixture in the evaporator tubes or the steam flow in the boiler superheater, the pressure drop along the flow path and the local changes in pressure are not big enough to be taken into account in the energy balance Eq. (2.142) and it is generally assumed that $Dp/Dt = 0$. Also, the rate of energy dissipated due to friction $(\dot{\phi})$ is usually omitted, i.e. $\dot{\phi} = 0$. It is only in the case of supersonic flows that terms Dp/Dt and $\dot{\phi}$ can have more significance. Energy dissipation should also be taken into consideration for the flow of fluids characterized by very high viscosity. In this situation, the dissipated energy rate $\dot{\phi}$ caused by friction can take considerable values and must not be omitted. The left side of Eq. (2.142) can be expressed as a function of specific heat at constant pressure c_p and of temperature T. Considering that for pure substances the following relation occurs [167, 346]:

$$di = \left(\frac{\partial i}{\partial T}\right)_p dT + \left(\frac{\partial i}{\partial p}\right)_T dp = c_p dT + v(1 - \beta\,T)dp, \qquad (2.143)$$

equation (2.142) can be written as:

$$\rho\, c_p \frac{DT}{Dt} = \nabla \cdot (\lambda \nabla T) - \nabla \cdot \dot{\mathbf{q}}_r + \dot{q}_v + \beta\, T \frac{Dp}{Dt} + \dot{\phi}, \qquad (2.144)$$

where:

$$c_p = \left(\frac{\partial i}{\partial T} \right)_p, \qquad (2.145)$$

$$\beta = \frac{1}{v} \left(\frac{\partial v}{\partial T} \right)_p = -\frac{1}{\rho} \left(\frac{\partial \rho}{\partial T} \right)_p. \qquad (2.146)$$

Quantity β defined by formula (2.146) is the volume thermal expansion coefficient.

For liquids and solid bodies, which can be treated as incompressible, it can be assumed that:

$$c_p \cong c_v = c \text{ and } \beta = 0 \qquad (2.147)$$

In the case of a perfect gas satisfying the Clapeyron equation $pv = RT$, coefficient β is defined as:

$$\beta = \frac{1}{T} \qquad (2.148)$$

and the energy balance Eq. (2.144) takes the following form:

$$\rho\, c_p \frac{DT}{Dt} = \nabla \cdot (\lambda \nabla T) - \nabla \cdot \dot{\mathbf{q}}_r + \dot{q}_v + \frac{Dp}{Dt} + \dot{\phi}. \qquad (2.149)$$

Equation (2.149) is the general energy balance equation for a solid body, gas or liquid moving with velocity $\mathbf{w}$. The thermo-physical properties λ, c and ρ can depend on location or temperature. Similarly, the unit thermal power of heat sources $\dot{q}_v$ can vary depending on temperature, location or time.

In Appendix C, the energy balance Eq. (2.144) is written in a Cartesian, cylindrical and spherical system of coordinates assuming that the fluid is incompressible.

2.8   Mass, Momentum, and Energy Conservation Equations for One-Dimensional Flows

The equations for one-dimensional flows in the direction of axis x constitute a special case of the general equations derived in the previous subsections.

However, they will be derived again to illustrate derivation of basic equations in the differential form.

The mass, momentum and energy balance is done for the control area (control volume) with length Δx and the cross-section surface area A (Fig. 2.11).

2.8.1   Mass Conservation Equation (Continuity Equation)

In the general case, mass rates $A\,\rho\,w_x|_x$ and $A\,\rho\,w_x|_{x+\Delta x}$ flowing into and out of the control volume, respectively, are not equal to each other because in time Δt, the mass of the substance contained in the control volume changes by $\Delta m = \Delta(\Delta V \cdot \rho)$:

$$A\,\rho\,w_x|_x\Delta t = A\,\rho\,w_x|_{x+\Delta x}\Delta t + \Delta(\Delta V \cdot \rho). \qquad (2.150)$$

Considering in Eq. (2.151) that $\Delta V = A|_s \cdot \Delta x$, the following is obtained:

$$\frac{\Delta(A\,\rho)}{\Delta t} = \frac{(A\,\rho\,w_x)|_{x+\Delta x} - (A\,\rho\,w_x)|_x}{\Delta x}. \qquad (2.151)$$

If $\Delta t \to 0$ and $\Delta x \to 0$, Eq. (2.151) takes the following form:

$$\frac{\partial(A\,\rho)}{\partial t} = \frac{\partial(A\,\rho\,w_x)}{\partial x}. \qquad (2.152)$$

If the cross-section surface area depends on location, i.e., $A = A(x)$, Eq. (2.152) can be transformed into the following forms:

$$\frac{\partial\rho}{\partial t} + w_x\frac{\partial\rho}{\partial x} = -\rho\left(\frac{w_x}{A}\frac{\partial A}{\partial x} + \frac{\partial w_x}{\partial x}\right), \qquad (2.153)$$

$$\frac{\partial\rho}{\partial t} = -\frac{1}{A}\frac{\partial\dot{m}}{\partial t}, \qquad (2.154)$$

where $\dot{m} = A\,w_x\rho$ denotes the fluid mass flow rate in the duct. Mass conservation equations are also referred to as continuity equations.

Fig. 2.11 Control volume diagram for one-dimensional flows

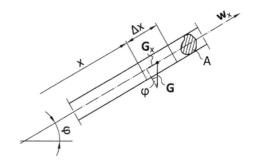

2.8.2 Momentum Conservation Equation

If Newton's second law is applied to the control volume presented in Fig. 2.11, the result is:

$$
\frac{\partial(\Delta V \cdot \rho\, w_x)}{\partial t}\bigg|_{x+\Delta x/2} = A\,\rho w_x w_x|_x - A\,\rho w_x w_x|_{x+\Delta x} - \left(A\,p|_{x+\Delta x} - A\,p|_x\right)
$$
$$
+ \left(p\frac{\partial A}{\partial x}\right)\bigg|_{x+\Delta x/2} \Delta x - \tau\,U|_{x+\Delta x/2}\cdot\Delta x - (\rho \cdot \Delta V \cdot g\,\sin\varphi)|_{x+\Delta x/2},
\tag{2.155}
$$

where $\Delta V = A \cdot \Delta x$, τ—shear stress on the duct wall, U—the duct inner perimeter. If $\Delta x \to 0$, Eq. (2.155) gives:

$$
\frac{1}{A}\frac{\partial(A\,\rho\,w_x)}{\partial t} = -\frac{1}{A}\cdot\frac{\partial(A\,\rho\,w_x w_x)}{\partial x}
$$
$$
-\frac{1}{A}\frac{\partial(p A)}{\partial x} + p\frac{\partial A}{\partial x} - \frac{\tau\,U}{A} - \rho g\,\sin\varphi\;.
\tag{2.156}
$$

If the cross-section surface area A is constant and location-independent, Eq. (2.156) is reduced to the following form:

$$
\frac{\partial(\rho\,w_x)}{\partial t} = -\frac{\partial(\rho\,w_x w_x)}{\partial x} - \frac{\partial p}{\partial x} - \frac{\tau\,U}{A} - \rho g\,\sin\varphi\;.
\tag{2.157}
$$

Transforming Eq. (2.156) to:

$$
\rho\frac{\partial w_x}{\partial t} + \frac{w_x}{A}\frac{\partial(\rho A)}{\partial t} = -\left(\rho w_x\frac{\partial(w_x)}{\partial x} + \frac{w_x}{A}\frac{\partial(\rho w_x A)}{\partial x}\right)
$$
$$
-\frac{\tau\,U}{A} - \frac{\partial p}{\partial x} - \frac{p}{A}\frac{\partial A}{\partial x} - \rho g\,\sin\varphi
\tag{2.158}
$$

and taking account of the continuity Eq. (2.153), the following is obtained from (2.158):

$$
\rho\frac{\partial w_x}{\partial t} + \rho\,w_x\frac{\partial w_x}{\partial x} = -\frac{\partial p}{\partial x} - \frac{\tau\,U}{A} - \rho g\,\sin\varphi\;.
\tag{2.159}
$$

Introducing the substantial derivative (2.46), which in this case is expressed as:

$$
\frac{D}{Dt} = \frac{\partial}{\partial t} + w_x\frac{\partial}{\partial x},
\tag{2.160}
$$

equation (2.159) can be written as follows:

$$\rho \frac{Dw_x}{Dt} = -\frac{\partial p}{\partial x} - \frac{\tau U}{A} - \rho g \sin \varphi \ . \tag{2.161}$$

The structures of Eq. (2.161) and of Newton's second law are identical. The first component is the force arising from pressure, the second—from friction forces on the duct wall, and the third component is the gravitational force.

All the three components are related to a unit of volume. It can be seen that if $(\partial p / \partial x) < 0$, i.e., if pressure p drops along the flow direction, velocity w_x rises. Friction force $\tau U / A$ always reduces the flow velocity. Gravitational force $(\rho g \sin \varphi)$ delays upward flows $(0 < \varphi < \pi)$ and accelerates downward flows $(\pi < \varphi < 2\pi)$.

The relation between tangential stress τ on the wall and the pressure drop caused by frictional resistance Δp_t is obtained from the following equation:

$$\tau U \Delta x = A \Delta p_t, \tag{2.162}$$

from which it follows that:

$$\frac{\tau U}{A} = \frac{\Delta p_t}{\Delta x}. \tag{2.163}$$

If $\Delta x \rightarrow 0$, then:

$$\frac{\tau U}{A} = \frac{\partial p_t}{\partial x}. \tag{2.164}$$

Pressure losses in a circular duct with inner diameter d_h or in a duct with a cross-section other than circular with equivalent hydraulic diameter d_h and length x are calculated using the following relation:

$$p_t = \lambda \frac{x}{d_h} \frac{\rho w_x^2}{2}. \tag{2.165}$$

Pressure losses per a unit of length are then expressed as:

$$\frac{\tau U}{A} = \frac{\partial p_t}{\partial x} \frac{\lambda}{d_h} \frac{\rho w_x^2}{2}. \tag{2.166}$$

In practical calculations, mass flow rate $\dot{m} = A \rho w_x$ is more often calculated than velocity w_x. Equation (2.156) gives:

$$\frac{\partial \dot{m}}{\partial t} = -\frac{\partial}{\partial x} \left(\frac{\dot{m}^2}{\rho A} \right) - \frac{\partial (p A)}{\partial x} - A \frac{\partial p_t}{\partial x} - A \rho g \sin \varphi \ . \tag{2.167}$$

If the channel cross-section is constant and independent of coordinate x, Eq. (2.167) can be written in the following form:

$$\frac{\partial \dot{m}}{\partial t} = -\frac{1}{A}\frac{\partial}{\partial x}\left(\frac{\dot{m}^2}{\rho}\right) - A\left(\frac{\partial p}{\partial x} + \frac{\partial p_t}{\partial x} + \rho g \sin\varphi\right). \tag{2.168}$$

The next to be derived is the energy conservation equation.

2.8.3 Energy Conservation Equation

According to formula (2.16), the energy of 1 kg of a fluid contained in a control volume is expressed as:

$$e = u + \frac{w_x^2}{2} + V_p = u + \frac{w_x^2}{2} + g x \sin\varphi, \tag{2.169}$$

where $V_p = g x \sin\varphi$ is potential energy.

In the fluid area only the heat transfer through conduction and convection will be taken into account when formulating the energy conservation equation. Radiation heat transfer between the hot fluid and the channel surface is taken into account by adding a radiation heat transfer coefficient to the convection heat transfer coefficient.

The energy balance for the control volume presented in Fig. 2.11 has the following form:

$$\begin{aligned}\frac{\partial}{\partial t}(\Delta V \rho e) &= \left(A\rho e w_x|_x - A\rho e w_x|_{x+\Delta x}\right) + \dot{q}\,U\cdot\Delta x \\ &+ \left[-A\lambda\frac{\partial T}{\partial x}\big|_x - \left(-A\lambda\frac{\partial T}{\partial x}\big|_{x+\Delta x}\right)\right] \\ &+ \left(A p w_x|_x - A p w_x|_{x+\Delta x}\right) + \dot{q}_v A\cdot\Delta x.\end{aligned} \tag{2.170}$$

Dividing Eq. (2.170) by $A\cdot\Delta x$ and assuming that $\Delta x \to 0$, the following is obtained:

$$\begin{aligned}\frac{1}{A}\frac{\partial}{\partial t}(A\rho e) &= -\frac{1}{A}\frac{\partial}{\partial x}(A\rho e w_x) + \frac{\dot{q}U}{A} + \frac{1}{A}\frac{\partial}{\partial x}\left(A\lambda\frac{\partial T}{\partial x}\right) \\ &- \frac{1}{A}\frac{\partial}{\partial x}(A p w_x) + \dot{q}_v.\end{aligned} \tag{2.171}$$

The term on the left side of the energy balance equation represents changes in energy accumulated in the fluid unit of volume over time. The first term on the right side represents the convective flux of total energy. The second term is the heat flux supplied to (or carried away from) the flowing fluid through the channel inner surface: if $\dot{q} > 0$, heat is supplied to the fluid; if $\dot{q} < 0$, heat flows away from the fluid to the outside through the inner surface of the channel (duct). The heat flux $\dot{q}$ is the sum of convection and radiation heat flux. The third term on the right side is the heat flux transferred in the fluid by means of convection, and the fourth—the power used to overcome pressure forces. The fifth term represents the heat flux generated in the fluid by heat sources giving up heat uniformly across the fluid entire volume.

All the heat fluxes are related to the fluid unit of volume. All terms in Eq. (2.171) are expressed in W/m^3. Equation (2.171) will be transformed to a simpler form. Substitution of (2.169) in (2.171) results in:

$$\frac{1}{A}\frac{\partial}{\partial t}\left[A\,\rho\left(u+\frac{w_x}{2}+g\,x\,\sin\varphi\right)\right] = -\frac{1}{A}\frac{\partial}{\partial x}\left[A\,\rho\,w_x\left(u+\frac{w_x}{2}+g\,x\,\sin\varphi\right)\right]$$
$$+\frac{\dot{q}\,U}{A}+\frac{1}{A}\frac{\partial}{\partial x}\left(A\lambda\frac{\partial T}{\partial x}\right)-\frac{1}{A}\frac{\partial}{\partial x}(A\,p\,w_x)+\dot{q}_v\;.$$

$$(2.172)$$

The expressions on the left and on the right side with term $g\,x\,\sin\varphi$ will be transformed first:

$$\frac{1}{A}\frac{\partial}{\partial t}[A\,\rho\,g\,x\,\sin\varphi]+\frac{1}{A}\frac{\partial}{\partial x}(A\,\rho\,w_x g\,x\,\sin\varphi)$$
$$=\frac{1}{A}A\,\rho\,g\,\sin\varphi\frac{\partial x}{\partial t}+\frac{1}{A}A\,g\,x\,\sin\varphi\frac{\partial\rho}{\partial t}$$
$$+\frac{1}{A}A\,\rho\,w_x g\,\sin\varphi+\frac{1}{A}\frac{\partial}{\partial x}(A\,\rho\,w_x)g\,x\,\sin\varphi\;.$$

$$(2.173)$$

Considering that $\partial x/\partial t = 0$, and taking account of continuity Eq. (2.152), expression (2.173) can be written in the following form:

$$\frac{1}{A}\frac{\partial}{\partial t}(A\,\rho\,g\,\sin\varphi)+\frac{1}{A}\frac{\partial}{\partial x}\left(A\,\rho\,w\,g\,\sin\varphi\right)\;.$$

$$(2.174)$$

The term on the left side and the first term on the right side of Eq. (2.172) will be transformed in a similar way with respect to the remaining part of the fluid energy $\left(u+w_x^2/2\right)$:

$$\frac{1}{A}\frac{\partial}{\partial t}\left[A\,\rho\left(u+\frac{w_x^2}{2}\right)\right]+\frac{1}{A}\frac{\partial}{\partial x}\left[A\,\rho\,w_x\left(u+\frac{w_x^2}{2}\right)\right]=$$
$$\frac{1}{A}\,\rho\frac{\partial}{\partial t}\left(u+\frac{w_x^2}{2}\right)+\frac{1}{A}\left(u+\frac{w_x^2}{2}\right)\frac{\partial(A\,\rho)}{\partial t}+\frac{1}{A}A\,\rho\,w_x\frac{\partial}{\partial x}\left(u+\frac{w_x^2}{2}\right)$$
$$+\left(u+\frac{w_x^2}{2}\right)\frac{1}{A}\frac{\partial}{\partial x}(A\,\rho\,w_x)=\rho\frac{\partial}{\partial t}\left(u+\frac{w_x^2}{2}\right)+\rho\,w_x\frac{\partial}{\partial x}\left(u+\frac{w_x^2}{2}\right)+$$
$$\frac{1}{A}\left(u+\frac{w_x^2}{2}\right)\left[\frac{\partial(A\,\rho)}{\partial t}+\frac{\partial}{\partial x}(A\,\rho\,w_x)\right]=\rho\frac{\partial}{\partial t}\left(u+\frac{w_x^2}{2}\right)+\rho\,w_x\frac{\partial}{\partial x}\left(u+\frac{w_x^2}{2}\right)\;.$$

$$(2.175)$$

Equation (2.175) is transformed taking account of the continuity equation. Considering (2.174) and (2.175) in Eq. (2.172), the following is obtained:

$$
\rho \frac{\partial}{\partial t} \left(u + \frac{w_x^2}{2} \right) + \rho\, w_x \frac{\partial}{\partial x} \left(u + \frac{w_x^2}{2} \right) = -\rho\, g\, w_x \sin \varphi
$$
$$
- \frac{1}{A} \frac{\partial}{\partial x} (A\, p\, w_x) + \frac{1}{A} \frac{\partial}{\partial x} \left(A\lambda \frac{\partial T}{\partial x} \right) + \frac{\dot{q}\, U}{A} + \dot{q}_v \, .
\tag{2.176}
$$

Equation (2.176) will be transformed using the mechanical energy balance equation obtained by multiplying the momentum conservation Eq. (2.159) by w_x:

$$
\rho\, w_x \frac{\partial w_x}{\partial t} = -\rho\, w_x^2 \frac{\partial w_x}{\partial x} - w_x \frac{\partial p}{\partial x} - w_x \frac{\tau\, U}{A} - \rho\, g\, w_x \sin \varphi \, .
\tag{2.177}
$$

It should be emphasized that Eq. (2.177) is not a mechanical energy conservation equation because mechanical energy $\left(w_x^2/2 + g x \sin \varphi \right)$ is only a part of energy e defined by formula (2.169) and is not conserved. Subtracting Eq. (2.177) from Eq. (2.176), the following is obtained:

$$
\rho \frac{\partial}{\partial t} \left(u + \frac{1}{2} w_x^2 \right) + \rho\, w_x \frac{\partial}{\partial t} \left(u + \frac{1}{2} w_x^2 \right) - \rho\, w_x \frac{\partial w_x}{\partial t}
$$
$$
= \frac{1}{A} \frac{\partial}{\partial x} \left(A\lambda \frac{\partial T}{\partial x} \right) + \frac{\dot{q}\, U}{A} + \dot{q}_v.
\tag{2.178}
$$

Equation (2.178) can be transformed into:

$$
\rho \frac{\partial u}{\partial t} + \rho\, w_x \frac{\partial w_x}{\partial t} + \rho\, w_x \frac{\partial u}{\partial x} + \rho\, w_x^2 \frac{\partial w_x}{\partial x} - \rho\, w_x \frac{\partial w_x}{\partial t}
$$
$$
= \rho\, w_x^2 \frac{\partial w_x}{\partial x} + w_x \frac{\tau\, U}{A} - \frac{1}{A} p \frac{\partial}{\partial x} (A\, w_x) - w_x \frac{\partial p}{\partial x} + w_x \frac{\partial p}{\partial x}
$$
$$
+ \frac{1}{A} \frac{\partial}{\partial x} \left(A\lambda \frac{\partial T}{\partial x} \right) + \frac{\dot{q}\, U}{A} + \dot{q}_v.
\tag{2.179}
$$

Simplifying and reducing Eq. (2.179), the following is obtained:

$$
\rho \frac{\partial u}{\partial t} + \rho\, w_x \frac{\partial u}{\partial x} = w_x \frac{\tau\, U}{A} - \frac{1}{A} p \frac{\partial}{\partial x} (A w_x) + \frac{1}{A} \frac{\partial}{\partial x} \left(A\lambda \frac{\partial T}{\partial x} \right) + \frac{\dot{q}\, U}{A} + \dot{q}_v.
\tag{2.180}
$$

Considering that:

$$
\frac{\partial u}{\partial t} + w_x \frac{\partial u}{\partial x} = \frac{Du}{Dt}
\tag{2.181}
$$

and taking account of formula (2.166), Eq. (2.180) can be written in a simpler form:

$$
\rho \frac{Du}{Dt} = w_x \frac{\partial p_t}{\partial x} - \frac{1}{A} p \frac{\partial}{\partial x} (A w_x) + \frac{1}{A} \frac{\partial}{\partial x} \left(A\lambda \frac{\partial T}{\partial x} \right) + \frac{\dot{q}\, U}{A} + \dot{q}_v.
\tag{2.182}
$$

which is the energy balance equation for a compressible medium unidirectional flow.

In the case of the flow of steam or gas, it is more convenient to write the energy balance equation not as a function of internal energy u but enthalpy i.

Considering that:

$$u = i - \frac{p}{\rho} \qquad (2.183)$$

equation (2.182) can be written as follows:

$$
\rho \frac{\partial \left(i - \frac{p}{\rho}\right)}{\partial t} + \rho w_x \frac{\partial \left(i - \frac{p}{\rho}\right)}{\partial x} = w_x \frac{\partial p_t}{\partial x} - \frac{1}{A} p \frac{\partial}{\partial x}(A\, w_x)
$$
$$
+ \frac{1}{A} \frac{\partial}{\partial x}\left(A\lambda \frac{\partial T}{\partial x}\right) + \frac{\dot{q}U}{A} + \dot{q}_v, \qquad (2.184)
$$

After transformations it gives:

$$
\rho \frac{\partial i}{\partial t} - \frac{\partial p}{\partial t} + \frac{p}{\rho} \frac{1}{A} A \frac{\partial p}{\partial t} = -\rho w_x \frac{\partial i}{\partial x} + w_x \frac{\partial p}{\partial x} - \frac{p w_x}{\rho} \frac{\partial \rho}{\partial x} - \frac{1}{A} \frac{p}{\rho} \frac{\partial}{\partial x}(A\,\rho\, w_x)
$$
$$
+ \frac{p w_x}{\rho} \frac{\partial \rho}{\partial x} - w_x \frac{\partial p_t}{\partial x} + \frac{1}{A} \frac{\partial}{\partial x}\left(A\lambda \frac{\partial T}{\partial x}\right) + \frac{\dot{q}U}{A} + \dot{q}_v. \qquad (2.185)
$$

Rearranging the terms in (2.185), the following is obtained:

$$
\rho \frac{\partial i}{\partial t} + \rho w_x \frac{\partial i}{\partial x} + \frac{p}{\rho} \frac{1}{A} \left[A \frac{\partial \rho}{\partial t} + \frac{\partial}{\partial x}(A\,\rho\, w_x) \right]
$$
$$
= \frac{\partial p}{\partial t} + w_x \frac{\partial p}{\partial x} + w_x \frac{\partial p_t}{\partial x} + \frac{1}{A} \frac{\partial}{\partial x}\left(A\lambda \frac{\partial T}{\partial x}\right) + \frac{\dot{q}U}{A} + \dot{q}_v. \qquad (2.186)
$$

Taking account of the continuity Eq. (2.152) and introducing the substantial derivative, Eq. (2.186) can be written as:

$$
\rho \frac{Di}{Dt} = \frac{Dp}{Dt} + w_x \frac{\partial p_t}{\partial x} + \frac{1}{A} \frac{\partial}{\partial x}\left(A\lambda \frac{\partial T}{\partial x}\right) + \frac{\dot{q}U}{A} + \dot{q}_v, \qquad (2.187)
$$

where:

$$
\frac{Di}{Dt} = \frac{\partial i}{\partial t} + w_x \frac{\partial i}{\partial t},
$$
$$
\frac{Dp}{Dt} = \frac{\partial p}{\partial t} + w_x \frac{\partial p}{\partial x}. \qquad (2.188)
$$

Equation (2.187) is frequently applied to model flows in tubes of the power boiler evaporators, steam superheaters and economizers.

In such cases, the main source of energy are flue gases, from which the heat flux is transferred to the medium flowing inside the tubes. The value of the heat flux is characterized by heat flux $\dot{q}$ set on the inner surface of a tube with perimeter U and the cross-section surface area A. While modelling flows through channels heated by flue gases or channels located on the furnace chamber waterwalls, it is a common practice to omit all the terms on the right side of Eq. (2.187) except $\dot{q}U/A$.

Equation (2.187) then takes the following form:

$$\rho \frac{\partial i}{\partial t} + \rho w_x \frac{\partial i}{\partial x} = \frac{\dot{q}U}{A}. \tag{2.189}$$

If the flue gas flow through channels is modeled as one-dimensional, considering the flue gas small heat capacity $c_p\rho$, the term $\rho \, \partial i/\partial t = c_p\rho \, \partial T/\partial t$ can be omitted and the flue gas flow can be modeled as a steady-state one. If flue gases with a specific chemical composition are cooled using tubes arranged in the flue gas duct, where water or steam flows through the tubes, the flue gas heat is transferred to the surfaces of the tubes through convection and radiation. Heat conduction in flue gases is usually omitted.

If the fuel combustion process is finished in the flue gases then $\dot{q}_v = 0$ and flue gases are only cooled. Equation (2.188) for the unidirectional flue gas flow can thus be reduced to the following expression:

$$\rho \, w_x \frac{\partial i}{\partial x} = \frac{\dot{q}U}{A}, \tag{2.190}$$

where $\dot{q}$ is the sum of convective and radiative heat flux.

Radiative heat flux is usually calculated using simplified formulae. If the fluid is heated, the heat flux $\dot{q}$ is positive and if it is cooled $\dot{q}$ is negative. If the hot optically active gas is cooled then the heat flux $\dot{q}$ in Eq. (2.190) is defined by the formula

$$\dot{q} = -(h_c + h_r)(T - T_w) \tag{2.191}$$

where the symbols h_c and h_r denote the convection and radiation heat transfer and T_w is the temperature of the channel surface.

Chapter 3
Laminar Flow of Fluids in Ducts

This chapter is devoted to the laminar flow of incompressible fluids with constant physical properties. The analysis will concern the developed laminar flow and the developing laminar flow in the inlet section (entrance length). Experimental correlations will be presented for the mean heat transfer coefficients on the wall of a channel with a finite length.

3.1 Developed Laminar Flow

The first part of the analysis is focused on the fluid axially symmetric laminar developed flow in ducts with a constant inner diameter d_w, for cases with the Reynolds number $\mathrm{Re} = w d_w / v$ higher than or equal to 2300. A flow which is fully developed hydraulically and thermally occurs at a large distance from the duct inlet, usually at $(x/d_w) \geq 200$. The velocity and temperature profiles are already developed—they depend on the radial coordinate only and do not change on the duct length. The fluid radial velocity profile $w_x(r)$, pressure changes $p(x)$ along the horizontal duct axis and radial temperature distribution $T(r)$ are determined based on the mass (2.56), momentum (B6) and energy (Appendix C) conservation equations. Considering that the fluid is incompressible ($\beta = 0$) and the flow is steady, i.e., time-independent, and assuming that the temperature field has no source ($\dot{q}_v = 0$) and there is no heat transfer using radiation ($\dot{q}_r = 0$), and omitting energy dissipation ($\dot{\phi} = 0$), the following system of equations is obtained:

$$\frac{1}{r}\frac{\partial}{\partial r}(r\,w_r) + \frac{\partial w_x}{\partial x} = 0, \tag{3.1}$$

$$\rho\left(w_r \frac{\partial w_x}{\partial r} + w_x \frac{\partial w_x}{\partial x}\right) = -\frac{\partial p}{\partial x} + \mu\left(\frac{\partial^2 w_x}{\partial r^2} + \frac{1}{r}\frac{\partial w_x}{\partial r}\right), \tag{3.2}$$

© Springer International Publishing AG, part of Springer Nature 2019
D. Taler, *Numerical Modelling and Experimental Testing of Heat Exchangers*,
Studies in Systems, Decision and Control 161,
https://doi.org/10.1007/978-3-319-91128-1_3

$$\rho\, c_p\left(w_r\frac{\partial T}{\partial r}+w_x\frac{\partial T}{\partial x}\right)=\lambda\left[\frac{1}{r}\frac{\partial}{\partial r}\left(r\frac{\partial T}{\partial r}\right)+\frac{\partial^2 T}{\partial x^2}\right]. \tag{3.3}$$

Equations (3.1)–(3.3) describe the axially symmetric laminar flow of the fluid, which, if the flow is developed, can be substantially simplified. As the fluid velocity w_x does not change on the tube length,

$$\frac{dw_x}{dx}=0. \tag{3.4}$$

Considering (3.4) in (3.1), the following equation is obtained:

$$\frac{\partial}{\partial r}(rw_r)=0, \tag{3.5}$$

from which it follows that:

$$rw_r=\text{const.} \tag{3.6}$$

Due to boundary condition $w_r(r=r_w)=0$,

$$w_r=0. \tag{3.7}$$

The energy conservation equation usually omits heat conduction in the fluid flow direction, i.e. it is assumed that:

$$\lambda\frac{\partial^2 T}{\partial x^2}=0. \tag{3.8}$$

Assumption (3.8) can be adopted if the following condition is satisfied:

$$\text{Pe}\gg 1, \tag{3.9}$$

where Pe is the Peclet number defined as:

$$\text{Pe}=\text{Re}\,\text{Pr}=\frac{\rho\, w_x d_w}{\mu}\frac{c_p\mu}{\lambda}=\frac{w_x d_w}{a},\quad a=\frac{\lambda}{\rho\, c_p}. \tag{3.10}$$

Considering conditions (3.4) and (3.8) and equality (3.7) in Eqs. (3.1)–(3.3), the following two-equation system is obtained:

$$\mu\left(\frac{d^2 w_x}{dr^2}+\frac{1}{r}\frac{dw_x}{dr}\right)=\frac{dp}{dx}, \tag{3.11}$$

$$\rho\, c_p w_x\frac{\partial T}{\partial x}=\frac{\lambda}{r}\frac{\partial}{\partial r}\left(r\frac{\partial T}{\partial r}\right). \tag{3.12}$$

Using the momentum conservation Eq. (3.11), a formula will be derived for the velocity distribution in the tube cross-section, and also—taking account of the Darcy-Weisbach friction factor definition—a formula for the pressure distribution along the tube length. Solving the energy conservation equation, the temperature distribution will be found in the tube cross-section, and the Nusselt number value will be determined for two different boundary conditions at the tube inner surface: a constant heat flux or a constant temperature. Then, using Duhamel's integral, the obtained solutions will be generalized to cover the case where the tube inner surface heat flux and temperature values change on the tube length.

3.1.1 Velocity Distribution and the Pressure Drop

The momentum conservation Eq. (3.11) written as:

$$\frac{1}{r}\frac{d}{dr}\left(r\frac{dw_x}{dr}\right) = \frac{1}{\mu}\frac{dp}{dx} \tag{3.13}$$

will be solved under the following boundary conditions:

$$\frac{dw_x}{dr}\Big|_{r=0} = 0, \tag{3.14}$$

$$w_x|_{r=r_w} = 0. \tag{3.15}$$

Integrating Eq. (3.13) twice, the result is:

$$w_x = \frac{1}{4\mu}\frac{dp}{dx}r^2 + C\ln r + C_1. \tag{3.16}$$

Substitution of solution (3.16) in conditions (3.14) and (3.15) gives:

$$C = 0, \quad C_1 = -\frac{1}{4\mu}\frac{dp}{dx}r_w^2. \tag{3.17}$$

Considering constants (3.17) in solution (3.16), the following velocity profile is obtained:

$$w_x = -\frac{1}{4\mu}\frac{dp}{dx}\left(r_w^2 - r^2\right). \tag{3.18}$$

The maximum velocity occurs in the duct axis for $r = 0$:

$$w_c = -\frac{1}{4\mu}\frac{dp}{dx}r_w^2. \tag{3.19}$$

Dividing velocity w_x by w_c, the result is:

$$\frac{w_x}{w_c} = \left(1 - \frac{r^2}{r_w^2}\right). \tag{3.20}$$

The fluid mean velocity is defined by the following formula:

$$w_m = \frac{2}{r_w^2}\int_0^{r_w} w_x r\,dr = -\frac{r_w^2}{8\mu}\frac{dp}{dx}. \tag{3.21}$$

Dividing expression (3.20) by (3.21), local velocity w_x can be expressed as a function of mean velocity and the radius:

$$w_x = 2w_m\left(1 - \frac{r^2}{r_w^2}\right). \tag{3.22}$$

It follows from formula (3.22) that the fluid velocity in the duct axis w_c is twice higher than mean velocity w_m.

The Darcy-Weisbach friction factor ξ is defined as:

$$\xi = \frac{d_w\left(-\frac{dp}{dx}\right)}{\frac{1}{2}\rho w_m^2}, \tag{3.23}$$

which gives:

$$\left(-\frac{dp}{dx}\right) = \frac{\xi}{d_w}\frac{\rho w_m^2}{2}. \tag{3.24}$$

The following is obtained from formula (3.21):

$$\left(-\frac{dp}{dx}\right) = \frac{8\mu}{r_w^2}w_m. \tag{3.25}$$

Comparing expressions (3.23) and (3.24), the following equation is obtained:

$$\frac{\xi}{d_w}\frac{\rho w_m^2}{2} = \frac{8\mu}{r_w^2}w_m, \tag{3.26}$$

from which the laminar flow friction factor is determined:

$$\xi = \frac{64}{\rho \, w_m d_w/\mu} = \frac{64}{\mathrm{Re}}. \tag{3.27}$$

If the fluid mass flow rate $\dot{m} = \pi \, r_w^2 \, \rho \, w_m$ or mean velocity w_m are known, the friction-related pressure drop on length L is found integrating expression (3.23):

$$\Delta p = p\Big|_{x_1=x} - p\Big|_{x_2=x+L} = \int_x^{x+L} \left(-\frac{\mathrm{d}p}{\mathrm{d}x}\right) \mathrm{d}x$$

$$= \int_x^{x+L} \frac{\xi}{d_w} \frac{\rho \, w_m^2}{2} \mathrm{d}x = \xi \frac{L}{d_w} \frac{\rho \, w_m^2}{2}. \tag{3.28}$$

Substituting the friction factor ξ defined by formula (3.27) in (3.28), the following is obtained:

$$\Delta p = p\Big|_{x_1=x} - p\Big|_{x_2=x+L} = \frac{64}{\mathrm{Re}} \frac{L}{d_w} \frac{\rho \, w_m^2}{2} = \frac{8 \, \mu L}{r_w^2} w_m. \tag{3.29}$$

An analysis of formula (3.29) indicates that the pressure drop in the tube on length L is proportional to the fluid flow mean velocity w_m and inversely proportional to the square of the inner surface radius r_w^2.

3.1.2 Temperature Distribution

If the flow is hydraulically developed, the velocity profile does not depend on the velocity profile at the tube inlet, and it is also independent of coordinate x (cf. Fig. 3.1a). Similarly, the fluid flow is thermally developed if the temperature profile is steady and does not change on the tube length (cf. Fig. 3.1b).

Both in laboratory testing and in many industrial installations, the heated tube section is preceded by an unheated one. The velocity profile at the heated section inlet will be developed if the unheated tube section is sufficiently long. On the heated section, the flow is already developed hydraulically, whereas, on the thermal entrance length L_T in the duct heated part, the temperature profile is still developing. Assuming that the tube unheated section length L_{un} is equal to the hydraulic entrance length L_H, it can be roughly calculated using the following formula:

$$L_H = \frac{L_T}{\mathrm{Pr}}, \tag{3.30}$$

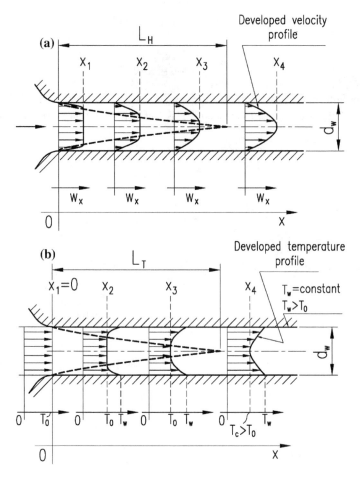

Fig. 3.1 Changes in velocity and temperature on the inlet section; **a** Development of the velocity profile in the tube, **b** development of the temperature profile in the tube; L_H—hydraulic entrance length, L_T—thermal entrance length

where the thermal entrance length L_T is calculated from the following expression:

$$L_T = 0.05\, d_w\,(\mathrm{Re\ Pr})\ . \tag{3.31}$$

Substitution of (3.31) in (3.30) results in:

$$L_H = 0.05\, d_w \mathrm{Re}\ . \tag{3.32}$$

The hydraulic entrance length defined by formula (3.30) depends on the Reynolds number only, whereas the thermal entrance length defined by formula (3.31) is determined by both the Reynolds and the Prandtl number. If the Prandtl number is smaller than one, the thermal entrance length is shorter than the hydraulic one (cf. Fig. 3.2).

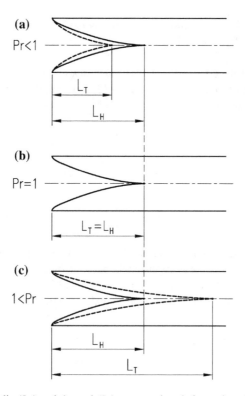

Fig. 3.2 The hydraulic (L_H) and thermal (L_T) entrance length for a given Reynolds number Re depending on the Prandtl number; **a** Pr < 1, **b** Pr = 1, **c** 1 < Pr

In the case of high Prandtl numbers, the thermal entrance length is longer than the hydraulic one.

It follows from the analysis of formula (3.32) that for the Reynolds number limit value Re = 2300, the hydraulic entrance length is 115 d_w. If the tube inner diameter is large, the length of the tube unheated section should be big. It should be noted that the temperature profile becomes developed only at the point with coordinate $x = L_H + L_T$. Coordinate x is counted from the unheated section inlet.

The velocity and temperature profiles will be fully developed at the distance of ($L_H + L_T$) from the tube inlet. A characteristic feature of the developed temperature profile is the constant temperature difference between any two points located in the same cross-section of the tube. This can be expressed as follows:

$$\frac{T_w - T}{T_w - T_m} = \psi\left(\frac{r}{r_w}\right).$$

(3.33)

Two cases of the boundary condition set on the tube heated section will be analysed. In the first, constant heat flux $\dot{q}_w$ is set on the tube inner surface; in the second, the temperature of the tube inner surface T_w is constant.

3.1.2.1   Temperature Distribution and the Nusselt Number at a Constant Heat Flux on the Tube Surface

The temperature distribution in fluid flow with steady and developed velocity and temperature profiles is described by Eq. (3.12) and the following boundary conditions (cf. Fig. 3.3):

$$T|_{x=0} = T_0(r) , \tag{3.34}$$

$$\lambda \frac{\partial T}{\partial r}|_{r=r_w} = \dot{q}_w, \tag{3.35}$$

$$\lambda \frac{\partial T}{\partial r}|_{r=0} = 0. \tag{3.36}$$

Due to two second kind boundary conditions set on the tube inner surface: $r = r_w$ and in the tube axis: $r = 0$, the constants in the solution of (3.12), (3.34)–(3.36) cannot be determined. The boundary condition defined in (3.36) will be replaced with a different boundary condition, which is presented in detail further on in this chapter.

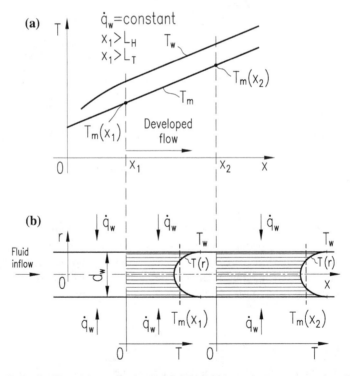

Fig. 3.3 Hydraulically and thermally developed fluid flow in a tube at a constant heat flux $\dot{q}_w$ set on the tube inner surface; **a** changes in the wall temperature T_w and in the fluid mass-averaged temperature T_m on the tube length, **b** radial distribution of fluid temperature

Multiplying both sides of Eq. (3.12) by r, and then integrating the equation along the radius from $r = 0$ to $r = r_w$, the following is obtained:

$$\int_0^{r_w} \rho\, c_p w_x \frac{\partial T}{\partial x} r\, \mathrm{d}r = \int_0^{r_w} \lambda \frac{\partial}{\partial r}\left(r \frac{\partial T}{\partial r}\right) \mathrm{d}r. \tag{3.37}$$

Considering that the fluid thermo-physical properties are constant, Eq. (3.37) gives:

$$\rho\, c_p \frac{\partial}{\partial x} \int_0^{r_w} w_x T\, r\, \mathrm{d}r = \lambda\left(r \frac{\partial T}{\partial r}\right)\Big|_{r=r_w} - \lambda\left(r \frac{\partial T}{\partial r}\right)\Big|_{r=0}. \tag{3.38}$$

Taking account of boundary conditions (3.35) and (3.36) and the mean velocity definition (3.21), and introducing mass-averaged temperature T_m, expressed as:

$$T_m = \frac{\int_{A_w} w_x T\, \mathrm{d}A}{\int_{A_w} w_x \mathrm{d}A} = \frac{2\pi \int_0^{r_w} w_x T\, r\, \mathrm{d}r}{2\pi \int_0^{r_w} w_x r\, \mathrm{d}r} = \frac{2}{r_w^2 w_m} \int_0^{r_w} w_x T\, r\, \mathrm{d}r \tag{3.39}$$

the following is obtained from Eq. (3.38):

$$\frac{\mathrm{d}T_m}{\mathrm{d}x} = \frac{2\dot{q}_w}{r_w \rho\, c_p w_m}. \tag{3.40}$$

Integrating Eq. (3.40) from 0 to x and taking account of boundary condition (3.34), the result is:

$$T_m = T_{m0} + \frac{2\dot{q}_w}{r_w \rho\, c_p w_m} x, \tag{3.41}$$

where T_{m0} is the fluid mass-averaged temperature in the tube cross-section for coordinate $x = 0$:

$$T_{m0} = \frac{2}{r_w^2 w_m} \int_0^{r_w} w_x T_0\, r\, \mathrm{d}r. \tag{3.42}$$

Equation (3.40) can also be obtained from the energy balance done for a control area with thickness dx:

$$\pi\, r_w^2 w_m \rho\, T_m|_x + \dot{q}_w \cdot 2\pi\, r_w \mathrm{d}x = \pi\, r_w^2 w_m \rho\, T_m|_{x+\mathrm{d}x}. \tag{3.43}$$

After simple transformations, Eq. (3.43) gives Eq. (3.40):

The solution of Eq. (3.12) under conditions (3.34)–(3.36) has the following form:

$$T = T_m(x) + T_2(r). \tag{3.44}$$

Substitution of (3.44) in (3.12) results in:

$$\rho c_p w_x \frac{\mathrm{d} T_m}{\mathrm{d} x} = \frac{\lambda}{r} \frac{\mathrm{d}}{\mathrm{d} r} \left(r \frac{\mathrm{d} T_2}{\mathrm{d} r} \right). \tag{3.45}$$

Considering (3.40) in Eq. (3.45), the following is obtained:

$$\frac{1}{r} \frac{\mathrm{d}}{\mathrm{d} r} \left(r \frac{\mathrm{d} T_2}{\mathrm{d} r} \right) = \frac{4 \dot{q}_w}{\lambda \, r_w} \left(1 - \frac{r^2}{r_w^2} \right). \tag{3.46}$$

Multiplying Eq. (3.46) by r and integrating it twice along r, the following is obtained:

$$T_2 = \frac{4 \dot{q}_w}{\lambda \, r_w} \left(\frac{1}{4} r^2 - \frac{1}{16 r_w^2} r^4 \right) + C_1 \ln r + C_2. \tag{3.47}$$

Considering that temperature T_2 has a finite value for $r = 0$, constant C_1 is equal to zero. Instead of boundary condition (3.36), the following condition will be used to determine constant C_2:

$$\frac{2}{r_w^2 w_m} \int_0^{r_w} w_x [T_m(x) + T_2(r)] r \, \mathrm{d} r = T_m(x), \tag{3.48}$$

which gives:

$$\frac{2}{r_w^2 w_m} \int_0^{r_w} w_x T_2(r) r \, \mathrm{d} r = 0. \tag{3.49}$$

It follows from condition (3.49) that mass-averaged temperature T_{2m}, defined as:

$$T_{2m} = \frac{2}{r_w^2 w_m} \int_0^{r_w} w_x T_2(r) r \, \mathrm{d} r \tag{3.50}$$

should be equal to zero:

$$\frac{2}{r_w^2} \int_0^{r_w} 2 \left(1 - \frac{r^2}{r_w^2} \right) \left[\frac{4 \dot{q}_w}{\lambda \, r_w} \left(\frac{1}{4} r^2 - \frac{1}{16 r_w^2} r^4 \right) + C_2 \right] r \, \mathrm{d} r = 0. \tag{3.51}$$

Condition (3.49) results in:

$$C_2 = -\frac{7}{24}\frac{\dot{q}_w r_w}{\lambda}. \tag{3.52}$$

Substituting $C_1 = 0$ and C_2 defined by formula (3.52) in (3.47), the following is obtained:

$$T_2 = \frac{\dot{q}_w r_w}{\lambda}\left[\left(\frac{r}{r_w}\right)^2 - \frac{1}{4}\left(\frac{r}{r_w}\right)^4 - \frac{7}{24}\right]. \tag{3.53}$$

The distribution of temperature $T(x, r)$ is obtained substituting temperature $T_m(x)$ defined by formula (3.41) and temperature $T_2(r)$ defined by formula (3.53) in formula (3.44):

$$T = T_{m0} + \frac{2\dot{q}_w}{\rho\, c_p w_m}\frac{x}{r_w} + \frac{\dot{q}_w r_w}{\lambda}\left[\left(\frac{r}{r_w}\right)^2 - \frac{1}{4}\left(\frac{r}{r_w}\right)^4 - \frac{7}{24}\right]. \tag{3.54}$$

Introducing dimensionless values:

$$R = \frac{r}{r_w}, \quad X_d = \frac{x}{r_w}, \quad \theta = \frac{T - T_{m0}}{\frac{\dot{q}_w r_w}{\lambda}}, \quad \mathrm{Re} = \frac{\rho\, w_m(2r_w)}{\mu},$$

$$\mathrm{Pr} = \frac{c_p\mu}{\lambda}, \quad \mathrm{Pe} = \mathrm{Re}\,\mathrm{Pr} = \frac{2r_w\rho\, c_p w_m}{\lambda} \tag{3.55}$$

expression (3.54) for the temperature distribution can be written as:

$$\theta = \frac{4X_d}{\mathrm{Pe}} + R^2 - \frac{R^4}{4} - \frac{7}{24}. \tag{3.56}$$

The heat transfer coefficient on the tube inner surface α can be found using the following formula:

$$\alpha = \frac{\dot{q}_w}{T|_{r=r_w} - T_m} = \frac{\dot{q}_w}{T_2|_{r=r_w}}. \tag{3.57}$$

The following is obtained from formula (3.53):

$$T|_{r=r_w} = \frac{11}{24}\frac{\dot{q}_w r_w}{\lambda}. \tag{3.58}$$

The heat transfer coefficient α can be determined substituting (3.58) in formula (3.57):

$$\alpha = \frac{\dot{q}_w}{T_2|_{r=r_w}} = \frac{24\,\lambda}{11\,r_w}. \tag{3.59}$$

The Nusselt number Nu is:

$$\mathrm{Nu} = \frac{\alpha \cdot 2r_w}{\lambda} = \frac{48}{11} \approx 4.364. \tag{3.60}$$

3.1.2.2 Temperature Distribution and the Nusselt Number at the Tube Wall Constant Temperature

The fluid temperature distribution is described by Eq. (3.12), boundary conditions (3.34), (3.36) and the following condition (cf. Fig. 3.4):

$$T|_{r=r_w} = T_w, \tag{3.61}$$

where T_w = const. is the wall constant temperature which does not change on the tube length.

The developed temperature distribution at the wall temperature set as constant is determined in [26, 37, 171, 265, 276] using the variables separation method.

The heat transfer coefficient on the tube wall and the Nusselt number are calculated using the following formulae:

$$\alpha = \frac{\lambda \frac{\partial T}{\partial r}|_{r=r_w}}{T_w - T_m}, \tag{3.62}$$

$$\mathrm{Nu} = \frac{\alpha \cdot (2r_w)}{\lambda}. \tag{3.63}$$

Fig. 3.4 Hydraulically and thermally developed laminar fluid flow in a tube at constant temperature T_w set on the tube inner surface

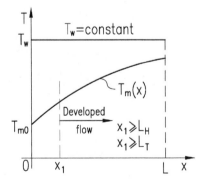

The Nusselt number determined in [26, 37, 171, 265, 276] is:

$$\mathrm{Nu} = \frac{\mu_0^2}{2} = \frac{2.70436^2}{2} = 3.657, \tag{3.64}$$

where λ_0 is the first eigenvalue of a respective Sturm-Liouville boundary value problem.

The history of mean temperature T_m as a function of coordinate x can be determined solving Eq. (3.40), which, considering the heat flux dependent on coordinate x and defined by the following formula:

$$\dot{q}_w = \alpha[T_w - T_m(x)] \tag{3.65}$$

can be transformed into:

$$\frac{dT_m}{dx} = \frac{2\alpha(T_w - T_m)}{r_w \, \rho \, c_p w_m}. \tag{3.66}$$

Equation (3.66) can be written as follows:

$$\frac{d(T_w - T_m)}{dx} = -\frac{2\alpha(T_w - T_m)}{r_w \rho \, c_p w_m}. \tag{3.67}$$

Integrating Eq. (3.67) under the following boundary condition:

$$T_m|_{x=0} = T_{m0} \tag{3.68}$$

the result is:

$$T_m(x) = T_w - (T_w - T_{m0}) \exp\left(\frac{-\alpha P x}{\dot{m} c_p}\right), \tag{3.69}$$

where P and $\dot{m}$ are the tube inner perimeter and the fluid mass flow rate, respectively.

The temperature distribution as a function of coordinates x and r in the fluid developed laminar flow can be found from expression (3.33):

$$T = T_w - \psi(T_w - T_m). \tag{3.70}$$

To determine function $\psi(R)$, the temperature distribution defined by (3.70) will be substituted in Eq. (3.12) and boundary conditions (3.36) and (3.61).

$$\frac{d^2\psi}{dR^2} + \frac{1}{R}\frac{d\psi}{dR} = -2\mathrm{Nu}\left(1 - R^2\right)\psi, \tag{3.71}$$

$$\frac{d\psi}{dR}\Big|_{R=0} = 0, \tag{3.72}$$

$$\psi|_{R=1} = 0. \tag{3.73}$$

Equations (3.71) and (3.72) constitute a two-point boundary value problem, because boundary condition (3.72) is written for $R = 0$, and condition (3.73) for $R = 1$.

The boundary value problem (3.71)–(3.73) will be replaced by the initial-value problem for a system of ordinary differential equations. A new variable ϕ will be introduced for this purpose. It is defined as:

$$\frac{d\psi}{dR} = \phi. \tag{3.74}$$

Equation (3.71) can thus be written as:

$$\frac{d\phi}{dR} + \frac{1}{R}\phi = -2\,\mathrm{Nu}\left(1 - R^2\right)\psi. \tag{3.75}$$

The initial conditions for the system of Eqs. (3.74) and (3.75) will be written for $R = 1$. Considering that the heat flux on the tube inner surface is defined as:

$$\lambda \frac{\partial T}{\partial r}\Big|_{r=r_w} = \alpha(T_w - T_m) \tag{3.76}$$

and taking account of the definition of variable ψ expressed by formula (3.33), the boundary condition defined by (3.76) can be transformed into the following form:

$$\lambda \left(\frac{\partial T}{\partial R}\frac{\partial R}{\partial r}\right)\Big|_{r=r_w} = \alpha(T_w - T_m), \tag{3.77}$$

$$\lambda \left[-(T_w - T_m)\frac{d\psi}{dR}\frac{1}{r_w}\right]\Big|_{R=1} = \alpha(T_w - T_m), \tag{3.78}$$

$$\frac{d\psi}{dR}\Big|_{R=1} = -\frac{\alpha \cdot (2r_w)}{2\lambda}. \tag{3.79}$$

Considering the definitions of ϕ and the Nusselt number $\mathrm{Nu} = \alpha \cdot (2r_w)/\lambda$, condition (3.79) can be written as:

$$\phi|_{R=1} = -\frac{\mathrm{Nu}}{2}. \tag{3.80}$$

The boundary-value problem for the system of ordinary differential Eqs. (3.74) and (3.75) and for boundary conditions (3.72)–(3.80) was solved iteratively using the shooting method [38, 39, 98]. According to it, the initial-value problem given by the equation system (3.74) and (3.75) with initial conditions (3.73) and (3.80) is solved iteratively. The Nusselt number in condition (3.80) was adjusted to satisfy condition (3.72), which, considering that $d\psi/dR = \phi$, can be written as:

$$\phi|_{R=0} = 0. \tag{3.81}$$

In every iteration step, the system of Eqs. (3.74) and (3.75) was solved using the explicit finite difference method. Interval [0, 1] was divided into $(N - 1)$ subintervals. The coordinate of the dimensionless radius $R_i = r_i/r_w$ is calculated using the following formula:

$$R_i = (i - 1)\Delta R, \quad i = 1, \ldots, N, \tag{3.82}$$

where ΔR is defined by the following expression:

$$\Delta R = \frac{1}{N - 1}. \tag{3.83}$$

Using the explicit finite difference method to approximate Eqs. (3.74) and (3.75), the result is:

$$\frac{\psi_i - \psi_{i-1}}{\Delta R} = \phi_i, \quad i = N, \ldots, 2, \tag{3.84}$$

$$\frac{\phi_i - \phi_{i-1}}{\Delta R} + \frac{1}{R}\phi_i = -2\mathrm{Nu}\left(1 - R_i^2\right)\psi_i, \quad i = N, \ldots, 2. \tag{3.85}$$

Solving Eq. (3.84) with respect to ψ_{i-1} and Eq. (3.85) with respect to ϕ_{i-1}, the following is obtained:

$$\psi_{i-1} = \psi_i - \Delta R \cdot \phi_i, \quad i = N, \ldots, 2, \tag{3.86}$$

$$\phi_{i-1} = \left(1 + \frac{\Delta R}{R_i}\right)\phi_i + 2 \cdot \Delta R \cdot \mathrm{Nu}\left(1 - R_i^2\right)\psi_i, \quad i = N, \ldots, 2. \tag{3.87}$$

Boundary conditions (3.73) and (3.80), which are now treated as initial conditions, take the following form:

$$\psi_N = 0, \tag{3.88}$$

$$\phi_N = -\frac{\mathrm{Nu}}{2}. \tag{3.89}$$

The value of the Nusselt number Nu in the condition (3.89) is found using the interval searching method to satisfy condition (3.81), which in this case takes the following form:

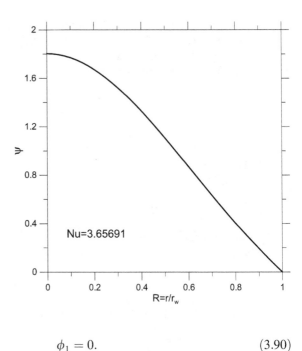

Fig. 3.5 Dimensionless temperature distribution $\psi = (T_w - T)/(T_w - T_m)$ as a function of dimensionless radius $R = r/r_w$ for the tube wall temperature $T|_{r=r_w} = T_w$

$$\phi_1 = 0. \tag{3.90}$$

Assuming the initial value of the Nusselt number as $\mathrm{Nu}^{(0)}$, the initial-value problem (3.86)–(3.89) is solved increasing subsequent Nusselt numbers by $\Delta\mathrm{Nu}$, according to the formula:

$$\mathrm{Nu}^{(k+1)} = \mathrm{Nu}^{(k)} + \Delta\mathrm{Nu}, \quad k = 0, 1, 2, \ldots. \tag{3.91}$$

The $\phi_1^{(0)}$ value for $\mathrm{Nu}^{(0)}$ is either negative or positive. The Nusselt number is increased until function $\phi_1^{(k+1)}$ changes its sign. The calculations are performed assuming that $\mathrm{Nu}^{(0)} = 3.65$ and $\Delta\mathrm{Nu} = 0.00001$. For this $\mathrm{Nu}^{(0)}$ value, the value of $\phi_1^{(0)}$ is negative. The Nusselt number was raised until $\phi_1^{(k+1)}$ was equal to or higher than zero. Assuming $N = 30001$, the result is $\mathrm{Nu} = 3.65691$. The determined temperature distribution is presented in Fig. 3.5.

Knowing the value of function $\psi_i, i = 1, \ldots N$, Eq. (3.70) enables determination of the distribution of temperature $T(R_i, x)$.

3.2   Laminar Heat Transfer in the Inlet Section

The distributions of the fluid velocity and temperature along the radius as it flows into a tube can be different. The velocity and temperature profiles become fully developed only at a considerable distance from the tube inlet. The entrance length of the hydrodynamically and thermally developing flow can be estimated by formulae (3.31) and (3.32). A comparison between the formulae indicates that if Pr = 1, the hydrodynamic and the thermal entrance lengths are equal. If Pr > 1, the thermal entrance length is longer than the hydrodynamic one. If Pr < 1, the opposite is the case, i.e. the thermal entrance length is shorter than the hydrodynamic one. Considering that the Prandtl number is the product of kinematic viscosity v and thermal diffusivity a, i.e. $Pr = v/a = c_p\mu/\lambda$, it can be seen that for liquids characterized by high viscosity, the thermal entrance length is bigger compared to the hydrodynamic one. Considering also that for liquids the Prandtl number is higher than one, the thermal entrance length for a liquid is higher compared to gases, for which the Prandtl number is generally smaller than one, e.g. for air Pr = 0.7.

3.2.1 Laminar Plug Flow at a Constant Heat Flux on the Tube Surface

In the case of the plug flow, it is assumed that the fluid velocity is uniform in the tube entire cross-section. Such a velocity profile at the inlet can be obtained by rounding the tube inlet edges. If valves and knees occur on the tube length, the velocity profile is far from a fully developed parabolic flow, and the assumption of constant velocity in the tube entire cross-section can be a good approximation of real conditions.

Assuming the fluid constant velocity $w_m = \dot{m}/(\rho \pi r_w^2)$, the energy balance equation for the flowing fluid takes the following form:

$$\rho c_p w_m \frac{\partial T}{\partial x} = \frac{\lambda}{r} \frac{\partial}{\partial r} \left(r \frac{\partial T}{\partial r} \right). \tag{3.92}$$

Equation (3.92) was solved under the following boundary conditions:

$$T|_{x=0} = T_0, \tag{3.93}$$

$$\lambda \frac{\partial T}{\partial r}|_{r=r_w} = \dot{q}_w, \tag{3.94}$$

$$\lambda \frac{\partial T}{\partial r}|_{r=0} = 0. \tag{3.95}$$

Equation (3.92) and boundary conditions (3.95) will be written in a dimensionless form.

For this purpose, the following dimensionless variables will be derived first:

$$\theta = \frac{T - T_0}{\dot{q}_w r_w / \lambda},$$
(3.96)

$$R = r/r_w,$$
(3.97)

$$X = \frac{at}{r_w^2} = \frac{ax}{r_w^2 w_m} = 4\frac{x}{d_w}\frac{1}{\mathrm{Re\,Pr}} = 4\frac{x}{d_w}\frac{1}{\mathrm{Pe}},$$
(3.98)

where:

$$a = \frac{\lambda}{\rho c_p}, \quad t = \frac{x}{w_m}, \quad d_w = 2r_w.$$

Considering (3.96)–(3.98) in (3.92)–(3.95), the following is obtained:

$$\frac{\partial \theta}{\partial X} = \frac{1}{R}\frac{\partial}{\partial R}\left(R\frac{\partial \theta}{\partial R}\right).$$
(3.99)

Equation (3.92) will be solved under the following boundary conditions:

$$\theta|_{x=0} = 0,$$
(3.100)

$$\frac{\partial \theta}{\partial R}\Big|_{R=1} = 1,$$
(3.101)

$$\frac{\partial \theta}{\partial R}\Big|_{R=0} = 0.$$
(3.102)

Analysing (3.99)–(3.102), it can be noticed that an identical equation and the same boundary conditions describe the unsteady-state temperature field in an infinitely long cylinder with the outer surface radius r_w and initial temperature T_0, which for $t > 0$ is heated with a constant heat flux $\dot{q}_w$. The dimensionless coordinate X in Eq. (3.99) corresponds to the Fourier number Fo in the heat conduction equation. The solution of the problem expressed by formulae (3.99)–(3.102) can be found in books [43] and [326] (p. 203 and pp. 430–434, respectively), where details are given of the procedure for obtaining the solution using the Laplace transform with respect to time. The solution is of the following form:

$$\theta = 2X + \frac{R^2}{2} - \frac{1}{4} - 2\sum_{n=1}^{\infty}\frac{J_0(\mu_n R)}{\mu_n^2 J_0(\mu_n)}\exp(-\mu_n^2 X),$$
(3.103)

where μ_n are the n-th roots of the following characteristic equation:

$$J_1(\mu) = 0. \tag{3.104}$$

Functions J_0 and J_1 denote the first-kind zero- and first-order Bessel functions, respectively [137, 338]. The values of μ for which $J_0(\mu) = 0$ or $J_1(\mu) = 0$ are listed in Table 3.1.

The fluid mean temperature in the tube cross-section determined from formula:

$$\theta_m = 2\int_0^1 \theta R\,dR \tag{3.105}$$

is:

$$\theta_m = 2X. \tag{3.106}$$

Analysing formula (3.106), it can be seen that the fluid mean temperature θ_m rises linearly with a rise in coordinate X, which is the effect of the constant heat flux on the tube inner surface. The tube surface dimensionless temperature θ_w is:

$$\theta_w = \theta|_{R=1} = 2X + \frac{1}{4} - 2\sum_{n=1}^{\infty}\frac{1}{\mu_n^2}\exp\left(-\mu_n^2 X\right). \tag{3.107}$$

The Nusselt number on the tube inner surface found from formula:

$$Nu = \frac{\alpha\,d_w}{\lambda} = \frac{2\dot{q}_w r_w}{\lambda(T_w - T_m)} = \frac{2}{(\theta_w - \theta_m)} \tag{3.108}$$

Table 3.1 Zeroes of function $J_0(\mu_i)$ and $J_1(\mu_i)$

μ_n	Roots of equation $J_0(\mu) = 0$	Roots of equation $J_1(\mu) = 0$
μ_1	2.4048	3.8317
μ_2	5.5201	7.0156
μ_3	8.6537	10.1735
μ_4	11.7915	13.3237
μ_5	14.9309	16.4706
μ_6	18.0711	19.6159
μ_7	21.2116	22.7601
μ_8	24.3525	25.9037
μ_9	27.4935	29.0468
μ_{10}	30.6346	32.1897

is:

$$\text{Nu} = \frac{2}{\left[2X + \frac{1}{4} - 2\sum_{n=1}^{\infty} \frac{1}{\mu_n^2} \exp\left(-\mu_n^2 X\right) - 2X\right]}$$

$$= \frac{8}{1 - 8\sum_{n=1}^{\infty} \frac{1}{\mu_n^2} \exp\left(-\mu_n^2 X\right)}. \tag{3.109}$$

For a thermally developed flow, if $X \to \infty$, the Nusselt number is 8, i.e.

$$\lim \text{Nu}|_{X \to \infty} = 8. \tag{3.110}$$

Mean heat transfer coefficient $\bar{\alpha}$ on the length from the tube inlet cross-section $X = 0$ to the cross-section with coordinate X is found using the following formula:

$$\int_0^X \dot{q}_w dX = \int_0^X \alpha(T_w - T_m)dX = \bar{\alpha} \int_0^X (T_w - T_m)dX, \tag{3.111}$$

which gives:

$$\bar{\alpha} = \frac{\dot{q}_w}{\frac{1}{X}\int_0^X (T_w - T_m)dX} = \frac{\dot{q}_w}{\bar{T}_w - \bar{T}_m}, \tag{3.112}$$

where mean temperatures $\bar{T}_w$ and $\bar{T}_m$ are defined as:

$$\bar{T}_w = \frac{1}{X}\int_0^X T_w dX, \quad \bar{T}_m = \frac{1}{X}\int_0^X T_m dX. \tag{3.113}$$

Introducing dimensionless temperatures, the Nusselt number mean value is defined as:

$$\text{Nu}_m = \frac{\alpha_m d_w}{\lambda} = \frac{2}{\bar{\theta}_w - \bar{\theta}_m}. \tag{3.114}$$

Determination of mean temperatures from the following formulae:

$$\bar{\theta}_w = \frac{1}{X}\int_0^X \theta_w dX, \quad \bar{\theta}_m = \frac{1}{X}\int_0^X \theta_m dX \tag{3.115}$$

and substitution thereof in (3.114) results in:

$$\mathrm{Nu}_m = \frac{8}{1 - 8 \sum_{n=1}^{\infty} \frac{1}{\mu_n^4 X} \left[1 - \exp\left(-\mu_n^2 X\right)\right]}. \qquad (3.116)$$

Table 3.2 presents the local and the mean Nusselt numbers for selected values of X.

The results presented in Table 3.2 indicate that at a small distance from the tube inlet, the values of the Nusselt number, both local and mean, are high. For $X > 0.4$, the Nusselt number is very close to $\mathrm{Nu} = 8$, i.e., to the Nusselt number value occurring in the fluid fully developed flow. The Nusselt number mean value is approaching the steady-state value only for $X > 2$.

3.2.2 Laminar Plug Flow at Constant Temperature on the Tube Surface

Assuming the constant fluid velocity $w_m = \dot{m}/(\rho \pi r_w^2)$, the energy balance equation for the flowing fluid (3.92) will be solved under boundary conditions (3.93) and (3.95), and condition (3.94) will be replaced by:

$$T|_{r=r_w} = T_w. \qquad (3.117)$$

The following dimensionless temperature will be derived first:

$$\theta = \frac{T - T_w}{T_0 - T_w}. \qquad (3.118)$$

The other dimensionless variables are defined by formulae (3.97) and (3.98).

Table 3.2 Values of the local and the mean Nusselt number (Nu and Nu_m, respectively) for selected values of X, in the case of constant heat flux on the tube inner surface	$X = 4\dfrac{x}{d_w}\dfrac{1}{\mathrm{Re}\,\mathrm{Pr}}$	Nu	Nu_m
	0.004	30.6	41.5
	0.01	20.4	28.6
	0.02	15.3	21.5
	0.04	11.9	16.3
	0.1	9.16	11.9
	0.2	8.24	9.95
	0.4	8.01	8.92
	1	8	8.34
	2	8	8.17
	4	8	8.08
	∞	8	8

Differential equation (3.92) and boundary conditions (3.93), (3.95) and (3.117) will be written in the following dimensionless form.

$$\frac{\partial \theta}{\partial X} = \frac{1}{R}\frac{\partial}{\partial R}\left(R\frac{\partial \theta}{\partial R}\right),$$ (3.119)

$$\theta|_{x=0} = 1,$$ (3.120)

$$\theta|_{R=1} = 0,$$ (3.121)

$$\frac{\partial \theta}{\partial R}|_{R=0} = 0.$$ (3.122)

Analysing (3.119)–(3.122), it can be noticed that an identical equation and the same boundary conditions describe the unsteady-state temperature field in an infinitely long cylinder with the outer surface radius r_w and initial temperature T_i, on the surface of which for $t > 0$ the heat flux is constant. The dimensionless coordinate X in Eq. (3.119) corresponds to the Fourier number (Fo) in the heat conduction equation. The solution of the problem expressed by formulae (3.119)–(3.122) can be found in books [43] and [326] (p. 199 and pp. 389–399, respectively), where details are given of the procedure for obtaining the solution using the variables separation method. The solution is of the following form:

$$\theta = 2\sum_{n=1}^{\infty} \frac{J_0(\mu_n R)}{\mu_n J_1(\mu_n)}\exp\left(-\mu_n^2 X\right),$$ (3.123)

where μ_n are the n-th roots of the following characteristic equation:

$$J_0(\mu) = 0.$$ (3.124)

The first ten roots of Eq. (3.124) are listed in Table 3.1.
The fluid mean temperature θ_m is calculated from the following formula:

$$\theta_m = 2\int_0^1 \theta R\, dR.$$ (3.125)

Substituting (3.123) in (3.125) and performing relevant operations, the following is obtained:

$$\theta_m = \sum_{n=1}^{\infty} \frac{4}{\mu_n^2}\exp\left(-\mu_n^2 X\right).$$ (3.126)

The heat transfer coefficient is calculated from:

$$\alpha = \frac{\lambda}{T_w - T_m} \frac{\partial T}{\partial r}\Big|_{r=r_w} = \frac{\lambda}{(\theta_w - \theta_m)r_w} \frac{\partial \theta}{\partial R}\Big|_{R=1}. \tag{3.127}$$

Considering that:

$$\frac{dJ_0(\mu_n R)}{dR} = -\mu_n J_1(\mu_n R) \tag{3.128}$$

derivative $\partial\theta/\partial R$ can be calculated:

$$\frac{\partial\theta}{\partial R}\Big|_{R=1} = -2\sum_{n=1}^{\infty}\exp(-\mu_n^2 X). \tag{3.129}$$

The local Nusselt number is found from the following formula:

$$\mathrm{Nu} = \frac{\alpha \cdot (2r_w)}{\lambda}, \tag{3.130}$$

which, considering formula (3.127), gives:

$$\mathrm{Nu} = \frac{2}{\theta_w - \theta_m} \frac{\partial\theta}{\partial R}\Big|_{R=1}. \tag{3.131}$$

Substituting mean temperature θ_m defined by formula (3.126) and derivative $(\partial\theta/\partial R)|_{R=1}$ defined by formula (3.129) in (3.131), and considering that $\theta_w = 0$, the following is obtained:

$$\mathrm{Nu} = \frac{\sum_{n=1}^{\infty}\exp(-\mu_n^2 X)}{\sum_{n=1}^{\infty}\frac{1}{\mu_n^2}\exp(-\mu_n^2 X)}. \tag{3.132}$$

For a fully developed flow, if X is high, it is enough to take account only of the first terms in the series in formula (3.132), which gives:

$$\mathrm{Nu}_\infty = \mathrm{Nu}|_{x\to\infty} = \mu_1^2 = 2.4048^2 = 5.7831. \tag{3.133}$$

The next to derive is the formula for the mean Nusselt number in the interval from $X = 0$ to X. The Nusselt number mean value will be determined using formula (3.67), which—if written in a dimensionless form—is expressed as:

$$\frac{d\theta_m}{\theta_m} = -\mathrm{Nu}\, dX. \tag{3.134}$$

Integrating Eq. (3.135) under the following condition:

$$\theta_m|_{X=0} = 1 \tag{3.135}$$

the result is:

$$\theta_m = \exp\left(-\int_0^X \mathrm{Nu}\, dX\right). \tag{3.136}$$

Taking the logarithm of both sides of Eq. (3.136) and dividing it by X, the following is obtained:

$$\mathrm{Nu}_m = \frac{\alpha_m d_w}{\lambda} = -\frac{\ln\theta_m}{X}, \tag{3.137}$$

where the fluid mean temperature θ_m is defined by formula (3.126).

Table 3.3 presents the Nusselt number local and mean values calculated by formulae (3.132) and (3.137), respectively, for selected values of X.

It follows from the analysis of the results presented in Figs. 3.6 and 3.7 that the Nusselt number both local and mean values are higher at a constant heat flux on the tube surface. It can be seen from Fig. 3.7 that the entrance length ends at $X \cong 0.2$. Assuming the values of the Reynolds number Re = 2000 and the Prandtl number Pr = 3.0 (laminar flow of water), the dimensionless entrance length is: $x/d_w = 0.2\,\mathrm{Re}\,\mathrm{Pr}\,/4 = 300$. It should be noted that the thermal entrance length is bigger for higher values of the Reynolds and the Prandtl numbers.

Table 3.3 Values of the local and the mean Nusselt number (Nu and Nu_m, respectively) for selected values of X, in the case of a constant temperature on the tube inner surface

$X = 4\dfrac{x}{d_w}\dfrac{1}{\mathrm{Re}\,\mathrm{Pr}}$	Nu	Nu_m
0.004	19.5	37.322
0.01	13.1	24.3
0.02	9.88	17.7
0.04	7.74	13.2
0.1	6.18	9.31
0.2	5.82	7.62
0.4	5.79	6.71
1	5.783	6.15
2	5.783	5.97
4	5.783	5.88
∞	5.783	5.783

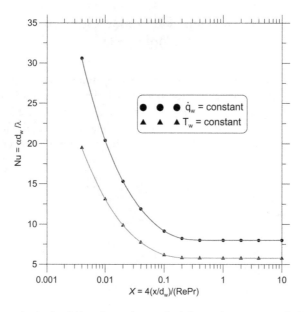

Fig. 3.6 Changes in the local Nusselt number on the inlet section—constant fluid flow velocity w_m in the tube cross-section

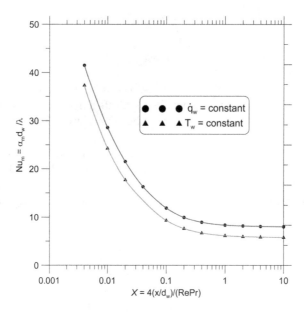

Fig. 3.7 Changes in the mean Nusselt number in the interval from 0 to X on the inlet section—constant fluid flow velocity w_m in the tube cross-section

3.3 Hydraulically Developed Laminar Flow and the Thermally Developing Flow

Before the formulae for the local and mean Nusselt number at a constant heat flux or temperature on the channel surface are derived, determination of the mean heat transfer coefficient α_m on the length $0–x$ and of the mean Nusselt number Nu_m corresponding to it will be discussed. The mean heat transfer coefficient α_m or the mean Nusselt number Nu_m are usually calculated using the following formulae:

$$\alpha_m = \frac{1}{x}\int_0^x \alpha\,dx, \quad \mathrm{Nu}_m = \frac{1}{x}\int_0^x \mathrm{Nu}\,dx, \tag{3.138}$$

where: $\mathrm{Nu} = \alpha\,d_w/\lambda$ and $\mathrm{Nu}_m = \alpha_m d_w/\lambda$.

However, it should be emphasized that such averaging is correct only if the difference between the temperature of the wall and the fluid $(T_w - T_m)$ is constant. In this case:

$$\int_0^x \dot{q}_w(x)dx = \int_0^x \alpha(x)(T_w - T_m)dx$$

$$= (T_w - T_m)\int_0^x \alpha(x)dx = x\alpha_m(T_w - T_m), \tag{3.139}$$

$$(T_w - T_m) = \text{const},$$

where the heat transfer coefficient α_m is defined as (3.138). Despite the fact that condition $(T_w - T_m) = \text{const}$. is generally not satisfied in practice, formulae (3.138) are often used to find the mean value of the heat transfer coefficient or of the mean Nusselt number.

If the tube surface temperature T_w is constant, mean heat transfer coefficient α_m can be found using formula (3.67): The procedure can be followed through taking formulae (3.134)–(3.137) as an example.

The mean heat transfer coefficient α_m can be also be found using the following formula:

$$\alpha_m = \frac{\dot{q}_w|_{0-x}}{\Delta T_{\ln}} = \frac{\dot{q}_w|_{0-x}}{\dfrac{(T_w - T_{m0}) - (T_w - T_m|_x)}{\ln\dfrac{(T_w - T_{m0})}{(T_w - T_m|_x)}}}, \tag{3.140}$$

where the mean heat flux is determined from the following:

$$\dot{Q} = \dot{m}c_p(T_m|_x - T_{m0}) = \dot{q}_w|_{0-x}(\pi \, d_w x),$$ (3.141)

which gives:

$$\dot{q}_w|_{0-x} = \frac{\dot{m}c_p(T_m|_x - T_{m0})}{(\pi \, d_w x)}.$$ (3.142)

Procedures (3.67) and (3.140) both give the same results.

If heat flux $\dot{q}_w$ on the tube inner surface is constant, the mean heat transfer coefficient calculated by means of formula (3.138) can be expressed as:

$$\alpha_m = \frac{1}{x}\int_0^x \frac{\dot{q}_w}{T_w - T_m}dx = \dot{q}_w \frac{1}{x}\int_0^x \frac{1}{T_w - T_m}dx = \frac{\dot{q}_w}{\Delta T_m},$$ (3.143)

where mean temperature difference ΔT_m is defined by the following formula:

$$\Delta T_m = \left[\frac{1}{x}\int_0^x \frac{1}{T_w(x) - T_m(x)}dx\right]^{-1}.$$ (3.144)

Formula (3.143) was applied in [270] to determine the mean heat transfer coefficient α_m and the mean Nusselt number Nu_m on length 0–x in a tube and on the surface of a flat slot.

It should be added that a different way of finding α_m can also be used [40] if heat flux $\dot{q}_w$ at the wall is set as constant.

Considering that:

$$\int_0^x \dot{q}_w dx = \int_0^x \alpha(T_w - T_m)dx = \alpha_m \int_0^x (T_w - T_m)dx$$ (3.145)

the following is obtained:

$$\alpha_m = \frac{\dot{q}_w}{\frac{1}{x}\int_0^x (T_w - T_m)dx} = \frac{\dot{q}_w}{\bar{T}_w - \bar{T}_m},$$ (3.146)

where mean temperatures $\bar{T}_w$ and $\bar{T}_m$ are defined by the following formulae:

$$\bar{T}_w = \frac{1}{x}\int_0^x T_w dx, \quad \bar{T}_m = \frac{1}{x}\int_0^x T_m dx.$$ (3.147)

Formulae (3.143) and (3.146) give different results.

This will be illustrated using an example of the mean heat transfer (the mean Nusselt number) determination on the surface of a plate transferring a constant heat flux $\dot{q}_w =$ const. to a fluid with temperature T_∞ flowing over the plate surface with velocity $w_{x\infty}$. In this case, the Nusselt number is defined by the following formula [106, 258]:

$$\text{Nu} = 0.462\text{Re}^{1/2}\text{Pr}^{1/3}, \quad \text{Pr} \geq 0.7, \tag{3.148}$$

where: $\text{Nu} = \alpha x/\lambda = \dot{q}_w x/[\lambda(T_w - T_\infty)]$, $\text{Re} = w_{x\infty}x/\nu$, $\text{Pr} = c_p\mu/\lambda$.

Calculating the mean temperature difference:

$$\bar{T}_w - \bar{T}_m = \frac{1}{x}\int_0^x (T_w - T_m)dx = \frac{1}{x}\int_0^x \frac{\dot{q}_w x}{\lambda \, \text{Nu}}dx$$

$$= \frac{1}{x}\int_0^x \frac{\dot{q}_w x}{\lambda\left[0.462(w_{x\infty}x/\nu)^{1/2}\text{Pr}^{1/3}\right]}dx = \frac{\dot{q}_w x/\lambda}{0.693\text{Re}^{1/2}\,\text{Pr}^{1/3}} \tag{3.149}$$

and substituting (3.149) in expression (3.146), the formula for the mean Nusselt number Nu_m on length 0–x can be determined:

$$\text{Nu}_m = \frac{\alpha_m x}{\lambda} = 0.693\text{Re}^{1/2}\text{Pr}^{1/3}. \tag{3.150}$$

If the first formula (3.138) is used for the averaging of α, considering (3.148), the following is obtained:

$$\text{Nu}_m = 0.924\text{Re}^{1/2}\text{Pr}^{1/3}, \quad \text{Pr} \geq 0.7. \tag{3.151}$$

Comparing formulae (3.150) and (3.151), it can be seen that the relative difference between the mean Nusselt numbers determined using the two different methods is rather big and totals:

$$\varepsilon = 100 \cdot (0.924 - 0.693)/0.924 = 25\%.$$

However, it should be noted that the heat flux absorbed by the fluid on length 0–x has to be calculated using the mean heat transfer coefficient α_m together with the corresponding mean difference between the temperature of the wall and the mass-averaged temperature of the fluid on length 0–x. Such a procedure always gives the same heat flux regardless of the way in which the mean heat transfer coefficient is determined. This has to be borne in mind in experimental determination of the heat transfer coefficient. For correct calculation of the heat flux transferred to the flowing fluid from the tube wall, the mean heat transfer coefficient should be multiplied by the mean temperature difference between the solid body surface and the fluid that corresponds to the definition of the mean heat transfer coefficient.

3.3.1 Hydraulically Developed Laminar Flow and the Thermally Developing Flow at Constant Heat Flux at the Tube Inner Surface

The energy balance equation at a parabolic distribution of velocity defined by formula (3.22) has the following form:

$$2w_m \left[1 - \left(\frac{r}{r_w} \right)^2 \right] \rho\, c_p \frac{\partial T}{\partial x} = \frac{\lambda}{r} \frac{\partial}{\partial r} \left(r \frac{\partial T}{\partial r} \right). \tag{3.152}$$

The fluid mass-averaged temperature at the tube inlet is set as $T_0 = T_{m0}$:

$$T|_{x=0} = T_0. \tag{3.153}$$

The heat flux on the tube inner surface is set as constant:

$$\lambda \frac{\partial T}{\partial r}\Big|_{r=r_w} = \dot{q}_w. \tag{3.154}$$

Due to the temperature field symmetry, the following condition has to be satisfied:

$$\frac{\partial T}{\partial r}\Big|_{r=0} = 0. \tag{3.155}$$

Introducing the following dimensionless variables:

$$\theta = \frac{T - T_0}{\dot{q}_w r_w / \lambda}, \tag{3.156}$$

$$R = r/r_w, \tag{3.157}$$

$$X^* = \frac{x}{r_w} \frac{1}{\mathrm{Re}\,\mathrm{Pr}} = \frac{x}{r_w} \frac{1}{\mathrm{Pe}} \tag{3.158}$$

Equation (3.152) and boundary conditions (3.153)–(3.155) take the following form:

$$(1 - R^2) \frac{\partial \theta}{\partial X^*} = \frac{1}{R} \frac{\partial}{\partial R} \left(R \frac{\partial \theta}{\partial R} \right), \tag{3.159}$$

$$\theta|_{X^*=0} = 0, \tag{3.160}$$

$$\frac{\partial \theta}{\partial R}\Big|_{R=1} = 1, \tag{3.161}$$

$$\frac{\partial \theta}{\partial R}\Big|_{R=0} = 0. \tag{3.162}$$

The problem expressed by (3.159)–(3.162) is linear and it will be solved using the superposition method.

$$\theta = \theta_1 + \theta_2, \tag{3.163}$$

where θ_1 is the solution for a flow which is fully developed hydraulically and thermally. Considering that temperature T_1 is determined by Eq. (3.54), the dimensionless temperature θ_1 can be written as:

$$\theta_1 = 4X^* + R^2 - \frac{1}{4}R^4 - \frac{7}{24}. \tag{3.164}$$

Temperature θ_2 satisfies the following equation and the following boundary conditions:

$$(1 - R^2)\frac{\partial \theta_2}{\partial X^*} = \frac{1}{R}\frac{\partial}{\partial R}\left(R\frac{\partial \theta_2}{\partial R}\right), \tag{3.165}$$

$$\theta_2|_{X^*=0} = -\left(R^2 - \frac{1}{4}R^4 - \frac{7}{24}\right), \tag{3.166}$$

$$\frac{\partial \theta_2}{\partial R}\Big|_{R=1} = 0, \tag{3.167}$$

$$\frac{\partial \theta_2}{\partial R}\Big|_{R=0} = 0. \tag{3.168}$$

Siegel et al. [274] solved the problem defined by (3.165)–(3.168). Assuming the solution of (3.165)–(3.168) in the following form:

$$\theta_2 = \sum_{n=1}^{\infty} C_n G_n \exp\left(-\mu_n^2 X^*\right) \tag{3.169}$$

and then substituting (3.169) in (3.165) and (3.167)–(3.168), the following Sturm-Liouville problem is obtained:

$$\frac{d^2 G_n}{dR^2} + \frac{1}{R}\frac{dG_n}{dR} + G_n\mu_n^2(1 - R^2) = 0, \tag{3.170}$$

$$\frac{dG_n}{dR}\Big|_{R=1} = 0, \tag{3.171}$$

$$\frac{dG_n}{dR}\Big|_{R=0} = 0, \tag{3.172}$$

where μ_n and G_n are eigenvalues and eigenfunctions of the problem expressed by formulae (3.170)–(3.172).

Constants C_n are determined from condition (3.166), which—after substitution of solution (3.169)—gives:

$$\sum_{n=1}^{\infty} C_n G_n = -\left(R^2 - \frac{1}{4}R^4 - \frac{7}{24}\right). \tag{3.173}$$

The solution of Eq. (3.173) has the following form:

$$C_n = \frac{-\int_0^1 (1 - R^2)\left(R^2 - \frac{1}{4}R^4 - \frac{7}{24}\right)G_n R dR}{\int_0^1 (1 - R^2)G_n^2 R\, dR}. \tag{3.174}$$

The Nusselt number will be found from the following formula:

$$Nu = \frac{\alpha\, d_w}{\lambda} = \frac{2\dot{q}_w r_w}{\lambda(T_w - T_m)} = \frac{2}{(\theta_w - \theta_m)}. \tag{3.175}$$

considering that the inner surface dimensionless temperature θ_w and mean dimensionless temperature θ_m are defined as follows:

$$\theta_w = \theta_1|_{R=1} + \theta_2\Big|_{R=1} = 4X^* + \frac{11}{24} + \sum_{n=1}^{\infty} C_n G_n(1)\exp\left(-\mu_n^2 X^*\right), \tag{3.176}$$

$$\theta_m = \theta_{1m} + \theta_{2m} = 4X^* + 0 = 4X^*. \tag{3.177}$$

Eigenvalues μ_n, eigenfunctions $G_n(1)$ and constants C_n are listed in Table 3.4 [128].

Substituting (3.163) and (3.164) in (3.162), the following formula for the Nusselt number is obtained:

$$Nu = \frac{2}{\frac{11}{24} + \sum_{n=1}^{\infty} C_n G_n(1)\exp\left(-\mu_n^2 X^*\right)}. \tag{3.178}$$

Using eigenvalues μ_n, eigenfunctions $G_n(1)$ and constants C_n listed in Table 3.4, the Nusselt number Nu was found for selected values of X^* (cf. Table 3.5).

For small values of $X^* \leq 0.01$, the series in the expression for the Nusselt number (3.178) is slow-convergent and even if twenty terms are taken into consideration in it, it is still not enough to achieve a satisfactory level of accuracy. For this reason, the Nusselt numbers for small values of X^* can be calculated using the formula:

Table 3.4 Eigenvalues μ_n, eigenfunctions $G_n(1)$ and constants C_n for a constant heat flux on the tube inner surface [128]

n	μ_n^2	$G_n(1)$	C_n
1	25.67961	−0.4925166	0.4034832
2	83.86175	0.3955085	−0.1751099
3	174.16674	−0.3458737	0.1055917
4	296.53630	0.31404646	−0.0732824
5	450.94720	−0.2912514	0.05503648
6	637.38735	0.2738069	−0.04348435
7	855.84953	−0.2598529	0.03559508
8	1106.32903	0.2483319	−0.02990845
9	1388.82260	−0.2385902	0.02564010
10	1703.32780	0.2301990	−0.02233368
11	2049.84300	−0.2228628	0.01970692
12	2438.3668	0.2163703	−0.01757646
13	2838.8981	−0.2105659	0.01581844
14	3281.4362	0.2053319	−0.01434637
15	3755.9803	−0.200577	0.01309817

Table 3.5 The Nusselt number Nu as a function of parameter $X^* = \dfrac{x}{r_w}\dfrac{1}{\mathrm{Re\,Pr}} = \dfrac{x}{r_w}\dfrac{1}{\mathrm{Pe}}$

$X^* = \dfrac{x}{r_w}\dfrac{1}{\mathrm{Re\,Pr}} = \dfrac{x}{r_w}\dfrac{1}{\mathrm{Pe}}$	$\mathrm{Nu} = \alpha d_w/\lambda$	$X^* = \dfrac{x}{r_w}\dfrac{1}{\mathrm{Re\,Pr}} = \dfrac{x}{r_w}\dfrac{1}{\mathrm{Pe}}$	$\mathrm{Nu} = \alpha d_w/\lambda$
0.010	7.46	0.08	4.62
0.012	7.08	0.09	4.56
0.014	6.78	0.10	4.51
0.016	6.53	0.12	4.45
0.018	6.32	0.14	4.42
0.020	6.15	0.16	4.39
0.03	5.55	0.20	4.37
0.04	5.20	0.30	4.36
0.05	4.97	0.50	4.36
0.06	4.82	1.00	4.36
0.07	4.70		

$$\mathrm{Nu} = \frac{1.6393}{(X^*)^{1/3}}, \quad X^* \leq 0.001$$

presented in [265]. The formula is obtained by solving the Lévêque problem [176], approximating the tube wall using a flat wall in the vicinity of which velocity changes linearly as a function of the distance from the wall surface. The solution of the Lévêque problem is discussed in more detail in Sect. 3.3.3.

The mean value of the Nusselt number can be determined from formula (3.114), where mean temperatures of the wall and of the fluid are defined as:

$$\bar{\theta}_w = \frac{1}{X^*} \int_0^{X^*} \theta_w dX^*, \quad \bar{\theta}_m = \frac{1}{X^*} \int_0^{X^*} \theta_m dX^*. \tag{3.179}$$

Substituting expressions (3.176) and (3.177) in formulae (3.179) and finding the mean temperatures, a formula is obtained for the mean Nusselt number in the interval from $X^* = 0$ to X^*:

$$Nu_m = \frac{\alpha_m d_w}{\lambda} = \frac{2}{\bar{\theta}_w - \bar{\theta}_m} = \frac{2}{\frac{11}{24} - \frac{1}{X^*} \sum_{n=1}^{\infty} \frac{C_n G_n(1)}{\mu_n^2} \left[\exp\left(-\mu_n^2 X^*\right) - 1 \right]}. \tag{3.180}$$

The analysis of (3.178) and (3.180) indicates that for a fully developed laminar flow, for $x \to 0$, the following value of the Nusselt number is obtained: $Nu = Nu_m = 48/11 = 4.3636$. Figure 3.8 presents changes in the local Nusselt number Nu as a function of parameter X^*, and Fig. 3.9—changes in the mean Nusselt number Nu_m.

The calculations were performed using formulae (3.178) and (3.180), respectively, and taking account of the first 20 terms in the series and the values of μ_n, $G_n(1)$ and C_n listed in Table 3.4 [128].

Fig. 3.8 Changes in the local Nusselt number Nu on the tube inner surface depending on parameter X^* for a constant heat flux $\dot{q}_w$ or constant temperature T_w set on the tube inner surface

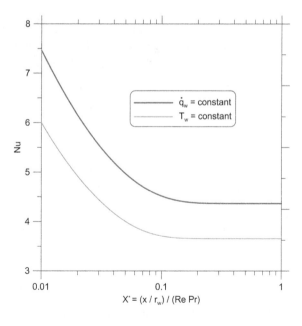

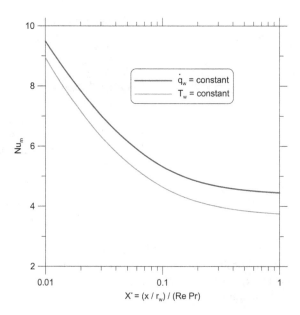

Fig. 3.9 Changes in the mean Nusselt number Nu_m on the tube inner surface depending on parameter X^* for a constant heat flux $\dot{q}_w$ or constant temperature T_w set on the tube inner surface

The mean Nusselt numbers Nu_m for selected values of X^* are also listed in Table 3.6. It can be seen that Nu_m approaches $Nu_m = 4.364$ only for $X^* > 1$, i.e., the entrance length is very long.

Because the analyzed problem is linear, the fluid temperature can also be found for the value of heat flux $\dot{q}_w(x)$ varying on the tube length (cf. Fig. 3.10) using Duhamel's integral. Duhamel's integral has the following form [324, 326]:

$$T - T_0 = \int_0^x \dot{q}_w(\xi) \frac{\partial u(x - \xi)}{\partial x} d\xi, \qquad (3.181)$$

where function $u(x, R)$, referred to as the impact function, is the temperature distribution (3.150) in the tube at $T_0 = 0$ for $x = 0$ and $\dot{q}_w = 1 \ Wm^2$ for $x > 0$. Impact function $u(x, R)$ has the following form:

$$
\begin{aligned}
\frac{u}{r_w/\lambda} &= 4X^* + R^2 - \frac{1}{4}R^4 - \frac{7}{24} + \sum_{n=1}^{\infty} C_n G_n \exp\left(-\mu_n^2 X^*\right) \\
&= 4\frac{x}{r_w}\frac{1}{Re\,Pr} + R^2 - \frac{1}{4}R^4 - \frac{7}{24} + \sum_{n=1}^{\infty} C_n G_n \exp\left(-\mu_n^2 \frac{x}{r_w}\frac{1}{Re\,Pr}\right).
\end{aligned}
\qquad (3.182)
$$

A derivative of the impact function expressed in (3.182) is defined by the following formula:

Table 3.6 Mean Nusselt number Nu_m depending on parameter $X^* = \frac{x}{r_w}\frac{1}{Re\,Pr} = \frac{x}{r_w}\frac{1}{Pe}$ for a constant heat flux set on the tube inner surface

$X^* = \frac{x}{r_w}\frac{1}{Re\,Pr} = \frac{x}{r_w}\frac{1}{Pe}$	$Nu_m = \alpha_m d_w/\lambda$		$X^* = \frac{x}{r_w}\frac{1}{Re\,Pr} = \frac{x}{r_w}\frac{1}{Pe}$	$Nu_m = \alpha_m d_w/\lambda$	
	Formula (3.180)	Shah and London [270]		Formula (3.180)	Shah and London [270]
0.010	9.49	11.08	0.06	6.14	6.49
0.012	9.03	10.45	0.08	5.78	6.04
0.014	8.65	9.95	0.10	5.53	5.75
0.016	8.34	9.54	0.20	4.97	5.08
0.020	7.84	8.90	0.30	4.77	4.85
0.040	6.52	7.24	0.40	4.67	4.72

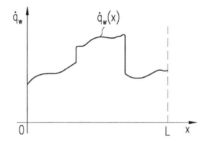

Fig. 3.10 Changes in heat flux $\dot{q}_w(x)$ set on the tube inner surface along the fluid flow direction

$$\frac{\partial u}{\partial x} = \frac{1}{\lambda}\left[\frac{4}{Re\,Pr} - \frac{1}{Re\,Pr}\sum_{n=1}^{\infty} C_n G_n \mu_n^2 \exp\left(-\mu_n^2 \frac{x}{r_w}\frac{1}{Re\,Pr}\right)\right]. \qquad (3.183)$$

Substitution of (3.183) in (3.181) results in:

$$T - T_0 = \frac{1}{\lambda}\int_0^x \dot{q}_w(\xi)\left[\frac{4}{Re\,Pr} - \frac{1}{Re\,Pr}\sum_{n=1}^{\infty} C_n G_n \mu_n^2 \exp\left(-\mu_n^2 \frac{x-\xi}{r_w}\frac{1}{Re\,Pr}\right)\right]d\xi.$$
$$(3.184)$$

Formula (3.184) enables determination of the temperature distribution at a change in $\dot{q}_w(x)$, for which the first derivative $\partial \dot{q}_w/\partial x$ is continuous, and also in the case where the first derivative $\partial \dot{q}_w/\partial x$ is discontinuous, i.e., when step changes occur in heat flux $\dot{q}_w(x)$ on the tube length (cf. Fig. 3.10).

3.3.2   Hydraulically Developed Laminar Flow and the Thermally Developing Flow at Constant Temperature on the Tube Inner Surface

In the case of cocurrent heat exchangers or if condensation or boiling occurs on the tube outer surface, the wall temperature can be assumed as constant. The fluid mass-averaged temperature $T_0 = T_{m0}$ and a developed parabolic distribution of velocity are set on the tube inlet.

Introducing the following dimensionless variables:

$$\theta = \frac{T - T_w}{T_0 - T_w},$$

(3.185)

$$R = r/r_w,$$

(3.186)

$$X^* = \frac{x}{r_w} \frac{1}{\text{Re Pr}} = \frac{x}{r_w} \frac{1}{\text{Pe}}$$

(3.187)

the energy balance equation and the boundary conditions take the following form:

$$\left(1 - R^2\right) \frac{\partial \theta}{\partial X^*} = \frac{1}{R} \frac{\partial}{\partial R} \left(R \frac{\partial \theta}{\partial R} \right),$$

(3.188)

$$\theta|_{X^*=0} = 0,$$

(3.189)

$$\theta|_{R=1} = 0,$$

(3.190)

$$\frac{\partial \theta}{\partial R}|_{R=0} = 0.$$

(3.191)

The distribution of temperature $\theta(X^*, R)$ will be found using the variables separation method, according to which the following solution is assumed:

$$\theta(X^*, R) = F(X^*)G(R).$$

(3.192)

Substitution of (3.192) in (3.188) results in:

$$F \frac{d^2 G}{dR^2} + \frac{1}{R} F \frac{dG}{dR} = \left(1 - R^2\right) G \frac{dF}{dX^*},$$

(3.193)

which, after the variables are separated, can be written as:

$$\frac{\frac{dF}{dx^*}}{F} = \frac{\frac{d^2G}{dR^2} + \frac{1}{R}\frac{dG}{dR}}{(1 - R^2)G} = -\mu^2.$$

(3.194)

The right side of Eq. (3.194) is a function of R only, whereas the left side depends only on X^*. The left and the right side of Eq. (3.192) can be equal to each other only if they are equal to $(-\mu^2)$. Equality (3.192) gives two ordinary differential equations:

$$\frac{dF}{dX^*} + \mu^2 F = 0, \tag{3.195}$$

$$\frac{d^2G}{dR^2} + \frac{1}{R}\frac{dG}{dR} + \mu^2\left(1 - R^2\right)G = 0. \tag{3.196}$$

Solving Eq. (3.195) using the variables separation method, the following is obtained:

$$F = C \exp\left(-\mu^2 X^*\right), \tag{3.197}$$

where C is a constant.

Equation (3.196) will be written in the following form:

$$\frac{d^2G}{d(\mu R)^2} + \frac{1}{\mu R}\frac{dG}{d(\mu R)} + \left(1 - \frac{(\mu R)^2}{\mu^2}\right)G = 0. \tag{3.198}$$

Substitution of solution (3.192) in boundary conditions (3.190) and (3.191) gives the following two boundary conditions for function G:

$$G|_{R=1} = 0, \tag{3.199}$$

$$\frac{\partial G}{\partial R}\Big|_{R=0} = 0. \tag{3.200}$$

Equation (3.198) combined with conditions (3.199) and (3.200) constitutes the Sturm-Liouville boundary value problem, which is a two-point value boundary problem. Eigenvalues μ_n are real numbers, whereas eigenfunctions $G_n(R)$ make up a set of orthogonal functions in interval (0, 1):

$$\int_0^1 R(1 - R^2)G_m(R)G_n(R)dR = 0, \quad \text{jeżeli } m \neq n, \tag{3.201}$$

$$\int_0^1 R(1 - R^2)G_m(R)G_n(R)dR = \frac{1}{2\mu_n}\left(\frac{dR_n}{d\mu_n}\frac{dR_n}{dR}\right)\Big|_{R=1}, \quad \text{jeżeli } m = n. \tag{3.202}$$

One of the methods that can be applied to solve the problem expressed by formulae (3.198)–(3.200) is the exponential series method, which is commonly used to solve ordinary differential equations. According to it, the solution can be presented in the form of the following exponential series:

$$G(\mu R) = \sum_{n=0}^{\infty} b_{2n}(\mu R)^{2n}.$$ (3.203)

Substitution of series (3.203) in Eq. (3.198) results in:

$$b_0 = 1, \quad n = 0,$$ (3.204)

$$b_2 = -\frac{1}{4}, \quad n = 1,$$ (3.205)

$$b_{2n} = \frac{1}{(2n)^2}\left(\frac{1}{\mu^2}b_{2n-4} - b_{2n-2}\right), \quad n \geq 2.$$ (3.206)

Substituting (3.203) in boundary condition (3.199) and taking account of coefficients (3.204)–(3.206), the following algebraic equation is obtained:

$$1 - \frac{1}{4}\mu^2 + \frac{1}{16}\left(\frac{1}{\mu^2} + \frac{1}{4}\right)\mu^4 - \frac{1}{36}\left[\frac{1}{4\mu^2} + \frac{1}{16}\left(\frac{1}{\mu^2} + \frac{1}{4}\right)\right]\mu^6, \ldots = 0.$$ (3.207)

This is an equation with an infinite number of roots μ_n, which are eigenvalues of the problem described by (3.198)–(3.200). For each eigenvalue there is a corresponding eigenfunction:

$$G_n(R) = G(\mu_n R, \mu_n), \quad n = 0, 1, 2, \ldots.$$ (3.208)

According to the variables separation method, the solution of Eq. (3.188) will be a sum of partial solutions obtained using (3.192) after substitution of the n-th eigenvalue and eigenfunction:

$$\theta_n(X^*, R) = C_n \exp\left(-\mu_n^2 X^*\right)G_n(R).$$ (3.209)

The solution of Eq. (3.188), being a sum of partial solutions of (3.209), can be written as:

$$\theta(X^*, R) = \sum_{n=0}^{\infty} C_n G_n(R) \exp\left(-\mu_n^2 X^*\right).$$ (3.210)

Constant C_n is determined from boundary condition (3.189), which—after substitution of (3.210)—gives:

$$1 = \sum_{n=0}^{\infty} C_n G_n(R).$$ (3.211)

Using orthogonality conditions (3.201) and (3.202), and relation:

$$\int_0^1 R(1 - R^2)G_n(R)dR = -\frac{1}{\mu_n^2}\left(\frac{dG_n}{dR}\right)|_{R=1}, \tag{3.212}$$

the following is obtained:

$$C_n = \frac{\int_0^1 R(1 - R^2)G_n(R)dR}{\int_0^1 R(1 - R^2)G_n^2(R)dR} = -\frac{2}{\mu_n\left(\frac{dG_n}{d\mu_n}\right)|_{R=1}}. \tag{3.213}$$

The local heat transfer coefficient on the tube inner surface will be found using the following formula:

$$\alpha = \frac{\lambda \frac{\partial T}{\partial r}|_{r=r_w}}{T_w - T_m} = -\frac{\lambda}{r_w}\frac{1}{\theta_m}\frac{\partial \theta}{\partial R}|_{R=1}. \tag{3.214}$$

Therefore, the local Nusselt number is defined as:

$$Nu = \frac{\alpha(2r_w)}{\lambda} = -\frac{2}{\theta_m}\frac{\partial \theta}{\partial R}|_{R=1}. \tag{3.215}$$

The fluid mass-averaged temperature θ_m is defined by the following formula:

$$\theta_m = \frac{T_m - T_w}{T_0 - T_w} = 2\int_0^1 \frac{w_x}{w_m}\theta R\, dR = 4\int_0^1 (1 - R^2)\theta R\, dR. \tag{3.216}$$

Substituting (3.210) in (3.216) and performing relevant operations, the following is obtained:

$$\theta_m = 8\sum_{n=0}^\infty \frac{B_n}{\mu_n^2}\exp(-\mu_n^2 X^*), \tag{3.217}$$

where:

$$B_n = \frac{1}{2}C_n\mu_n^2\int_0^1 G_n(R)(1 - R^2)R\, dR. \tag{3.218}$$

Considering relation (3.213) in expression (3.218), the result is:

$$B_n = -\frac{1}{2}C_n\left(\frac{dG_n}{dR}\right)|_{R=1}. \tag{3.219}$$

After derivative $\frac{\partial \theta}{\partial R}|_{R=1}$ is calculated, and taking account of relation (3.219):

$$\frac{\partial \theta}{\partial R}\Big|_{R=1} = \sum_{n=0}^{\infty} C_n \frac{dG_n}{dR}\Big|_{R=1} \exp\left(-\mu_n^2 X^*\right) = -2\sum_{n=0}^{\infty} B_n \exp\left(-\mu_n^2 X^*\right) \qquad (3.220)$$

and after substitution of expressions (3.220) and (3.219) in (3.216), the following formula is obtained which enables determination of the local Nusselt number:

$$\mathrm{Nu} = \frac{\sum_{n=0}^{\infty} B_n \exp\left(-\mu_n^2 X^*\right)}{2\sum_{n=0}^{\infty} \frac{B_n}{\mu_n^2} \exp\left(-\mu_n^2 X^*\right)}. \qquad (3.221)$$

In the case of a fully developed flow, if $x \to \infty$, formula (3.221) gives $\mathrm{Nu}_\infty = \mu_0^2/2 = 2.70436^2/2 = 3.6568$, i.e. a value which is the same as the one produced by formula (3.64).

The formula for the mean Nusselt number can be determined from formula (3.67) using the same procedure as for derivation of formulae (3.135)–(3.138). The result is:

$$\mathrm{Nu}_m = -\frac{\ln \theta_m(X^*)}{2X^*}. \qquad (3.222)$$

The series in the expressions for the distribution of temperature θ, mean temperature θ_m and the Nusselt number Nu_m are slow-convergent if $X^* \leq 0.01$. Even if the number of terms in the series is raised to more than twenty, the accuracy of the obtained results cannot be improved. It is only after the number of the terms in the series in Eq. (3.221) is raised to more than 100 that results are obtained with good accuracy.

Sellars et al. [265], based on the solution of the Lévêque problem [179], put forward the following formula that makes it possible to calculate the local Nusselt number:

$$\mathrm{Nu} = \left(\frac{16}{9}\right)^{1/3} \frac{1}{\Gamma\left(\frac{4}{3}\right)(X^*)^{1/3}} = 1.356597(X^*)^{-1/3}, \quad X^* \leq 0.01, \qquad (3.223)$$

where $\Gamma(4/3) = 0.8929795$.

They also put forward the following approximate formula for $n > 2$:

$$\mu_n = 4n + \frac{8}{3}, \qquad (3.224)$$

$$C_n = \frac{2.8461(-1)^n}{\mu_n^{2/3}}, \qquad (3.225)$$

$$B_n = -\frac{1}{2}C_n\left(\frac{dG_n}{dR}\right)\Big|_{R=1} = \frac{2.0256}{2\mu_n^{1/3}}. \qquad (3.226)$$

Eigenvalues μ_n and coefficients B_n and C_n for the Graetz problem are listed in Table 3.7. The first eleven coefficients are adopted from [26, 228]; the rest are calculated using approximate formulae (3.224)–(3.226). The Nu and the Nu_m values listed in Table 3.8 are calculated taking account of 200 terms in the series in expression (3.221).

If the tube inner surface temperature T_w depends on coordinate x (cf. Fig. 3.11), using Duhamel's integral [324, 326]:

Table 3.7 Eigenvalues μ_n and coefficients B_n and C_n for the Graetz problem [26, 110, 228]

N	μ_n	B_n	C_n	N	μ_n	B_n	C_n
1	2.70436	0.74877	1.47643	16	62.66667	0.25498	−0.18040
2	6.67903	0.54383	−0.80612	17	66.66667	0.24978	0.17311
3	10.67338	0.46286	0.58876	18	70.66667	0.24497	−0.16651
4	14.67108	0.41542	−0.47585	19	74.66667	0.24052	0.16051
5	18.66987	0.38292	0.40502	20	78.66667	0.23637	−0.15502
6	22.66914	0.35869	−0.35575	21	82.66667	0.23249	0.14998
7	26.66866	0.33962	0.31917	22	86.66667	0.22886	−0.14533
8	30.66832	0.32406	−0.29074	23	90.66667	0.22545	0.14102
9	34.66807	0.31110	0.26789	24	94.66667	0.22222	−0.13702
10	38.66788	0.29984	−0.24906	25	98.66667	0.21918	0.13329
11	42.66762	0.28984	0.23322	26	102.6667	0.21630	−0.12981
12	46.66667	0.28131	−0.21957	27	106.6667	0.21355	0.12654
13	50.66667	0.27371	0.20786	28	110.6667	0.21095	−0.12347
14	54.66667	0.26686	−0.19759	29	114.6667	0.20847	0.12059
15	58.66667	0.26065	0.18851	30	118.6667	0.20610	−0.11786

Table 3.8 Local and mean Nusselt numbers (Nu and Nu_m, respectively) depending on parameter $X^* = \dfrac{x}{r_w}\dfrac{1}{\mathrm{Re\,Pr}} = \dfrac{x}{r_w}\dfrac{1}{\mathrm{Pe}}$ for a constant temperature set on the tube inner surface

$X^* = \dfrac{x}{r_w}\dfrac{1}{\mathrm{Pe}}$	$\mathrm{Nu} = \alpha d_w/\lambda$	$\mathrm{Nu}_m = \alpha_m d_w/\lambda$	$X^* = \frac{x}{r_w}\frac{1}{\mathrm{Pe}}$	$\mathrm{Nu} = \alpha d_w/\lambda$	$\mathrm{Nu}_m = \alpha_m d_w/\lambda$
0.00001	61.86	93.66	0.020	4.92	7.16
0.0001	28.25	42.84	0.040	4.17	5.81
0.001	12.82	19.50	0.07	3.8197	5.02
0.010	6.00	8.94	0.10	3.7100	4.64
0.012	5.68	8.43	0.20	3.6581	4.16
0.014	5.43	8.01	0.30	3.6568	3.99
0.016	5.23	7.68	0.50	3.6568	3.86

$$T - T_0 = \int_0^x [T_w(\xi) - T_0] \frac{\partial u(x - \xi)}{\partial x} \, d\xi \tag{3.227}$$

it is possible to determine the temperature distribution in the flowing fluid.

Impact function $u(x, R)$ is the distribution of temperature (3.210) in the tube at $T_0 = 0\,°\mathrm{C}$ for $x = 0$ and at the tube inner surface temperature $T_w = 1\,°\mathrm{C}$ for $x > 0$:

$$u = 1 - \sum_{n=0}^{\infty} C_n G_n(R) \exp(-\mu_n^2 X^*). \tag{3.228}$$

A derivative of the impact function described by (3.224) is defined by the following formula:

$$\frac{\partial u}{\partial x} = \frac{1}{r_w} \frac{1}{\mathrm{Re}\,\mathrm{Pr}} \sum_{n=1}^{\infty} C_n G_n(R) \mu_n^2 \exp\left(-\mu_n^2 \frac{x}{r_w} \frac{1}{\mathrm{Re}\,\mathrm{Pr}}\right). \tag{3.229}$$

Substitution of (3.229) in (3.227) results in:

$$T - T_0 = \frac{1}{r_w} \frac{1}{\mathrm{Re}\,\mathrm{Pr}} \int_0^x \left\{ [T_w(\xi) - T_0] \sum_{n=1}^{\infty} C_n G_n(R) \mu_n^2 \exp\left(-\mu_n^2 \frac{x - \xi}{r_w} \frac{1}{\mathrm{Re}\,\mathrm{Pr}}\right) \right\} d\xi. \tag{3.230}$$

Formula (3.230) enables determination of the temperature distribution at a change in $T_w(x)$, for which the first derivative $\partial T_w/\partial x$ is continuous, and also in the case where the first derivative $\partial T_w/\partial x$ is discontinuous, i.e. when step changes occur in the tube inner surface temperature $T_w(x)$ on the tube length (cf. Fig. 3.11).

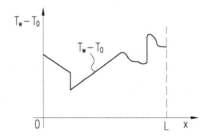

Fig. 3.11 Changes in temperature $T_w(x)$ set on the tube inner surface along the fluid flow direction

3.3.3 *Hydraulically Developed Laminar Flow and the Thermally Developing Flow at Constant Heat Flux on the Flat Slot Inner Surface*

First, the velocity distribution in a flat slot with width $2b$ will be determined (cf. Fig. 3.12).

In this case, the momentum conservation equation has the following form:

$$\frac{d^2 w_x}{dy^2} = \frac{1}{\mu}\frac{dp}{dx},\tag{3.231}$$

and it was solved under the following boundary conditions:

$$\frac{dw_x}{dy}\Big|_{y=0} = 0,\tag{3.232}$$

$$w_x\big|_{y=b} = 0.\tag{3.233}$$

Integrating Eq. (3.231) twice, the result is:

$$w_x = \frac{1}{2\mu}\frac{dp}{dx}y^2 + C_1 y + C_2.\tag{3.234}$$

Substitution of solution (3.324) in conditions (3.232) and (3.233) gives:

$$C_1 = 0, \quad C_1 = -\frac{1}{2\mu}\frac{dp}{dx}b^2.\tag{3.235}$$

Considering constants (3.235) in solution (3.234), the following velocity profile is obtained:

$$w_x = \frac{1}{2\mu}\frac{dp}{dx}\left(y^2 - b^2\right).\tag{3.236}$$

Fig. 3.12 Developed laminar flow in a flat slot with width $2b$

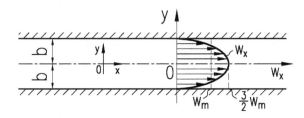

The maximum velocity occurs in the flat channel axis for $y = 0$:

$$w_c = -\frac{1}{2\mu}\frac{dp}{dx}b^2. \tag{3.237}$$

Dividing velocity w_x by w_c, the result is:

$$\frac{w_x}{w_c} = \left(1 - \frac{y^2}{b^2}\right). \tag{3.238}$$

The following formula defines the mean fluid velocity:

$$w_m = \frac{1}{b}\int\limits_0^{r_w} w_x\,dy = -\frac{b^2}{3\mu}\frac{dp}{dx}. \tag{3.239}$$

Dividing expression (3.236) by (3.239), local velocity w_x can be expressed as a function of mean velocity and the radius:

$$w_x = \frac{3}{2}w_m\left(1 - \frac{y^2}{b^2}\right). \tag{3.240}$$

It follows from formula (3.240) that the fluid velocity in the duct axis w_c is one and a half times higher than mean velocity w_m.

The Darcy-Weisbach friction factor ξ is defined as:

$$\xi = \frac{d_h\left(-\frac{dp}{dx}\right)}{\frac{1}{2}\rho w_m^2}, \tag{3.241}$$

which gives:

$$\left(-\frac{dp}{dx}\right) = \frac{\xi}{d_h}\frac{\rho w_m^2}{2}. \tag{3.242}$$

The following is obtained from formula (3.239):

$$\left(-\frac{dp}{dx}\right) = \frac{3\mu}{b^2}w_m. \tag{3.243}$$

Comparing expressions (3.242) and (3.243), the following equation is obtained:

$$\frac{\xi}{d_h}\frac{\rho w_m^2}{2} = \frac{3\mu}{b^2}w_m, \tag{3.244}$$

from which the laminar flow friction factor is determined:

$$\xi = \frac{96}{\rho \, w_m d_h / \mu} = \frac{96}{\text{Re}},$$ (3.245)

where: $\text{Re} = \rho w_m d_h / \mu$, $d_h = 4b$.

If the fluid mean velocity w_m is known, the friction-related pressure drop on length L is found integrating expression (3.242):

$$\Delta p = p \big|_{x_1=x} - p \big|_{x_2=x+L} = \int_{x}^{x+L} \left(-\frac{dp}{dx} \right) dx$$

$$= \int_{x}^{x+L} \frac{\xi}{d_h} \frac{\rho \, w_m^2}{2} dx = \xi \frac{L}{d_h} \frac{\rho \, w_m^2}{2}.$$ (3.246)

Substituting the friction factor ξ defined by formula (3.245) in (3.246), the following is obtained:

$$\Delta p = p \big|_{x_1=x} - p \big|_{x_2=x+L} = \frac{96}{\text{Re}} \frac{L}{d_h} \frac{\rho \, w_m^2}{2} = \frac{3 \, \mu L}{b^2} w_m.$$ (3.247)

An analysis of formula (3.247) indicates that the pressure drop in the tube on length L is proportional to the fluid flow mean velocity w_m and inversely proportional to the square of half of the distance between the parallel plates making up the slot.

The fluid flow temperature field was determined by Cess and Shaffer [48], who, like Siegel, Sparrow, and Hallman [274] for the tube, applied the variables separation method for the flat slot at the surfaces of which a constant value of heat flux $\dot{q}_w$ is set. Considering expression (3.240), the energy balance Eq. (3.12) takes the following form:

$$\frac{3}{2} \rho \, c_p w_m \left(1 - \frac{y^2}{b^2} \right) \frac{\partial T}{\partial x} = \lambda \frac{\partial^2 T}{\partial y^2}.$$ (3.248)

Equation (3.248) will be solved under the following boundary conditions:

$$T \big|_{x=0} = T_0,$$ (3.249)

$$\lambda \frac{\partial T}{\partial y} \big|_{y=b} = \dot{q}_w,$$ (3.250)

$$\lambda \frac{\partial T}{\partial y} \big|_{y=0} = 0.$$ (3.251)

Introducing the following dimensionless variables:

$$\eta = \frac{y}{b}, \quad \text{Re} = \frac{w_m d_h}{\nu}, \quad d_h = 4b, \quad X^* = \frac{1}{\text{Re Pr}}\frac{x}{b} = \frac{1}{\text{Pe}}\frac{x}{b}, \quad \theta = \frac{T - T_0}{\frac{\dot{q}_w b}{\lambda}} \quad (3.252)$$

the problem (3.248)–(3.251) can be written as:

$$\frac{3}{8}\left(1 - \eta^2\right)\frac{\partial\theta}{\partial X^*} = \frac{\partial^2\theta}{\partial\eta^2}. \quad (3.253)$$

Equation (3.248) will be solved under the following boundary conditions:

$$\theta|_{X^*=0} = 0, \quad (3.254)$$

$$\frac{\partial\theta}{\partial\eta}\Big|_{\eta=1} = 1, \quad (3.255)$$

$$\frac{\partial\theta}{\partial\eta}\Big|_{\eta=0} = 0. \quad (3.256)$$

The problem expressed by (3.253)–(3.256) will be solved using the superposition method, according to which dimensionless temperature θ can be presented as a sum of two solutions θ_1 and θ_2:

$$\theta = \theta_1 + \theta_2. \quad (3.257)$$

Substituting (3.257) in Eq. (3.253) and in boundary conditions (3.254)–(3.256), two boundary value problems are obtained, whose solutions are dimensionless temperatures θ_1 and θ_2. The differential equations and the boundary conditions for θ_1 have the following form:

$$\frac{3}{8}\left(1 - \eta^2\right)\frac{\partial\theta_1}{\partial X^*} = \frac{\partial^2\theta_1}{\partial\eta^2}, \quad (3.258)$$

$$\theta_1|_{X^*=0} = 0, \quad (3.259)$$

$$\frac{\partial\theta_1}{\partial\eta}\Big|_{\eta=1} = 1, \quad (3.260)$$

$$\frac{\partial\theta_1}{\partial\eta}\Big|_{\eta=0} = 0. \quad (3.261)$$

Dimensionless temperature θ_1 represents the developed temperature profile if $x \to \infty$.

In such a situation, the velocity and the temperature profiles are fully formed. The solution of problem (3.258)–(3.261) can be found in the same way as in the case of the developed laminar flow inside a tube, which is presented in detail in Sect. 3.1.2.1. Solution θ_1 is then of the following form:

$$\theta_1 = 4X^* + \frac{3}{4}\eta^2 - \frac{1}{8}\eta^4 - \frac{39}{280}. \tag{3.262}$$

Dimensionless temperature θ_2 is obtained by solving the following problem:

$$\frac{3}{8}(1 - \eta^2)\frac{\partial\theta_2}{\partial X^*} = \frac{\partial^2\theta_2}{\partial\eta^2}, \tag{3.263}$$

$$\theta_2\big|_{X^*=0} = -\left(\frac{3}{4}\eta^2 - \frac{1}{8}\eta^4 - \frac{39}{280}\right), \tag{3.264}$$

$$\frac{\partial\theta_2}{\partial\eta}\Big|_{\eta=1} = 0, \tag{3.265}$$

$$\frac{\partial\theta_2}{\partial\eta}\Big|_{\eta=0} = 0. \tag{3.266}$$

The solution of Eq. (3.263) with boundary conditions (3.264)–(3.266) has the following form [48]:

$$\theta_2 = \sum_{n=1}^{\infty} c_n Y_n(\eta)\exp\left(-\frac{8}{3}\mu_n^2 X^*\right), \tag{3.267}$$

where μ_n and $Y_n(\eta)$ are eigenvalues and eigenfunctions, respectively, of the Sturm-Liouville problem:

$$\frac{d^2Y}{d\eta^2} + \mu^2(1 - \eta^2)Y = 0 \tag{3.268}$$

with the following boundary conditions:

$$\frac{dY}{d\eta}\Big|_{\eta=0} = 0, \tag{3.269}$$

$$\frac{dY}{d\eta}\Big|_{\eta=1} = 1. \tag{3.270}$$

Constants c_n are determined using the following equation:

$$c_n = \frac{\int_0^1 \left(\frac{39}{280} - \frac{3}{4}\eta^2 + \frac{1}{8}\eta^4\right)(1 - \eta^2)Y_n d\eta}{\int_0^1 (1 - \eta^2)Y_n^2 d\eta}. \tag{3.271}$$

Constants c_n were determined in [48] using the method proposed by Graetz [110]:

$$c_n = \frac{2}{\mu_n \left(\frac{\partial^2 Y}{\partial \mu \partial \eta}\right)_{\mu=\mu_n, \eta=1}}. \tag{3.272}$$

The Nusselt number will be found from the following formula:

$$\text{Nu} = \frac{\alpha \, d_h}{\lambda} = \frac{\dot{q}_w}{(T_w - T_m)} \frac{4b}{\lambda} = \frac{4}{\theta_w - \theta_m}. \tag{3.273}$$

Formula (3.273) takes account of the fact that hydraulic diameter d_h for a flat slot is $4b$. For a channel with a rectangular cross-section, with dimensions $s \times 2b$, the hydraulic diameter is defined as follows:

$$d_h = \frac{4s \cdot 2b}{2s + 2 \cdot 2b} = \frac{4bs}{s + 2b}, \tag{3.274}$$

from which, if $s \to \infty$, the following is obtained:

$$d_h = \lim_{s \to \infty} \left(\frac{4bs}{s + 2b}\right) = 4b. \tag{3.275}$$

Considering expressions (3.262) and (3.264) in formula (3.257), the channel surface dimensionless temperature θ_w can be expressed as:

$$\theta_w = \frac{T_w - T_0}{\dot{q}_w b / \lambda} = 4X^* + \frac{17}{35} + \sum_{n=1}^{\infty} c_n Y_n(1) \exp\left(-\frac{8}{3} \mu_n^2 X^*\right). \tag{3.276}$$

Considering that mass-averaged temperature θ_m is defined by the following formula:

$$\theta_m = \frac{T_m - T_0}{\dot{q}_w b / \lambda} = \int_0^1 \frac{w_x}{w_m} \theta \, d\eta = 4X^*, \tag{3.277}$$

and temperature θ_w by expression (3.276), formula (3.273) gives:

$$\text{Nu} = \frac{4}{\frac{17}{35} + \sum_{n=1}^{\infty} c_n Y_n(1) \exp\left(-\frac{8}{3} \mu_n^2 X^*\right)}. \tag{3.278}$$

It can be seen that for the developed laminar flow, if $x \to \infty$, the Nusselt number Nu_∞ is: $\text{Nu}_\infty = 140/17 = 8.2353$.

The mean Nusselt number $\text{Nu}_m = \frac{1}{x}\int_0^x \text{Nu}\, dx$ can be found using formula (3.143).

If the heat flux changes on the channel length, the temperature on the channel inner surface can be determined using Duhamel's integral [324]:

$$
\begin{aligned}
T_w - T_0 &= \int_0^x \dot{q}_w(\xi) \frac{\partial u_w(x-\xi)}{\partial x} d\xi \\
&= \frac{1}{\lambda} \int_0^x \left\{ \frac{4}{\text{Pe}} - \frac{8}{3} \sum_{n=1}^{\infty} c_n Y_n(1) \frac{\mu_n^2}{\text{Pe}} \exp\left[-\frac{8}{3} \frac{\mu_n^2}{\text{Pe}} \frac{(x-\xi)}{b} \right] \right\} \dot{q}_w(\xi) d\xi,
\end{aligned}
$$

(3.279)

where u_w is the wall temperature at heat flux:

$$
\dot{q}_w = 1\,\text{W/m}^2,
$$

(3.280)

which is defined as:

$$
u_w = \frac{b}{\lambda} \left[4X^* + \frac{17}{35} + \sum_{n=1}^{\infty} c_n Y_n(1) \exp\left(-\frac{8}{3} \mu_n^2 X^* \right) \right].
$$

(3.281)

The series in expressions (3.276) and (3.279) are slow-convergent if $x \to 0$.

To achieve good accuracy, the number of terms should not be smaller than one hundred. For this reason, it is essential to know the values of μ_n and $Y_n(1)$ for high values of n, which are found from the solution of the problem expressed in (3.268)–(3.270). The problem can be solved using the exponential series method. Solution (3.268) satisfying boundary condition (3.269) has the following form:

$$
Y(\eta) = \sum_{m=0}^{\infty} a_{2m} \eta^{2m},
$$

(3.282)

where:

$$
a_0 = 1, \quad a_2 = -\frac{\mu^2}{2},
$$

(3.283)

$$
a_{2m} = \frac{\mu^2}{2m(2m-1)} (a_{2m-4} - a_{2m-2}), \quad m = 2,3,\ldots.
$$

(3.284)

Substitution of solution (3.282) in condition (3.270) gives the following equation:

$$
\sum_{m=0}^{\infty} m\, a_{2m} = 0,
$$

(3.285)

from which eigenvalues μ_n are found.

Table 3.9 Eigenvalues μ_n and constants c_n and $Y_n(1)$ [101]	n	μ_n	c_n	$Y_n(1)$
	1	4.287224	0.17503	−1.2697
	2	8.30372	−0.051727	1.4022
	3	12.3106	0.025053	−1.4916
	4	16.3145	−0.014924	1.5601
	5	20.3171	0.0099692	−1.6161
	6	24.3189	−0.0072637	1.6638
	7	28.3203	0.0054147	−1.7054
	8	32.3214	−0.0042475	1.7425
	9	36.3223	0.0034280	−1.7760
	10	40.3231	−0.0028294	1.8066

Table 3.9 presents the first ten values of μ_n, c_n, $Y_n(1)$ [101].

For high values of μ and η, approximate eigenvalues μ_n and coefficients $Y(1)$ and c_n can be calculated using the following formulae:

$$\mu_n = 4n + \frac{1}{3}, \tag{3.286}$$

$$Y_n(1) = Y_n(\eta = 0) = \frac{(-1)^n \pi^{1/2}}{6^{1/6} \Gamma\left(\frac{2}{3}\right)} \mu_n^{1/6} = (-1)^n 0.97103\, \mu_n^{1/6}. \tag{3.287}$$

$$c_n = (-1)^{n+1} \frac{2^{11/6} \Gamma\left(\frac{4}{3}\right) 3^{4/3}}{\pi^{3/2}} \mu_n^{-11/6} = (-1)^{n+1} 2.4727\, \mu_n^{-11/6} \tag{3.288}$$

Formulae (3.287) and (3.288) take account of the fact that $\Gamma(2/3) = 1.354117939$ and $\Gamma(4/3) = 0.892979512$, respectively. The Nusselt number defined by formula (3.278) decreases with a rise in x. For:

$$x/b \geq 0.046\, \text{Pe} \tag{3.289}$$

the difference between the Nusselt numbers Nu and $\text{Nu}_\infty = 140/17$ is less than 5%.

3.3.4 Hydraulically Developed Laminar Flow and the Thermally Developing Flow at Constant Temperature at the Flat Slot Inner Surface

The temperature field in a flowing fluid was determined by Cess, Shaffer, and Sellars [48, 265]. The energy conservation Eq. (3.248) will be solved under the following boundary conditions:

$$T|_{x=0} = T_0, \tag{3.290}$$

$$T|_{y=b} = T_w, \tag{3.291}$$

$$\lambda \frac{\partial T}{\partial y}\Big|_{y=0} = 0. \tag{3.292}$$

Introducing the following dimensionless variables:

$$\eta = \frac{y}{b}, \quad \text{Re} = \frac{w_m d_h}{\nu}, \quad d_h = 4b, \quad X^* = \frac{1}{\text{Re Pr}}\frac{x}{b} = \frac{1}{\text{Pe}}\frac{x}{b}, \quad \theta = \frac{T - T_w}{T_0 - T_w}, \tag{3.293}$$

the problem expressed in (3.248), (3.290) and (3.291) can be written as:

$$\frac{3}{8}(1 - \eta^2)\frac{\partial\theta}{\partial X^*} = \frac{\partial^2\theta}{\partial\eta^2}. \tag{3.294}$$

Equation (3.248) will be solved under the following boundary conditions:

$$\theta|_{X^*=0} = 1, \tag{3.295}$$

$$\theta|_{\eta=1} = 0, \tag{3.296}$$

$$\frac{\partial\theta}{\partial\eta}\Big|_{\eta=0} = 0. \tag{3.297}$$

The solution of problem (3.294)–(3.297) will be found using the following form:

$$\theta = \sum_{n=0}^{\infty} c_n Y_n(\eta)\exp\left(-\frac{8}{3}\mu_n^2 X^*\right), \tag{3.298}$$

where μ_n and $Y_n(\eta)$ are eigenvalues and eigenfunctions, respectively, of the Sturm-Liouville problem:

$$\frac{d^2 Y}{d\eta^2} + \mu^2(1 - \eta^2)Y = 0 \tag{3.299}$$

with the following boundary conditions:

$$\frac{dY}{d\eta}\Big|_{\eta=0} = 0, \tag{3.300}$$

$$Y|_{\eta=1} = 1. \tag{3.301}$$

Constants c_n are determined from the following expression:

$$c_n = \frac{\int_0^1 Y_n(1 - \eta^2)d\eta}{\int_0^1 Y_n^2(1 - \eta^2)d\eta}.$$ (3.302)

The heat flux on the channel inner surface is found from:

$$\dot{q}_w = \lambda \frac{\partial T}{\partial y}\Big|_{y=b} = \frac{\lambda}{b}(T_0 - T_w)\frac{\partial \theta}{\partial \eta}\Big|_{\eta=0}$$

$$= \frac{\lambda}{b}(T_0 - T_w)\sum_{n=0}^{\infty} c_n \frac{dY_n(1)}{d\eta} Y_n(\eta) \exp\left(-\frac{8}{3}\mu_n^2 X^*\right).$$ (3.303)

The fluid mass-averaged temperature θ_m is defined by the following formula:

$$\theta_m = 3\sum_{n=0}^{\infty} \frac{G_n}{\mu_n^2} \exp\left(-\frac{8}{3}\mu_n^2 X^*\right).$$ (3.304)

The Nusselt number Nu is defined as:

$$\mathrm{Nu} = \frac{\dot{q}_w d_h}{\lambda(T_w - T_m)} = -\frac{4\dot{q}_w b}{\lambda\theta_m(T_0 - T_w)}.$$ (3.305)

Substitution of (3.303) and (3.304) in (3.305) results in:

$$\mathrm{Nu} = \frac{8\sum_{n=0}^{\infty} G_n \exp\left(-\frac{8}{3}\mu_n^2 X^*\right)}{3\sum_{n=0}^{\infty} \frac{G_n}{\mu_n^2} \exp\left(-\frac{8}{3}\mu_n^2 X^*\right)},$$ (3.306)

where:

$$G_n = -\frac{1}{2}c_n \frac{dY_n}{d\eta}(1).$$ (3.307)

Table 3.10 presents the first ten values of μ_n, c_n, $Y_n(1)$ and G_n [37, 269, 270]. For $n > 9$, the following formulae can be used [265]:

$$\mu_n = 4n + \frac{5}{3},$$ (3.308)

$$c_n \approx 2.28(-1)^n \mu_n^{-7/6},$$ (3.309)

$$-c_n \frac{dY_n}{d\eta}\Big|_{\eta=1} = 2.02557458\mu_n^{-1/3}.$$ (3.310)

For small values of X^* in formula (3.306), the satisfactory accuracy of the Nusselt number calculations cannot be achieved unless more than one hundred

Table 3.10 First ten values of μ_n, c_n, $Y_n(1)$ and G_n

n	μ_n	c_n	$Y_n(1)$	G_n
0	1.6815953222	1.200830	−1.4292	0.858086674
1	5.6698573459	−0.299160	3.8071	0.569462850
2	9.6682424625	0.160826	−5.9202	0.476065463
3	13.6676614426	−0.107437	7.8925	0.423973730
4	17.6673735653	0.079646	−9.7709	0.389108706
5	21.6672053243	−0.062776	11.5798	0.363465044
6	25.6670964863	0.051519	−13.3339	0.343475506
7	29.6670210447	−0.043511	15.0430	0.327265745
8	33.6669660687	0.037542	−16.7141	0.313739318
9	37.6669244563	−0.032933	18.3525	0.302204200

terms are taken into consideration in each of the two series. If the values of X^* are very small, e.g., if $X^* < 1$, it is better to calculate the local and the mean Nusselt numbers (Nu and Nu_m, respectively) using the formulae obtained from the solution of the Lévêque problem, as presented further below in Sect. 3.4. For a hydraulically and thermally developed laminar flow, if $X^* \to \infty$, it is sufficient to take account of only one term in the series in expression (3.307). Considering that $\mu_0 = 1.6815953222$, formula (3.307) gives:

$$\mathrm{Nu}_\infty = \frac{8}{3}\mu_0^2 = \frac{8}{3}1.6815953222^2 = 7.5407. \tag{3.311}$$

The mean Nusselt number Nu_m is found transforming formula (3.69), which results in:

$$\mathrm{Nu}_m = -\frac{1}{X^*}\ln\theta_m, \tag{3.312}$$

where:

$$\mathrm{Nu}_m = \frac{\alpha_m d_h}{\lambda}, \quad \mathrm{Re} = \frac{w_m d_h}{\nu}, \quad \mathrm{Pr} = \frac{c_p\mu}{\lambda}, \quad d_h = 4b,$$

$$X^* = \frac{1}{\mathrm{Re}\,\mathrm{Pr}}\frac{x}{b} = \frac{1}{\mathrm{Pe}}\frac{x}{b}, \quad \theta_m = \frac{T_m - T_w}{T_0 - T_w}.$$

If:

$$\frac{x}{b} \geq 0.032\mathrm{Re}\,\mathrm{Pr}, \tag{3.313}$$

the fluid flow in a slot at a constant temperature of the channel surface can be considered as thermally developed.

3.4   Asymptotic Solutions for Small Values of Coordinate x

The series in the formulae for the temperature distribution and the local Nusselt numbers presented so far are very slow-convergent. For small values of x, i.e. close to the tube inlet, more than 100 terms have to be considered to obtain an appropriate solution. For $X^* \leq 10^{-4}$, where $X^* = (x/r_w)[1/(\text{Re Pr})]$, Lévêque [176] put forward a simple flow model where the tube surface is replaced by a flat surface (a flat plate) over which the fluid flows. The fluid temperature at the inlet is T_0. Constant values of temperature T_w or heat flux $\dot{q}_w$ can be set on the plate surface. The fluid flow velocity over the plate should be constant and equal to velocity w_m (plug flow), or it should change linearly. The latter case is of more practical significance.

3.4.1   Constant Fluid Flow Velocity Over a Flat Surface

The energy conservation equation for a fluid flowing at a constant velocity w_m has the following form:

$$c_p \rho\, w_m \frac{\partial T}{\partial x} = \lambda \frac{\partial^2 T}{\partial y^2}. \qquad (3.314)$$

The equation will be solved setting the channel surface constant temperature T_w or a constant heat flux on the channel surface $\dot{q}_w$.

3.4.1.1   Constant Temperature of the Channel Surface

Equation (3.314) will be solved first assuming that the temperature of the surface over which the fluid flows is constant and totals T_w, i.e. the boundary conditions are as follows:

$$T\big|_{x=0} = T_0, \qquad (3.315)$$

$$T\big|_{y=0} = T_w, \qquad (3.316)$$

$$T\big|_{y=\infty} = T_0. \qquad (3.317)$$

The problem described by Eq. (3.314) and the boundary conditions presented above is the same as the transient-state heat conduction problem in a half space with a constant value of initial temperature T_w, where the temperature of the surface rises in jumps to temperature T_w for time $t > 0$.

In the problem under analysis, coordinate x and product $c_p \rho w_m$ correspond to time t and product $c_p \rho$ in the transient-state heat conduction problem, respectively. Using the solution of the transient-state heat conduction problem presented in [326], the heat flux on the channel surface is defined by the following formula:

$$\dot{q}_w = (T_w - T_0)\sqrt{\frac{\lambda\, c_p \rho\, w_m}{\pi\, x}}.$$ (3.318)

The heat transfer coefficient α is obtained from:

$$\alpha = \frac{\dot{q}_w}{T_w - T_0} = \sqrt{\frac{\lambda\, c_p \rho\, w_m}{\pi\, x}}.$$ (3.319)

Introducing the following dimensionless numbers:

$$\mathrm{Re} = \frac{w(2\, r_w)}{\nu} = \frac{w\, d_w}{\nu}, \quad \mathrm{Pr} = \frac{\nu}{a} = \frac{c_p \rho\, \nu}{\lambda} = \frac{c_p \mu}{\lambda},$$ (3.320)

the Nusselt number is expressed as follows:

$$\mathrm{Nu} = \frac{\alpha(2\, r_w)}{\lambda} = \sqrt{\frac{4\, r_w^2\, c_p \rho\, w_m}{\pi\, \lambda\, x}} = \frac{1}{\sqrt{\pi}}\left(\mathrm{Re}\ \mathrm{Pr}\,\frac{d_w}{x}\right)^{1/2}.$$ (3.321)

The mean Nusselt number on length 0–x is defined as:

$$\mathrm{Nu}_m = \frac{\alpha_m(2\, r_w)}{\lambda} = \frac{1}{x}\int_0^x \mathrm{Nu}\, dx = \frac{2}{\sqrt{\pi}}\left(\mathrm{Re}\ \mathrm{Pr}\,\frac{d_w}{x}\right)^{1/2}.$$ (3.322)

It should be noted that the exponent of the product $\left(\mathrm{Re}\ \mathrm{Pr}\,\dfrac{d_w}{x}\right)$ is 1/2, whereas it totals 1/3 if the fluid flow velocity over the channel surface has a linear distribution.

The same form of formulae for the local and the mean Nusselt numbers (Nu and Nu_m, respectively) is obtained for a flat slot with width $2b$. The tube inner diameter d_w should be replaced with hydraulic diameter d_h, which—for a flat slot—is: $d_h = 4b$.

3.4.1.2 Constant Heat Flux at the Channel Surface

Like in the case with the channel surface constant temperature, the solution will be applied of the initial-boundary problem for a half space with a uniform initial temperature T_0 which is heated abruptly by a constant heat flux $\dot{q}_w$. The fluid temperature at the channel inlet is T_0. At a big distance from the channel wall, it also equals zero. The boundary conditions have the following form:

$$T|_{x=0} = T_0, \tag{3.323}$$

$$-\lambda \frac{\partial T}{\partial y} = \dot{q}_w, \tag{3.324}$$

$$T|_{y=\infty} = T_0. \tag{3.325}$$

Using the solution of the problem of transient-state heat conduction in a half space presented in [326], the difference between the temperatures of the channel surface and of the fluid at a big distance from the wall $(T_w - T_0)$ is defined by the following formula:

$$T_w - T_0 = \frac{2\dot{q}_w}{\sqrt{\pi}} \sqrt{\frac{x}{\lambda c_p \rho w_m}}. \tag{3.326}$$

The local heat transfer coefficient on the channel surface is calculated using the following formula:

$$\alpha = \frac{\dot{q}_w}{T_w - T_0} = \frac{\sqrt{\pi}}{2} \sqrt{\frac{\lambda c_p \rho w_m}{x}}. \tag{3.327}$$

According to the definition, the Nusselt number is found from the following expression:

$$\begin{aligned}
\mathrm{Nu} &= \frac{\alpha(2\,r_w)}{\lambda} = \frac{\sqrt{\pi}}{2} \sqrt{\frac{c_p \rho\, w_m (2\,r_w)^2}{\lambda\, x}} \\
&= \frac{\sqrt{\pi}}{2} \sqrt{\left(\frac{w_m\, 2\,r_w}{\nu}\right) \frac{\nu\, 2\,r_w}{a\; x}} = \frac{\sqrt{\pi}}{2} \sqrt{\mathrm{Re}\, \mathrm{Pr} \frac{d_w}{x}}.
\end{aligned} \tag{3.328}$$

The mean Nusselt number on length 0–x is calculated as:

$$\mathrm{Nu}_m = \frac{\alpha_m(2\,r_w)}{\lambda} = \frac{1}{x} \int_0^x \mathrm{Nu}\, dx = \sqrt{\pi} \left(\mathrm{Re}\, \mathrm{Pr} \frac{d_w}{x}\right)^{1/2}. \tag{3.329}$$

It can be noticed that the mean Nusselt number for coordinate x is twice larger than the local Nusselt number for the same coordinate, i.e. $\mathrm{Nu}_m(x) = 2\,\mathrm{Nu}(x)$. The Nusselt number at a constant heat flux on the channel surface is $\pi/2$ times bigger than the Nusselt number at the surface constant temperature. It should be emphasized that the formula for the heat transfer coefficient calculation (3.329) is also valid for a flat slot. The expression for the Nusselt number for a flat slot with width $2b$ has the following form:

$$Nu = \frac{\alpha(4b)}{\lambda} = \frac{\sqrt{\pi}}{2} \sqrt{\frac{c_p \rho \, w_m (4b)^2}{\lambda x}}$$

$$= \frac{\sqrt{\pi}}{2} \sqrt{\left(\frac{w_m \, 4b}{\nu}\right) \frac{\nu \, 4b}{a \, x}} = \frac{\sqrt{\pi}}{2} \sqrt{Re \, Pr \frac{d_h}{x}}. \tag{3.330}$$

It can be seen that the formulae for the slot and for the tube have an identical form. If inner diameter d_w is replaced by hydraulic diameter $d_h = 4b$ in the formulae valid for the tube, formulae are obtained for the flat slot.

3.4.2 Linear Change in the Fluid Flow Velocity Over a Flat Surface

Two cases will be analyzed—with constant temperature T_w or with constant heat flux $\dot{q}_w$ set on the channel flat surface

3.4.2.1 Constant Temperature of the Channel Surface

For a laminar flow, the developed distribution of velocity in the tube is defined as follows:

$$w = 2 w_m \left[1 - \left(\frac{r}{r_w}\right)^2\right]. \tag{3.331}$$

Close to the tube surface, velocity $w(r)$ can be expressed as a Taylor series in which all terms of the higher order except the linear one are omitted:

$$w(r) = w|_{r=r_w} + \frac{dw}{dr}|_{r=r_w}(r - r_w) + \cdots. \tag{3.332}$$

Considering that the fluid velocity on the tube surface equals zero, i.e. $w|_{r=r_w} = 0$, calculating derivative $\frac{dw}{dr}|_{r=r_w} = -\frac{4 w_m}{r_w}$ and introducing coordinate $y = r_w - r$ from formula (3.332), the following is obtained:

$$w = \frac{4 w_m}{r_w} y. \tag{3.333}$$

The same procedure can be used for a channel formed by two infinitely large plates which are parallel to each other. Considering that the velocity distribution in a flat channel with width $2b$ is defined as:

$$w = \frac{3}{2}w_m \left[1 - \left(\frac{b-y}{b}\right)^2\right],$$ (3.334)

the velocity distribution close to the flat channel wall can be expressed using the following formula:

$$w = \frac{3\,w_m}{b}y,$$ (3.335)

where coordinate y is measured from the channel surface, i.e. for the channel surface $y = 0$ and for the channel axis $y = b$.

Expressions (3.333) and (3.335) can be written using the following single formula (cf. Fig. 3.13):

$$w = B_w y,$$ (3.336)

where:

$$B_w = \frac{4\,w_m}{r_w} \text{ for the tube,}$$ (3.337)

$$B_w = \frac{3\,w_m}{b} \text{ for the flat channel.}$$ (3.338)

Linear approximations of the fluid velocity close to the tube and the channel inner surface are presented in Fig. 3.13.

The energy conservation equation for very small values of coordinate x can be written for the tube and the flat channel using a single formula as presented below:

$$c_p \rho\, B_w y \frac{\partial T}{\partial x} = \lambda \frac{\partial^2 T}{\partial y^2}.$$ (3.339)

Expression (3.339) is a differential energy balance equation for a fluid flowing over a flat surface.

Introducing the following variables:

$$\theta = \frac{T - T_0}{T_w - T_0},$$ (3.340)

$$\eta = \frac{y}{C}x^{-1/3},$$ (3.341)

$$C = \left(\frac{9a}{B_w}\right)^{1/3},$$ (3.342)

Fig. 3.13 Linear approximation of the real change in the fluid velocity close to the channel wall; **a** Developed laminar flow in a tube with diameter $d_w = 2r_w$, **b** developed laminar flow in a slot with width $2b$

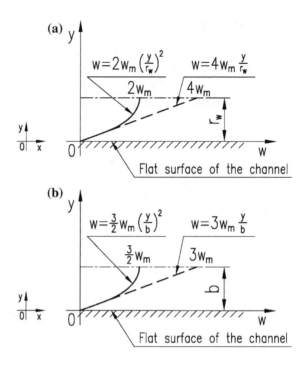

equation (3.339) can be written as follows:

$$\frac{d^2\theta}{d\eta^2} + 3\eta^2 \frac{d\theta}{d\eta} = 0. \tag{3.343}$$

If constant temperature T_w is set on the channel surface, the boundary conditions take the following form:

$$\theta\big|_{\eta=0} = 1, \tag{3.344}$$

$$\theta\big|_{\eta=\infty} = 0. \tag{3.345}$$

Assuming that

$$\frac{d\theta}{d\eta} = p \tag{3.346}$$

and substituting it in the differential Eq. (3.343), the result is as follows:

$$\frac{dp}{d\eta} + 3\eta^2 p = 0. \tag{3.347}$$

Variables separation and integration give:

$$\frac{d\theta}{d\eta} = C_1 \exp(-\eta^3). \tag{3.348}$$

Another integration results in:

$$\theta = C_1 \int_0^\eta \exp(-\eta^3)d\eta + C_2. \tag{3.349}$$

Substituting (3.349) in the first boundary condition (3.344), the following is obtained:

$$C_2 = 0. \tag{3.350}$$

Condition (3.345) gives:

$$C_1 = -\frac{1}{\int_0^\infty \exp(-\eta^3)d\eta}. \tag{3.351}$$

Substituting C_1 and C_2 in (3.349), the solution takes the following form:

$$\theta = 1 - \frac{\int_0^\eta \exp(-\eta^3)d\eta}{\int_0^\infty \exp(-\eta^3)d\eta} = \frac{\int_\eta^\infty \exp(-\eta^3)d\eta}{\int_0^\infty \exp(-\eta^3)d\eta}. \tag{3.352}$$

Noting that:

$$\int_0^\infty \exp(-\eta^3)d\eta = \Gamma\left(\frac{4}{3}\right), \tag{3.353}$$

the solution can be written as follows:

$$\theta = 1 - \frac{1}{\Gamma\left(\frac{4}{3}\right)} \int_0^\eta \exp(-\eta^3)d\eta = \frac{1}{\Gamma\left(\frac{4}{3}\right)} \int_\eta^\infty \exp(-\eta^3)d\eta. \tag{3.354}$$

The heat flux on the channel surface is:

$$\dot{q}_w = -\lambda \frac{\partial T}{\partial y}\Big|_{y=0} = -\lambda(T_w - T_0)\frac{\partial \theta}{\partial y}\Big|_{y=0} = -\lambda(T_w - T_0)\frac{\partial \theta}{\partial \eta}\frac{\partial \eta}{\partial y}\Big|_{\eta=0}$$

$$= -\lambda(T_w - T_0)C_1\frac{\partial \eta}{\partial y}\Big|_{\eta=0} = \frac{\lambda}{\Gamma(\frac{4}{3})}\frac{x^{-1/3}}{C}(T_w - T_0).$$

(3.355)

In this case, $\frac{\partial \theta}{\partial \eta}\Big|_{\eta=0}$ is determined using the expression (3.348) and taking account of the fact that constant C_1 is defined by formula (3.351). The derivative can also be found differentiating function (3.354). The Leibniz formula should be used in this case because the integration limit in Eq. (3.354) is a function of η. According to the Leibniz rule:

$$\frac{d}{dx}\left[\int_{a(x)}^{b(x)} f(x, \xi)d\xi\right] = \int_{a(x)}^{b(x)} \frac{\partial f(x, \xi)}{\partial x} dx + f(x, b)\frac{db}{dx} - f(x, a)\frac{da}{dx},$$

(3.356)

derivative $\frac{\partial \theta}{\partial y}\Big|_{y=0}$ in formula (3.355) can be found in the following way:

$$\frac{\partial \theta}{\partial y}\Big|_{y=0} = \left[\exp(-\eta^3)\frac{\partial \eta}{\partial y}\right]\Big|_{y=0} = \frac{\partial \eta}{\partial y}\Big|_{y=0}.$$

(3.357)

The heat transfer coefficient is calculated from:

$$\alpha = \frac{\dot{q}_w}{T_w - T_0} = \frac{\lambda}{\Gamma(\frac{4}{3})}\frac{x^{-1/3}}{C}.$$

(3.358)

The Nusselt number is defined as:

$$Nu = \frac{\alpha d_h}{\lambda} = \frac{1}{\Gamma(\frac{4}{3})}\frac{d_h x^{-1/3}}{C},$$

(3.359)

where d_h denotes the hydraulic diameter. For a tube $d_h = 2r_w$, whereas for a flat channel (flat slot) with width $2b$ the diameter is $d_h = 4b$.

Substituting C defined by formula (3.342) in (3.359) and considering B_w defined by formulae (3.337) and (3.338), the following is obtained:

- for a tube with diameter d_w

$$\text{Nu} = \frac{\alpha\,d_w}{\lambda} = \frac{\left(\frac{8}{9}\right)^{1/3}}{\Gamma(4/3)}\left(\frac{1}{\text{Re Pr}}\frac{x}{d_w}\right)^{-1/3} = 1.0767\left(\text{Re Pr}\frac{d_w}{x}\right)^{1/3}, \qquad (3.360)$$

where:

$$\text{Re} = \frac{w_m d_w}{\nu}, \quad \text{Pr} = \frac{c_p\mu}{\lambda}, \quad d_w = 2r_w, \quad \Gamma(4/3) = 0.892\,979\,512;$$

- for a flat channel with width $2b$

$$\text{Nu} = \frac{\alpha\,d_h}{\lambda} = \frac{\left(\frac{4}{3}\right)^{1/3}}{\Gamma(4/3)}\left(\frac{1}{\text{Re Pr}}\frac{x}{d_h}\right)^{-1/3} = 1.2326\left(\text{Re Pr}\frac{d_h}{x}\right)^{1/3}, \qquad (3.361)$$

where:

$$\text{Re} = \frac{w_m d_h}{\nu}, \quad \text{Pr} = \frac{c_p\mu}{\lambda}, \quad d_h = 4b.$$

The mean Nusselt number on length 0–x is found from the following formula:

$$\text{Nu}_m = \frac{1}{x}\int_0^x \text{Nu}\,dx = \frac{3}{2}\text{Nu}(x). \qquad (3.362)$$

Substitution of expressions (3.360) and (3.361), respectively, results in:

$$\text{Nu}_m = \frac{\alpha_m d_w}{\lambda} = \frac{3^{1/3}}{\Gamma(4/3)}\left(\frac{1}{\text{Re Pr}}\frac{x}{d_w}\right)^{-1/3} = 1.6151\left(\text{Re Pr}\frac{d_w}{x}\right)^{1/3} \qquad (3.363)$$

for the tube and

$$\text{Nu}_m = \frac{\alpha_m d_h}{\lambda} = \frac{\left(\frac{9}{2}\right)^{1/3}}{\Gamma(4/3)}\left(\frac{1}{\text{Re Pr}}\frac{x}{d_h}\right)^{-1/3} = 1.8489\left(\text{Re Pr}\frac{d_h}{x}\right)^{1/3} \qquad (3.364)$$

for the flat channel.

3.4.2.2   Constant Heat Flux at the Channel Surface

The energy conservation equation for the tube can be written as:

$$c_p \rho \, w_a \frac{y}{r_w} \frac{\partial T}{\partial x} = \lambda \frac{\partial^2 T}{\partial y^2}, \tag{3.365}$$

where $w_a = B_w \, r_w = 4 \, w_m$ is the fluid velocity in the duct axis ($r = 0$) at a linear distribution of velocity.

Considering that the heat flux component in the direction of axis y is:

$$\dot{q}_y = -\lambda \frac{\partial T}{\partial y} \tag{3.366}$$

and dividing Eq. (3.365) by y and differentiating both sides of the equation in relation to coordinate y, the following is obtained:

$$\frac{w_a}{r_w} \frac{\partial \dot{q}_y}{\partial x} = a \frac{\partial}{\partial y} \left(\frac{1}{y} \frac{\partial \dot{q}_y}{\partial y} \right). \tag{3.367}$$

Introducing the following dimensionless coordinates:

$$\psi = \frac{\dot{q}_y}{\dot{q}_w}, \quad \eta = \frac{y}{r_w}, \quad \tau = \frac{a \, x}{w_a \, r_w^2}, \tag{3.368}$$

Eq. (3.367) can be transformed into:

$$\frac{\partial \psi}{\partial \tau} = \frac{\partial}{\partial \eta} \left(\frac{1}{\eta} \frac{\partial \psi}{\partial \eta} \right). \tag{3.369}$$

Equation (3.369) will be solved under the following boundary conditions:

$$\psi|_{\tau=0} = 0, \tag{3.370}$$

$$\psi|_{\eta=0} = 1, \tag{3.371}$$

$$\psi|_{\eta=\infty} = 0. \tag{3.372}$$

Introducing the following new variable:

$$\chi = \frac{\eta}{(9\tau)^{1/3}} \tag{3.373}$$

Eq. (3.374) can be written as follows:

$$\chi \frac{d^2\psi}{d\chi^2} + (3\chi^3 - 1)\frac{d\psi}{d\chi} = 0. \tag{3.374}$$

Introducing new variable p:

$$\frac{d\psi}{d\chi} = p, \tag{3.375}$$

Eq. (3.369) can be written as follows:

$$\chi \frac{dp}{d\chi} + (3\chi^3 - 1)p = 0. \tag{3.376}$$

After Eq. (3.376) is solved, $p(\chi)$ is determined, and then, substituting $p(\chi)$ in Eq. (3.375), Equation (3.375) is solved using boundary conditions (3.371) and (3.372), which results in:

$$\psi(\chi) = \frac{\int_\chi^\infty \bar{\chi} \exp(-\bar{\chi}^3) d\bar{\chi}}{\int_0^\infty \bar{\chi} \exp(-\bar{\chi}^3) d\bar{\chi}} = \frac{3}{\Gamma\left(\frac{2}{3}\right)} \int_\chi^\infty \bar{\chi} \exp(-\bar{\chi}^3) d\bar{\chi}, \tag{3.377}$$

where the gamma function is defined as [137, 338]: $\Gamma(z) = \int_0^\infty t^{z-t}e^{-t}dt$.

The temperature distribution is found from the Fourier law (3.366):

$$\int_T^{T_0} d\bar{T} = -\frac{1}{\lambda} \int_y^\infty \dot{q}_{\bar{y}} d\bar{y}. \tag{3.378}$$

Considering that $\dot{q}_y = \dot{q}_w \psi$, formula (3.378) gives the distribution of temperature:

$$T - T_0 = \frac{\dot{q}_w}{\lambda} \int_y^\infty \psi d\bar{y}, \tag{3.379}$$

which can be expressed in the following dimensionless form:

$$\theta(\eta, \tau) = \frac{T - T_0}{\dot{q}_w r_w / \lambda} = (9\tau)^{1/3} \int_\chi^\infty \psi(\bar{\chi}) d\bar{\chi}. \tag{3.380}$$

Substitution of expression $\psi(\chi)$, defined by formula (3.377), in (3.380) results in:

$$\theta(\eta,\tau) = (9\tau)^{1/3} \int_{\chi}^{\infty} \left[\frac{3}{\Gamma(2/3)} \int_{\tilde{\chi}}^{\infty} \tilde{\chi} \exp(-\tilde{\chi}^3) d\tilde{\chi} \right] d\tilde{\chi}. \qquad (3.381)$$

After integration, the following is obtained:

$$\theta(\eta,\tau) = \frac{(9\tau)^{1/3}}{\Gamma(2/3)} \left[\exp(-\chi^3) - 3\chi \int_{\chi}^{\infty} \tilde{\chi} \exp(-\tilde{\chi}^3) d\tilde{\chi} \right]. \qquad (3.382)$$

Noting that the incomplete gamma function is defined by the formula $\Gamma(a,z) = \int_{z}^{\infty} t^{a-1} e^{-t} dt$ [137, 338], expression (3.382) can be written as:

$$\theta(\eta,\tau) = \frac{(9\tau)^{1/3}}{\Gamma(2/3)} \left[\exp(-\chi^3) - \chi \Gamma\left(\frac{2}{3},\chi^3\right) \right]. \qquad (3.383)$$

Assuming in formula (3.383) that $\chi = 0$, the following expression is obtained for the temperature difference between the wall surface T_w and the fluid at a big distance from the wall T_0:

$$T_w - T_0 = \frac{\dot{q}_w r_w}{\lambda} \frac{(9\tau)^{1/3}}{\Gamma(2/3)}. \qquad (3.384)$$

The heat transfer coefficient is defined as:

$$\alpha = \frac{\dot{q}_w}{T_w - T_0} = \frac{\lambda \Gamma(2/3)}{r_w (9\tau)^{1/3}}. \qquad (3.385)$$

Knowing the heat transfer coefficient, the expression for the Nusselt number can be found:

$$Nu = \frac{\alpha(2r_w)}{\lambda} = \frac{2\Gamma(2/3)}{(9\tau)^{1/3}}. \qquad (3.386)$$

Considering that for the tube $w_a = 4w_m$, the expression describing τ takes the following form:

$$\tau = X_d = \frac{ax}{w_a r_w^2} = \frac{ax}{4w_m r_w^2} = \frac{1}{\frac{w_m(2r_w)}{\nu} \frac{\nu}{a}} \frac{x}{2r_w} = \frac{1}{\mathrm{Re}\,\mathrm{Pr}} \frac{x}{d_w}. \qquad (3.387)$$

The expression for the Nusselt number takes the following form:

$$\text{Nu} = \left(\frac{8}{9}\right)^{1/3} \Gamma(2/3) \left(\text{Re Pr}\frac{d_w}{x}\right)^{1/3}. \tag{3.388}$$

Considering that $\Gamma(2/3) = 1.354\,117\,939$, expression (3.388) can be written as:

$$\text{Nu} = 1.3020 \left(\text{Re Pr}\frac{d_w}{x}\right)^{1/3}. \tag{3.389}$$

Sellars et al. [265] give a slightly different form of the formula for the Nusselt number Nu:

$$\text{Nu} = \frac{2^{1/3}9^{2/3}}{3} \Gamma(5/3) \left(\text{Re Pr}\frac{r_w}{x}\right)^{1/3} = 3^{1/3}\Gamma(5/3) \left(\text{Re Pr}\frac{d_w}{x}\right)^{1/3}. \tag{3.390}$$

Noting that $\Gamma(z+1) = z\Gamma(z)$ [137, 338], i.e. $\Gamma(\frac{5}{3}) = \Gamma(\frac{2}{3}+1) = \frac{2}{3}\Gamma(\frac{2}{3})$, expression (3.390) can be written in the following form:

$$\text{Nu} = \left(\frac{16}{9}\right)^{1/3} \Gamma\left(\frac{2}{3}\right) \left(\text{Re Pr}\frac{r_w}{x}\right)^{1/3} = \left(\frac{8}{9}\right)^{1/3} \Gamma\left(\frac{2}{3}\right) \left(\text{Re Pr}\frac{d_w}{x}\right)^{1/3} \tag{3.391}$$

or as:

$$\text{Nu} = 1.6404 \left(\text{Re Pr}\frac{r_w}{x}\right)^{1/3} = 1.3020 \left(\text{Re Pr}\frac{d_w}{x}\right)^{1/3}. \tag{3.392}$$

It can be seen that both formulae: (3.388) and (3.391) lead to the same results. The mean Nusselt number on length 0–x is defined as:

$$\text{Nu}_m = \frac{\alpha_m d_w}{\lambda} = \frac{1}{x}\int_0^x \text{Nu}\,dx. \tag{3.393}$$

Substituting (3.388) in (3.393) and performing integration, the following is obtained:

$$\text{Nu}_m = \frac{3}{2}\text{Nu}(x) = 3^{1/3}\Gamma(2/3) \left(\text{Re Pr}\frac{d_w}{x}\right)^{1/3} = 1.9530 \left(\text{Re Pr}\frac{d_w}{x}\right)^{1/3}. \tag{3.394}$$

Using a similar procedure, an expression can be obtained for the Nusselt number for a flat channel with width $2b$, on the surface of which a constant heat flux is set. Considering that in this case $w_a = 3w_m$ and $d_h = 4b$, the expression describing τ takes the following form:

$$\tau = X_d = \frac{ax}{w_a b^2} = \frac{ax}{3 w_m b^2} = \frac{16}{3 \frac{w_m(4b)}{\nu} \frac{\nu}{a}} \frac{x}{4b} = \frac{16}{3} \frac{1}{\text{Re Pr}} \frac{x}{d_h}. \tag{3.395}$$

The expression for the Nusselt number has the following form:

$$\begin{aligned}
\text{Nu} &= \frac{\alpha \, d_h}{\lambda} = \frac{4\Gamma(2/3)}{(9\tau)^{1/3}} = \left(\frac{4}{3}\right)^{1/3} \Gamma(2/3) \left(\text{Re Pr} \frac{d_h}{x}\right)^{1/3} \\
&= 1.4904 \left(\text{Re Pr} \frac{d_h}{x}\right)^{1/3}.
\end{aligned} \tag{3.396}$$

The following formula defines the mean Nusselt number:

$$\begin{aligned}
\text{Nu}_m &= \frac{\alpha_m d_h}{\lambda} = \frac{3}{2} \text{Nu}(x) = \left(\frac{9}{2}\right)^{1/3} \Gamma(2/3) \left(\text{Re Pr} \frac{d_h}{x}\right)^{1/3} \\
&= 2.2356 \left(\text{Re Pr} \frac{d_h}{x}\right)^{1/3},
\end{aligned} \tag{3.397}$$

where $d_h = 4b$.

3.4.3 Formulae for Determination of the Nusselt Number in Tubes and Flat Slots Valid in the Initial Part of the Inlet Section

The formulae for the Nusselt number derived in Sect. 3.4 are often used in practice [106, 209, 258]. For this reason, the formulae for the local and mean Nusselt numbers will be listed in tables.

The previous subsections present derivation of the formulae needed to calculate the local and mean Nusselt numbers in tubes and flat channels (slots) for the thermal entrance length assuming that the velocity profile is developed. Formulae for the Nusselt number are also derived for the fluid plug flow through a tube at a constant temperature or a constant heat flux on the tube surface. The formulae for the Nusselt number derived based on the solution of the Lévêque problem and presented in Tables 3.11 and 3.12 are valid if:

$$\frac{1}{\text{Re Pr}} \frac{x}{d_h} \le 0.0005. \tag{3.398}$$

For higher values of expression $\frac{1}{\text{Re Pr}} \frac{x}{d_h}$, the Nusselt number is calculated using the formulae obtained from the solution of the Graetz problem and containing infinite series. However, the formulae are rather complex—they contain infinite series, which

Table 3.11 Nusselt numbers for the developed laminar flow and local Nusselt numbers for the initial part of the inlet section for a tube at a constant temperature set on the tube inner surface (asymptotic solutions for small values of x)

Flow type	Local Nusselt number Nu	Nusselt number Nu_∞ for the developed flow
Plug flow	$$Nu = \frac{\alpha\, d_w}{\lambda} = \frac{1}{\sqrt{\pi}}\left(Re\, Pr\, \frac{d_w}{x}\right)^{1/2},$$ $$Re = \frac{w_m d_w}{\nu},\, Pr = \frac{\nu}{a} = \frac{c_p \mu}{\lambda}$$	$Nu_\infty = 5.7831$
Laminar flow, $w_x = \dfrac{4\,w_m}{r_w}y$	$$Nu = \frac{\alpha\, d_w}{\lambda} = \frac{\left(\frac{8}{9}\right)^{1/3}}{\Gamma(4/3)}\left(\frac{1}{Re\, Pr}\frac{x}{d_w}\right)^{-1/3}$$ $$= 1.0767\left(Re\, Pr\, \frac{d_w}{x}\right)^{1/3},$$ $$Re = \frac{w_m d_w}{\nu},\, Pr = \frac{\nu}{a} = \frac{c_p \mu}{\lambda}$$	$Nu_\infty = 3.6568$

Table 3.12 Mean Nusselt numbers for the initial part of the inlet section and mean Nusselt numbers for the developed laminar flow in a tube at a constant temperature set on the tube inner surface (asymptotic solutions for small values of x)

Flow type	Mean Nusselt number Nu_m	Nusselt number $Nu_{m\infty}$ for the developed flow
Plug flow, $w_x = w_m = $ const.	$$Nu_m = \frac{\alpha_m d_w}{\lambda} = \frac{2}{\sqrt{\pi}}\left(Re\, Pr\frac{d_w}{x}\right)^{1/2},$$ $$Re = \frac{w_m d_w}{\nu},\, Pr = \frac{\nu}{a} = \frac{c_p \mu}{\lambda}$$	$Nu_{m\infty} = 5.7831$
Laminar flow, $w_x = \dfrac{4\,w_m}{r_w}y$	$$Nu_m = \frac{\alpha_m d_w}{\lambda} = \frac{3^{1/3}}{\Gamma(4/3)}\left(\frac{1}{Re\, Pr}\frac{x}{d_w}\right)^{-1/3}$$ $$= 1.6151\left(Re\, Pr\frac{d_w}{x}\right)^{1/3},$$ $$Re = \frac{w_m d_w}{\nu},\, Pr = \frac{\nu}{a} = \frac{c_p \mu}{\lambda}$$	$Nu_{m\infty} = 3.6568$

for smaller values of x require considering a few hundred terms. For this reason, in practice, asymptotic formulae are applied for the Nusselt numbers which are valid for $\dfrac{1}{Re\, Pr}\dfrac{x}{d_h} \rightarrow 0$, and formulae valid for $x \rightarrow \infty$, i.e., for the developed laminar flow. The Nusselt number values determined in a wide range of changes in parameter $\dfrac{1}{Re\, Pr}\dfrac{x}{d_h}$ are approximated using various simple functions using asymptotic solutions.

Tables 3.13 and 3.14 present the formulae needed to calculate local and mean Nusselt numbers for small values of the parameter $\dfrac{1}{Re\, Pr}\dfrac{x}{d_w}$ for the laminar flow in a tube with a constant heat flux on the tube inner surface.

Table 3.13 Nusselt numbers for the developed laminar flow and local Nusselt numbers for the initial part of the inlet section for a tube at a constant heat flux set on the tube inner surface (asymptotic solutions for small values of x)

Flow type	Local Nusselt number Nu	Nusselt number Nu_∞ for the developed flow
Plug flow, $w_x = w_m = \text{const.}$	$\mathrm{Nu} = \dfrac{\alpha\, d_w}{\lambda} = \dfrac{\sqrt{\pi}}{2}\sqrt{\mathrm{Re}\,\mathrm{Pr}\,\dfrac{d_w}{x}},$ $\mathrm{Re} = \dfrac{w_m d_w}{\nu},\ \mathrm{Pr} = \dfrac{\nu}{a} = \dfrac{c_p \mu}{\lambda}$	$\mathrm{Nu}_\infty = 8$
Laminar flow, $w_x = \dfrac{4\,w_m}{r_w}\,y$	$\mathrm{Nu} = \dfrac{\alpha\, d_w}{\lambda} = \left(\dfrac{8}{9}\right)^{1/3}\Gamma(2/3)\left(\mathrm{Re}\,\mathrm{Pr}\,\dfrac{d_w}{x}\right)^{1/3}$ $= 1.3020\left(\mathrm{Re}\,\mathrm{Pr}\,\dfrac{d_w}{x}\right)^{1/3},$ $\mathrm{Re} = \dfrac{w_m d_w}{\nu},\ \mathrm{Pr} = \dfrac{\nu}{a} = \dfrac{c_p \mu}{\lambda}$	$\mathrm{Nu}_\infty = \dfrac{48}{11} = 4.3636$

Table 3.14 Mean Nusselt numbers for the developed laminar flow and mean Nusselt numbers for the initial part of the inlet section for a tube at a constant heat flux set on the tube inner surface (asymptotic solutions for small values of x)

Flow type	Mean Nusselt number Nu_m	Nusselt number $\mathrm{Nu}_{m\infty}$ for the developed flow
Plug flow, $w_x = w_m = \text{const.}$	$\mathrm{Nu}_m = \dfrac{\alpha_m d_w}{\lambda} = \sqrt{\pi}\left(\mathrm{Re}\,\mathrm{Pr}\,\dfrac{d_w}{x}\right)^{1/2},$ $\mathrm{Re} = \dfrac{w_m d_w}{\nu},\ \mathrm{Pr} = \dfrac{\nu}{a} = \dfrac{c_p \mu}{\lambda}$	$\mathrm{Nu}_m = 8$
Laminar flow, $w_x = \dfrac{4\,w_m}{r_w}\,y$	$\mathrm{Nu}_m = \dfrac{\alpha_m d_w}{\lambda} = 3^{1/3}\Gamma(2/3)\left(\mathrm{Re}\,\mathrm{Pr}\,\dfrac{d_w}{x}\right)^{1/3}$ $= 1.9530\left(\mathrm{Re}\,\mathrm{Pr}\,\dfrac{d_w}{x}\right)^{1/3},$ $\mathrm{Re} = \dfrac{w_m d_w}{\nu},\ \mathrm{Pr} = \dfrac{\nu}{a} = \dfrac{c_p \mu}{\lambda}$	$\mathrm{Nu}_m = \dfrac{48}{11} = 4.3636$

Tables 3.15 and 3.16 present the formulae needed to calculate local and mean Nusselt numbers for small values of the parameter $\dfrac{1}{\mathrm{Re}\,\mathrm{Pr}}\dfrac{x}{d_w}$ for the laminar flow in a flat slot with a constant temperature and a constant heat flux, respectively, at the slot inner surface.

The formulae presented in Tables 3.11, 3.12, 3.13, 3.14, 3.15 and 3.16 can be used to calculate the local and the mean Nusselt number (Nu and Nu_m, respectively) for small values of a parameter $\dfrac{1}{\mathrm{Re}\,\mathrm{Pr}}\dfrac{x}{d_h}$ that satisfy condition (3.398).

Figure 3.14 presents changes in the local and the mean Nusselt number (Nu and Nu_m) for the thermally developing flow in the tube if the fluid flow is laminar. The velocity profile is already developed and defined by a parabolic function.

Table 3.15 Nusselt numbers for the developed laminar flow and local and mean Nusselt numbers for the initial part of the inlet section for a flat slot with width $2b$ at a constant temperature set on the slot inner surface (asymptotic solutions for small values of x)

Velocity distribution	Nusselt number Nu or Nu_m	Nusselt numbers Nu_∞ and $\mathrm{Nu}_{m\infty}$ for the developed flow
$w_x = \dfrac{3\,w_m}{b}y$	$\mathrm{Nu} = \dfrac{\alpha\,d_h}{\lambda} = \dfrac{\left(\frac{4}{3}\right)^{1/3}}{\Gamma(4/3)}\left(\mathrm{Re\,Pr}\dfrac{d_h}{x}\right)^{1/3}$ $= 1.2326\left(\mathrm{Re\,Pr}\dfrac{d_h}{x}\right)^{1/3},$ $d_h = 4b, \mathrm{Re} = \dfrac{w_m d_h}{\nu}, \mathrm{Pr} = \dfrac{\nu}{a} = \dfrac{c_p\mu}{\lambda}$	$\mathrm{Nu}_\infty = \mathrm{Nu}_{m\infty} = 7.541$
$w_x = \dfrac{3\,w_m}{b}y$	$\mathrm{Nu}_m = \dfrac{\alpha_m d_h}{\lambda} = \dfrac{\left(\frac{9}{2}\right)^{1/3}}{\Gamma(4/3)}\left(\mathrm{Re\,Pr}\dfrac{d_h}{x}\right)^{1/3}$ $= 1.8489\left(\mathrm{Re\,Pr}\dfrac{d_h}{x}\right)^{1/3},$ $d_h = 4b, \mathrm{Re} = \dfrac{w_m d_h}{\nu}, \mathrm{Pr} = \dfrac{\nu}{a} = \dfrac{c_p\mu}{\lambda}$	$\mathrm{Nu}_\infty = \mathrm{Nu}_{m\infty} = 7.541$

Table 3.16 Nusselt numbers for the developed laminar flow and local and mean Nusselt numbers for the initial part of the inlet section for a flat slot with width $2b$ at a constant heat flux set on the slot inner surface (asymptotic solutions for small values of x)

Velocity distribution	Nusselt number Nu or Nu_m	Nusselt numbers Nu_∞ and $\mathrm{Nu}_{m\infty}$ for the developed flow
$w_x = \dfrac{3\,w_m}{b}y$	$\mathrm{Nu} = \dfrac{\alpha\,d_h}{\lambda} = \left(\dfrac{4}{3}\right)^{1/3}\Gamma(2/3)\left(\mathrm{Re\,Pr}\dfrac{d_h}{x}\right)^{1/3}$ $= 1.4904\left(\mathrm{Re\,Pr}\dfrac{d_h}{x}\right)^{1/3},$ $d_h = 4b, \mathrm{Re} = \dfrac{w_m d_h}{\nu}, \mathrm{Pr} = \dfrac{\nu}{a} = \dfrac{c_p\mu}{\lambda}$	$\mathrm{Nu}_\infty = \mathrm{Nu}_{m\infty} = \dfrac{140}{17} = 8.235$
$w_x = \dfrac{3\,w_m}{b}y$	$\mathrm{Nu}_m = \dfrac{\alpha_m d_h}{\lambda} = \left(\dfrac{9}{2}\right)^{1/3}\Gamma(2/3)\left(\mathrm{Re\,Pr}\dfrac{d_h}{x}\right)^{1/3}$ $= 2.2356\left(\mathrm{Re\,Pr}\dfrac{d_h}{x}\right)^{1/3},$ $d_h = 4b, \mathrm{Re} = \dfrac{w_m d_h}{\nu}, \mathrm{Pr} = \dfrac{\nu}{a} = \dfrac{c_p\mu}{\lambda}$	$\mathrm{Nu}_\infty = \mathrm{Nu}_{m\infty} = \dfrac{140}{17} = 8.235$

Figure 3.15 illustrates changes in the local and the mean Nusselt number (Nu and Nu_m) for the thermal entrance length in a flat slot at the laminar fluid flow and developed velocity profile. Both figures are made based on the table data presented in [269, 270]. The histories of the Nusselt numbers Nu and Nu_m depending on parameter $\dfrac{1}{\mathrm{Re\,Pr}}\dfrac{x}{d_h}$ presented in the works mentioned above are found using solutions of the Lévêque problem, and the Graetz problem analyzed in the works of

Fig. 3.14 Local and mean Nusselt number (Nu and Nu_m, respectively) at the fluid flow in a tube with a constant heat flux $\dot{q}_w$ or constant temperature T_w as a function of the parameter

$$X^* = \frac{1}{\text{Re Pr}} \frac{x}{d_w}$$

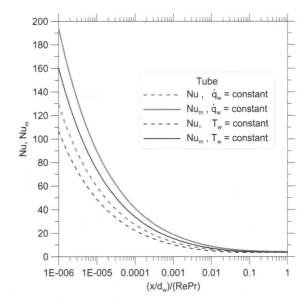

Fig. 3.15 Local and mean Nusselt number (Nu and Nu_m, respectively) at the fluid flow in a flat slot with width $2b$ with a constant heat flux $\dot{q}_w$ or constant temperature T_w as a function of parameter $\frac{1}{\text{Re Pr}} \frac{x}{d_h}$; symbol $d_h = 4b$ denotes the hydraulic slot diameter

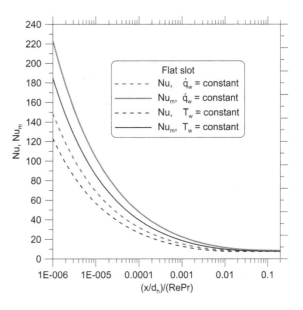

Mercer [199], Worsøe-Schmidt [364], Newman [211] and Nunge et al. [216]. They improved the Lévêque problem solution obtained under the assumption of linear changes in the fluid velocity by taking account of the real velocity profiles near the channel wall and the variations in the surface area of the tube cross-section depending on the radius. Using the solutions obtained by Worsøe-Schmidt [364] and Newman [211], Shah [270] determined the Nusselt numbers for small values of the parameter $\dfrac{1}{\operatorname{Re}\operatorname{Pr}}\dfrac{x}{d_h}$. For the parameter higher values, the Nusselt numbers Nu and Nu_m are calculated using the Graetz problem solutions, which are discussed in detail in this chapter.

The expressions for the Nusselt number obtained from the Lévêque problem solution find application in the determination of correlations for the Nusselt number in plate and tube heat exchangers [191–193].

It follows from the analysis of the results presented in Figs. 3.14 and 3.15 that for:

$$0.04 \leq \frac{1}{\operatorname{Re}\operatorname{Pr}}\frac{x}{d_h} \tag{3.399}$$

there are only slight changes in the Nusselt numbers Nu and Nu_m, and the fluid flow is already thermally developed. Furthermore, considering that the fluid velocity profile is assumed to be developed, the flow is also developed hydraulically.

3.5   Laminar Fluid Flow and Heat Transfer in the Inlet Section—Formulae Used in Engineering Practice

The formulae derived in this chapter for the Nusselt number determination for the laminar fluid flow in the inlet section (on the entrance length) are complex, and the infinite series in them are slow-convergent, which makes the calculation of both the local and the mean Nusselt number values difficult. For this reason, approximation formulae are usually used to calculate the Nusselt numbers presented in Figs. 3.14 and 3.15 [106, 209, 258, 269, 270]. Good results are obtained if the Nusselt numbers are approximated in the entire range of changes in the parameter $\dfrac{1}{\operatorname{Re}\operatorname{Pr}}\dfrac{x}{d_h}$ as a function of the Nusselt numbers $\mathrm{Nu}|_{x\to 0}$ obtained based on the solution of the Lévêque problem for small values of the parameter, and using the Nusselt numbers $\mathrm{Nu}|_{x\to\infty}$ for the developed laminar flow if the parameter is high. The changes in the Nusselt number in the interval:

$$1 \times 10^{-6} \leq \frac{1}{\operatorname{Re}\operatorname{Pr}}\frac{x}{d_h} \leq 1.0 \tag{3.400}$$

are approximated using the following function:

$$\mathrm{Nu} = \left[(\mathrm{Nu}|_{x\to\infty})^{c_1} + (\mathrm{Nu}|_{x\to 0})^{c_1}\right]^{1/c_1}, \tag{3.401}$$

where c_1 is a constant determined using the least squares method.

The Nusselt numbers found using exact formulae are approximated using a different function, applied in [106, 258]:

$$\text{Nu} = \left[(\text{Nu}|_{x \to \infty})^3 + c_2^3 + (\text{Nu}|_{x \to 0} - c_2)^3 \right]^{1/3}, \qquad (3.402)$$

where c_2 is a constant determined by means of the least squares method.

Functions (3.401) and (3.402) are also used to approximate the mean Nusselt number for tubes and flat slots.

The following formulae are obtained, which approximate changes in the Nusselt number on the inlet section length where the flow is hydraulically developed but thermally developing:

- **tube, constant heat flux on the wall**

$$\text{Nu} = \left\{ 4.364^{6.8332} + \left[1.302 \left(\text{Re} \, \text{Pr} \frac{d_w}{x} \right)^{1/3} \right]^{6.8332} \right\}^{(1/6.8332)}, \qquad (3.403)$$

$r^2 = 0.99961, \quad |\varepsilon_{\max}| \le 3.9\%,$

$$\text{Nu}_m = \left\{ 4.364^{4.8290} + \left[1.953 \left(\text{Re} \, \text{Pr} \frac{d_w}{x} \right)^{1/3} \right]^{4.8290} \right\}^{(1/4.8290)}, \qquad (3.404)$$

$r^2 = 0.99979, \quad |\varepsilon_{\max}| \le 3.5\%,$

$$\text{Nu} = \left\{ 4.364^3 + 0.8245^3 + \left[1.302 \left(\text{Re} \, \text{Pr} \frac{d_w}{x} \right)^{1/3} - 0.8245 \right]^3 \right\}^{1/3}, \qquad (3.405)$$

$r^2 = 0.99999, \quad |\varepsilon_{\max}| \le 4.0\%,$

$$\text{Nu}_m = \left\{ 4.364^3 + 0.7999^3 + \left[1.953 \left(\text{Re} \, \text{Pr} \frac{d_w}{x} \right)^{1/3} - 0.7999 \right]^3 \right\}^{1/3}, \qquad (3.406)$$

$r^2 = 0.99999, \quad |\varepsilon_{\max}| \le 3.1\%;$

- **tube, constant wall temperature**

$$\text{Nu} = \left\{ 3.66^{10.5649} + \left[1.077 \left(\text{Re} \, \text{Pr} \frac{d_w}{x} \right)^{1/3} \right]^{10.5649} \right\}^{(1/10.5649)}, \qquad (3.407)$$

$r^2 = 0.99998, \quad |\varepsilon_{\max}| \le 6.4\%,$

$$\text{Nu}_m = \left\{ 3.66^{5.8965} + \left[1.615 \left(\text{Re} \, \text{Pr} \frac{d_w}{x} \right)^{1/3} \right]^{5.8965} \right\}^{(1/5.8965)}, \qquad (3.408)$$

$$r^2 = 0.99958, \quad |\varepsilon_{\max}| \leq 5.7\%,$$

$$\text{Nu} = \left\{ 3.66^3 + 0.9705^3 + \left[1.077 \left(\text{Re} \, \text{Pr} \frac{d_w}{x} \right)^{1/3} - 0.9705 \right]^3 \right\}^{1/3}, \qquad (3.409)$$

$$r^2 = 0.99998, \quad |\varepsilon_{\max}| \leq 3.6\%,$$

$$\text{Nu}_m = \left\{ 3.66^3 + 0.9301^3 + \left[1.615 \left(\text{Re} \, \text{Pr} \frac{d_w}{x} \right)^{1/3} - 0.9301 \right]^3 \right\}^{1/3}, \qquad (3.410)$$

$$r^2 = 0.99999, \quad |\varepsilon_{\max}| \leq 3.1\%;$$

- **flat slot, constant heat flux on the wall**

$$\text{Nu} = \left\{ 8.235^{4.2011} + \left[1.4904 \left(\text{Re} \, \text{Pr} \frac{d_h}{x} \right)^{1/3} \right]^{4.2011} \right\}^{(1/4.2011)}, \qquad (3.411)$$

$$r^2 = 0.99998, \quad |\varepsilon_{\max}| \leq 3.3\%,$$

$$\text{Nu}_m = \left\{ 8.235^{3.3857} + \left[2.2356 \left(\text{Re} \, \text{Pr} \frac{d_h}{x} \right)^{1/3} \right]^{3.3857} \right\}^{(1/3.3857)}, \qquad (3.412)$$

$$r^2 = 0.99999, \quad |\varepsilon_{\max}| \leq 0.6\%,$$

$$\text{Nu} = \left\{ 8.235^3 + 0.3159^3 + \left[1.4904 \left(\text{Re} \, \text{Pr} \frac{d_w}{x} \right)^{1/3} - 0.3159 \right]^3 \right\}^{1/3}, \qquad (3.413)$$

$$r^2 = 0.99994, \quad |\varepsilon_{\max}| \leq 7.5\%,$$

$$\text{Nu}_m = \left\{ 8.235^3 + 0.2063^3 + \left[2.2356 \left(\text{Re} \, \text{Pr} \frac{d_w}{x} \right)^{1/3} - 0.2063 \right]^3 \right\}^{1/3}, \qquad (3.414)$$

$$r^2 = 0.99999, \quad |\varepsilon_{\max}| \leq 2.2\%;$$

- **flat slot, constant wall temperature**

$$\text{Nu} = \left\{ 7.541^{4.5261} + \left[1.2326 \left(\text{Re} \, \text{Pr} \frac{d_h}{x} \right)^{1/3} \right]^{4.5261} \right\}^{(1/4.5261)}, \tag{3.415}$$

$r^2 = 0.99997, \quad |\varepsilon_{\max}| \leq 3.0\%,$

$$\text{Nu}_m = \left\{ 7.541^{3.546} + \left[1.8489 \left(\text{Re} \, \text{Pr} \frac{d_h}{x} \right)^{1/3} \right]^{3.546} \right\}^{(1/3.546)}, \tag{3.416}$$

$r^2 = 0.99999, \quad |\varepsilon_{\max}| \leq 0.5\%,$

$$\text{Nu} = \left\{ 7.541^3 + 0.3724^3 + \left[1.2326 \left(\text{Re} \, \text{Pr} \frac{d_w}{x} \right)^{1/3} - 0.3724 \right]^3 \right\}^{1/3}, \tag{3.417}$$

$r^2 = 0.99992, \quad |\varepsilon_{\max}| \leq 8.0\%,$

$$\text{Nu}_m = \left\{ 7.541^3 + 0.2643^3 + \left[1.8489 \left(\text{Re} \, \text{Pr} \frac{d_w}{x} \right)^{1/3} - 0.2643 \right]^3 \right\}^{1/3}, \tag{3.418}$$

$r^2 = 0.99999, \quad |\varepsilon_{\max}| \leq 2.4\%.$

Relative errors ε_i are calculated using the following formula:

$$\varepsilon_i = \frac{\text{Nu}_i^{\text{data}} - \text{Nu}_i^{\text{apr}}}{\text{Nu}_i^{\text{data}}} 100, \; \%, \tag{3.419}$$

where $\text{Nu}_i^{\text{data}}$ denotes data calculated from the solution of the Lévêque or the Graetz problem [269, 270], and Nu_i^{apr} is the Nusselt number calculated using a function approximating the data. The relative errors in the approximation of the mean Nusselt number Nu_m are calculated in the same manner. It should be emphasized that the maximum values of relative errors given next to the approximation formulae occur in a few points only. For most points, the errors are several times smaller compared to the presented maximum ones. Symbol r^2 is the determination coefficient. Figures 3.16 and 3.17 present a comparison of data taken from the work of Shah and London [270] with exponential approximation functions. Figure 3.16 presents exponential functions (3.403) and (3.407) approximating the local Nusselt

Fig. 3.16 Nu approximation
for a tube at a constant heat
flux $\dot{q}_w$ or constant
temperature T_w set on the tube
inner surface using functions
(3.403) and (3.407),
respectively

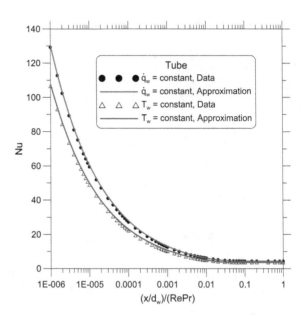

Fig. 3.17 Nu_m
approximation for a tube at a
constant heat flux $\dot{q}_w$ or
constant temperature T_w set
on the tube inner surface by
means of functions (3.404)
and (3.408), respectively

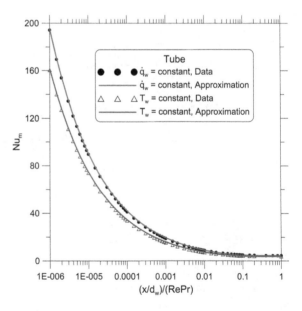

number for a constant heat flux or a constant temperature set on the tube inner
surface, and Fig. 3.17—functions (3.404) and (3.408) approximating the mean
Nusselt number for a constant heat flux or a constant temperature set at the tube
inner surface.

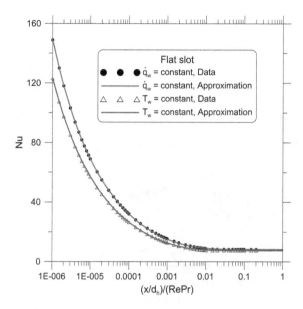

Fig. 3.18 Nu approximation for a flat slot at a constant heat flux or a constant temperature set at the inner slot surface using functions (3.411) and (3.415), respectively

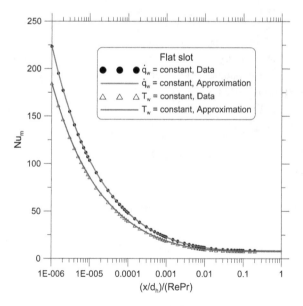

Fig. 3.19 Nu_m approximation for a flat slot at a constant heat flux or a constant temperature set at the slot inner surface by means of functions (3.412) and (3.415), respectively

The analysis of the results presented in Figs. 3.16 and 3.17 indicates that exponential functions approximate data obtained using the solution of the Lévêque or the Graetz problem very well [269, 270].

A similar comparison is made in Figs. 3.18 and 3.19 for a slot with width $2b$.

Also, in this case, the accuracy of data approximation using exponential functions is excellent.

Considering the simpler form of function (3.401) and very good accuracy of the obtained formulae approximating the local and mean Nusselt numbers for the tube and the flat slot, it is recommended that they should be used in practice.

3.6 Hydrodynamically and Thermally Developing Flow in the Inlet Section

In heat exchangers, when fluid flows from the header (manifold) into a parallel system of tubes, the fluid flow in the initial section is not developed hydraulically or thermally, i.e., both velocity and temperature profiles differ from those observed in the developed laminar flow. In this situation, the fluid velocity and temperature values in the inlet cross-section are usually assumed as constant.

Because on the tube length the velocity profile in the tube cross-section varies, the problem could not be solved analytically. The approximation formulae presented in the literature are derived based on numerical calculation results. This subsection presents formulae put forward in [106, 258], which are simpler and characterized by a form similar to function (3.402). For small values of x, the fluid flow with a constant velocity and temperature at the inlet is similar to the flow over a flat surface with a constant temperature or heat flux. For this reason, function (3.402) takes additional account of the expression for the Nusselt number, which—for a problem formulated in this way—is known [209, 232].

3.6.1 Flow in a Tube with a Constant Temperature of the Inner Surface

The local Nusselt number is defined as [106]:

$$\mathrm{Nu}_{x,T} = \left[\mathrm{Nu}_{x,T,1}^3 + 0.7^3 + \left(\mathrm{Nu}_{x,T,2} - 0.7\right)^3 + \mathrm{Nu}_{x,T,3}^3\right]^{1/3}, \quad (3.420)$$

where:

$$\mathrm{Nu}_{x,T,1} = 3.66 \quad (3.421)$$

is the Nusselt number for the developed laminar flow. The methods of the finding $\mathrm{Nu}_{x,T,1} = 3.66$ are presented in Sects. 3.1.2.2 and 3.3.2.

The Nusselt number:

$$\mathrm{Nu}_{x,T,2} = 1.077 \left(\mathrm{Re} \ \mathrm{Pr} \frac{d_i}{x} \right)^{1/3} \qquad (3.422)$$

is obtained by solving the Lévêque problem. Formula (3.422) is derived in Sect. 3.4.2.1 (formula 3.360). Formula (3.420) contains the Nusselt number $\mathrm{Nu}_{x,T,3}$, which is a generalization of the formula obtained by Pohlhausen [21, 40, 106, 141, 232] for the fluid flow with a constant velocity and temperature in the inlet cross-section over a flat surface with a constant temperature:

$$\mathrm{Nu}_{x,T} = 0.332 \, \mathrm{Pr}^{1/3} \left(\mathrm{Re} \frac{d_i}{x} \right)^{1/2}. \qquad (3.423)$$

Formula (3.423) is generalized to the form presented below [106, 258] to cover fluids with different values of the Prandtl number:

$$\mathrm{Nu}_{x,T,3} = \frac{1}{2} \left(\frac{2}{1 + 22 \, \mathrm{Pr}} \right)^{1/6} (\mathrm{Re} \ \mathrm{Pr} \ d_i/x)^{1/2}. \qquad (3.424)$$

Calculating the mean values for the Nusselt numbers defined by formulae (3.421), (3.422) and (3.424), the formula $\mathrm{Nu}_m = \frac{1}{x} \int_0^x \mathrm{Nu} \, dx$ makes it possible to determine the relation for the mean Nusselt number on length 0–x (length $l = x$) [106]:

$$\mathrm{Nu}_{m,T} = \left[\mathrm{Nu}_{m,T,1}^3 + 0.7^3 + \left(\mathrm{Nu}_{m,T,2} - 0.7 \right)^3 + \mathrm{Nu}_{m,T,3}^3 \right]^{1/3}, \qquad (3.425)$$

where:

$$\mathrm{Nu}_{m,T,1} = 3.66, \qquad (3.426)$$

$$\mathrm{Nu}_{m,T,2} = 1.615 \left(\mathrm{Re} \ \mathrm{Pr} \frac{d_i}{l} \right)^{1/3}, \qquad (3.427)$$

$$\mathrm{Nu}_{m,T,3} = \left(\frac{2}{1 + 22 \, \mathrm{Pr}} \right)^{1/6} (\mathrm{Re} \ \mathrm{Pr} \ d_i/l)^{1/2}. \qquad (3.428)$$

The dimensionless numbers are defined as follows:

$$\mathrm{Nu} = \frac{\alpha \, d_w}{\lambda}, \qquad \mathrm{Re} = \frac{w d_w}{\nu}, \qquad \mathrm{Pr} = \frac{c_p \mu}{\lambda}. \qquad (3.429)$$

3.6.2 Flow in a Tube with a Constant Heat Flux at the Inner Surface

The Nusselt local number value $\mathrm{Nu}_{x,q}$ can be found using the formula recommended in [106]:

$$\mathrm{Nu}_{x,q} = \left[\mathrm{Nu}^3_{x,q,1} + 1 + \left(\mathrm{Nu}_{x,q,2} - 1\right)^3 + \mathrm{Nu}^3_{x,q,3}\right]^{1/3}, \qquad (3.430)$$

where $\mathrm{Nu}_{x,q,1}$ is the Nusselt number for the developed laminar flow and equals:

$$\mathrm{Nu}_{x,q,1} = 4.364. \qquad (3.431)$$

Formula (3.431) is derived in Sect. 3.1.2.1.

The Nusselt number $\mathrm{Nu}_{x,q,2}$ is a solution of the Lévêque problem valid for small values of the parameter $\dfrac{1}{\mathrm{Re}\,\mathrm{Pr}}\dfrac{x}{d_w}$, and it is defined by the following formula:

$$\mathrm{Nu}_{x,q,2} = 1.302\left(\mathrm{Re}\,\mathrm{Pr}\frac{d_i}{x}\right)^{1/3}. \qquad (3.432)$$

Formula (3.432) is derived in Sect. 3.4.2.2. $\mathrm{Nu}_{x,q,3}$ is the local Nusselt number on the surface of a plate (on a flat surface) over which the fluid flows with a constant velocity and temperature in the inlet cross-section [84, 106]:

$$\mathrm{Nu}_{x,q,3} = 0.462\,\mathrm{Pr}^{1/3}\left(\mathrm{Re}\frac{d_i}{x}\right)^{1/2} \text{ for } \mathrm{Pr} \geq 0.7. \qquad (3.433)$$

The same procedure can be used to calculate the mean Nusselt number values on length 0–x [106]:

$$\mathrm{Nu}_{m,q} = \left[\mathrm{Nu}^3_{m,q,1} + 0.6^3 + \left(\mathrm{Nu}_{m,q,2} - 0.6\right)^3 + \mathrm{Nu}^3_{m,q,3}\right]^{1/3}, \qquad (3.434)$$

where:

$$\mathrm{Nu}_{m,q,1} = 4.364, \qquad (3.435)$$

$$\mathrm{Nu}_{m,q,2} = 1.953\left(\mathrm{Re}\,\mathrm{Pr}\frac{d_i}{l}\right)^{1/3}, \qquad (3.436)$$

$$\mathrm{Nu}_{m,q,3} = 0.924\mathrm{Pr}^{1/3}\left(\mathrm{Re}\frac{d_i}{l}\right)^{1/2}. \qquad (3.437)$$

Formulae (3.436) and (3.437) are obtained calculating mean Nusselt numbers using formulae (3.432) and (3.433), respectively.

Mean values of the heat transfer coefficient are used most often to calculate heat exchangers, both in the method based on the logarithmic mean temperature difference and in the ε—NTU method. Due to the different ways of calculating the mean temperature difference between the tube inner surface and the fluid mass-averaged temperature for the wall constant temperature and a constant heat flux value, the calculation results can differ a little from those obtained if the heat exchanger calculations take account of changes in the local heat transfer coefficient on the length of a tube or channel with a different cross-section. For this reason, if the temperature distributions of mediums and the heat flux transferred from the hot to the cold medium are calculated using numerical methods, e.g., the finite volume method, it is better to perform the calculations using the local heat transfer coefficient.

The inlet section (the entrance length) contributes substantially to the intensification of the heat transfer in heat exchangers. Car radiators, or heat exchangers in

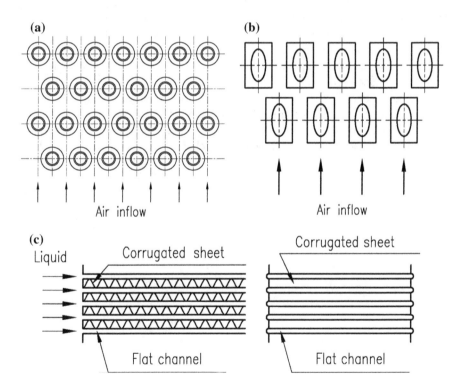

Fig. 3.20 Evolution of the design of air-cooled steam condensers used to condense the turbine exhaust steam in what is referred to as dry cooling systems; **a** First generation—four rows of finned tubes with a circular cross-section, **b** the 1980s—two rows of individually finned elliptical or oval tubes; **c** now—flattened, very wide and long channels with corrugated sheets of steel in between acting as fins

power plants used to carry away the steam condensation heat in what is referred to as dry cooling systems, make use of air-cooled finned exchangers. Fins can be separate or continuous (lamellae). Air flows through flat slots created by the fins. The largest heat flux is transferred from hot water to air in the first tube row [305, 314, 315, 318, 319, 328]. This is mainly the effect of the big heat flux transferred from the surfaces of the lamellae to air in the air inflow area. The heat transfer coefficient on the surface of the fins (lamellae) on the side of the air inflow is very high.

As the distance from the air inflow area gets bigger, the heat transfer coefficient on the surface of the fins gets smaller and thermal effectiveness of the second and further rows of the tubes decreases. For this reason, modern finned-tube heat exchangers, where a hot liquid flows inside the tubes and the cooling air stream is perpendicular to the tube axis, are built as single-row devices (cf. Fig. 3.20).

The formulae derived in this chapter can be generalized to cover the laminar fluid flow in micro-channels with diameter $d_w = 10–100$ μm and in mini-channels with diameter $d_w = 100$ μm–1 mm. At such dimensions, a slip occurs in the fluid flow on the channel surface, i.e., the fluid velocity is not equal to zero, and the fluid temperature is different from the temperature of the channel surface. Similar issues also arise in the case of flows of diluted gases in channels with bigger diameters. Such flows were investigated intensively in the early 1960s.

Due to the advancement in micro-exchangers characterized by high thermal power values achieved at very small dimensions, extensive and detailed research is now being conducted on issues related to the fluid flow in micro-channels.

The methods for determination of the velocity and temperature distribution in micro-channels are similar to those presented in this chapter. The survey of literature and solutions related to the fluid flow in micro-channels can be found in [101, 143].

Chapter 4
Turbulent Fluid Flow

Turbulent flows involve irregular fluctuations in velocity, pressure and temperature. The momentary velocity vector varies from mean velocity in terms of both value and direction. All parameters characterizing the turbulent flow, such as velocity, pressure and temperature, oscillate around the mean value over time (cf. Fig. 4.1).

Any $\varphi(x, y, z, t)$ can be presented as a sum of mean value $\bar{\varphi}(x, y, z, t)$ and random fluctuation $\varphi'(x, y, z, t)$:

$$\varphi = \bar{\varphi} + \varphi', \tag{4.1}$$

where:

$$\bar{\varphi}(x, y, z, t) = \frac{1}{t_0} \int_{t-t_0/2}^{t+t_0/2} \varphi(x, y, z, t)\, dt. \tag{4.2}$$

Compared to the period of random fluctuations, time interval t_0 must be sufficiently long but also small enough to avoid substantial changes in mean value $\bar{\varphi}(x, y, z, t)$ in the analysed interval of time t_0. Considering that fluctuations φ' are random, the following is obtained:

$$\bar{\varphi}' = \frac{1}{t_0} \int_{t-t_0/2}^{t+t_0/2} \varphi'\, dt = 0. \tag{4.3}$$

The turbulence level (turbulence intensity) is expressed using parameter T.

$$T = \frac{\sqrt{\frac{1}{3}\left(\overline{w_x'^2} + \overline{w_y'^2} + \overline{w_z'^2}\right)}}{|\bar{\mathbf{w}}|}, \tag{4.4}$$

© Springer International Publishing AG, part of Springer Nature 2019
D. Taler, *Numerical Modelling and Experimental Testing of Heat Exchangers*,
Studies in Systems, Decision and Control 161,
https://doi.org/10.1007/978-3-319-91128-1_4

Fig. 4.1 Changes in φ
(velocity, pressure or
temperature) for the turbulent
flow

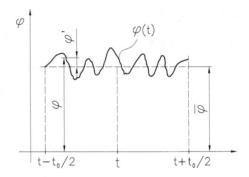

where $\bar{\mathbf{w}}$ is the velocity vector.

$$\bar{\mathbf{w}} = \mathbf{i}\,\bar{w}_x + \mathbf{j}\,\bar{w}_y + \mathbf{k}\,\bar{w}_z. \tag{4.5}$$

Not only the Reynolds number Re but also turbulence intensity T need to be known in experimental testing and numerical simulations to characterize the turbulent flow.

4.1   Averaged Reynolds Equations

The turbulent fluid flow will be analyzed based on an example of a low-velocity flow, which is usually the case in practice, assuming constant location- and temperature-independent physical properties. Time-averaged mass, momentum, and energy conservation equations can be derived considering that:

$$w_x = \bar{w}_x + w'_x, \tag{4.6}$$

$$w_y = \bar{w}_y + w'_y, \tag{4.7}$$

$$w_z = \bar{w}_z + w'_z, \tag{4.8}$$

$$p = \bar{p} + p', \tag{4.9}$$

$$T = \bar{T} + T', \tag{4.10}$$

$$\rho = \bar{\rho} + \rho' \approx \bar{\rho}. \tag{4.11}$$

Equation (4.11) is the Boussinesq approximation.

Expressions (4.6)–(4.11) will be substituted in the mass (2.55), momentum (2.78)–(2.80) and energy (2.144) conservation equations. The following assumptions also have to be taken into consideration: the fluid physical properties are constant, the fluid is incompressible, i.e., $\beta = 0$, no radiative heat transfer occurs, i.e., $\dot{q}_r = 0$ and the dissipation term in the energy balance Eq. (2.144) equals zero, i.e., $\dot{\phi} = 0$. The equations will be transformed taking account of the Reynolds principles of averaging two variables f and g. Assuming that

$$f = \bar{f} + f', \tag{4.12}$$

$$g = \bar{g} + g', \tag{4.13}$$

Reynolds put forward the following principles of averaging:

$$\overline{f + g} = \bar{f} + \bar{g}, \tag{4.14}$$

$$\overline{Cf} = C\bar{f}, \quad \text{gdzie } C = \text{const}, \tag{4.15}$$

$$\overline{f'} = \overline{g'} = \overline{f'\bar{f}} = \overline{g'\bar{f}} = 0, \tag{4.16}$$

$$\bar{\bar{f}} = \bar{f}, \tag{4.17}$$

$$\overline{\bar{f}\bar{g}} = \bar{f}\,\bar{g}, \tag{4.18}$$

$$\overline{f\,g} = \bar{f}\,\bar{g} + \overline{f'\,g'}, \tag{4.19}$$

$$\overline{\left(\frac{\partial f}{\partial s}\right)} = \frac{\partial \bar{f}}{\partial s}, \tag{4.20}$$

$$\overline{\left(\int f\,ds\right)} = \int \bar{f}\,ds. \tag{4.21}$$

Considering (4.3)–(4.8) in formula (2.55), the following is obtained:

$$\frac{\partial \bar{w}_x}{\partial x} + \frac{\partial \bar{w}'_x}{\partial x} + \frac{\partial \bar{w}_y}{\partial y} + \frac{\partial \bar{w}'_y}{\partial y} + \frac{\partial \bar{w}_z}{\partial z} + \frac{\partial \bar{w}'_z}{\partial z} = 0. \tag{4.22}$$

Integrating Eq. (4.22) with respect to time and using formula (4.3), the result is:

$$\frac{\partial \bar{w}_x}{\partial x} + \frac{\partial \bar{w}_y}{\partial y} + \frac{\partial \bar{w}_z}{\partial z} = 0. \tag{4.23}$$

The same procedure can be applied to transform the components of the momentum balance Eqs. (2.78)–(2.80):

$$\frac{\partial \bar{w}_x}{\partial t} + \bar{w}_x \frac{\partial \bar{w}_x}{\partial x} + \bar{w}_y \frac{\partial \bar{w}_x}{\partial y} + \bar{w}_z \frac{\partial \bar{w}_x}{\partial z} = -\frac{1}{\rho}\frac{\partial \bar{p}}{\partial x} + \nu \nabla^2 \bar{w}_x - \frac{\partial}{\partial x}\left(\overline{w_x'^2}\right)$$
$$- \frac{\partial}{\partial y}\left(\overline{w_x' w_y'}\right) - \frac{\partial}{\partial z}\left(\overline{w_x' w_z'}\right), \tag{4.24}$$

$$\frac{\partial \bar{w}_y}{\partial t} + \bar{w}_x \frac{\partial \bar{w}_y}{\partial x} + \bar{w}_y \frac{\partial \bar{w}_y}{\partial y} + \bar{w}_z \frac{\partial \bar{w}_y}{\partial z} = -\frac{1}{\rho}\frac{\partial \bar{p}}{\partial y} + \nu \nabla^2 \bar{w}_y - \frac{\partial}{\partial x}\left(\overline{w_x' w_y'}\right)$$
$$- \frac{\partial}{\partial y}\left(\overline{w_y'^2}\right) - \frac{\partial}{\partial z}\left(\overline{w_y' w_z'}\right), \tag{4.25}$$

$$\frac{\partial \bar{w}_z}{\partial t} + \bar{w}_x \frac{\partial \bar{w}_z}{\partial x} + \bar{w}_y \frac{\partial \bar{w}_z}{\partial y} + \bar{w}_z \frac{\partial \bar{w}_z}{\partial z} = -\frac{1}{\rho}\frac{\partial \bar{p}}{\partial z} + \nu \nabla^2 \bar{w}_z - \frac{\partial}{\partial x}\left(\overline{w_x' w_z'}\right)$$
$$- \frac{\partial}{\partial y}\left(\overline{w_y' w_z'}\right) - \frac{\partial}{\partial z}\left(\overline{w_z'^2}\right). \tag{4.26}$$

Equations (4.24)–(4.26) can be written in the following alternative form:

$$\rho\left(\frac{\partial \bar{w}_x}{\partial t} + \bar{w}_x \frac{\partial \bar{w}_x}{\partial x} + \bar{w}_y \frac{\partial \bar{w}_x}{\partial y} + \bar{w}_z \frac{\partial \bar{w}_x}{\partial z}\right) = -\frac{\partial \bar{p}}{\partial x} + \frac{\partial}{\partial x}\left(\mu \frac{\partial \bar{w}_x}{\partial x} - \rho \overline{w_x'^2}\right)$$
$$+ \frac{\partial}{\partial y}\left(\mu \frac{\partial \bar{w}_x}{\partial y} - \rho \overline{w_x' w_y'}\right)$$
$$+ \frac{\partial}{\partial z}\left(\mu \frac{\partial \bar{w}_x}{\partial z} - \rho \overline{w_x' w_z'}\right), \tag{4.27}$$

$$\rho\left(\frac{\partial \bar{w}_y}{\partial t} + \bar{w}_x \frac{\partial \bar{w}_y}{\partial x} + \bar{w}_y \frac{\partial \bar{w}_y}{\partial y} + \bar{w}_z \frac{\partial \bar{w}_y}{\partial z}\right) = -\frac{\partial \bar{p}}{\partial y} + \frac{\partial}{\partial x}\left(\mu \frac{\partial \bar{w}_y}{\partial x} - \rho \overline{w_y' w_x'}\right)$$
$$+ \frac{\partial}{\partial y}\left(\mu \frac{\partial \bar{w}_y}{\partial y} - \rho \overline{w_y'^2}\right)$$
$$+ \frac{\partial}{\partial z}\left(\mu \frac{\partial \bar{w}_y}{\partial z} - \rho \overline{w_y' w_z'}\right), \tag{4.28}$$

$$\rho\left(\frac{\partial \bar{w}_z}{\partial t} + \bar{w}_x \frac{\partial \bar{w}_z}{\partial x} + \bar{w}_y \frac{\partial \bar{w}_z}{\partial y} + \bar{w}_z \frac{\partial \bar{w}_z}{\partial z}\right) = -\frac{\partial \bar{p}}{\partial z} + \frac{\partial}{\partial x}\left(\mu \frac{\partial \bar{w}_z}{\partial x} - \rho \overline{w_z' w_x'}\right)$$
$$+ \frac{\partial}{\partial y}\left(\mu \frac{\partial \bar{w}_z}{\partial y} - \rho \overline{w_z' w_y'}\right)$$
$$+ \frac{\partial}{\partial z}\left(\mu \frac{\partial \bar{w}_z}{\partial z} - \rho \overline{w_z'^2}\right). \tag{4.29}$$

Substituting expressions (4.6)–(4.11) in the energy balance Eq. (2.144) and considering that $\beta = 0$, $\dot{q}_r = 0$ and $\phi = 0$ the following is obtained after transformations:

$$\frac{\partial \bar{T}}{\partial t} + \bar{w}_x \frac{\partial \bar{T}}{\partial x} + \bar{w}_y \frac{\partial \bar{T}}{\partial y} + \bar{w}_z \frac{\partial \bar{T}}{\partial z} = a \nabla^2 \bar{T} - \frac{\partial}{\partial x} \left(\overline{w'_x T'} \right)$$

$$- \frac{\partial}{\partial y} \left(\overline{w'_y T'} \right) - \frac{\partial}{\partial z} \left(\overline{w'_z T'} \right). \tag{4.30}$$

Equation (4.30) can also be written as:

$$\rho c_p \left(\frac{\partial \bar{T}}{\partial t} + \bar{w}_x \frac{\partial \bar{T}}{\partial x} + \bar{w}_y \frac{\partial \bar{T}}{\partial y} + \bar{w}_z \frac{\partial \bar{T}}{\partial z} \right) = \frac{\partial}{\partial x} \left(\lambda \frac{\partial \bar{T}}{\partial x} - \rho c_p \overline{w'_x T'} \right) + \frac{\partial}{\partial y} \left(\lambda \frac{\partial \bar{T}}{\partial y} - \rho c_p \overline{w'_y T'} \right)$$

$$+ \frac{\partial}{\partial z} \left(\lambda \frac{\partial \bar{T}}{\partial z} - \rho c_p \overline{w'_z T'} \right). \tag{4.31}$$

It follows that the turbulent flow of a fluid with constant thermophysical properties is described by the mass conservation Eq. (4.23), the momentum conservation Eqs. (4.24)–(4.26) or (4.27)–(4.29) and the energy balance Eq. (4.30) or (4.31), and by relevant boundary and initial conditions. However, an analysis of the equations mentioned above indicates that the number of unknowns in the system of Eqs. (4.23), (4.27)–(4.29) and (4.31) is larger than the number of equations themselves, i.e., the system is underdetermined. The excess variables include terms resulting from the turbulent fluid flow. Additional relations need to be introduced so that the products of fluctuations in the velocity vector components w'_x, w'_y, w'_z and the products of fluctuations in the velocity vector components w'_x, w'_y, w'_z and fluctuations in temperature T' should be expressed by means of time-averaged velocity vector components $\bar{w}_x$, $\bar{w}_y$, $\bar{w}_z$, time-averaged temperature $\bar{T}$, the fluid thermophysical properties and, possibly, by means of time t and coordinates x, y, z. In order to simplify the momentum conservation Eqs. (4.27)–(4.29), Reynolds introduced turbulent stresses, which later on were renamed Reynolds stresses [287].

In the averaged energy conservation Eq. (4.31), the number of variables is reduced by introducing the heat flux vector turbulent components.

4.2   Turbulent Viscosity and Diffusivity

It is assumed that the fluid flows over a flat surface in the direction of axis x. The fluid flow is steady, i.e., time-independent. The fluid free flow velocity and temperature at a large distance from the solid body surface are $w_{x\infty}$ and T_∞, respectively. Velocity and temperature changes in the direction of axis z can be omitted. Considering that w'_x and w'_y are velocity fluctuations caused by a circular vortex carried with velocity $\bar{w}_x$, it can be assumed that velocities w'_x and w'_y are of the same order. It can, therefore, be assumed that:

$$\frac{\partial w_x'^2}{\partial x} << \frac{\partial \left(\overline{w_x' w_y'} \right)}{\partial y} \tag{4.32}$$

and

$$\frac{\partial \left(\overline{w_x' T'} \right)}{\partial x} << \frac{\partial \left(\overline{w_y' T'} \right)}{\partial y}. \tag{4.33}$$

Equations (4.23), (4.28) and (4.31) are reduced to the following form:

$$\frac{\partial \bar{w}_x}{\partial x} + \frac{\partial \bar{w}_y}{\partial y} = 0, \tag{4.34}$$

$$\rho \left(\bar{w}_x \frac{\partial \bar{w}_x}{\partial x} + \bar{w}_y \frac{\partial \bar{w}_x}{\partial y} \right) = -\frac{\partial \bar{p}}{\partial x} + \frac{\partial}{\partial y} \left(\mu \frac{\partial \bar{w}_x}{\partial y} - \rho \overline{w_x' w_y'} \right), \tag{4.35}$$

$$\rho c_p \left(\bar{w}_x \frac{\partial \bar{T}}{\partial x} + \bar{w}_y \frac{\partial \bar{T}}{\partial y} \right) = \frac{\partial}{\partial y} \left(\lambda \frac{\partial \bar{T}}{\partial y} - \rho c_p \overline{w_y' T'} \right). \tag{4.36}$$

Analysing Eq. (4.35), it can be seen that in the parentheses on the right side, apart from molecular tangential stress τ_m, defined as:

$$\tau_m = \mu \frac{\partial \bar{w}_x}{\partial x} = \rho \nu \frac{\partial \bar{w}_x}{\partial x}, \tag{4.37}$$

turbulent tangential stresses appear as well.

$$\tau_t = -\rho \overline{w_x' w_y'} = \rho \varepsilon_\tau \frac{\partial \bar{w}_x}{\partial y}, \tag{4.38}$$

where turbulent momentum diffusivity ε_τ is defined as:

$$\varepsilon_\tau = \frac{-\overline{w_x' w_y'}}{\frac{\partial \bar{w}_x}{\partial y}}. \tag{4.39}$$

Tangential stress τ can, therefore, be expressed as a sum of molecular stress τ_m and turbulent stress τ_t:

$$\tau = \tau_m + \tau_t = (\mu + \rho \varepsilon_\tau) \frac{\partial \bar{w}_x}{\partial y} = \rho(\nu + \varepsilon_\tau) \frac{\partial \bar{w}_x}{\partial y}. \tag{4.40}$$

Similarly, heat flux can be expressed as a sum of heat flux $\dot{q}_m$ caused by molecular heat conduction and defined by the Fourier law and heat flux $\dot{q}_t$ resulting due to the turbulent fluid flow:

$$\dot{q} = \dot{q}_m + \dot{q}_t = -\left(\lambda + \rho\, c_p\, \varepsilon_q\right) \frac{\partial \bar{T}}{\partial y}, \tag{4.41}$$

where turbulent thermal diffusivity ε_q, also referred to as eddy diffusivity for heat transfer, is defined by the following formula:

$$\varepsilon_q = -\frac{\overline{w'_y T'}}{\frac{\partial \bar{T}}{\partial y}}. \tag{4.42}$$

Considering relations (4.39) and (4.42) in the momentum and energy conservation Eqs. (4.35) and (4.36), respectively, the following is obtained:

$$\bar{w}_x \frac{\partial \bar{w}_x}{\partial x} + \bar{w}_y \frac{\partial \bar{w}_x}{\partial y} = -\frac{1}{\rho}\frac{\partial \bar{p}}{\partial x} + \frac{\partial}{\partial y}\left[(\nu + \varepsilon_\tau)\frac{\partial \bar{w}_x}{\partial y}\right], \tag{4.43}$$

$$\bar{w}_x \frac{\partial \bar{T}}{\partial x} + \bar{w}_y \frac{\partial \bar{T}}{\partial y} = \frac{\partial}{\partial y}\left[(a + \varepsilon_q)\frac{\partial \bar{T}}{\partial y}\right], \tag{4.44}$$

where $a = \lambda/(\rho\, c_p)$ is the molecular thermal diffusivity, which is also called thermal diffusivity.

Considering relations (4.37) and (4.38), the tangential turbulent stress τ_t to molecular stress τ_m ratio is defined as follows:

$$\frac{\tau_t}{\tau_m} = \frac{\varepsilon_\tau}{\nu}. \tag{4.45}$$

The ratio between the turbulent and the molecular heat flux ($\dot{q}_t$ to $\dot{q}_m$) can be determined taking account of (4.41):

$$\frac{\dot{q}_t}{\dot{q}_m} = \frac{-\rho\, c_p\, \varepsilon_q \frac{\partial \bar{T}}{\partial y}}{-\lambda \frac{\partial \bar{T}}{\partial y}} = \frac{\rho\, c_p\, \varepsilon_q}{\lambda} = \frac{\mu\, c_p\, \varepsilon_q\, \varepsilon_\tau}{\lambda\, \varepsilon_\tau\, \nu}. \tag{4.46}$$

Introducing the molecular and the turbulent Prandtl number defined, respectively, by the following formulae:

$$\mathrm{Pr} = \frac{\mu\, c_p}{\lambda}, \quad \mathrm{Pr}_t = \frac{\varepsilon_\tau}{\varepsilon_q} \tag{4.47}$$

and equality (4.45), formula (4.46) can be written in the following form:

$$\frac{\dot{q}_t}{\dot{q}_m} = \frac{\varepsilon_q}{a} = \frac{\mathrm{Pr}\,\tau_t}{\mathrm{Pr}_t\,\tau_m} = \frac{\mathrm{Pr}\,\varepsilon_\tau}{\mathrm{Pr}_t\,\nu}. \tag{4.48}$$

In the case of the boundary layer, the system of Eqs. (4.34), (4.43) and (4.44) should be solved using appropriate boundary conditions. As a result, three equations are obtained with the following five unknowns: $\bar{w}_x$, $\bar{w}_y$, $\bar{T}$, ε_τ and ε_q. Using experimental tests, eddy diffusivity for momentum transfer (turbulent momentum diffusivity) ε_τ and eddy diffusivity for heat transfer (turbulent thermal diffusivity) are determined as a function of the distance from the solid body surface y, the Reynolds number Re and the molecular Prandtl number Pr. In this way, the number of unknowns is reduced to three, which means that the numbers of equations and unknowns are now the same. Usually, experimental tests are used to find ε_τ depending on the dimensionless distance from the solid body surface y^+, and instead of the direct determination of ε_q, the turbulent Prandtl number Pr_t is found as a function of the Reynolds number Re, the Prandtl number Pr and dimensionless distance y^+. Knowing experimental values and Pr_t, formula (4.47) can be used to determine turbulent thermal diffusivity $\varepsilon_q = \varepsilon_\tau / Pr_t$. The search for experimental relations describing ε_τ and Pr_t is just one of numerous attempts to model the flow turbulence. Although there are several methods of the phenomenon mathematical modeling, the problem has not been solved completely despite the efforts made by many researchers. The results of the turbulent flow modeling performed using Computational Fluid Dynamics (CFD) programs based on different turbulence models differ from each other considerably. This especially concerns the heat transfer coefficient values on the surface of a solid body.

4.3   Mixing Path Model

The mixing path concept proposed by Prandtl [235–237] is the first attempt to describe the turbulent fluid flow. Turbulent tangential stress τ_t and eddy diffusivity for momentum transfer (momentum diffusivity) ε_τ can be expressed as a function of the mixing path length l. Based on the concept of the mixing path length l, it is also possible to determine the fluid velocity distribution in the turbulent core. An analysis will be conducted for the fluid flow in the direction of axis x over an infinitely long plate. Considering that mean velocities of the fluid flow in the direction of axes y and z equal zero, i.e., $\bar{w}_y = \bar{w}_z$, the momentary fluid velocities are as follows:

$$w_x = \bar{w}_x + w'_x, \quad w_y = w'_y, \quad w_z = w'_z. \tag{4.49}$$

The molecular momentum and heat transfers are analyzed using the concept of the mean free path, i.e., the distance covered by a particle before it collides with another. Prandtl introduced a similar concept of the mixing path to describe the turbulent fluid flow. The mixing path length l introduced by Prandtl is the mean distance covered by a turbulent "lump" of fluid in the direction perpendicular to the direction of the fluid flow. Assume that the turbulent "lump" is located over or

Fig. 4.2 Diagram illustrating
the mixing path concept

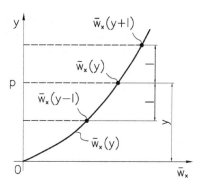

under a horizontal plane P–P at distance l (cf. Fig. 4.2). The fluid turbulent lumps
move up and down passing through plane P–P. This movement of the lumps in the
direction perpendicular to the plane involves the momentum transfer, which in turn
results in fluctuations in velocity w'_x:

$$w'_x = \bar{w}_x(y+l) - \bar{w}_x(y) = \bar{w}_x(y) - \bar{w}_x(y-l). \tag{4.50}$$

Velocities $\bar{w}_x(y+l)$ and $\bar{w}_x(y-l)$ can be determined by approximation using a
Taylor series that takes account of first-order derivatives only:

$$\bar{w}_x(y+l) \cong \bar{w}_x(y) + l\frac{\partial \bar{w}_x}{\partial y}, \tag{4.51}$$

$$\bar{w}_x(y-l) \cong \bar{w}_x(y) - l\frac{\partial \bar{w}_x}{\partial y}. \tag{4.52}$$

Substitution of (4.51) and (4.52) in (4.50) results in:

$$w'_x \cong l\frac{\partial \bar{w}_x}{\partial y}. \tag{4.53}$$

Prandtl named length l the mixing length. According to Prandtl, the fluctuation
in velocity w'_y can be expressed using a formula which is similar to expression
(4.53). Turbulent tangential stress τ_t can thus be expressed as:

$$\tau_t = -\rho w'_x w'_y \cong \rho l^2 \left(\frac{\partial \bar{w}_x}{\partial y}\right)^2 = \rho \varepsilon_\tau \frac{\partial \bar{w}_x}{\partial y}. \tag{4.54}$$

It follows from equality (4.54) that:

$$\varepsilon_\tau = l^2 \left|\frac{\partial \bar{w}_x}{\partial y}\right|, \quad \mu_t = \rho l^2 \left|\frac{\partial \bar{w}_x}{\partial y}\right|. \tag{4.55}$$

Close to a flat wall in the viscous sublayer, the turbulent momentum diffusivity is zero, i.e., $\varepsilon_\tau = 0$. It follows from relation (4.55) that $l = 0$.

In [237], Prandtl suggested that in the turbulent sublayer near the wall, the change in mixing length l depending on distance y from the wall should be approximated using a linear function.

$$l = \kappa y, \tag{4.56}$$

where κ is a constant.

Based on experimental testing results, Nikuradse [213–214, 256] proposed the following function for the turbulent core mixing length at the turbulent fluid flow through a tube with a circular cross-section for $\mathrm{Re} \geq 10^5$:

$$\frac{l}{r_w} = 0.14 - 0.08\left(1 - \frac{y}{r_w}\right)^2 - 0.06\left(1 - \frac{y}{r_w}\right)^4 = 0.14 - 0.08\left(\frac{r}{r_w}\right)^2 - 0.06\left(\frac{r}{r_w}\right)^4. \tag{4.57}$$

Expanding function (4.57) into a Taylor series and using only the first two terms in the series:

$$\frac{l}{r_w} = \left(\frac{l}{r_w}\right)\bigg|_{y=0} + \frac{d}{dy}\left(\frac{l}{r_w}\right)_{y=0} y \tag{4.58}$$

the result is $\kappa = 0.4$.

The velocity distribution near the wall is usually presented as a function of dimensionless coordinates. First, friction velocity u_τ will be introduced:

$$\frac{\tau_w}{\rho} = \frac{f\, w_{x\infty}^2}{2} = u_\tau^2, \tag{4.59}$$

where f is the Fanning friction factor.

Equality (4.59) gives:

$$u_\tau = \sqrt{\frac{\tau_w}{\rho}}. \tag{4.60}$$

The following dimensionless variables will be introduced next:

$$u^+ = \frac{\bar{w}_x}{u_\tau} = \frac{\bar{w}_x/w_{x\infty}}{\sqrt{f/2}} = \frac{\bar{w}_x/w_{x\infty}}{\sqrt{\xi/8}}, \tag{4.61}$$

$$v^+ = \frac{\bar{w}_y}{u_\tau} = \frac{\bar{w}_y/w_{y\infty}}{\sqrt{f/2}} = \frac{\bar{w}_y/w_{y\infty}}{\sqrt{\xi/8}}, \tag{4.62}$$

$$x^+ = \frac{x u_\tau}{v} = \frac{x\sqrt{\tau_w/\rho}}{v} = \frac{x w_{x\infty}\sqrt{f/2}}{v} = \frac{x w_{x\infty}\sqrt{\xi/8}}{v},$$ (4.63)

$$y^+ = \frac{y u_\tau}{v} = \frac{y\sqrt{\tau_w/\rho}}{v} = \frac{y w_{x\infty}\sqrt{f/2}}{v} = \frac{y w_{x\infty}\sqrt{\xi/8}}{v},$$ (4.64)

where u^+ and v^+ are dimensionless velocities in the direction of axes x an y, and v is kinematic viscosity.

Expression (4.40) describing shear stress τ can be expressed as a function of dimensionless quantities using the following formula:

$$\frac{\tau}{\tau_w} = \left(1 + \frac{\varepsilon_\tau}{v}\right)\frac{du^+}{dy^+}.$$ (4.65)

It follows from Eq. (4.61) that using it, if velocity profile $u^+ = u^+(y^+)$ or derivative du^+/dy^+ is known, it is easy to determine ratio ε_τ/v:

$$\frac{\varepsilon_\tau}{v} = \left(\frac{\tau}{\tau_w}\right)\Big/\left(\frac{du^+}{dy^+}\right) - 1 = \frac{\tau}{\tau_w}\frac{dy^+}{du^+} - 1.$$ (4.66)

Equation (4.65) also enables determination of velocity profile $u^+(y^+)$, provided that the dependence of the ε_τ/v and the τ/τ_w ratios on dimensionless coordinate y^+ is known from experimental testing. Separating the variables in Eq. (4.65) and integrating from 0 to y^+, considering that $u^+(y^+ = 0) = 0$, the following is obtained:

$$u^+(y^+) = \int\limits_0^{y^+} \frac{(\tau/\tau_w)}{1 + \frac{\varepsilon_\tau}{v}}\,dy^+.$$ (4.67)

The two-layer model of the near-wall area proposed by Prandtl and Taylor is based on the assumption that $\tau/\tau_w = 1$ both for the laminar and the turbulent sublayer. If the flow in the tube is fully developed, in formula (4.67) it should be taken into account that $\tau/\tau_w = r/r_w$.

4.4 Universal Velocity Profiles

An analysis will be conducted of an incompressible viscous fluid developed flow over a stationary flat surface. The surface is hydraulically smooth, i.e., it is not rough. The fluid free flow velocity at a large distance from the wall is $w_{x\infty}$.

4.4.1 Prandtl Velocity Profile

The boundary layer arising in the turbulent fluid flow can be divided into two areas: the viscous sublayer, where $v \gg \varepsilon_\tau$, and the turbulent sublayer, where $v \ll \varepsilon_\tau$. The velocity distribution in the turbulent sublayer will be determined using the theory of the mixing path length proposed by Prandtl [235–237].

4.4.1.1 Velocity Profile in the Viscous Sublayer

The turbulent fluid flow in channels or over a solid body surface is characterized by zero velocity $\bar{w}_x$ on the solid body surface and disappearance velocity fluctuations which are characteristic for the turbulent flow. Equation (4.43) written in the following form:

$$\bar{w}_x \frac{\partial \bar{w}_x}{\partial x} + \bar{w}_y \frac{\partial \bar{w}_x}{\partial y} = -\frac{1}{\rho}\frac{\partial \bar{p}}{\partial x} + \frac{\partial \tau}{\partial y} \tag{4.68}$$

can be reduced, considering that in the case of the viscous sublayer the terms on the left side can be omitted. It is further assumed that the viscous sublayer is the Couette flow area [69, 167], for which derivative $\partial \bar{p}/\partial x$ equals zero. It should be reminded that the Couette flow occurs between two plates, one of which is stationary and the other moves in parallel with a constant velocity. The Couette flow characteristic features are the zero pressure gradient, i.e., $\partial \bar{p}/\partial x = 0$, the linear velocity distribution in between the plates and the constancy of tangential stresses on the channel width in between the plates. Considering all the assumptions: $\bar{w}_y = 0$, $\partial \bar{w}_x/\partial y = 0$ and $\partial \bar{p}/\partial x = 0$ in Eq. (4.68), the following is obtained:

$$\frac{\partial \tau}{\partial y} = 0. \tag{4.69}$$

Integrating Eq. (4.69) from 0 to y, the result is:

$$\tau(y) = \tau_w, \tag{4.70}$$

where τ_w is tangential stress on the wall.

Considering that Reynolds stress $\tau_t = -\rho\, w'_x w'_y$ in the viscous sublayer totals zero, tangential stress τ defined by formula (4.37) is reduced to the following form:

$$\tau = \mu \frac{d\bar{w}_x}{dy} = \rho\, v \frac{d\bar{w}_x}{dy}. \tag{4.71}$$

Separating variables in Eq. (4.71) and taking account of (4.70), the following is obtained:

$$d\bar{w}_x = \frac{\tau_w}{\rho\nu}dy = \frac{\tau_w}{\mu}dy. \tag{4.72}$$

Integration of both sides of Eq. (4.72):

$$\int_0^{\bar{w}_x} dw_x = \frac{\tau_w}{\mu}\int_0^y dy, \tag{4.73}$$

results in:

$$\bar{w}_x = \frac{\tau_w}{\mu}y. \tag{4.74}$$

Considering dimensionless coordinates (4.61)–(4.64), expression (4.74) can be written as:

$$u^+ = y^+. \tag{4.75}$$

The same result is obtained from formula (4.67) assuming that $\tau/\tau_w = 1$ and ε_τ/ν.

The linear velocity profile occurs in the interval $0 \le y^+ \le y_{l-t}^+$, where y_{l-t}^+ is the coordinate of the boundary between the laminar and the turbulent sublayer.

4.4.1.2 Velocity Profile in the Turbulent Sublayer

The turbulent sublayer velocity distribution will be found using the mixing length concept. Considering that in the turbulent sublayer $\varepsilon_\tau \gg \nu$, i.e., $\varepsilon_\tau/\nu \gg 1$ and $\tau/\tau_w = 1$, taking account of (4.55), the following is obtained from Eq. (4.65):

$$\kappa^2 (y^+)^2 \left(\frac{du^+}{dy^+}\right)^2 = 1. \tag{4.76}$$

Solving Eq. (4.76) with respect to du^+/dy^+, the following is obtained:

$$\frac{du^+}{dy} = \frac{1}{\kappa y^+}. \tag{4.77}$$

Integrating Eq. (4.77) within limits from the laminar sublayer edge coordinate y_{l-t}^+ to y^+, the result is:

$$u^+(y^+) - u^+(y_{l-t}^+) = \frac{1}{\kappa}\ln y^+ - \frac{1}{\kappa}\ln y_{l-t}^+, \quad y^+ \ge y_{l-t}^+. \tag{4.78}$$

Considering in (4.78) that the fluid velocity on the boundary between the laminar and the turbulent sublayer y^+_{l-t} determined from formula (4.75) is:

$$u^+\left(y^+_{l-t}\right) = y^+_{l-t}, \qquad (4.79)$$

the dimensionless distribution of velocity in the laminar sublayer has the following form:

$$u^+ = \frac{1}{\kappa} \ln y^+ + y^+_{l-t} - \frac{1}{\kappa} \ln y^+_{l-t}, \quad y^+ \ge y^+_{l-t}. \qquad (4.80)$$

Introducing constants:

$$A = \frac{1}{\kappa}, \qquad (4.81)$$

$$B = y^+_{l-t} - \frac{1}{\kappa} \ln y^+_{l-t}, \qquad (4.82)$$

formula (4.80) can be written as:

$$u^+ = A \ln y^+ + B, \quad y^+ \ge y^+_{l-t}. \qquad (4.83)$$

According to Prandtl, the constants are: $\kappa = 0.4$ and $y^+_{l-t} = 11.6$. Substituting the constants in formulae (4.81) and (4.82), the following is obtained:

$$A = 2.5 \quad \text{i} \quad B = 5.5. \qquad (4.84)$$

The velocity distribution in the laminar and the turbulent boundary layer is described in the Prandtl-Taylor model using the following functions:

$$u^+ = y^+, \quad 0 \le y^+ \le 11.6, \qquad (4.85)$$

$$u^+ = 2.5 \ln y^+ + 5.5, \quad y^+ \ge 11.6. \qquad (4.86)$$

The two-layer Prandtl model described by expressions (4.85)–(4.86) is the basis for determination of the Prandtl number formula for the fluid flow over a flat solid body.

4.4.2   von Kármán Velocity Profile

von Kármán [145] split the boundary layer into three subareas (cf. Fig. 4.3):

- the laminar sublayer in the area $0 \le y^+ \le 5$, which adjoins the solid body surface,
- the buffer (transition) sublayer in area $5 < y^+ \le 30$,
- the turbulent sublayer (the turbulent core) in the area $30 < y^+$.

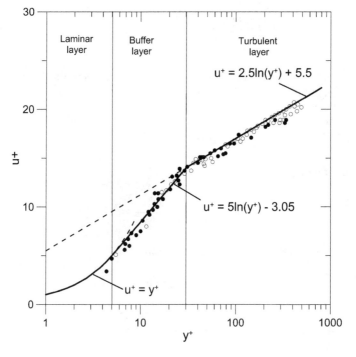

Fig. 4.3 Comparison between the velocity distribution obtained using the von Kármán formulae and the Nikuradse (○) and Reichardt (●) experimental data [245, 357]

In each of the areas mentioned above, von Kármán proposed the following functions describing the velocity distribution:

$$u^+ = y^+, \quad 0 \le y^+ \le 5, \tag{4.87}$$

$$u^+ = 5 \ln\left(\frac{y^+}{5}\right) + 5 = 5 \ln y^+ - 3.05, \quad 5 < y^+ \le 30, \tag{4.88}$$

$$u^+ = 2.5 \ln y^+ + 5.5, \quad 30 < y^+. \tag{4.89}$$

The experimental testing results indicate that the upper limit for the application of formula (4.89) y^+ depends on the Reynolds number.

4.4.3   Deissler Velocity Profile

Based on experimental data, Deissler [75, 76] found that turbulent momentum diffusivity (eddy diffusivity for momentum transfer) ε_τ decreases exponentially with a rise in the distance from the wall:

$$\frac{\varepsilon_\tau}{\nu} = n^2 u^+ y^+ \left[1 - \exp\left(-n^2 u^+ y^+\right)\right], \quad 0 \leq y^+ \leq 26, \tag{4.90}$$

where $n = 0.124$.

The universal velocity profile near the wall can be determined using formula (4.65), considering that near the wall $\tau = \tau_w$:

$$\frac{du^+}{dy^+} = \frac{1}{1 + \frac{\varepsilon_\tau}{\nu}}. \tag{4.91}$$

Substitution of (4.90) in (4.91) results in:

$$\frac{du^+}{dy^+} = \frac{1}{1 + n^2 u^+ y^+ \left[1 - \exp(-n^2 u^+ y^+)\right]}, \quad 0 \leq y^+ \leq 26. \tag{4.92}$$

The initial condition has the following form:

$$u^+\big|_{y^+=0} = 0. \tag{4.93}$$

The velocity distribution can be found integrating Eq. (4.92) and taking account of the initial condition defined by expression (4.93):

$$u^+ = \int\limits_0^{y^+} \frac{1}{1 + n^2 u^+ y^+ \left[1 - \exp(-n^2 u^+ y^+)\right]}\, dy^+, \quad 0 \leq y^+ \leq 26. \tag{4.94}$$

The integral in solution (4.94) can be determined using the trapezoidal rule. Velocity profile $u^+(y^+)$ can also be found solving the first-order differential Eq. (4.92) under initial condition (4.93) using one of the many numerical methods which are available, e.g., the Euler method or the Runge-Kutta fourth-order method. Applying the trapezoidal rule to solve the initial-value problem expressed in (4.92)–(4.93), the following difference equation is obtained:

$$\frac{u^+_{i+1} - u^+_i}{\Delta y^+} = \frac{1}{2}\left\{ \frac{1}{1 + n^2 u^+_i y^+_i \left[1 - \exp(-n^2 u^+_i y^+_i)\right]} \right.$$
$$\left. + \frac{1}{1 + n^2 u^+_{i+1} y^+_{i+1} \left[1 - \exp(-n^2 u^+_{i+1} y^+_{i+1})\right]} \right\}, \quad 0 \leq y^+ \leq 26$$
$$\tag{4.95}$$

from which u_{i+1}^+ is determined:

$$
\begin{aligned}
u_{i+1}^+ = u_i^+ + \frac{\Delta y^+}{2} \Bigg\{ & \frac{1}{1 + n^2 u_i^+ y_i^+ \left[1 - \exp(-n^2 u_i^+ y_i^+)\right]} \\
& + \frac{1}{1 + n^2 u_{i+1}^+ y_{i+1}^+ \left[1 - \exp(-n^2 u_{i+1}^+ y_{i+1}^+)\right]} \Bigg\}, \quad 0 \le y^+ \le 26,
\end{aligned}
\tag{4.96}
$$

where:

$$
y_i^+ = (i-1)\Delta y^+, \quad u_i^+ = u^+\left(y_i^+\right), \quad u_1^+ = u^+\big|_{y^+=0} = 0, \quad i = 1,\dots,n-1.
\tag{4.97}
$$

Symbol n denotes the number of the difference mesh equidistant points in the integration interval $0 \le y^+ \le 26$. The number of steps Δy^+ is $(n-1)$. Equation (4.96) is solved using simple iterations with respect to u_{i+1}^+:

$$
\begin{aligned}
\left(u_{i+1}^+\right)^{(k+1)} = u_i^+ + \frac{\Delta y^+}{2} \Bigg\{ & \frac{1}{1 + n^2 u_i^+ y_i^+ \left[1 - \exp(-n^2 u_i^+ y_i^+)\right]} \\
& + \frac{1}{1 + n^2 \left(u_{i+1}^+\right)^{(k)} y_{i+1}^+ \left[1 - \exp\left(-n^2 \left(u_{i+1}^+\right)^{(k)} y_{i+1}^+\right)\right]} \Bigg\}, \quad 0 \le y^+ \le 26,
\end{aligned}
\tag{4.98}
$$

where $k = 0, 1, \dots$ is the iteration number.

$\left(u_{i+1}^+\right)^{(0)} = u_i^+$ is adopted as the first approximation. Ten iterations are enough to obtain a convergent solution which does not change up to four decimal points with a rise in the number of iterations.

Like Prandtl, Dreissler assumed that the velocity distribution in the area $26 \le y^+$ had the form as presented by formula (4.83):

$$
u^+ = \frac{1}{0.36} \ln \frac{y^+}{26} + 12.85 = 2.78 \ln y^+ + 3.8, \quad 26 \le y^+.
\tag{4.99}
$$

The formula describing eddy viscosity at the fluid flow through the tube can be derived from formula (4.65) considering that tangential stress τ is a linear function of radius r

$$
\frac{\tau}{\tau_w} = \frac{r}{r_w} = \frac{r_w - y}{r_w} = 1 - \frac{y}{r_w} = 1 - \frac{y^+}{r_w^+},
\tag{4.100}
$$

where $r_w^+ = r_w u_\tau / \nu$.

r_w is the inner radius of the tube. Substitution of (4.100) in (4.65) results in:

$$1 - \frac{y^+}{r_w^+} = \left(1 + \frac{\varepsilon_\tau}{\nu}\right)\frac{du^+}{dy^+}. \tag{4.101}$$

Considering (4.99) and performing operations from formula (4.101), the following is obtained:

$$\frac{\varepsilon_\tau}{\nu} = 0.36\left(1 - \frac{y^+}{r_w^+}\right)y^+ - 1 = \frac{y^+}{2.78}\left(1 - \frac{y^+}{r_w^+}\right) - 1, \quad 26 \leq y^+. \tag{4.102}$$

The formulae presented so far are valid only for a specific y^+ interval. Reichardt [243–245] put forward a velocity profile which is valid for any y^+.

4.4.4 Reichardt Velocity Profile

In his research [243–245], Reichardt aimed to find a simple velocity profile formula that would hold for the entire area of the fluid flow. Substituting velocity profile (4.83) in expression (4.101), the following formula is obtained for the area near the wall if $(y^+/r_w^+) << 1$:

$$\frac{\varepsilon_\tau}{\nu} = \kappa y^+. \tag{4.103}$$

The same formula was obtained by Nikuradse [213, 214] based on experimental testing. If $y^+ \to 0$, eddy viscosity in the viscous sublayer approaches zero. Reichardt [245] proposed a different function which is free from this disadvantage:

$$\frac{\varepsilon_\tau}{\nu} = \kappa\left[y^+ - y_n^+ \tanh\left(\frac{y^+}{y_n^+}\right)\right], \quad y^+ \leq 50, \tag{4.104}$$

where: $\kappa = 0.4$ and $y_n^+ = 11$. It can be seen that if $y^+ = 0$, formula (4.104) gives $\varepsilon_\tau/\nu = 0$.

Substituting formula (4.104) in Eq. (4.101) and assuming that $(y^+/r_w^+) << 1$ the following velocity distribution is obtained after integration:

$$u^+ = \frac{1}{\kappa}\ln(1 + \kappa y^+) + C\left[1 - \exp\left(-\frac{y^+}{y_n^+}\right) - \frac{y^+}{y_n^+}\exp\left(-\frac{y^+}{3}\right)\right], \quad y^+ \leq 50, \tag{4.105}$$

where constant C is expressed as:

$$C = 5.5 - \frac{1}{\kappa} \ln \kappa. \tag{4.106}$$

Considering that $\kappa = 0.4$ and $y_n^+ = 11$, velocity profile (4.105) can be expressed as:

$$u^+ = 2.5 \ln(1 + 0.4 y^+) + 7.8 \left[1 - \exp\left(-\frac{y^+}{11}\right) - \frac{y^+}{11} \exp\left(-\frac{y^+}{3}\right) \right], \tag{4.107}$$
$$y^+ \leq 50.$$

Near the wall, the velocity values determined from formula (4.107) show very good agreement with experimental data [245].

Close to the tube axis, molecular viscosity ν can be omitted due to its very small value compared to momentum turbulent diffusivity ε_τ.

Equation (4.101) can be written in cylindrical coordinates as:

$$\frac{r}{r_w} = \frac{r_w u_\tau}{\nu} \frac{\varepsilon_\tau}{\nu} \frac{du^+}{d\left(\frac{r}{r_w}\right)}, \tag{4.108}$$

where r_w is the tube inner radius.

Equation (4.108) can also be written in a slightly different form:

$$\frac{du^+}{dR} = \frac{R}{r_w^+ \frac{\varepsilon_\tau}{\nu}}, \tag{4.109}$$

where $R = r/r_w$, $r_w^+ = r_w u_\tau/\nu$.

Based on experimental data, Reichardt proposed the following form of a function for turbulent viscosity (turbulent momentum diffusivity) determination in the tube central part:

$$\frac{\varepsilon_\tau}{\nu} = \frac{\kappa r_w^+}{3} (1 - R^2) \left(\frac{1}{2} + R^2\right), \quad y^+ > 50. \tag{4.110}$$

Considering dimensionless coordinate y^+ defined by formula (4.64) and noting that $y = r_w - r$ and $\kappa = 0.4$, formula (4.110) can be written as:

$$\frac{\varepsilon_\tau}{\nu} = 0.1333 \, y^+ (1 + R) \left(\frac{1}{2} + R^2\right), \quad y^+ > 50. \tag{4.111}$$

Substituting (4.109) in Eq. (4.108) and integrating the equation from r to r_w, the following is obtained [245]:

$$u^+ = 2.5 \ln\left[y^+ \; \frac{1.5\,(1+R)}{1+2\,R^2}\right] + 5.5, \quad y^+ > 50. \tag{4.112}$$

Based on the velocity profiles expressed by (4.107) and (4.112), Reichardt [245] proposed a universal velocity distribution for a circular tube which is valid for the tube entire cross-section (for any y^+).

$$u^+ = \frac{1}{\kappa}\ln\left[(1+\kappa y^+)\;\frac{3\,(1+R)}{2\,(1+2\,R^2)}\right] + C\left[1 - \exp\left(-\frac{y^+}{y_n^+}\right) - \frac{y^+}{y_n^+}\exp\left(-\frac{y^+}{3}\right)\right].$$
$$\tag{4.113}$$

Substituting relevant constants, formula (4.113) can be written in the following form:

$$u^+ = 2.5 \ln\left[(1+0.4\,y^+)\;\frac{1.5\,(1+R)}{1+2\,R^2}\right] + 7.8\left[1 - \exp\left(-\frac{y^+}{11}\right) - \frac{y^+}{11}\exp\left(-\frac{y^+}{3}\right)\right].$$
$$\tag{4.114}$$

The velocities determined using formula (4.114) are an excellent approximation of the experimental data obtained by Nikuradse [213, 214] and Reichardt [245]. Two variables appear in relations (4.112)–(4.113)—y^+ and R. It should be noted that the following relation holds between y^+ and R:

$$y^+ = \frac{u_\tau\, y}{\nu} = \frac{u_\tau\,(r_w - r)}{\nu} = \frac{u_\tau\, r_w}{\nu}(1 - R) = r_w^+\,(1-R), \tag{4.115}$$

from which it follows that:

$$R = 1 - \frac{y^+}{r_w^+}\,. \tag{4.116}$$

Determination of velocity distribution $u^+(y^+)$ for the fluid flow in a tube makes it necessary to establish the following relations:

$$r_w^+ = r_w\, u_\tau/\nu = \frac{\mathrm{Re}}{2}\sqrt{\frac{\xi}{8}}, \tag{4.117}$$

where ξ is the Darcy-Weisbach friction factor.

Reichardt's formulae (4.104), (4.111) and (4.114) enable very accurate modeling of turbulent flows in tubes with a circular cross-section.

4.4.5 van Driest Velocity Profile

van Driest [140, 148, 354] put forward the following mixing path length formula:

$$l = \kappa y \left[1 - \exp\left(-\frac{y^+}{A} \right) \right],$$ (4.118)

where $\kappa = 0.4$ and $A = 26$ or $A = 25$ [140, 148, 354]. It follows from the analysis of formula (4.118) that the mixing path length l increases as the distance from the wall gets bigger.

$$u_\tau^2 = (\nu + \varepsilon_\tau) \frac{d\bar{w}_x}{dy} = \left(\nu + l^2 \left| \frac{d\bar{w}_x}{dy} \right| \right) \frac{d\bar{w}_x}{dy}.$$ (4.119)

Writing Eq. (4.119) in a dimensionless form, the following is obtained:

$$\left[1 + (l^+)^2 \left| \frac{du^+}{dy^+} \right| \right] \frac{du^+}{dy^+} = 1.$$ (4.120)

The solution of the quadratic algebraic Eq. (4.120) has the following form:

$$\frac{du^+}{dy^+} = \frac{\left[1 + 4(l^+)^2 \right]^{1/2} - 1}{2(l^+)^2}.$$ (4.121)

Multiplying the numerator and the denominator by:

$$\left[1 + 4(l^+)^2 \right]^{1/2} + 1$$ (4.122)

the following is obtained by simple transformations:

$$\frac{du^+}{dy^+} = \frac{2}{1 + \left[1 + 4(l^+)^2 \right]^{1/2}}.$$ (4.123)

Considering (4.118) in (4.121), the following first-order differential equation is obtained:

$$\frac{du^+}{dy^+} = \frac{2}{1 + \left\{ 1 + 4\kappa^2 (y^+)^2 [1 - \exp(-y^+ /A)]^2 \right\}^{1/2}}, \quad A = 26,$$ (4.124)

which, to find the velocity distribution, is solved under the following initial condition:

$$u^+ \big|_{y^+ = 0} = 0.$$ (4.125)

The solution of Eq. (4.124) considering initial condition (4.125) can be written as follows:

$$u^+(y^+) = \int_0^{y^+} \frac{2}{1 + \left\{1 + 4\kappa^2(y^+)^2[1 - \exp(-y^+/A)]^2\right\}^{1/2}} dy^+, \quad A = 26.$$

(4.126)

Integral (4.124) can be calculated numerically, e.g., using the rectangle method or the trapezoidal rule.

The velocity distribution can also be determined by solving the initial-value problem (4.124)–(4.125) numerically using one of the many existing methods, e.g., the Euler method or the Runge-Kutta method.

Applying the trapezoidal rule to solve Eq. (4.124), the following is obtained:

$$\frac{u_{i+1}^+ - u_i^+}{\Delta y^+} = \frac{1}{2} \left\{ \frac{2}{1 + \left\{1 + 4\kappa^2(y_i^+)^2[1 - \exp(-y_i^+/A)]^2\right\}^{1/2}} \right. $$
$$\left. + \frac{2}{1 + \left\{1 + 4\kappa^2(y_{i+1}^+)^2[1 - \exp(-y_{i+1}^+/A)]^2\right\}^{1/2}} \right\}.$$

(4.127)

Difference Eq. (4.124) is used to determine u_{i+1}:

$$u_{i+1}^+ = u_i^+ + \Delta y^+ \left\{ \frac{1}{1 + \left\{1 + 4\kappa^2(y_i^+)^2[1 - \exp(-y_i^+/A)]^2\right\}^{1/2}} \right.$$
$$\left. + \frac{1}{1 + \left\{1 + 4\kappa^2(y_{i+1}^+)^2[1 - \exp(-y_{i+1}^+/A)]^2\right\}^{1/2}} \right\}, \quad i = 1, \ldots, n-1,$$

(4.128)

where:

$$y_i^+ = (i-1)\Delta y^+, \quad u_i^+ = u^+(y_i^+), \quad u_1^+ = u^+\big|_{y^+=0} = 0, \quad i = 1, \ldots, n.$$

(4.129)

Symbol n is the number of nodes where velocity u^+ is found. The entire integration interval is divided into $(n-1)$ subintervals.

The formulae for universal velocity distributions are listed in Table 4.1.

Table 4.1 List of universal velocity profiles

$u^+(y^+)$	Interval/range	Publication
$u^+ = y^+$ $u^+ = 2.5 \ln y^+ + 5.5$	$0 < y^+ \leq 11.6$ $y^+ \geq 11.6$	Prandtl and Taylor [234–237, 337]
$u^+ = y^+$ $u^+ = 5 \ln y^+ - 3.05$ $u^+ = 2.5 \ln y^+ + 5.5$	$0 < y^+ \leq 5$ $5 \leq y^+ \leq 30$ $y^+ \geq 30$	von Kármán [145]
$u^+ = 14.53 \tanh(y^+/14.53)$ $u^+ = 2.5 \ln y^+ + 5.5$	$0 < y^+ \leq 27.5$ $y^+ \geq 27.5$	Rannie [354]
$\dfrac{du^+}{dy^+} = \dfrac{2}{1 + \left\{1 + 4\kappa^2 y^{+2}\,[1 - \exp(-y^+/A^+)]^2\right\}^{1/2}}$ $\kappa = 0.4,\ A^+ = 26$	for any y^+	van Driest [40, 140, 148, 354]
$u^+ = 2.5 \ln(1 + 0,4\,y^+)$ $\quad + 7.8[1 - \exp(-y^+/11) - (y^+/11)\exp(-0.33\,y^+)]$	for any y^+	Reichardt [245]
$\dfrac{du^+}{dy^+} = \dfrac{1}{1 + n^2 u^+ y^+ [1 - \exp(-n^2 u^+ y^+)]}$ $n = 0.124,\ u^+ = 2.78 \ln y^+ + 3.8$	$0 < y^+ \leq 26$	Deissler [75, 76]
$y^+ = u^+ + A\left[\exp Bu^+ - 1 - Bu^+ - \frac{1}{2}(Bu^+)^2 - \frac{1}{6}(Bu^+)^3 - \frac{1}{24}(Bu^+)^4\right]$	for any y^+ $A = 0.1108$ $B = 0.4$	Spalding [40, 140, 148, 354]

Figure 4.4 presents a comparison between velocity distributions determined by means of the Prandtl formulae, experimental data [148] and the velocity profile obtained from integration of the van Driest equation at $A = 26$.

Reynolds number $\mathrm{Re}_{\delta_2} = w_{x\infty}\delta_2/\nu$ is based on what is referred to as momentum thickness δ_2 being a measure of the reduction in the momentum flux due to the boundary layer. The definition of δ_2 is discussed in detail in Chap. 5 of the book [148]. It should be noted that instead of the formula given by Prandtl (4.86):

$$u^+ = 2.5 \ln y^+ + 5.5$$

the following relation is assumed:

$$u^+ = 2.44 \ln y^+ + 5 \qquad (4.130)$$

which is applied in [64, 148, 240, 354].

Results closer to experimental data can be obtained assuming in van Driest's formulae that $A = 23$ (cf. Fig. 4.4). Figure 4.5 presents another comparison of formulae (4.85) and (4.1) with Johnson's [148] and Wieghardt's [148] experimental data obtained for $\mathrm{Re}_{\delta_2} = 1500$ and $\mathrm{Re}_{\delta_2} = 15,000$, respectively.

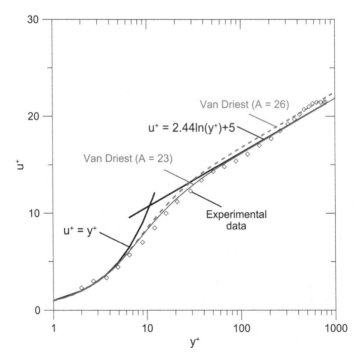

Fig. 4.4 Comparison of the Prandtl and van Driest universal velocity distributions with Johnson's experimental data for $\mathrm{Re}_{\delta_2} = 2500$ [148]

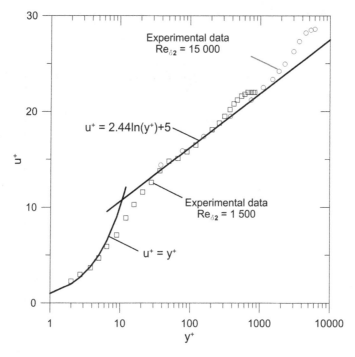

Fig. 4.5 Comparison of selected universal velocity distributions with Johnson's [148] and Wieghardt's [148] experimental data for $Re_{\delta_2} = 1500$ and $Re_{\delta_2} = 15,000$, respectively

Figure 4.6 presents a comparison of the universal velocity profiles listed in Table 4.1.

The velocity distribution obtained from Deissler's and van Driest's formulae was determined using step $\Delta y^+ = 1000/30,000 = 1/30$. The number of iterations k performed to solve the nonlinear algebraic Eq. (4.98) was 10. The analysis of the results presented in Fig. 4.6 indicates that all velocity distributions are very similar to each other, except the profile proposed by Prandtl. In Figs. 4.7, 4.8 and 4.9, respectively, Reichardt's, van Driest's and Deissler's velocity profiles are compared to the profile proposed by von Kármán.

An additional comparison of velocity profiles proposed by different authors is presented in Table 4.2.

It follows from the analysis of the results presented in Figs. 4.3, 4.4, 4.5, 4.6 and in Table 4.2 that the profile proposed by Reichardt is the best approximation of the von Kármán distribution, which was put forward the earliest and which shows good agreement with experimental data (cf. Fig. 4.3). The Reichardt formula has a simple form, and its essential advantage is the fact that it is valid for the entire near-wall area without the need to divide the boundary layer into the laminar and buffer sublayers and the turbulent core. This facilitates numerical modeling of turbulent flows based on a variant of the k-ε turbulence model where boundary layer

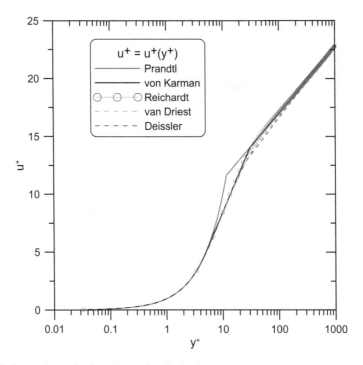

Fig. 4.6 Comparison of universal velocity distributions

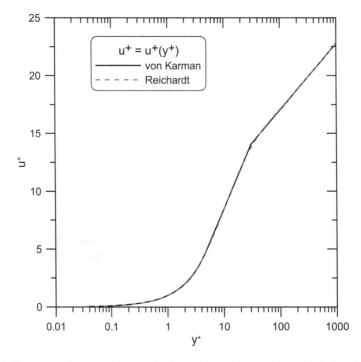

Fig. 4.7 Comparison between the von Kármán and the Reichardt universal velocity distribution

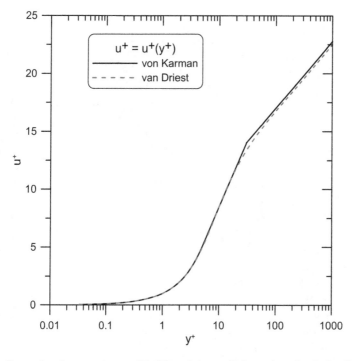

Fig. 4.8 Comparison between the von Kármán and the van Driest universal velocity distribution

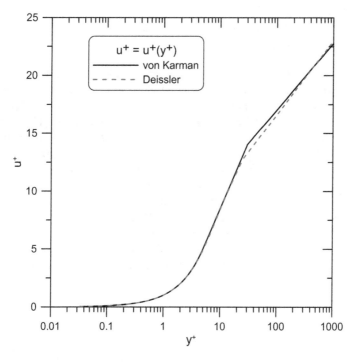

Fig. 4.9 Comparison between the von Kármán and the Deissler universal velocity distribution

Table 4.2 Comparison of velocity distributions $u^+(y^+)$ calculated according to formulae proposed by different authors

y^+	Prandtl	von Kármán	Reichardt	van Driest	Deissler
0	0	0	0	0	0
1	1	1	1.0109	1	1
2	2	2	2.0381	1.9986	1.9985
3	3	3	3.0504	2.9897	2.9892
4	4	4	4.0190	3.9596	3.9573
5	5	5	4.9259	4.8880	4.8816
10	10	8.4629	8.4281	8.4632	8.3858
15	12.2701	10.4903	10.5984	10.5188	10.3936
20	12.9893	11.9287	12.0089	11.7944	11.7091
25	13.5472	13.0444	12.9868	12.6727	12.6818
30	14.0030	14.0030	13.7013	13.3256	13.2475
50	15.2801	15.2801	15.3285	14.9147	14.6665
100	17.0129	17.0129	17.0831	16.7481	16.5919
200	18.7458	18.7458	18.7861	18.4766	18.5173
300	19.7595	19.7595	19.7895	19.4852	19.6436
400	20.4787	20.4787	20.5035	20.2018	20.4427
500	21.0365	21.0365	21.0583	20.7581	21.0625
600	21.4923	21.4923	21.512	21.2129	21.5690
700	21.8777	21.8777	21.8959	21.5975	21.9972
800	22.2115	22.2115	22.2286	21.9308	22.3681
900	22.5060	22.5060	22.5222	22.2249	22.6953
1000	22.7694	22.7694	22.7849	22.4879	22.9879

functions describing the universal velocity distribution and turbulent viscosity (diffusivity of momentum transfer) are used close to the surface of a solid body instead of increasing the finite volume or the finite element mesh density. An example application of the Reichardt formulae for numerical modeling of the turbulent fluid flow between two flat surfaces using the FVM-FEM method is presented in [103]. The comparison between selected results of numerical simulations using the k-ε turbulence model and the results obtained by means of Direct Numerical Simulations (DNS), consisting in solving the mass, momentum and energy conservation equations directly, without introducing an additional turbulence model, indicates higher accuracy of the Reichardt formulae compared to the formulae proposed by van Driest and used in the standard Fluent package. The velocity profile determination from the solution of a differential equation, as proposed by Deissler and van Driest, is more complicated and laborious.

Chapter 5
Analogies Between the Heat and the Momentum Transfer

Heat exchangers, boilers, and many other devices are commonly calculated using the heat transfer coefficient. For several dozens of years, a search has been going on for simple $Nu = f(Re, Pr)$ relations that could be applied in practice to calculate the heat transfer coefficient α on the solid body surface as a function of the fluid velocity, temperature, and physical properties.

One of the ways to determine the correlations for the Nusselt number depending on the Reynolds and the Prandtl numbers is to make use of the analogies between the momentum and the heat transfer. Written in a dimensionless form, momentum and energy conservation equations have an identical form.

The solution of the momentum conservation equation provides a velocity distribution that can be used to find friction factor ξ depending on the Reynolds number, and the solution of the energy conservation equation gives a temperature distribution based on which the heat transfer coefficient can be determined as a function of the Reynolds number. A correlation exists between the friction factor and the heat transfer coefficient. For the same Reynolds number—the higher the friction factor, the higher the heat transfer coefficient. It should be added that the friction factor can also be found experimentally based on measurements of the pressure drop on a given length and using the flow mean velocity known from measurements. The friction factor correlations are known for smooth and rough tubes. Many different functions describing the friction factor dependence on the Reynolds number—$\xi(Re)$—can be found in the literature.

This chapter presents an analysis of the turbulent flow of an incompressible fluid with constant thermophysical properties over a flat surface. The fluid flows in the direction parallel with axis x. The heat flux on the solid body surface ($y = 0$) is set as constant (cf. Fig. 5.1). It is further assumed that the fluid mean velocity and mean temperature are equal to the fluid velocity and temperature at a large distance from the flat surface, i.e.:

© Springer International Publishing AG, part of Springer Nature 2019 157
D. Taler, *Numerical Modelling and Experimental Testing of Heat Exchangers*,
Studies in Systems, Decision and Control 161,
https://doi.org/10.1007/978-3-319-91128-1_5

$$\bar{w}_{x,m} = \bar{w}_{x,\infty}, \quad \bar{T}_{x,m} = \bar{T}_{x,\infty}. \tag{5.1}$$

Of all the analogies put forward so far, only four will be presented: the Reynolds, the Chilton-Colburn, the Prandtl and the von Kármán analogy. A survey of the analogies proposed so far is presented in [125, 140, 142, 182, 187, 335, 353, 357, 358].

The forms of the formulae resulting from analogies between the momentum and the heat transfer can be used to approximate mean heat transfer coefficients in heat exchangers. The numerical factors present in the formulae usually differ from the coefficients resulting from the analogies and are determined using the least squares method.

5.1 Reynolds and Chilton-Colburn Analogy

The Reynolds analogy is the oldest. As early as in 1874, Reynolds published his work [248] suggesting that the momentum and the heat transfer processes occur similarly. Reynolds ignores the laminar and the buffer sublayer and assumes that the flow over a flat surface is turbulent. The formula describing tangential stress:

$$\tau = \rho(v + \varepsilon_\tau)\frac{d\bar{w}_x}{dy} \tag{5.2}$$

is similar to the heat flux definition:

$$\dot{q} = -\rho c_p(a + \varepsilon_q)\frac{d\bar{T}}{dy} = -\rho c_p\left(\frac{v}{\text{Pr}} + \frac{\varepsilon_\tau}{\text{Pr}_t}\right)\frac{d\bar{T}}{dy}, \tag{5.3}$$

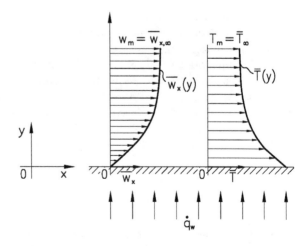

Fig. 5.1 Velocity and temperature distributions for the fluid flow over a flat surface used in the Reynolds analogy

where

$$\Pr = \frac{\nu}{a}, \quad \Pr_t = \frac{\varepsilon_\tau}{\varepsilon_q}. \tag{5.4}$$

The assumption that the momentum and the heat transfer processes, both molecular and turbulent, occur in the same manner results in the following equalities:

$$a = \nu, \quad \varepsilon_q = \varepsilon_\tau. \tag{5.5}$$

It follows from (5.5) that:

$$\Pr = 1, \quad \Pr_t = 1 \tag{5.6}$$

The other assumption, confirmed by experimental testing, is the constancy of the ratio between tangential stress and the heat flux.

$$\frac{\tau}{\dot{q}} = \text{const.} \tag{5.7}$$

The assumption expressed in (5.7) implies the following equality:

$$\frac{\tau}{\dot{q}} = \frac{\tau_w}{\dot{q}_w} = \text{const.} \tag{5.8}$$

The formulae resulting from the Reynolds analogy can be used for gases due to the assumption that Pr = 1.

Considering (5.3) in (5.5), the following is obtained:

$$\dot{q} = -\rho\, c_p (\nu + \varepsilon_\tau) \frac{d\bar{T}}{dy}. \tag{5.9}$$

Dividing Eqs. (5.9) and (5.2) by sides, the result is:

$$\frac{d\bar{T}}{d\bar{w}_x} = -\frac{\dot{q}}{\tau}\frac{1}{c_p}. \tag{5.10}$$

Considering equality (5.8), Eq. (5.10) can be transformed into:

$$\frac{d\bar{T}}{d\bar{w}_x} = -\frac{\dot{q}_w}{\tau_w}\frac{1}{c_p}. \tag{5.11}$$

Separating the variables in Eq. (5.11):

$$d\bar{T} = -\frac{\dot{q}_w}{c_p \tau_w} d\bar{w}_x, \tag{5.12}$$

and performing integration on both sides, the following is obtained:

$$\int_{T_w}^{\bar{T}_m} d\bar{T} = -\int_{0}^{\bar{w}_{x,m}} \frac{\dot{q}_w}{c_p \tau_w} d\bar{w}_x, \tag{5.13}$$

where $\bar{T}_m = T_\infty$ and $\bar{w}_{x,m} = \bar{w}_{x,\infty}$ (cf. Fig. 5.1).

After integration, Eq. (5.13) takes the following form:

$$\bar{T}_m - T_w = -\frac{\dot{q}_w \bar{w}_{x,m}}{c_p \tau_w}, \tag{5.14}$$

which, considering that:

$$\tau_w = \frac{\xi}{8} \rho \bar{w}_{x,m}^2, \tag{5.15}$$

$$\dot{q}_w = \alpha(T_w - \bar{T}_m), \tag{5.16}$$

results in:

$$\frac{\alpha}{\rho\, c_p \bar{w}_{x,m}} = \frac{\xi}{8}. \tag{5.17}$$

The left side of Eq. (5.17) can be transformed by introducing the following dimensionless numbers:

$$\mathrm{Nu} = \frac{\alpha x}{\lambda}, \quad \mathrm{Pr} = \frac{c_p \mu}{\lambda}, \quad \mathrm{Re} = \frac{\rho \bar{w}_{x,m} x}{\mu}, \tag{5.18}$$

where x is the analyzed point distance from the plate inflow edge.

Equation (5.17) can then be written as:

$$\frac{\mathrm{Nu}}{\mathrm{Re}\,\mathrm{Pr}} = \frac{\xi}{8}. \tag{5.19}$$

Equation (5.19), referred to as the Reynolds analogy, enables determination of the heat transfer coefficient for the fluid fully developed turbulent flow over a flat surface based on the known empirical relation for the friction factor: $\xi(\mathrm{Re})$. As mentioned above, if $\mathrm{Pr} \cong 1$, relation (5.19) can be used for gases. For high values of the Reynolds number, the friction factor can be expressed using the following formula [208]:

$$\xi = \frac{0.2328}{Re^{0.2}}, \quad 5 \times 10^5 \leq Re < 10^7. \tag{5.20}$$

Substitution of (5.20) in (5.19) results in:

$$Nu = 0.0291 Re^{0.8} Pr, \quad 5 \times 10^5 \leq Re < 10^7. \tag{5.21}$$

A generalization of the Reynolds analogy is the Chilton-Colburn analogy [40, 102, 148]:

$$\frac{Nu}{Re\,Pr^{1/3}} = \frac{\xi}{8}, \tag{5.22}$$

which can be used at any Prandtl number value, i.e., both for liquids and gases. Formula (5.22) is of great practical significance because it can be used for any turbulent flow, e.g., for the fully developed turbulent flow through a tube.

Considering that in this case the friction factor is defined as [66, 128, 208, 246]:

$$\xi = \frac{0.184}{Re^{0.2}}, \quad 10^4 \leq Re \leq 10^6 \tag{5.23}$$

the following is obtained from formula (5.22) [128, 142]:

$$Nu = 0.023 Re^{0.8} Pr^{1/3}. \tag{5.24}$$

The Dittus-Boelter formula [65, 79, 126, 142, 196] frequently used in engineering calculations has the following form

$$Nu = 0.023 Re^{0.8} Pr^{n}, \quad 10^4 \leq Re < 1.2 \times 10^5, \quad 0.7 < Pr < 120, \tag{5.25}$$

where for the heated and the cooled tube, respectively, $n = 0.4$ and $n = 0.3$.

The Nusselt and the Reynolds numbers are calculated from the following formulae:

$$Nu = \frac{\alpha\, d_w}{\lambda}, \quad Re = \frac{\rho\, w_m d_w}{\mu}, \tag{5.26}$$

where $d_w = 2r_w$ and w_m are the tube inner diameter and the fluid flow mean velocity in the tube, respectively.

The Colburn (5.24) and the Dittus-Boelter (5.25) formulae are commonly used in practice. They are still used successfully to approximate the experimental results when the Prandtl number varies in a narrow range. Power-type correlations for wide ranges of Reynolds and Prandtl number were proposed by Taler [326]:

$$Nu = 0.02155\, Re^{0.8018}\, Pr^{0.7095} \quad 3 \times 10^3 \leq Re \leq 10^6, \quad 0.1 \leq Pr \leq 1 \tag{5.27}$$

$$Nu = 0.01253 Re^{0.8413} Pr^{0.6179} \quad 3 \times 10^3 \leq Re \leq 10^6, \quad 1 < Pr \leq 3 \qquad (5.28)$$

$$Nu = 0.00881 Re^{0.8991} Pr^{0.3911} \quad 3 \times 10^3 \leq Re \leq 10^6, \quad 3 < Pr \leq 1000 \qquad (5.29)$$

Heat transfer correlations (5.27)–(5.29) were compared with experimental results available in the literature [326]. The performed comparisons confirmed the good accuracy of the proposed correlations (5.27)–(5.29). Heat transfer relationships (5.27)–(5.29) can be used in broader ranges of Reynolds and Prandtl numbers compared with a widely used correlation of Dittus-Boelter. They are also much more straightforward in comparison to the relationship of Gnielinski, which is also widely used in the heat transfer calculations.

5.2   Prandtl Analogy

The assumptions concerning the profiles of velocity and temperature and the conditions adopted in the Prandtl analogy [234–238] are presented in Fig. 5.2.

The velocity distributions in the laminar and the turbulent layer are defined by formulae (4.85) and (4.86), respectively. In the interval $0 \leq y \leq y_{l-t}$, the fluid velocity distribution is linear and described by formula (4.85), and in the turbulent flow region $y_{l-t} \leq y \leq y_{fs}$ the velocity distribution is described by the logarithmic function presented in (4.86). In the region $y_{fs} \leq y$, the fluid velocity is constant and equal to the free flow velocity. No turbulent flow occurs in the laminar layer, i.e., momentum diffusivity ε_τ and thermal diffusivity a_t are equal to zero. Tangential stress τ and heat flux $\dot{q}$ are therefore defined as:

$$\tau = \rho v \frac{d\bar{w}_x}{dy}, \qquad (5.30)$$

Fig. 5.2 Velocity and temperature distributions for the fluid flow over a flat surface used in the Prandtl analogy

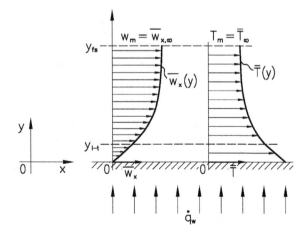

$$\dot{q} = -\rho c_p a \frac{d\bar{T}}{dy}. \tag{5.31}$$

Due to the boundary layer small thickness y_{l-t}, it is assumed that both tangential stress and the heat flux are constant and equal to their respective values on the solid body surface, i.e.:

$$\tau = \tau_w, \tag{5.32}$$

$$0 \le y \le y_{l-t},$$

$$\dot{q} = \dot{q}_w. \tag{5.33}$$

Integrating Eq. (5.30) from 0 to y_{l-t} and taking account of assumption (5.32) and boundary condition $\bar{w}_x|_{y=0} = 0$, the following is obtained:

$$\frac{\tau_w}{\rho \nu} y_{l-t} = \bar{w}_{x,l-t}. \tag{5.34}$$

Equation (5.34) can be transformed into:

$$\frac{y_{l-t} u_\tau}{\nu} = \frac{\bar{w}_{x,l-t}}{u_\tau}, \tag{5.35}$$

where $u_\tau = \sqrt{\tau_w/\rho}$.

Taking account of the definitions of the dimensionless quantities expressed in (4.61) and (4.64), Eq. (5.35) can be written as:

$$u_{l-t}^+ = y_{l-t}^+. \tag{5.36}$$

Integrating Eq. (5.31) from 0 to y_{l-t} and taking account of assumption (5.33) and boundary condition $\bar{T}|_{y=0} = T_w$, the following is obtained:

$$\frac{\dot{q}_w}{\rho c_p a} = T_w - T|_{l-t}. \tag{5.37}$$

Dividing Eq. (5.37) by Eq. (5.34), the following is obtained:

$$\frac{T_w - T|_{l-t}}{\bar{w}_{x,l-t}} = \frac{\dot{q}_w}{\tau_w} \frac{\nu}{a c_p}. \tag{5.38}$$

Using the same procedure, it is possible to determine temperature and velocity differences in the turbulent area thickness. Due to the turbulent flow nature, molecular viscosity and diffusivity are omitted, i.e., $\nu = 0$ and $a = 0$. In this case, the tangential stress and the heat flux formulae take the following form:

$$\tau = \rho \varepsilon_\tau \frac{d\bar{w}_x}{dy}, \tag{5.39}$$

$$\dot{q} = -\rho \varepsilon_q c_p \frac{d\bar{T}}{dy}. \tag{5.40}$$

Additionally, identical assumptions are adopted as in the Reynolds analogy:

$$\varepsilon_q = \varepsilon_\tau, \quad \tfrac{\tau}{\dot{q}} = \tfrac{\tau_{l-t}}{\dot{q}_{l-t}} = \text{const.} \tag{5.41}$$

Integrating Eqs. (5.39) and (5.40) from y_{l-t} to y_d, the result is:

$$w_m - \bar{w}_{x,l-t} = \int_{y_{l-t}}^{y_d} \frac{\tau}{\rho \varepsilon_\tau} dy, \tag{5.42}$$

$$\bar{T}_{l-t} - T_m = \int_{y_{l-t}}^{y_d} \frac{\dot{q}}{\rho c_p \varepsilon_q} dy, \tag{5.43}$$

where

$$w_m = \bar{w}_{x,\infty} \quad \text{and} \quad T_m = T_\infty. \tag{5.44}$$

Dividing Eq. (5.43) by Eq. (5.42), and taking account of the assumptions expressed in (5.41), the following equation is obtained:

$$\frac{\bar{T}_{l-t} - T_m}{w_m - \bar{w}_{x,l-t}} = \frac{\dot{q}_{l-t}}{c_p \tau_{l-t}} = \frac{\dot{q}_w}{c_p \tau_w}. \tag{5.45}$$

Finding T_{l-t} from Eq. (5.38) and substituting it in (5.45), the following is obtained:

$$\frac{\dot{q}_w}{c_p \tau_w} = \frac{T_w - T_m}{w_m} \frac{1}{1 + \frac{\bar{w}_{x,l-t}}{w_m}(\text{Pr} - 1)}. \tag{5.46}$$

Considering that the heat transfer coefficient is defined as:

$$\alpha = \frac{\dot{q}_w}{T_w - T_m} \tag{5.47}$$

and introducing the following dimensionless numbers:

$$\text{Nu} = \frac{\alpha L}{\lambda}, \quad \text{Re} = \frac{w_m L}{\nu}, \quad \text{Pr} = \frac{c_p \nu \rho}{\lambda} = \frac{c_p \eta}{\lambda}, \tag{5.48}$$

Eq. (5.46) can be written as follows:

$$\text{Nu} = \frac{\frac{\xi}{8} \text{Re} \, \text{Pr}}{1 + \frac{\bar{w}_{x,l-t}}{w_m} (\text{Pr} - 1)}. \tag{5.49}$$

Symbol L in the dimensionless numbers defined in (5.48) denotes any adopted dimension. Equation (5.40) can be transformed, taking account of equality (5.36) written as:

$$\frac{\bar{w}_{x,l-t}}{\sqrt{\frac{\tau_w}{\rho}}} = y_{l-t}^+, \tag{5.50}$$

which gives:

$$\bar{w}_{x,l-t} = y_{l-t}^+ \sqrt{\frac{\tau_w}{\rho}}. \tag{5.51}$$

Dividing both sides of Eq. (5.51) by w_m:

$$\frac{\bar{w}_{x,l-t}}{w_m} = y_{l-t}^+ \sqrt{\frac{\tau_w}{\rho w_m^2}} \tag{5.52}$$

and considering that tangential stress on the wall is defined as:

$$\tau_w = \frac{\xi}{8} \rho w_m^2, \tag{5.53}$$

from which it follows that:

$$\frac{\tau_w}{\rho w_m^2} = \frac{\xi}{8}, \tag{5.54}$$

expression (5.52) can be written in the following form:

$$\frac{\bar{w}_{x,l-t}}{w_m} = y_{l-t}^+ \sqrt{\frac{\xi}{8}}. \tag{5.55}$$

Considering equality (5.55) in formula (5.49), the following formula for the Nusselt number is obtained:

$$Nu = \frac{\frac{\xi}{8} \operatorname{Re} \operatorname{Pr}}{1 + y_{l-t}^+ \sqrt{\frac{\xi}{8}}(\operatorname{Pr} - 1)}. \tag{5.56}$$

The coordinate of the point of intersection of the velocity distributions described in (4.85) and (4.86) y_{l-t}^+ is 11.6. Equation (5.56) then takes the following form:

$$Nu = \frac{\frac{\xi}{8} \operatorname{Re} \operatorname{Pr}}{1 + 11.6 \sqrt{\frac{\xi}{8}}(\operatorname{Pr} - 1)}. \tag{5.57}$$

If the modified form of the velocity distribution formula (4.129) proposed in [148] is adopted, then:

$$y_{l-t}^+ = 10.8 \sqrt{\frac{\xi}{8}}. \tag{5.58}$$

As a result, the Prandtl formula takes the following form:

$$Nu = \frac{\frac{\xi}{8} \operatorname{Re} \operatorname{Pr}}{1 + 10.8 \sqrt{\frac{\xi}{8}}(\operatorname{Pr} - 1)}. \tag{5.59}$$

Based on experimental testing results, Prandtl [236, 237] suggested that the following should be adopted: $y_{l-t}^+ = 8.7$. Equation (5.56) then takes the following form:

$$Nu = \frac{\frac{\xi}{8} \operatorname{Re} \operatorname{Pr}}{1 + 8.7 \sqrt{\frac{\xi}{8}}(\operatorname{Pr} - 1)}. \tag{5.60}$$

Friction factor ξ can be calculated using different relations describing the turbulent flow over a smooth surface. Relations having the same form as (5.56) are used to calculate the heat transfer coefficient α on the inner surface of tubes. Due to the assumptions presented in (5.44), this can produce inaccurate results because the fluid temperature in the tube axis is not equal to the mass-averaged temperature, and the fluid velocity in the tube axis differs from the mean velocity value. Analysis of relation (5.60) indicates that for high Prandtl numbers $\operatorname{Pr}/(\operatorname{Pr} - 1) \to 1$, i.e., in such cases the Nusselt number does not depend on the Prandtl number. This conclusion does not agree with experimental testing results. For this reason, formula (5.60) proposed by Prandtl and defining the Nusselt number can be used for smaller values of the Prandtl number.

5.3 Von Kármán Analogy

The assumptions concerning the profiles of velocity and temperature and the conditions adopted in the Kármán analogy are presented in Fig. 5.3.

The velocity distributions in the laminar, the buffer (transition) and the turbulent layer, respectively, are defined by formulae (4.87)–(4.89). In the interval $0 \leq y \leq y_s$, the fluid velocity distribution is linear and described by formula (4.87), in the transient flow region $y_s \leq y \leq y_b$, it is defined by formula (4.88), and in the turbulent flow region, the velocity distribution is described by a logarithmic function (4.89). In the region $y_{fs} \leq y$ the fluid velocity is constant and equal to the free flow velocity. Tangential stress τ and heat flux $\dot{q}$ are therefore defined as:

$$\tau = \rho(\nu + \varepsilon_\tau)\frac{d\bar{w}_x}{dy}, \tag{5.61}$$

$$\dot{q} = -\rho\,c_p\left(a + \varepsilon_q\right)\frac{d\bar{T}}{dy}. \tag{5.62}$$

Equations (5.61)–(5.62) will be written in a dimensionless form using the variables defined by formulae (4.61)–(4.64) and dimensionless temperature T^+ defined as:

$$T^+ = \frac{\bar{T} - T_w}{\frac{\dot{q}_w}{\rho\,c_p\,u_\tau}}. \tag{5.63}$$

Expressions (5.61)–(5.62) are then written as follows:

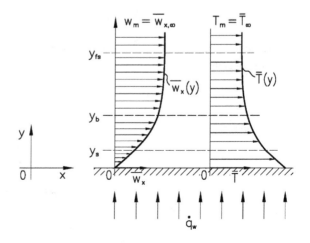

Fig. 5.3 Velocity and temperature distributions for the fluid flow over a flat surface used in the von Kármán analogy

$$\frac{\tau}{\tau_w} = \left(1 + \frac{\varepsilon_\tau}{\nu}\right)\frac{du^+}{dy^+},\tag{5.64}$$

$$-\frac{\dot{q}}{\dot{q}_w} = \left(\frac{1}{Pr} + \frac{\varepsilon_\tau/\nu}{Pr_t}\right)\frac{dT^+}{dy^+}.\tag{5.65}$$

No turbulent flow occurs in the laminar layer, i.e., momentum diffusivity ε_τ and thermal diffusivity a_t are equal to zero. Moreover, in the laminar layer, like in the Prandtl analogy, tangential stress τ and heat flux $\dot{q}$ are equal to the values of τ_w and $\dot{q}_w$, respectively. Equation (5.65) then takes the following form:

$$\frac{dT^+}{dy^+} = -Pr, \quad 0 \le y^+ \le 5.\tag{5.66}$$

Separating the variables and integrating Eq. (5.66) from $y^+ = 0$ to y^+, considering that $T^+\big|_{y^+=0} = 0$, the following is obtained:

$$T^+ = -Pr\, y^+, \quad 0 \le y^+ \le 5.\tag{5.67}$$

The temperature drop in the laminar layer is expressed as:

$$\Delta T_s^+ = T^+\big|_{y^+=0} - T^+\big|_{y^+=5} = 0 - T_s^+ = Pr\, y_s^+ = 5\,Pr, \quad 0 \le y^+ \le 5.\tag{5.68}$$

Like in the laminar layer, the following is assumed in the buffer layer $5 \le y^+ \le 30$:

$$\tau = \tau_w, \quad \dot{q} = \dot{q}_w, \quad 5 \le y^+ \le 30.\tag{5.69}$$

Considering that in the buffer layer the universal velocity profile is defined by formula (4.88), the following expression defining derivative du^+/dy^+ is obtained:

$$\frac{du^+}{dy^+} = \frac{5}{y^+}, \quad 5 \le y^+ \le 30.\tag{5.70}$$

Substituting (5.70) into (5.64) and taking account of (5.69), the following equation is obtained:

$$1 = \left(1 + \frac{\varepsilon_\tau}{\nu}\right)\frac{5}{y^+}, \quad 5 \le y^+ \le 30,\tag{5.71}$$

from which it follows that:

$$\frac{\varepsilon_\tau}{\nu} = \frac{y^+}{5} - 1, \quad 5 \le y^+ \le 30.\tag{5.72}$$

Substituting (5.72) in (5.65) and taking account of (5.69), the following differential equation is obtained:

$$\frac{\mathrm{d}T^+}{\mathrm{d}y^+} = -\frac{1}{\frac{1}{\mathrm{Pr}} + \frac{1}{\mathrm{Pr}_t}\left(\frac{y^+}{5} - 1\right)}, \quad 5 \leq y^+ \leq 30, \tag{5.73}$$

which, after integration from $y^+ = y_s^+ = 5$ to y^+, enables determination of the temperature distribution in the buffer sublayer:

$$T^+ = T_s^+ - 5\,\mathrm{Pr}_t\,\ln\left[1 + \frac{\mathrm{Pr}}{\mathrm{Pr}_t}\left(\frac{y^+}{5} - 1\right)\right], \quad 5 \leq y^+ \leq 30. \tag{5.74}$$

The temperature drop ΔT_b^+ in the buffer layer obtained by using formula (5.74) is:

$$\Delta T_b^+ = T_s^+ - T_b^+ = T^+\big|_{y^+=5} - T^+\big|_{y^+=30} = 5\,\mathrm{Pr}_t\,\ln\left(1 + 5\frac{\mathrm{Pr}}{\mathrm{Pr}_t}\right), 5 \leq y^+ \leq 30. \tag{5.75}$$

In the turbulent core, molecular viscosity v and molecular diffusivity a can be omitted because the momentum transfer and the heat transfer are affected by momentum diffusivity ε_τ and turbulent diffusivity a_t, respectively. Equations (5.64) and (5.65) now take the following form:

$$\frac{\tau}{\tau_w} = \frac{\varepsilon_\tau}{v}\frac{\mathrm{d}u^+}{\mathrm{d}y^+}, \tag{5.76}$$

$$\frac{\dot{q}}{\dot{q}_w} = -\frac{\varepsilon_\tau/v}{\mathrm{Pr}_t}\frac{\mathrm{d}T^+}{\mathrm{d}y^+}. \tag{5.77}$$

Considering relation (5.8), which results from the analogy between the momentum and the heat transfer, the right side of Eqs. (5.76) and (5.77) are equal to each other:

$$\frac{\varepsilon_\tau}{v}\frac{\mathrm{d}u^+}{\mathrm{d}y^+} = -\frac{\varepsilon_\tau}{v}\frac{1}{\mathrm{Pr}_t}\frac{\mathrm{d}T^+}{\mathrm{d}y^+}. \tag{5.78}$$

Equation (5.78) gives:

$$\frac{\mathrm{d}T^+}{\mathrm{d}y^+} = -\mathrm{Pr}_t\frac{\mathrm{d}u^+}{\mathrm{d}y^+}. \tag{5.79}$$

Integrating Eq. (5.79) from $y^+ = y_b^+ = 30$ to y_{fs}^+, the following equality is obtained:

$$T_\infty^+ - T_b^+ = \mathrm{Pr}_t\left(u_b^+ - u_\infty^+\right). \tag{5.80}$$

Considering that:

$$u_b^+ = 5 \ln\left(\frac{y_b^+}{5}\right) + 5 = 5 \ln\left(\frac{30}{5}\right) + 5 = 5(1 + \ln 6), \qquad (5.81)$$

$$T_b^+ - T_\infty^+ = \Pr{}_t\left[u_\infty^+ - 5(1 + \ln 6)\right]. \qquad (5.82)$$

Adding the sides of Eqs. (5.65), (5.72) and (5.79), the following is obtained:

$$-T_\infty^+ = \left\{5 \Pr + 5 \Pr{}_t \ln\left(1 + 5\frac{\Pr}{\Pr_t}\right) + \Pr{}_t\left[u_\infty^+ - 5(1 + \ln 6)\right]\right\}, \qquad (5.83)$$

which, taking account of the dimensionless temperature definition expressed in (5.63), can be written as:

$$(T_w - T_\infty)\frac{\rho c_p u_\tau}{\dot{q}_w} = 5\left[\Pr - \Pr{}_t + \Pr{}_t \ln\left(\frac{5 \Pr / \Pr_t + 1}{6}\right)\right] + \Pr{}_t u_\infty^+. \qquad (5.84)$$

Considering assumptions (5.44) and the heat transfer coefficient definition (5.47), Eq. (5.84) can be expressed in the following form:

$$\frac{\mathrm{Nu}}{\mathrm{Re}\,\Pr} = \frac{1}{(u_m^+)^2}\frac{1}{\Pr_t + 5/u_m^+\{\Pr - \Pr_t + \Pr_t \ln[(5 \Pr / \Pr_t + 1)/6]\}}. \qquad (5.85)$$

Considering also that:

$$u_m^+ = \sqrt{\frac{8}{\xi}}, \qquad (5.86)$$

Equation (5.85) can be written in a form that enables an easy calculation of the heat transfer coefficient on a flat surface at the fluid developed and turbulent flow:

$$\mathrm{Nu} = \frac{\frac{\xi}{8}\mathrm{Re}\,\Pr}{\Pr_t + 5\sqrt{\frac{\xi}{8}}\{\Pr - \Pr_t + \Pr_t \ln[(5 \Pr / \Pr_t + 1)/6]\}}. \qquad (5.87)$$

In the case of many fluids for which the Prandtl number is higher than ~ 0.1, it can be assumed that the turbulent Prandtl number is 1, i.e. $\Pr_t = 1$. Equation (5.87) then takes the following form:

$$\mathrm{Nu} = \frac{\frac{\xi}{8}\mathrm{Re}\,\Pr}{1 + 5\sqrt{\frac{\xi}{8}}\{\Pr - 1 + \ln[(5 \Pr + 1)/6]\}}, \qquad 0.1 < \Pr. \qquad (5.88)$$

The Nusselt number Nu, the Reynolds number Re and the Prandtl number Pr are defined by formulae (5.48). The structure of formula (5.88) obtained above is similar to that of formula (5.54) obtained by Prandtl. Instead of the number 11.7 in the Prandtl formula, term $\sqrt{\xi/8}$ in formula (5.88) is preceded by the number 5. Additionally, the curly bracket in the denominator of formula (5.88) includes the following term:

$$\ln\left(\frac{5\,\text{Pr}+1}{6}\right), \tag{5.89}$$

which is small. It can be demonstrated that the Prandtl formula (5.54) and the von Kármán formula produce almost identical results if the number 11.6 in the Prandtl formula (5.57) is replaced with the number 5, which is equal to the laminar sublayer dimensionless thickness.

The term included in the curly bracket in the denominator of formula (5.88) describing the Nusselt number can be approximated in the following way:

$$5\left[\text{Pr}-1+\ln\left(\tfrac{5\,\text{Pr}+1}{6}\right)\right] \cong C(\text{Pr}-1), \quad 1 \le \text{Pr} \le 200, \tag{5.90}$$

where constant C determined using the least squares method is 5.1739, at the determination coefficient $r^2 = 0.9994$. The 95% confidence interval of constant C is $5.1653 \le C \le 5.1825$.

Other analogies can also be found in literature, e.g., the Martinelli analogy, which enables determination of the heat transfer coefficient on the tube inner surface. It is assumed in it [40, 142, 357–358] that the heat flux is a linear function of the radius. In the case of a heated tube, the largest heat flux occurs on the inner surface of the wall, whereas in the tube axis it equals zero. More accurate results obtained by solving the momentum and energy conservation equations using numerical methods indicate that the heat flux is not a linear function of the radius, especially in the case of smaller Reynolds numbers. A considerable advantage resulting from the Prandtl and the von Kármán analogy is the form of the formulae indicating that the Nusselt number Nu is a function of the Reynolds and the Prandtl numbers (Re and Pr) and the friction factor ξ on the tube inner surface. Formulae with an identical or modified form can be used to approximate results of numerical calculations or experimental data where unknown coefficients are found using the least squares method. The Pietukhov-Kirillov formula [228] and its modifications [226–227, 229–230], as well as the Gnielinski formula [104–108] can serve as an example here.

Chapter 6
Developed Turbulent Fluid Flow in Ducts with a Circular Cross-Section

This chapter presents derivation of the Nusselt number correlations for fluid flows over a flat surface. The formulae are also used to calculate the heat transfer coefficient on the inner surface of tubes. However, the coefficient values obtained in this manner should be treated as an approximation only, due to the fact that the fluid flow velocity distributions in a tube and over a flat surface differ from each other. The fluid mean temperature used to calculate the heat transfer coefficient is also different.

This chapter presents determination of velocity, temperature and heat flux distributions for fully developed turbulent flows. The Nusselt number correlations will also be determined for the fluid turbulent flow and in the transitional range of $2300 \leq \mathrm{Re} \leq 10000$.

The conservation equations for turbulent flows in a tube with a circular cross-section are of the following form [140, 148, 227]:

- the mass conservation equation (continuity equation):

$$\frac{1}{r}\frac{\partial}{\partial r}(r\bar{w}_r) + \frac{\partial \bar{w}_x}{\partial x} = 0, \tag{6.1}$$

- the momentum conservation equation (in the direction of x and r):

$$\rho\left(\bar{w}_x\frac{\partial \bar{w}_x}{\partial x} + \bar{w}_r\frac{\partial \bar{w}_x}{\partial r}\right) = -\frac{dp}{dx} + \frac{1}{r}\frac{\partial}{\partial r}\left[r\left(\mu\frac{\partial \bar{w}_x}{\partial r} - \rho\overline{w'_x w'_r}\right)\right]$$
$$+ \frac{\partial}{\partial x}\left[\mu\frac{\partial \bar{w}_x}{\partial x} - \rho\overline{w'_x w'_r}\right], \tag{6.2}$$

© Springer International Publishing AG, part of Springer Nature 2019
D. Taler, *Numerical Modelling and Experimental Testing of Heat Exchangers*,
Studies in Systems, Decision and Control 161,
https://doi.org/10.1007/978-3-319-91128-1_6

$$\rho\left(\bar{w}_x\frac{\partial\bar{w}_r}{\partial x}+\bar{w}_r\frac{\partial\bar{w}_r}{\partial r}\right)=-\frac{dp}{dr}+\frac{1}{r}\frac{\partial}{\partial r}\left[r\left(\mu\frac{\partial\bar{w}_r}{\partial r}-\rho\overline{w'_r w'_r}\right)\right]$$
$$+\frac{\partial}{\partial x}\left[\mu\frac{\partial\bar{w}_r}{\partial x}-\rho\overline{w'_r w'_x}\right],$$

(6.3)

- the energy conservation equation:

$$\rho c_p\left(\bar{w}_x\frac{\partial\bar{T}}{\partial x}+w_r\frac{\partial\bar{T}}{\partial r}\right)=\frac{1}{r}\frac{\partial}{\partial r}\left[r\left(\lambda\frac{\partial\bar{T}}{\partial r}-\rho c_p\overline{w'_r T'}\right)\right]$$
$$+\frac{\partial}{\partial x}\left[\lambda\frac{\partial\bar{T}}{\partial x}-\rho c_p\overline{w'_x T'}\right].$$

(6.4)

Conservation Eqs. (6.1)–(6.4) make it possible to find the velocity and temperature profiles, the pressure distribution and the friction factor as a function of the Reynolds number.

6.1 Hydromechanics of the Fluid Flow in Channels

The fluid flow characteristics, such as the velocity distribution, the friction factor or the pressure drop, are determined based on the mass (6.1) and the momentum (6.2)–(6.3) conservation equations. The analysis concerns a steady-state stabilized flow where velocity $\bar{w}_x$ is independent of time t or coordinate x, i.e.:

$$\frac{\partial\bar{w}_x}{\partial t}=0,\quad\frac{\partial\bar{w}_x}{\partial x}=0.$$

(6.5)

Taking account of the second condition in the continuity Eq. (6.1), the following is obtained:

$$\bar{w}_r=0.$$

(6.6)

Considering that the radial component of velocity from Eq. (6.3) is zero, it follows that $d\bar{p}/dr=0$, i.e. pressure is only a function of coordinate x. Considering (6.5) and (6.6) in Eq. (6.2), the following is obtained:

$$\frac{1}{r}\frac{d}{dr}\left[r\left(\mu\frac{d\bar{w}_x}{dr}-\rho\overline{w'_x w'_r}\right)\right]=\frac{d\bar{p}}{dx},$$

(6.7)

because velocity $\bar{w}_x$ is a function of radius r only.

The following marking will be introduced:

$$\tau = \tau_m + \tau_t = -\left(\mu \frac{d\bar{w}_x}{dr} - \rho\overline{w'_x w'}_r\right), \tag{6.8}$$

where τ_m is shear stress due to the fluid molecular viscosity:

$$\tau_m = -\mu \frac{d\bar{w}_x}{dr} \tag{6.9}$$

and τ_t—turbulent shear stress due to the momentum transfer caused by velocity fluctuations:

$$\tau_t = \rho\overline{w'_x w'}_r. \tag{6.10}$$

Defining the momentum turbulent diffusivity ε_τ as:

$$\varepsilon_\tau = -\frac{\overline{w'_x w'}_r}{\dfrac{d\bar{w}_x}{dr}}, \tag{6.11}$$

Turbulent shear stress τ_t can be expressed using the following formula:

$$\tau_t = -\rho\varepsilon_\tau \frac{d\bar{w}_x}{dr}. \tag{6.12}$$

Substitution of (6.9) and (6.12) in (6.8) gives:

$$\tau = -(\mu + \rho\varepsilon_\tau)\frac{d\bar{w}_x}{dr} = -\rho(\nu + \varepsilon_\tau)\frac{d\bar{w}_x}{dr} = -\mu\left(1 + \frac{\varepsilon_\tau}{\nu}\right)\frac{d\bar{w}_x}{dr}. \tag{6.13}$$

Considering (6.11), Eq. (6.7) can be expressed in the following form:

$$\frac{1}{r}\frac{d}{dr}\left[r\rho(\nu + \varepsilon_\tau)\frac{d\bar{w}_x}{dr}\right] = \frac{d\bar{p}}{dx}. \tag{6.14}$$

In order to integrate Eq. (6.14), the dependence of $d\bar{p}/dx$ on shear stress on the wall will be determined. Using the equation describing the equilibrium of forces for the control area (cf. Fig. 6.1), the following is obtained:

$$p(x)\pi r_w^2 = 2\pi r_w \Delta x \tau_w \left(x + \frac{\Delta x}{2}\right) + p(x + \Delta x)\pi r_w^2. \tag{6.15}$$

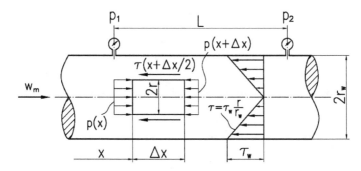

Fig. 6.1 Diagram illustrating shear stress determination

Finding $\tau_w(x + \Delta x/2)$ from Eq. (6.15):

$$\tau_w\left(x + \frac{\Delta x}{2}\right) = -\frac{p(x + \Delta x) - p(x)}{\Delta x}\frac{r_w}{2}. \tag{6.16}$$

If $\Delta x \to 0$, expression (6.16) takes the following form:

$$\tau_w = -\frac{r_w}{2}\frac{dp}{dx}, \tag{6.17}$$

which gives:

$$\frac{dp}{dx} = -\frac{2\tau_w}{r_w}. \tag{6.18}$$

Considering (6.13) and (6.18) in Eq. (6.14), the following can be written:

$$\frac{1}{r}\frac{d}{dr}(r\tau) = \frac{2\tau_w}{r_w}. \tag{6.19}$$

Multiplying both sides of Eq. (6.19) by r and integrating along r, the result is:

$$\tau = \tau_w\frac{r}{r_w} + \frac{C}{\ln r}. \tag{6.20}$$

Constant C in Eq. (6.20) equals zero because stress τ in the tube axis ($r = 0$) has a finite value. Therefore, Eq. (6.20) takes the following form:

$$\tau = \tau_w\frac{r}{r_w}, \tag{6.21}$$

which gives:

$$\frac{\tau}{\tau_w} = \frac{r}{r_w}.$$ (6.22)

Equation (6.22) indicates that the distribution of shear stress τ along radius r is linear (cf. Fig. 6.1). Formula (6.22) can also be derived based on the balance of forces for a cylindrical control area located in the fluid.

The equation describing the equilibrium of forces:

$$\pi r^2 p(x) = \pi r^2 p(x + \Delta x) + 2\pi r \Delta x \tau \left(x + \frac{\Delta x}{2} \right)$$ (6.23)

results in:

$$\tau \left(x + \frac{\Delta x}{2} \right) = -\frac{r}{2} \frac{p(x + \Delta x) - p(x)}{\Delta x}.$$ (6.24)

If $\Delta x \to 0$, Eq. (6.24) gives:

$$\tau = -\frac{r}{2} \frac{dp}{dx}.$$ (6.25)

Stress τ_w on the tube inner surface is found substituting $r = r_w$ in expression (6.25), which gives:

$$\tau_w = -\frac{r_w}{2} \frac{dp}{dx}.$$ (6.26)

Dividing Eqs. (6.25) and (6.26) by sides, the result is relation (6.22).

Stress τ_w can be found easily based on the measurement of the pressure drop on length L (cf. Fig. 6.1). Considering that:

$$\frac{dp}{dx} = \frac{p_2 - p_1}{L} = -\frac{p_1 - p_2}{L} = -\frac{\Delta p}{L}$$ (6.27)

and substituting (6.27) in (6.26), the following is obtained:

$$\tau_w = \frac{r_w}{2} \frac{p_1 - p_2}{L}.$$ (6.28)

Measuring the pressure difference $(p_1 - p_2)$ on length L, it is possible to determine shear stress from formula (6.28).

6.1.1 Determination of Fluid Velocity and Friction Factor—Integral Formulation

Velocity distribution $\bar{u}(r)$ is found from formula (6.13), which is first transformed into the following form:

$$\frac{d\bar{w}_x}{dr} = -\tau_w \frac{r}{r_w} \frac{1}{\mu\left(1 + \frac{\varepsilon_\tau}{\nu}\right)}.$$ (6.29)

Introducing dimensionless radius $R = r/r_w$, Eq. (6.29) can be written as:

$$\frac{d\bar{w}_x}{dR} = -\frac{\tau_w r_w}{\mu} \frac{R}{\left(1 + \frac{\varepsilon_\tau}{\nu}\right)}.$$ (6.30)

Integrating Eq. (6.29) from R to 1, the result is:

$$\bar{w}_x(R) = \frac{\tau_w r_w}{\mu} \int_R^1 \frac{R dR}{1 + \frac{\varepsilon_\tau}{\nu}}.$$ (6.31)

Next, mean velocity w_m is calculated from the following formula:

$$w_m = 2 \int_0^1 \bar{w}_x R dR.$$ (6.32)

Substitution of (6.31) in (6.32) results in:

$$w_m = \frac{2\tau_w r_w}{\mu} \int_0^1 \left(\int_R^1 \frac{R dR}{1 + \frac{\varepsilon_\tau}{\nu}} \right) R dR.$$ (6.33)

In order to derive the formula defining friction factor ξ, the relationship between pressure drop $\Delta p = p(x) - p(x + \Delta x)$ and friction velocity u_τ will be determined first. From the condition of the equilibrium of forces acting on the fluid element included in a control area with length Δx (cf. Fig. 6.1) it follows that:

$$\pi r_w^2 \Delta p = 2\pi r_w \Delta x \tau_w,$$ (6.34)

which, after simple transformations, gives:

$$\Delta p = \frac{2\Delta x \tau_w}{r_w}.$$ (6.35)

Using friction velocity as defined in (4.60), expression (6.35) can be transformed into the following form:

$$\Delta p = \frac{2\Delta x \rho u_\tau^2}{r_w}. \tag{6.36}$$

The pressure drop can also be expressed using the Darcy-Weisbach formula:

$$\Delta p = \xi \frac{\Delta x}{2 r_w} \frac{\rho w_m^2}{2}. \tag{6.37}$$

Comparing Eqs. (6.35) and (6.37), the following formula is obtained for shear stress τ_w:

$$\tau_w = \xi \frac{\rho w_m^2}{8}. \tag{6.38}$$

Comparing pressure drop relations (6.36) and (6.37), the following is obtained:

$$u_\tau = w_m \sqrt{\frac{\xi}{8}} = w_m \sqrt{\frac{f}{2}}. \tag{6.39}$$

Multiplying both sides of Eq. (6.38) by r_w/v, the result is:

$$r_w^+ = \frac{u_\tau r_w}{v} = \frac{w_m r_w}{v} \sqrt{\frac{\xi}{8}} = \frac{\text{Re}}{2} \sqrt{\frac{\xi}{8}} = \frac{\text{Re}}{2} \sqrt{\frac{f}{2}}, \tag{6.40}$$

where $\text{Re} = 2 w_m r_w / v$ is the Reynolds number.

It should be noted that r_w^+ depends on friction factor ξ.

Substituting τ_w defined by formula (6.38) in (6.33), the following expression is obtained for the calculation of friction factor ξ:

$$\xi = \frac{8}{\text{Re}} \left[\int\limits_0^1 \left(\int\limits_R^1 \frac{R dR}{1 + \frac{\varepsilon_\tau}{v}} \right) R dR \right]^{-1}. \tag{6.41}$$

It is more convenient to write formulae (6.31), (6.33) and (6.41) in a dimensionless form because the ε_τ/v ratio is found based on experimental testing as a function of dimensionless coordinate y^+. Considering that:

$$R = \frac{r_w - y}{r_w} = 1 - \frac{y^+}{r_w^+} \quad \text{i} \quad dR = -\frac{dy^+}{r_w^+}, \tag{6.42}$$

formula (6.31) can be written as:

$$u^+(y^+) = \frac{1}{r_w^+} \int_0^{y^+} \frac{(r_w^+ - y^+)\,dy^+}{1 + \frac{\varepsilon_\tau}{\nu}},$$

(6.43)

where $u^+ = \bar{w}_x/u_\tau$.

Also mean velocity defined by (6.33) can be written in a dimensionless form:

$$u_m^+ = \frac{w_m}{u_\tau} = 2\int_0^1 u^+ R\,dR = 2r_w^+ \int_0^{r_w^+} \left(\int_0^{y^+} \frac{\left(\frac{r_w^+ - y^+}{r_w^+}\right)d\left(\frac{r^+}{r_w^+}\right)}{1 + \frac{\varepsilon_\tau}{\nu}}\right)\left(\frac{r_w^+ - y^+}{r_w^+}\right)d\left(\frac{r^+}{r_w^+}\right).$$

(6.44)

The friction factor ξ definition (formula 6.41) can be written as:

$$\xi = \frac{8(r_w^+)^4}{\mathrm{Re}}\left[\int_0^{r_w^+}\left(\int_0^{y^+} \frac{(r_w^+ - y^+)\,dy^+}{1 + \frac{\varepsilon_\tau}{\nu}}\right)(r_w^+ - y^+)\,dy^+\right]^{-1}.$$

(6.45)

The friction factor can also be found using formula (6.39).

$$\xi = \frac{8}{(u_m^+)^2}.$$

(6.46)

If the experimental relation $u^+(y^+)$ is known, e.g. in the form of the Reichardt formula (4.114), mean velocity u_m^+ can be determined using formula (6.44) and then the friction factor can be found using formula (6.46). The velocity profile is determined iteratively. First, friction factor ξ is calculated using empirical formulae, e.g. the correlations proposed by Blasius, Filonenko or others. Then, the universal velocity distribution is found from formula (6.43) or from an empirical relation, such as the Reichardt formula for example. Having found mean velocity u_m^+, the friction factor is calculated using formula (6.46). So-determined friction factor ξ is then used to find velocity distribution $u^+(y^+)$, and the whole procedure for ξ determination is repeated until the ξ value does not change any more. The number of iterations needed to find the correct ξ value is usually smaller than 10. It should be noted that due to the fact that velocity values near the wall vary dramatically, the integrals in the derived formulae are usually calculated numerically, e.g. using the trapezoidal rule, dividing the tube radius into a few dozen thousand intervals.

6.1.2 Determination of Fluid Velocity and Friction Factor—Differential Formulation

Equation (4.101) can be written as follows:

$$\frac{du^+}{dy^+} = \frac{1}{r_w^+} \frac{r_w^+ - y^+}{1 + \varepsilon_\tau/\nu}. \tag{6.47}$$

Boundary condition:

$$u^+\big|_{y^+=0} = 0 \tag{6.48}$$

results from the fluid velocity falling to zero on the channel surface.

Equation (6.47) will be solved using the trapezoidal rule:

$$\frac{u_{i+1}^+ - u_i^+}{\Delta y^+} = \frac{1}{2r_w^+} \left[\frac{r_w^+ - y_i^+}{1 + (\varepsilon_\tau/\nu)\big|_{y_i^+}} + \frac{r_w^+ - y_{i+1}^+}{1 + (\varepsilon_\tau/\nu)\big|_{y_{i+1}^+}} \right], \quad i = 1, \ldots, n-1. \tag{6.49}$$

Equation (6.49) gives:

$$u_{i+1}^+ = u_i^+ + \frac{\Delta y^+}{2r_w^+} \left[\frac{r_w^+ - y_i^+}{1 + (\varepsilon_\tau/\nu)\big|_{y_i^+}} + \frac{r_w^+ - y_{i+1}^+}{1 + (\varepsilon_\tau/\nu)\big|_{y_{i+1}^+}} \right], \quad i = 1, \ldots, n-1. \tag{6.50}$$

Boundary condition (6.48) takes the following form:

$$u_1^+ = 0. \tag{6.51}$$

The marking in Eq. (6.49) is as follows:

$$\Delta y^+ = \frac{r_w^+}{n-1}, \quad r_w^+ = \frac{\mathrm{Re}}{2} \sqrt{\frac{\xi}{8}}, \tag{6.52}$$

$$y_i^+ = (i-1)\Delta y^+, \quad i = 1, \ldots, n, \quad y_n^+ = (n-1)\Delta y^+. \tag{6.53}$$

Mean velocity u_m^+ is defined as:

$$u_m^+ = \frac{1}{\pi (r_w^+)^2} \int_0^{r_w^+} 2\pi r^+ u^+ \, dr^+ = \frac{2}{(r_w^+)^2} \int_0^{r_w^+} r^+ u^+ \, dr^+. \tag{6.54}$$

Considering that $r^+ = r_w^+ - y^+$, $dr^+ = -dy^+$ and noting that if $r = r_w^+$, $y^+ = 0$ and if $r^+ = 0$, $y^+ = r_w^+$, formula (6.54) can be written as:

$$u_m^+ = \frac{2}{\left(r_w^+\right)^2} \int_0^{r_w^+} u^+ \left(r_w^+ - y^+\right) dy^+. \tag{6.55}$$

Integral (6.55) is calculated using the trapezoidal rule:

$$u_m^+ = \frac{2\Delta y^+}{\left(r_w^+\right)^2} \sum_{i=1}^{n-1} \frac{u_i^+ \left(r_w^+ - y_i^+\right) + u_{i+1}^+ \left(r_w^+ - y_{i+1}^+\right)}{2}. \tag{6.56}$$

Friction factor ξ can be calculated from formula (6.46).

6.2 Friction Factor for Smooth and Rough Tubes

6.2.1 Empirical Formulae for Friction Factor in Smooth and Rough Tubes

Pressure drop Δp in the fluid flow through a channel is calculated using the following formula:

$$\Delta p = \xi \frac{l}{d_h} \frac{\rho w_m^2}{2}, \tag{6.57}$$

where: ξ—friction factor, l—channel length, d_h—channel hydraulic diameter, ρ—density, w_m—fluid flow mean velocity.
The hydraulic diameter is defined as:

$$d_h = \frac{4A_w}{U_w}, \tag{6.58}$$

where: A_w—channel cross-section surface area, U_w—wetted perimeter.
If the fluid flows through the channel entire cross-section, U_w is the channel inner perimeter. For channels with a circular cross-section $d_h = d_w$, where d_w is the inner diameter.
If the fluid flow is laminar or if the channel inner surface is smooth, the friction factor is a function of the Reynolds number only. For turbulent flows through channels with a rough inner surface, the friction factor depends on the Reynolds number and on the channel surface relative roughness (cf. Fig. 6.2).

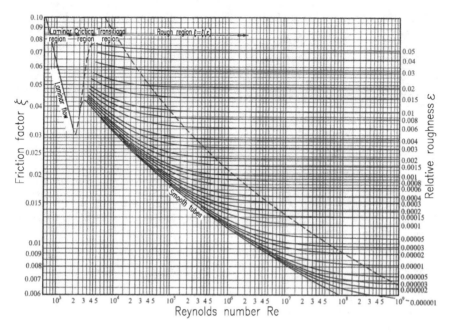

Fig. 6.2 Moody chart for friction factor determination for the fluid flow through circular tubes

Relative roughness ε is defined as:

$$\varepsilon = \frac{R_a}{d_w},\tag{6.59}$$

where R_a is the mean arithmetic deviation of the profile from the average line, i.e. the surface absolute roughness [188], and d_w is the duct inner diameter.

For the fluid laminar flow, friction factor ξ is calculated from the following formula:

$$\xi = \frac{64}{Re}, \quad Re \leq 2300,\tag{6.60}$$

where the Reynolds number is defined as:

$$Re = \frac{w_m d_h}{v} = \frac{\rho w_m d_h}{\mu},\tag{6.61}$$

where: μ—dynamic viscosity, v—kinematic viscosity.

6.2.2   Empirical Formulae for Friction Factor in Smooth and Rough Tubes

This subsection presents the formulae used to calculate the friction factor in engineering practice. They are simple relations determined based on experimental data.

6.2.2.1   Friction Factor for Turbulent Flows Through Channels with a Smooth Surface

In the case of fully developed turbulent flows through a tube with a smooth surface, the friction factor can be determined from the Prandtl-von Kármán-Nikuradse formula [214]:

$$\frac{1}{\sqrt{\xi}} = -0.8 + 2\log\left(\mathrm{Re}\sqrt{\xi}\right) = -2\log\left(\frac{2.51}{\mathrm{Re}\sqrt{\xi}}\right), \tag{6.62}$$

which can be written in the following alternative form:

$$\xi = \frac{1}{\left(-0.8 + 2\log\left(\mathrm{Re}\sqrt{\xi}\right)\right)^2} = \frac{1}{\left(2\log\left(\frac{2.51}{\mathrm{Re}\sqrt{\xi}}\right)\right)^2}. \tag{6.63}$$

The use of formula (6.62) is troublesome because, in order to find friction factor ξ, a nonlinear algebraic equation has to be solved. Equation (6.63) can be solved using the fixed-point iteration, the interval search method or one of the many methods of solving nonlinear algebraic equations which are available. Literature offers explicit expressions for the friction factor calculation which can be easily applied in engineering practice. The Blasius formula [31] is one of the earliest:

$$\xi = \frac{0.3164}{\mathrm{Re}^{0.25}}, \quad 3 \times 10^3 \leq \mathrm{Re} \leq 1 \times 10^5. \tag{6.64}$$

For higher Reynolds numbers, the Moody formula of the form presented below is more accurate [127, 160, 202, 208]:

$$\xi = \frac{0.184}{\mathrm{Re}^{0.2}}, \quad 10^5 \leq \mathrm{Re}. \tag{6.65}$$

Nevertheless, for high Reynolds numbers both the Blasius formula (6.64) and the Moody formula (6.65) give too low values of the friction factor.

In a wider range of the Reynolds numbers, the Konakov formula can be applied [157]:

$$\xi = \frac{1}{(1.8 \log \mathrm{Re} - 1.5)^2}, \quad 3 \times 10^3 \leq \mathrm{Re} \leq 10^6. \tag{6.66}$$

The Konakov formula (6.66) is a good approximation of the Prandtl-von Kármán-Nikuradse formula (6.62) [106, 324], and it is used in the Petukhov-Kirillov formula to calculate the Nusselt number in the case of the fluid turbulent flow through tubes with a smooth surface [106, 108]. The Filonenko formula is also commonly used [92]:

$$\xi = \frac{1}{(1.82 \log \mathrm{Re} - 1.64)^2}, \quad 3 \times 10^3 \leq \mathrm{Re} \leq 5 \times 10^7. \tag{6.67}$$

A different form of the formula defining the friction factor for the fluid turbulent flow through tubes with a circular cross-section is proposed by Allen and Eckert [8]:

$$\xi = 0.00556 + \frac{0.432}{\mathrm{Re}^{0.308}}, \quad 1.3 \times 10^4 \leq \mathrm{Re} \leq 1.2 \times 10^5. \tag{6.68}$$

Formula (6.68) was determined based on measurements of the pressure drop in circular tubes for water flows for the Prandtl numbers 7 and 8.

6.2.2.2 Friction Factor for Turbulent Flows Through Channels with a Rough Surface

For high Reynolds numbers, Prandtl and von Kármán put forward the following equation:

$$\frac{1}{\sqrt{\xi}} = 1.14 - 2 \log \varepsilon = -2 \log \frac{\varepsilon}{3.7}, \tag{6.69}$$

$$\xi = \left[2 \log \left(\frac{\varepsilon}{3.7} \right) \right]^{-2}. \tag{6.70}$$

Colebrook [60, 61] proposed the following nonlinear algebraic equation for determination of friction factor ξ:

$$\frac{1}{\sqrt{\xi}} = -2 \log \left(\frac{2.51}{\mathrm{Re}\sqrt{\xi}} + \frac{\varepsilon}{3.7} \right). \tag{6.71}$$

The structure of Eq. (6.71) results from the combination of Eqs. (6.62) and (6.69). Equation (6.71) can be written in the following alternative form:

$$\xi = \left[2\log\left(\frac{2.51}{\mathrm{Re}\sqrt{\xi}} + \frac{\varepsilon}{3.7}\right)\right]^{-2}. \tag{6.72}$$

Although nowadays Eqs. (6.71) or (6.72) can be solved without any difficulties, more than 20 explicit formulae approximating the results obtained from the solution of the Colebrook Eq. (6.71) have been put forward in the last several dozens of years. They enable an easy calculation of friction factor ξ without the need to solve Eqs. (6.71) or (6.72) iteratively.

Churchill [55, 56] suggested that the implicit Prandtl-von Kármán-Nikuradse Eq. (6.62) [214] in the Colebrook formula (6.72) should be replaced with the explicit expression proposed by Nikuradse [213, 214]:

$$\xi = \frac{1}{[1.8\log(\mathrm{Re}/7)]^2}. \tag{6.73}$$

Considering (6.73) in (6.72), Churchill [55] obtained the following explicit expression for the calculation of ξ:

$$\xi = \left\{2\log\left[\left(\frac{7}{\mathrm{Re}}\right)^{0.9} + \frac{\varepsilon}{3.7}\right]\right\}^{-2} = \left\{2\log\left[\frac{5.76}{\mathrm{Re}^{0.9}} + \frac{\varepsilon}{3.7}\right]\right\}^{-2}. \tag{6.74}$$

Three years later, Swamee and Jain [291] put forward a very similar formula for the calculation of friction factor ξ:

$$\xi = \left\{2\log\left[\frac{5.74}{\mathrm{Re}^{0.9}} + \frac{\varepsilon}{3.7}\right]\right\}^{-2}, \quad 5000 \leq \mathrm{Re} \leq 10^8. \tag{6.75}$$

In 1983, an expression characterized by a simple form and good accuracy was proposed by Haaland [117]:

$$\xi = \left\{1.8\log\left[\frac{6.9}{\mathrm{Re}} + \left(\frac{\varepsilon}{3.7}\right)^{1.11}\right]\right\}^{-2}, \quad 4000 \leq \mathrm{Re} \leq 10^8, \tag{6.76}$$

For natural gas pipelines with very small relative roughness, Haaland recommends a different formula:

$$\xi = \left\{\frac{1.8}{n}\log\left[\left(\frac{7.7}{\mathrm{Re}}\right)^n + \left(\frac{\varepsilon}{3.7}\right)^{1.11n}\right]\right\}^{-2}, \quad n = 3, \quad 4000 \leq \mathrm{Re} \leq 10^8. \tag{6.77}$$

In 1997, Manadilli [189] proposed the following simple and fairly accurate formula:

$$\xi = \left\{ 2\log\left[\frac{95}{\mathrm{Re}^{0.983}} - \frac{96.82}{\mathrm{Re}} + \frac{\varepsilon}{3.7}\right]\right\}^{-2}, \quad 5235 \le \mathrm{Re} \le 10^8, \tag{6.78}$$

which was modified by Fang et al. [89] and used to calculate ξ in smooth tubes:

$$\xi = \frac{1}{\left[2\log\left(\frac{150.39}{\mathrm{Re}^{0.98865}} - \frac{152.66}{\mathrm{Re}}\right)\right]^2}, \quad 3 \times 10^3 \le \mathrm{Re} \le 10^8. \tag{6.79}$$

The Chen formula [50] demonstrates good accuracy, but it is more complex:

$$\xi = \left\{ 2\log\left[\frac{\varepsilon}{3.7065} - \frac{5.0452}{\mathrm{Re}}\log\left(\frac{\varepsilon^{1.1098}}{2.8257} + \frac{5.8506}{\mathrm{Re}^{0.8981}}\right)\right]\right\}^{-2}, \tag{6.80}$$

$$4 \times 10^3 \le \mathrm{Re} \le 4 \times 10^8.$$

The expressions referred to above, which enable non-iterative calculation of friction factor ξ, are valid in the range of turbulent flows, i.e. for the Reynolds numbers higher than 3000. In 1977, Churchill [56] developed the following formula, which holds for all ranges: laminar, transitional and turbulent.

$$\xi = 8\left[\left(\frac{8}{\mathrm{Re}}\right)^{12} + \frac{1}{(A+B)^{3/2}}\right]^{1/12}, \tag{6.81}$$

where:

$$A = \left[2.457\ln\frac{1}{\left(\frac{7}{\mathrm{Re}}\right)^{0.9} + 0.27\varepsilon}\right]^{16}, \tag{6.82}$$

$$B = \left(\frac{37530}{\mathrm{Re}}\right)^{16}. \tag{6.83}$$

Formula (6.83) draws on expression (6.74) proposed by Churchill in 1973 and on the earlier publication of Churchill and Usagi [58].

Expression (6.81) also takes account of the transitional region data approximation proposed in 1975 by Wilson and Azad [366]. It can also be seen that the first term in the square bracket in expression (6.81) corresponds to the Hagen-Poiseuille equation defining the friction factor in the laminar flow, i.e. if the second term in the square bracket approaches zero, formula (6.81) gives:

$$\xi = 64/\text{Re}.$$

Rennels and Hudson modified the Churchill formula (6.81), pointing out that instead of the Nikuradse formula (6.73), approximating the friction factor for a smooth tube, a linearized form of the Hoerl function could be used [246]:

$$\frac{1}{\sqrt{\xi}} = -2\log\left[0.883\frac{(\ln \text{Re})^{1.282}}{\text{Re}^{1.007}}\right]. \qquad (6.84)$$

If in the definition of A expressed by formula (6.82) term $(7/\text{Re})^{0.9}$ is replaced with:

$$0.883\frac{(\ln \text{Re})^{1.282}}{\text{Re}^{1.007}}, \qquad (6.85)$$

the Churchill formula modified in this manner will be more accurate in the transitional region, especially for $0 \le \varepsilon \le 0.002$. Schroeder [260] pointed out that the Colebrook-White formula gave too high values of the friction factor from the transitional range up to the Reynolds number of about 10^5. In connection with that, Rennels and Hudson [246] suggested that the denominator in the definition of A expressed by formula (6.82) should be reduced by the value of $110\varepsilon/\text{Re}$. Ultimately, the Churchill formula (6.81) modified by Rennels and Hudson [246] has the following form:

$$\xi = \left[\left(\frac{64}{\text{Re}}\right)^{12} + \frac{1}{(A+B)^{3/2}}\right]^{1/12}, \qquad (6.86)$$

where:

$$A = \left[0.8687\ln\frac{1}{\frac{0.883(\ln \text{Re})^{1.282}}{\text{Re}^{1.007}} + 0.27\varepsilon - \frac{110\varepsilon}{\text{Re}}}\right]^{16}, \qquad (6.87)$$

$$B = \left(\frac{13269}{\text{Re}}\right)^{16}. \qquad (6.88)$$

The changes in the friction factor determined using formula (6.86) as a function of the Reynolds number Re and relative roughness ε are illustrated in Fig. 6.3.

Comparing the Moody chart in Fig. 6.2 with the chart presented in Fig. 6.3, very good agreement can be observed between friction factors ξ for different values of the Reynolds number Re and the surface relative roughness ε. In practical applications and also in mathematical modelling of the fluid flow in tubes, it is more convenient to use formula (6.86) than read the ξ value from the Moody chart presented in Fig. 6.2.

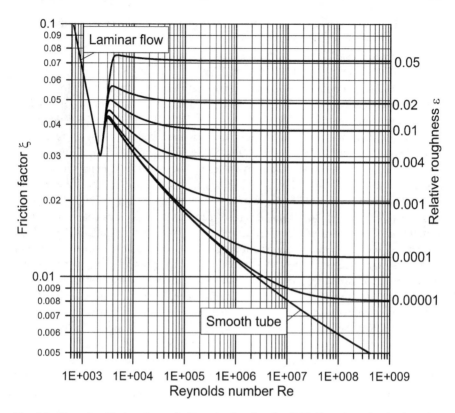

Fig. 6.3 Chart for friction factor ξ determination for the fluid flow through circular tubes developed based on formula (6.86)

6.2.3   Comparison of Friction Factors for the Fluid Turbulent Flows Through Channels with a Smooth Surface

Friction factors for the fluid turbulent flow, both in smooth and rough tubes, are usually found based on the measurement of the pressure drop on the tube section for which the length, the inner diameter and the inner surface roughness are known. Performing measurements for the fluid different velocities, the friction factor can be determined as a function of the Reynolds number.

The friction factor can also be found based on the velocity distribution in the tube cross-section using formula (6.46). The fluid flow mean universal velocity u_m^+ is determined from formula (6.55). The distribution of velocity can be found based on experimental testing or solving the momentum conservation equation. In the last case, velocity $u^+(y^+)$ can be found from formula (6.43), or numerically from formula (6.56) at the known ε_τ/ν ratio determined experimentally. The fluid flow through tubes with a circular cross-section is commonly analysed by means of the Reichardt formulae (4.104) and (4.111).

Both methods of friction factor $\xi(\text{Re})$ determination are based on experimentally determined relation $u^+(y^+)$ or relation $\frac{\varepsilon_t}{\nu}(y^+)$, and they can produce slightly different results. A comparison is made below between velocity distributions found from the Reichardt formula (4.114) and those determined numerically using formula (6.50) and taking account of formulae (4.104) and (4.111). Calculating velocity by means of formula (6.50), the integration interval from 0 to r_w^+ is divided into 30,000 subintervals. So many points of velocity determination are necessary due to the very big changes in the fluid velocity close to the tube inner surface. The calculation results obtained for the Reynolds numbers Re = 10000, Re = 30000, Re = 50000 and Re = 100000 are shown in Figs. 6.4, 6.5, 6.6 and 6.7. Analysing the results presented in the figures, it can be seen that the velocities determined from formula (4.114) are higher compared to those obtained from formula (6.50), which is based on the momentum conservation equation and turbulent viscosity ε_t found experimentally. The differences between the velocities obtained by means of the two procedures are bigger for small Reynolds numbers. For the Reynolds number Re = 100000 (cf. Fig. 6.7) the differences between the results are already slight, and they will be even smaller for higher Re values.

Having found the velocity distributions, the histories of changes in the values of friction factor ξ depending on the Reynolds number Re were determined using formula (6.46). Mean velocity was calculated from formula (6.56), dividing the integration interval into 30,000 subintervals. The calculation results and their comparison with experimental formulae are presented in Figs. 6.8 and 6.9.

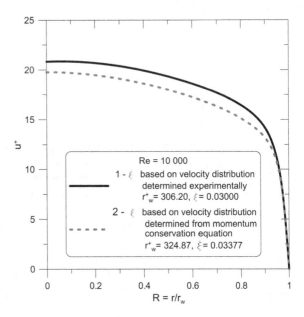

Fig. 6.4 Velocity distributions found from formula (4.114) and formula (6.50); Re = 10000

Fig. 6.5 Velocity
distributions found from
formula (4.114) and formula
(6.50); Re = 30000

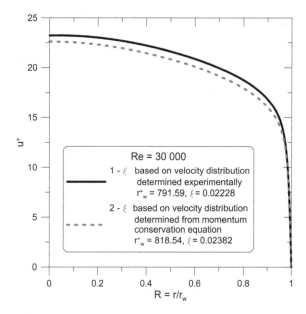

Fig. 6.6 Velocity
distributions found from
formula (4.114) and formula
(6.50); Re = 50000

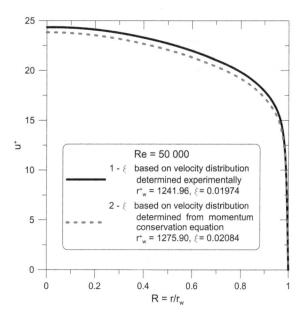

In order to assess the differences between the results of calculations and the
following experimental formulae: the Blasius formula (6.64), the Konakov formula
(6.66), the Filonenko formula (6.67) and the Allen-Eckert formula (6.68), a comparison is presented in Fig. 6.10 between relative differences calculated from the
following formula:

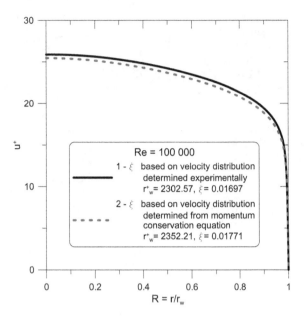

Fig. 6.7 Velocity distributions found from formula (4.114) and formula (6.50); Re = 100000

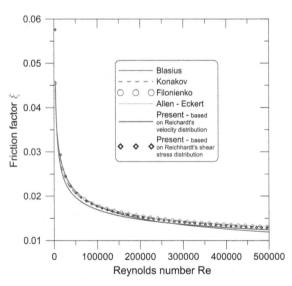

Fig. 6.8 Changes in friction factor ξ as a function of the Reynolds number—linear scale of Reynolds numbers Re

$$e = \frac{\xi - \xi_m}{\xi_m} \cdot 100 \ (\%), \tag{6.89}$$

where ξ_m is the arithmetic mean of the friction factors shown in Fig. 6.8 or 6.9.

The results presented in Figs. 6.8, 6.9 and 6.10 demonstrate considerable agreement. Only the friction factor determined based on the velocity distribution

Fig. 6.9 Changes in friction factor ζ as a function of the Reynolds number— logarithmic scale of Reynolds numbers Re

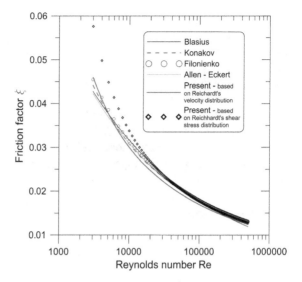

Fig. 6.10 Relative differences between the friction factor values determined by means of different formulae and methods

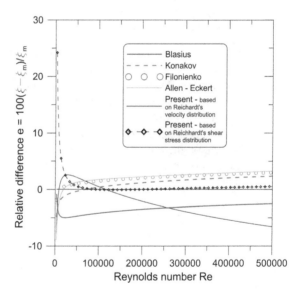

obtained from integration of the momentum conservation equation and using Reichardt's empirical formulae (4.104) and (4.111) for turbulent viscosity is in the Reynolds number range of $3000 \leq \mathrm{Re} \leq 20000$ a little higher compared to the other values. For Re > 20000, the factor shows the best agreement with the mean value $\xi_m(\mathrm{Re})$.

It follows from the analysis of the Moody chart (Fig. 6.2), and also of the chart presented in Fig. 6.3, that for tubes with a smooth inner surface the friction factor rises fast in the Reynolds number range from 2300 to 3000. In the range of

$3000 \leq \mathrm{Re} \leq 10000$, friction factor ξ can therefore be calculated from formulae valid for the turbulent flow. Considering the analogy between the transfer of momentum and heat, the heat transfer coefficients (the Nusselt numbers) can in the range of $3000 \leq \mathrm{Re} \leq 10000$ be approximated by means of formulae with forms used for the turbulent flow. It should be noted, however, that in this range of values the heat transfer coefficients are by 20–40% lower compared to the coefficient values found from the correlations used for the fully developed turbulent flow if the Reynolds number is higher than 10000 [67, 128, 202, 208, 227, 270]. The issue of determining the friction factors for turbulent fluid flow in pipes with a smooth inner surface is the subject of the papers by Taler [324, 325].

6.3 Heat Transfer

The analysis will concern the fully developed flow of a fluid in a circular duct at a constant heat flux on the wall.

Such flows are common in practice, both in circular tubes and in ducts where the cross-section has a different shape. Equations (4.23)–(4.31) will be reduced taking account of the following conditions:

(a) the fluid is incompressible and its physical properties are constant,
(b) the fluid flow is steady and hydrodynamically developed, which means that the analysis concerns flows where the velocity and temperature profiles in the duct are fully developed, i.e. $\bar{w}_x = \bar{w}_x(r)$, $\bar{T} = \bar{T}(r)$ and $\partial \bar{w}_x / \partial x = 0$; it follows from the mass conservation Eq. (4.23) that under such assumptions the velocity radial component equals zero, i.e. $\bar{w}_r = 0$,
(c) heat flux $\dot{q}_w$ on the duct inner surface is constant,
(d) heat conduction and the turbulent heat flow in the liquid in the axial direction are negligibly small compared to the same in the radial direction, i.e.:

$$\frac{\partial}{\partial x}\left[\lambda \frac{\partial \bar{T}}{\partial x} - \rho c_p \overline{w'_x T'}\right] \ll \frac{1}{r}\frac{\partial}{\partial r}\left[r\left(\lambda \frac{\partial \bar{T}}{\partial r} - \rho c_p \overline{w'_r T'}\right)\right]. \tag{6.90}$$

Taking account of inequality (6.90), the energy balance Eq. (4.31) takes the following form:

$$\rho c_p \bar{w}_x \frac{\partial \bar{T}}{\partial x} = \frac{1}{r}\frac{\partial}{\partial r}\left[r\left(\lambda \frac{\partial \bar{T}}{\partial r} - \rho c_p \overline{w'_r T'}\right)\right]. \tag{6.91}$$

Equation (6.90) can be written as follows:

$$\rho c_p \bar{w}_x \frac{\partial \bar{T}}{\partial x} = \frac{1}{r}\frac{\partial}{\partial r}(r\dot{q}), \tag{6.92}$$

where

$$\dot{q} = \dot{q}_m + \dot{q}_t = \lambda \frac{\partial \bar{T}}{\partial r} - \rho c_p \overline{w'_r T'} = \left(\lambda + \rho c_p \varepsilon_q\right) \frac{\partial \bar{T}}{\partial r}, \qquad (6.93)$$

$$\dot{q}_m = \lambda \frac{\partial \bar{T}}{\partial r}, \quad \dot{q}_t = -\rho c_p \overline{w'_r T'} = \rho c_p \varepsilon_q \frac{\partial \bar{T}}{\partial r}. \qquad (6.94)$$

It should be noted that in this case the heat flux is positive because heat flows from the duct wall to the fluid. Therefore function $T(r)$ is increasing and conductive (molecular) heat flux $\lambda \partial \bar{T}/\partial r$ is also positive.

In order to find the correlation for the Nusselt number: $\mathrm{Nu} = f(\mathrm{Re}, \mathrm{Pr})$, the heat transfer coefficient α will first be determined using the following formula:

$$\alpha = \frac{\dot{q}_w}{T_w - T_m}, \qquad (6.95)$$

where $(T_w - T_m)$ is the difference between the tube inner surface temperature T_w and the mass-averaged temperature of the medium.

The changes in mass-averaged temperature $T_m(x)$ on the duct length can be found from the heat balance equation. Assuming that heat flux $\dot{q}_w$ on the duct inner surface is constant and coordinate x is measured from the duct inlet, the heat balance is expressed as follows:

$$2\pi r_w \dot{q}_w x + \pi r_w^2 w_m \rho c_p T_m|_{x=0} = \pi r_w^2 w_m \rho c_p T_m(x). \qquad (6.96)$$

The fluid mass-averaged temperature $T_m(x)$ at constant density ρ and the fluid constant specific heat c_p is determined using the following equality:

$$\pi r_w^2 w_m \rho c_p T_m(x) = \int_0^{r_w} 2\pi r dr \rho c_p \bar{w}_x \bar{T}(x, r), \qquad (6.97)$$

which gives:

$$T_m(x) = \frac{2}{r_w^2 w_m} \int_0^{r_w} \bar{w}_x(r) \bar{T}(x, r) r dr. \qquad (6.98)$$

The fluid mass-averaged temperature $T_m(x)$ found according to formula (6.96) is defined by the following expression:

$$T_m(x) = T_m|_{x=0} + \frac{2\dot{q}_w}{\rho c_p w_m r_w} x. \qquad (6.99)$$

Differentiating both sides of Eq. (6.99) with respect to x, the following is obtained:

$$\frac{dT_m}{dx} = \frac{2\dot{q}_w}{\rho c_p w_m r_w}. \tag{6.100}$$

The solution of Eq. (6.92) will be found in the following form:

$$\bar{T}(x, r) = \bar{T}_1(x) + \bar{T}_2(r). \tag{6.101}$$

Substituting function (6.101) in differential Eq. (6.92) and taking account of Eq. (6.93), the following is obtained:

$$\frac{\partial \bar{T}_1}{\partial x} = \frac{1}{r\bar{w}_x} \frac{\partial}{\partial r} \left[r(a + \varepsilon_q) \frac{\partial \bar{T}_2}{\partial r} \right]. \tag{6.102}$$

Differentiating both sides of expression (6.101) with respect to x, the result is:

$$\frac{\partial \bar{T}}{\partial x} = \frac{d\bar{T}_1}{dx}. \tag{6.103}$$

It follows from the analysis of Eq. (6.102) that the left side depends on x only, whereas the right side—on r, which means that the two sides are equal to a certain constant.

Therefore, Eq. (6.103) takes the following form:

$$\frac{\partial \bar{T}}{\partial x} = \frac{d\bar{T}_1}{dx} = \text{const.} \tag{6.104}$$

Derivative $d\bar{T}_1/dx$ can be determined from Eq. (6.102). Multiplying the equation by $(r\bar{w}_x \rho c_p)$ and integrating from $r = 0$ to $r = r_w$, the result is:

$$\rho c_p \frac{d\bar{T}_1}{dx} \int_0^{r_w} \bar{w}_x r dr = \left[r(\lambda + \rho c_p \varepsilon_q) \frac{\partial \bar{T}_2}{\partial r} \right]\Big|_0^{r_w}. \tag{6.105}$$

Considering boundary conditions:

$$\left[(\lambda + \rho c_p \varepsilon_q) \frac{\partial \bar{T}_2}{\partial r} \right]\Big|_{r=r_w} = \dot{q}_w, \tag{6.106}$$

$$\left[\frac{\partial \bar{T}_2}{\partial r} \right]\Big|_{r=0} = 0 \tag{6.107}$$

and defining mean velocity as:

$$w_m = \frac{2}{r_w^2} \int_0^{r_w} \bar{w}_x r \, dr \qquad (6.108)$$

Equation (6.105) gives:

$$\frac{\rho c_p r_w^2 w_m}{2} \rho c_p \frac{d\bar{T}_1}{dx} = r_w \dot{q}_w. \qquad (6.109)$$

Reducing Eq. (6.109), the following is obtained:

$$\frac{d\bar{T}_1}{dx} = \frac{2\dot{q}_w}{\rho c_p r_w w_m}. \qquad (6.110)$$

Comparing expressions (6.100) and (6.110) it can be seen that

$$\bar{T}_1(x) = T_m(x). \qquad (6.111)$$

It is worth noting that Eq. (6.110) can also be obtained from the energy balance done for a control area with length Δx and diameter $2r_w$

$$\dot{m} c_p T_m|_x + 2\pi r_w \Delta x \dot{q}_w = \dot{m} c_p T_m|_{x+\Delta x}, \qquad (6.112)$$

where the fluid mass flow rate $\dot{m}$ is defined as:

$$\dot{m} = \rho \pi r_w^2 w_m. \qquad (6.113)$$

Transforming Eq. (6.112), the following is obtained:

$$\frac{T_m|_{x+\Delta x} - T_m|_x}{\Delta x} = \frac{2\dot{q}_w}{\rho c_p r_w w_m}. \qquad (6.114)$$

If it is assumed that $\Delta x \to 0$, considering (6.111) in Eq. (6.114), the result is Eq. (6.110).

Considering (6.104) and (6.111), the following is obtained:

$$\frac{\partial \bar{T}}{\partial x} = \frac{dT_m}{dx} = \frac{d\bar{T}_1}{dx} = \frac{2\dot{q}_w}{\rho c_p w_m r_w} = \text{const.} \qquad (6.115)$$

Multiplying Eq. (6.101) by $2w_x r / w_m r_w^2$ and integrating both sides, the result is:

$$\frac{2}{w_m r_w^2} \int_0^{r_w} \bar{T}(x,r) w_x r \, dr = \frac{2}{w_m r_w^2} \int_0^{r_w} \bar{T}_1(x) w_m r \, dr + \frac{2}{w_m r_w^2} \int_0^{r_w} \bar{T}_2(r) w_x r \, dr. \quad (6.116)$$

Taking account of mass-averaged temperature $T_m(x)$ as defined by (6.98), Eq. (6.116) can be written in the following form:

$$T_m(x) = \bar{T}_1(x) + T_{2m},\qquad(6.117)$$

where

$$T_{2m} = \frac{2}{w_m r_w^2} \int_0^{r_w} \bar{T}_2(r) \bar{w}_x r \, dr.\qquad(6.118)$$

Considering equality (6.111), expression (6.118) gives:

$$T_{2m} = 0.\qquad(6.119)$$

Considering relations (6.103) and (6.110) in Eq. (6.92), the following is obtained:

$$\rho c_p \bar{w}_x \frac{2\dot{q}_w}{\rho c_p w_m r_w} = \frac{1}{r}\frac{d}{dr}(r\dot{q}).\qquad(6.120)$$

By introducing non-dimensional coordinate $R = r/r_w$, Eq. (6.120) can be transformed into the following form:

$$2\frac{\bar{w}_x}{w_m} = \frac{1}{R}\frac{d}{dR}\left(R\frac{\dot{q}}{\dot{q}_w}\right).\qquad(6.121)$$

Taking account of Eq. (6.101), Eq. (6.93) gives:

$$\frac{d\bar{T}_2}{dr} = \frac{\dot{q}}{\lambda + \rho c_p \varepsilon_q}.\qquad(6.122)$$

Considering that

$$\frac{\rho c_p}{\lambda}\varepsilon_q = \frac{\varepsilon_q}{a} = \frac{\mathrm{Pr}}{\mathrm{Pr}_t}\frac{\varepsilon_\tau}{v},\qquad(6.123)$$

where symbols:

$$\mathrm{Pr} = \frac{v}{a} = \frac{c_p \mu}{\lambda} \quad \text{and} \quad \mathrm{Pr}_t = \frac{\varepsilon_\tau}{\varepsilon_q}\qquad(6.124)$$

denote the molecular and the turbulent Prandtl number, respectively, Eq. (6.122) can be written as:

$$\frac{d\bar{T}_2}{dr} = \frac{\dot{q}}{\lambda\left(1 + \frac{\mathrm{Pr}}{\mathrm{Pr}_t}\frac{\varepsilon_\tau}{v}\right)}.\qquad(6.125)$$

In order to solve the system of Eqs. (6.121) and (6.125), appropriate boundary conditions have to be set. For Eq. (6.121), the boundary condition has the following form resulting from the temperature field symmetry:

$$\dot{q}|_{r=0} = 0. \tag{6.126}$$

The distribution of temperature $\bar{T}_2(r)$ will be determined by selecting the tube surface temperature $T_{2w} = \bar{T}_2(r)|_{r=0}$ in such a manner that condition (6.119) is satisfied.

6.3.1 Determination of Temperature, the Heat Flux and the Nusselt Number—Integral Formulation

Integration of Eq. (6.121) along radius R from 0 to R:

$$\int_0^R 2\frac{\bar{w}_x}{w_m} R\,dR = \int_0^R \frac{1}{R}\frac{d}{dR}\left(R\frac{\dot{q}}{\dot{q}_w}\right) R\,dR \tag{6.127}$$

results in:

$$\frac{\dot{q}}{\dot{q}_w} = \frac{2}{R}\int_0^R \frac{\bar{w}_x}{w_m} R\,dR. \tag{6.128}$$

Introducing dimensionless coordinates (4.61), (6.40) and (6.42), solution (6.128) can be written as:

$$\frac{\dot{q}}{\dot{q}_w} = \frac{2}{r_w^+ (r_w^+ - y^+)} \int_{y^+}^{r_w^+} \frac{u^+}{u_m^+}\left(r_w^+ - y^+\right)dy^+. \tag{6.129}$$

The heat flux can be found from formula (6.128) or (6.129), calculating integrals by means of one of the available numerical methods. Using the trapezoidal rule to calculate the integral in formula (6.128), the following is obtained:

$$\frac{\dot{q}_i}{\dot{q}_w} = \frac{\Delta R}{R_i w_m}\sum_{j=2}^i \left(\bar{w}_{x,j-1}R_{j-1} + \bar{w}_{x,j}R_j\right), \quad i = 2,\ldots,n, \tag{6.130}$$

where $(n-1)$ is the number of subintervals with equal length $\Delta R = 1/(n-1)$, into which interval $0 \le R \le 1$ is divided.

Introducing dimensionless radius $R = r/r_w$ in Eq. (6.125) and considering the heat flux distribution (6.128), the following is obtained:

$$\frac{d\bar{T}_2}{dR} = \frac{2\dot{q}_w r_w}{\lambda} \frac{\int_0^R \frac{\bar{w}_x}{w_m} R dR}{\left(1 + \frac{Pr}{Pr_t}\frac{\varepsilon_\tau}{v}\right)R}. \tag{6.131}$$

The difference between the wall temperature $T_{2w} = \bar{T}_2|_{R=1}$ and the fluid temperature $\bar{T}_2(R)$ can be found by integrating Eq. (6.131) along R from R to $R = 1$.

$$T_{2w} - \bar{T}_2(R) = \frac{2\dot{q}_w r_w}{\lambda} \int_R^1 \frac{\int_0^R \frac{\bar{w}_x}{w_m} R dR}{\left(1 + \frac{Pr}{Pr_t}\frac{\varepsilon_\tau}{v}\right)R} dR. \tag{6.132}$$

Introducing dimensionless coordinates (4.61), (6.40) and (6.42), solution (6.128) can be written in the following form:

$$T_{2w} - \bar{T}_2(y^+) = \frac{2\dot{q}_w r_w}{\lambda} \int_0^{y^+} \frac{\int_{y^+}^{r_w^+} \frac{1}{(r_w^+)^2} \frac{u^+}{u_m^+}(r_w^+ - y^+) dy^+}{\left(1 + \frac{Pr}{Pr_t}\frac{\varepsilon_\tau}{v}\right)(r_w^+ - y^+)} dy^+. \tag{6.133}$$

Formulae (6.131) or (6.132) do not make it possible to determine the distribution of temperature $\bar{T}_2$ in the tube cross-section because the wall temperature T_{2w} is unknown.

However, it can be done after the heat transfer coefficient (the Nusselt number) is determined on the tube inner surface.

The integrals in formula (6.131) or (6.132) can be calculated numerically by means of the trapezoidal rule. If the heat flux is determined, formula (6.128) implies that:

$$\frac{2}{R} \int_0^R \frac{\bar{w}_x}{w_m} R dR = \frac{\dot{q}}{\dot{q}_w} \frac{R}{2}. \tag{6.134}$$

Substitution of (6.134) in (6.132) results in:

$$T_{2w} - \bar{T}_2(R) = \frac{\dot{q}_w r_w}{\lambda} \int_R^1 \frac{\frac{\dot{q}}{\dot{q}_w}}{\left(1 + \frac{Pr}{Pr_t}\frac{\varepsilon_\tau}{v}\right)} dR. \tag{6.135}$$

The following is obtained using the trapezoidal rule to calculate the integral in formula (6.135):

$$\bar{T}_{2,i-1} = T_{2w} - \frac{\dot{q}_w r_w}{\lambda} \sum_{j=i}^{n} \frac{\Delta R}{2} \left[\frac{\frac{\dot{q}_{j-1}}{\dot{q}_w}}{1 + \frac{\Pr}{\Pr_{t,j-1}} \left(\frac{\varepsilon_\tau}{\nu}\right)_{j-1}} + \frac{\frac{\dot{q}_j}{\dot{q}_w}}{1 + \frac{\Pr}{\Pr_{t,j}} \left(\frac{\varepsilon_\tau}{\nu}\right)_j} \right], \tag{6.136}$$

$$i = n, \ldots, 2.$$

In order to determine the Nusselt number, mass-averaged temperature T_{2m} in the tube cross-section, expressed by formula (6.118), will be written in the following dimensionless form:

$$T_{2m} = \frac{1}{\pi r_w^2 w_m} \int_0^{r_w} 2\pi r \bar{T}_2 \bar{w}_x dr = 2 \int_0^1 \bar{T}_2 \frac{\bar{w}_x}{w_m} R dR. \tag{6.137}$$

Using integration by parts for the second term on the right side of formula (6.137), mean temperature can be expressed as:

$$T_{2m} = 2 \left[\bar{T}_2 \int_0^R \left(\frac{\bar{w}_x}{w_m} R dR \right) \Big|_0^1 - \int_0^1 \frac{\partial \bar{T}_2}{\partial R} \left(\int_0^R \frac{\bar{w}_x}{w_m} R dR \right) dR \right]. \tag{6.138}$$

Considering that

$$\bar{T}_2 |_{R=1} = T_{2w}, \tag{6.139}$$

$$2 \int_0^R \left(\frac{\bar{w}_x}{w_m} R dR \right) \Big|_{R=1} = 2 \int_0^1 \frac{\bar{w}_x}{w_m} R dR = 1, \tag{6.140}$$

$$2 \int_0^R \left(\frac{\bar{w}_x}{w_m} R dR \right) \Big|_{R=0} = 2 \int_0^0 \frac{\bar{w}_x}{w_m} R dR = 0 \tag{6.141}$$

and substituting (6.131) in (6.138), the following is obtained:

$$T_{2w} - T_{2m} = \frac{2\dot{q}_w d_w}{\lambda} \int_0^1 \frac{\left(\int_0^R \frac{\bar{w}_x}{w_m} R dR \right)^2}{\left(1 + \frac{\Pr}{\Pr_t} \frac{\varepsilon_\tau}{\nu} \right) R} dR, \tag{6.142}$$

where $d_w = 2r_w$ is the tube inner diameter.

Considering that

$$T_w(x) = T_{2w} + T_m(x), \tag{6.143}$$

$$T_m(x) = T_{2m} + T_m(x) \tag{6.144}$$

expression (6.95) takes the following form:

$$\alpha = \frac{\dot{q}_w}{T_{2w} - T_{2m}} = \frac{\dot{q}_w}{T_{2w}}, \tag{6.145}$$

because according to formula (6.119), T_{2m} equals zero.

Considering the heat transfer coefficient definition (6.145), Eq. (6.142) can be written as:

$$\frac{1}{\mathrm{Nu}} = 2 \int_0^R \frac{\left(\int_0^R \frac{\bar{w}_x}{w_m} R \, dR \right)^2}{\left(1 + \frac{\mathrm{Pr}}{\mathrm{Pr}_t} \frac{\varepsilon_\tau}{v}\right) R} dR, \tag{6.146}$$

where Nu denotes the Nusselt number $\mathrm{Nu} = \alpha d_w / \lambda$. Formula (6.146) is known in the literature as the Lyon integral [186]. Introducing dimensionless coordinates (4.61), (6.40) and (6.42), expression (6.146) can be written in the following form:

$$\frac{1}{\mathrm{Nu}} = 2 \int_0^{r_w^+} \frac{\left[\frac{1}{(r_w^+)^2} \int_{y^+}^{r_w^+} \frac{u^+}{u_m^+} \left(r_w^+ - y^+\right) dy^+ \right]^2}{\left(1 + \frac{\mathrm{Pr}}{\mathrm{Pr}_t} \frac{\varepsilon_\tau}{v}\right) \left(r_w^+ - y^+\right)} dy^+. \tag{6.147}$$

Formula (6.146) or formula (6.147) can be expressed as a function of the heat flux, which will be demonstrated on the example of expression (6.146). Substitution of (6.134) in (6.146) results in:

$$\frac{1}{\mathrm{Nu}} = \frac{1}{2} \int_0^1 \frac{\left(\frac{\dot{q}}{\dot{q}_w}\right)^2 R}{\left(1 + \frac{\mathrm{Pr}}{\mathrm{Pr}_t} \frac{\varepsilon_\tau}{v}\right)} dR. \tag{6.148}$$

The integral in the formula can be calculated numerically using the trapezoidal rule, which gives:

$$\frac{1}{\mathrm{Nu}} = \frac{1}{2} \sum_{i=2}^n \frac{\Delta R}{2} \left[\frac{\left(\frac{\dot{q}_{i-1}}{\dot{q}_w}\right)^2 R_{i-1}}{1 + \frac{\mathrm{Pr}}{\mathrm{Pr}_{t,i-1}} \left(\frac{\varepsilon_\tau}{v}\right)_{i-1}} + \frac{\left(\frac{\dot{q}_{i-1}}{\dot{q}_w}\right)^2 R_i}{1 + \frac{\mathrm{Pr}}{\mathrm{Pr}_{t,i}} \left(\frac{\varepsilon_\tau}{v}\right)_i} \right]. \tag{6.149}$$

Finding the Nusselt number from formula (6.146) or (6.147), it is possible to calculate temperature T_{2w}.

$$T_{2w} = \frac{\dot{q}_w}{\alpha} = \frac{\dot{q}_w d_w}{\lambda} \frac{1}{\mathrm{Nu}}. \tag{6.150}$$

Substituting (6.148) in (6.132), an expression is obtained for the distribution of temperature $\bar{T}_2$ in the tube cross-section.

$$\bar{T}_2(R) = \frac{2\dot{q}_w r_w}{\lambda} \left[\frac{1}{\mathrm{Nu}} - \int_R^1 \frac{\int_0^R \frac{w_x}{w_m} R dR}{\left(1 + \frac{\mathrm{Pr}}{\mathrm{Pr}_t} \frac{\varepsilon_\tau}{v}\right) R} dR \right], \quad 0 \le R \le 1. \tag{6.151}$$

Introducing dimensionless coordinates (4.61), (6.40) and (6.42), expression (6.151) can be written in the following form:

$$\bar{T}_2(y^+) = \frac{2\dot{q}_w r_w}{\lambda} \left[\frac{1}{\mathrm{Nu}} - \int_0^{y^+} \frac{\int_{y^+}^{r_w^+} \frac{1}{(r_w^+)^2} \frac{u^+}{u_m^+}(r_w^+ - y^+) dy^+}{\left(1 + \frac{\mathrm{Pr}}{\mathrm{Pr}_t} \frac{\varepsilon_\tau}{v}\right)(r_w^+ - y^+)} dy^+ \right]. \tag{6.152}$$

The integrals in formulae: (6.132), (6.133), (6.137), (6.142), (6.146), (6.151) and (6.152) have to be calculated numerically due to the complex form of the expressions used to calculate u^+ and ε_τ/v. The integrals can be calculated using the rectangle method, the trapezoidal rule or other methods. In order to achieve a satisfactory accuracy of the calculations, interval $0 \le R \le 1$ or $0 \le y^+ \le r_w^+$ should be divided into a few dozen thousand subintervals. This is necessitated by the huge changes in velocity and temperature on the fluid small thickness near the wall.

The algorithm for determination of the Nusselt number and the fluid temperature distribution as a function of radius R or dimensionless distance from the tube wall y^+ for set values of the Reynolds and the Prandtl numbers is as follows:

- set numerical values of Re, Pr, $\dot{q}_w$, r_w and λ,
- divide interval $[0, r_w^+]$ into $(n-1)$ subintervals,
- calculate velocity u_i^+, $i = 1, \ldots, n$ using formula (4.113) or from integral (6.43),
- calculate mean velocity u_m^+ using formula (6.44),
- calculate heat flux $\dot{q}(y^+)$ using formula (6.129),
- calculate the Nusselt number Nu using formula (6.146) [or by means of formula (6.147) or (6.149)],
- calculate the distribution of temperature $\bar{T}_2(y^+)$ or $\bar{T}_2(R)$.

Alternatively, the algorithm presented above can be used to perform the calculations by determining the fluid velocity $w_x(R)$ from formula (6.31) and the mean velocity from formula (6.33), heat flux $\dot{q}$ form formula (6.130), the Nusselt number Nu from formula (6.151) and the distribution of temperature $\bar{T}_2(R)$.

Having found temperature $\bar{T}_2(y^+)$ or $\bar{T}_2(R)$, the fluid temperature $\bar{T}(x,r)$ can be calculated for set coordinates x and r using formula (6.101) for set values of: the tube inner surface radius r_w, the fluid flow mean velocity in the tube w_m, the heat flux on the tube inner surface $\dot{q}_w$ and the following physical properties of the fluid: density ρ, thermal conductivity λ, specific heat c_p, dynamic viscosity μ.

6.3.2 Determination of Temperature, the Heat Flux, and the Nusselt Number—Differential Formulation

Differential Eqs. (6.121) and (6.125) will be written as a function of dimensionless heat flux

$$q^* = \frac{\dot{q}}{\dot{q}_w} \tag{6.153}$$

moreover, dimensionless temperature

$$T_2^* = \frac{\bar{T}_2}{\dot{q}_w r_w / \lambda}. \tag{6.154}$$

Noting that $\bar{w}_x / w_m = u^+ / u_m^+$, Eqs. (6.121) and (6.125) will be written in the following form:

$$\frac{1}{R} \frac{d}{dR}(R\dot{q}^*) = 2\frac{u^+}{u_m^+}, \tag{6.155}$$

$$\frac{dT_2^*}{dR} = \frac{\dot{q}^*}{1 + \frac{Pr}{Pr_t} \frac{\varepsilon_\tau}{\nu}}. \tag{6.156}$$

The boundary conditions have the following form:

$$\dot{q}^*|_{R=1} = 1, \tag{6.157}$$

$$\dot{q}^*|_{R=0} = 0, \tag{6.158}$$

$$T_2^*|_{R=1} = T_{2w}^*, \tag{6.159}$$

$$T_{2m}^* = 0, \tag{6.160}$$

where the mean temperature T_{2m}^* is defined as:

$$T_{2m}^* = \frac{2}{u_m^+} \int\limits_0^1 T_2^* u^+ R dR. \tag{6.161}$$

The tube inner surface temperature T_{2w}^* is unknown and must be calculated iteratively to satisfy condition (6.160), i.e. to ensure that the fluid mass-averaged temperature T_{2m}^* is zero.

The problems expressed in (6.155)–(6.160) will be solved using the finite difference method. First, interval $0 \leq R \leq 1$ is divided into $(n - 1)$ subintervals ΔR.

Dimensionless radius R_i is defined by the following formula:

$$R_i = (i - 1)\Delta R, \quad i = 1, \ldots, n, \tag{6.162}$$

where

$$\Delta R = 1/(n - 1). \tag{6.163}$$

Universal velocity u^+ is usually expressed as a function of y^+. Coordinates y_i^+ corresponding to dimensionless radii R_i are found from the following formula:

$$y_i^+ = (1 - R_i)r_w^+, \tag{6.164}$$

where $r_w^+ = \frac{Re}{2} \sqrt{\frac{\xi}{8}}$.

Equation (6.149) will be approximated as follows:

$$\frac{R_i \dot{q}_i^* - R_{i-1} \dot{q}_{i-1}^*}{\Delta R} = \frac{1}{2} \left(2R_{i-1} \frac{u_{i-1}^+}{u_m^+} + 2R_i \frac{u_i^+}{u_m^+} \right), \quad i = 2, \ldots, n, \tag{6.165}$$

Mean velocity u_m^+ is calculated from the formula:

$$u_m^+ = 2\Delta R \sum_{i=2}^n \frac{R_{i-1}u^+\left(y_{i-1}^+\right) + R_iu^+\left(y_i^+\right)}{2}, \tag{6.166}$$

where the distribution of velocity $u^+(y^+)$ is defined by formula (4.113).

Equation (6.165) is used to determine q_i^*:

$$\dot{q}_i^* = \frac{R_{i-1}}{R_i} \dot{q}_{i-1}^* + \frac{\Delta R}{R_i} \left(R_{i-1} \frac{u_{i-1}^+}{u_m^+} + R_i \frac{u_i^+}{u_m^+} \right), \quad i = 2, \ldots, n, \tag{6.167}$$

where $\dot{q}_1^* = 0$.

Equation (6.150) is approximated in a similar way:

$$\frac{T_{2,i}^* - T_{2,i-1}^*}{\Delta R} = \frac{1}{2}\left[\frac{\dot{q}_i^*}{1+\frac{Pr}{Pr_{t,i}}\left(\frac{\varepsilon_\tau}{\nu}\right)_i} + \frac{\dot{q}_{i-1}^*}{1+\frac{Pr}{Pr_{t,i-1}}\left(\frac{\varepsilon_\tau}{\nu}\right)_{i-1}}\right], \quad i=n,\ldots,2. \quad (6.168)$$

Solving Eq. (6.168) with respect to $T_{2,i}^*$, the following is obtained:

$$T_{2,i-1}^* = T_{2,i}^* - \frac{\Delta R}{2}\left[\frac{\dot{q}_i^*}{1+\frac{Pr}{Pr_{t,i}}\left(\frac{\varepsilon_\tau}{\nu}\right)_i} + \frac{\dot{q}_{i-1}^*}{1+\frac{Pr}{Pr_{t,i-1}}\left(\frac{\varepsilon_\tau}{\nu}\right)_{i-1}}\right], \quad i=n,\ldots,2,$$

$$(6.169)$$

where

$$T_{2,n}^* = T_{2w}^*. \quad (6.170)$$

Temperature T_{2w}^* is selected to satisfy condition (6.160). Temperature T_{2w}^* can be determined by means of one of the many methods used to solve nonlinear algebraic equations with one unknown, e.g. the interval search method, the bisection (interval halving) method or the secant method. Mass-averaged temperature T_{2m}^* is found numerically from the following formula:

$$T_{2m}^* = \frac{2\Delta r}{u_m^+}\sum_{i=2}^n \frac{\left(T_{2,i-1}^* u_{i-1}^+ R_{i-1} + T_{2,i}^* u_i^+ R_i\right)}{2}. \quad (6.171)$$

Considering the heat transfer coefficient definition (6.145), the Nusselt number can be determined using the following expression:

$$Nu = \frac{2}{T_2^*|_{R=1} - T_{2m}^*} = \frac{2}{T_{2w}^*}. \quad (6.172)$$

Both the integral and the differential formulation give practically identical results. Tables 6.1 and 6.2 present a comparison of the Nusselt numbers determined using the integral and the differential formulation, respectively. In both cases, the Nusselt number is calculated from formula (6.172). The integration interval is divided into $n = 30,001$ subintervals.

Adopting the integral formulation, the temperature distribution and the mean temperature are determined using formulas (6.135) and (6.137), respectively. Practically identical values of the Nusselt number are obtained from formula (6.149).

Table 6.1 Comparison of the Nusselt numbers determined using the integral formulation and the Lyon integral; integral method $n = 30001$

	T_{2w}^*	T_{2m}^*	$\mathrm{Nu} = 2/(T_{2w}^* - T_{2m}^*)$	Lyon integral
Pr = 3 Re = 5000	5.622134×10^{-2}	-9.239543×10^{-6}	35.5678	35.5684
Pr = 3 Re = 10000	3.307378×10^{-2}	3.133279×10^{-7}	60.4714	60.4735
Pr = 3 Re = 50000	9.065129×10^{-3}	7.743707×10^{-7}	220.6445	220.6500
Pr = 3 Re = 100000	5.104993×10^{-3}	9.302475×10^{-7}	391.8447	391.8480
Pr = 3 Re = 300000	2.024791×10^{-3}	9.732795×10^{-9}	987.6292	987.6762
Pr = 3 Re = 1000000	7.233373×10^{-4}	6.569460×10^{-7}	2764.5250	2764.6250

Table 6.2 Comparison of the Nusselt numbers determined using the differential formulation and the Lyon integral; difference method $n = 30001$

	T_{2w}^*	T_{2m}^*	$\mathrm{Nu} = 2/(T_{2w}^* - T_{2m}^*)$	Lyon integral
Pr = 3 Re = 5000	5.623115×10^{-2}	9.110954×10^{-7}	35.56805	35.5684
Pr = 3 Re = 10000	3.307451×10^{-2}	8.060670×10^{-7}	60.4710	60.4735
Pr = 3 Re = 50000	9.065129×10^{-3}	7.881067×10^{-7}	220.6448	220.6500
Pr = 3 Re = 100000	5.104993×10^{-3}	7.876159×10^{-7}	391.8337	391.8480
Pr = 3 Re = 300000	2.025292×10^{-3}	2.397983×10^{-7}	987.6291	987.6762
Pr = 3 Re = 1000000	7.235433×10^{-4}	9.156552×10^{-8}	2764.5240	2764.6250

In the case of the differential formulation, temperatures T_{2w}^* and T_m^* in formula (6.172) are calculated by means of the formulae derived in Sect. 6.3.2.

The obtained Nusselt numbers are compared to the Nu values determined by means of the Lyon integral (6.152).

6.3.3 Distributions of the Fluid Flow Velocity, Heat Flux and Temperature

The fluid heat flux and temperature distributions depending on the dimensionless radius are found using the formulae presented in Sect. 6.3.2, which are obtained by

means of the finite difference method. Velocity profile $\bar{w}_x/w_m = (\bar{w}_x/u_\tau)/$
$(w_m/u_\tau) = u^+/u_m^+$ is determined from the Reichardt formula (4.114), and mean
velocity u_m^+ is found using formula (6.55) and calculating the integral by means of
the trapezoidal rule using formula (6.56). The calculation results are presented in
Figs. 6.11, 6.12 and 6.13.

The analysis of the results presented in Figs. 6.11, 6.12 and 6.13 indicates that
the distribution of heat flux $\dot{q}/\dot{q}_w$ as a function of dimensionless radius R is a
nonlinear function. For Re = 10000, dimensionless heat flux $\dot{q}/\dot{q}_w$ near the tube
wall rises to a value that slightly exceeds 1. For the laminar flow, the rise in heat
flux $\dot{q}/\dot{q}_w$ over one is even higher. This can be accounted for by the fluid small
velocity near the wall for the laminar and the turbulent flow at small values of the
Reynolds number. If the fluid was stationary, the heat transfer would proceed by
means of heat conduction. An increment in the heat flux would then occur due to
the decrease in the heat transfer surface area. If the fluid velocity near the wall rises,
which is the case for higher Reynolds numbers, heat is collected from the boundary
layer by the fluid flowing with a bigger velocity and the heat flux in the radial
direction decreases. For Re = 100000, as well as for the Reynolds number higher
values, no increase in heat flux $\dot{q}/\dot{q}_w$ to a value exceeding 1 is observed close to the
wall (cf. Fig. 6.13).

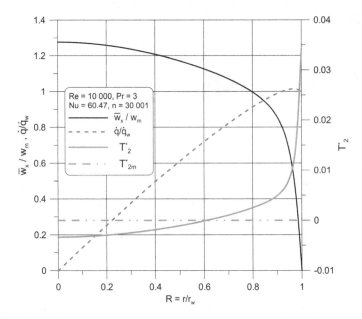

Fig. 6.11 Distribution of the fluid dimensionless velocity $\bar{w}_x/w_m$, heat flux $\dot{q}/\dot{q}_w$ and temperature
T_2^* as a function of dimensionless radius R and mean temperature T_{2m}^* for the Reynolds number
Re = 10000

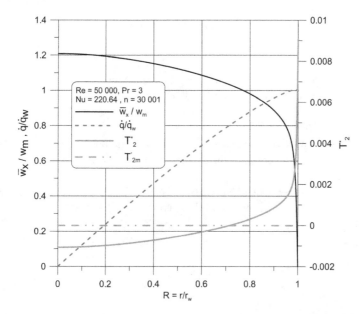

Fig. 6.12 Distribution of the fluid dimensionless velocity $\bar{w}_x/w_m$, heat flux $\dot{q}/\dot{q}_w$ and temperature T_2^* as a function of dimensionless radius R and mean temperature T_{2m}^* for the Reynolds number $Re = 50000$

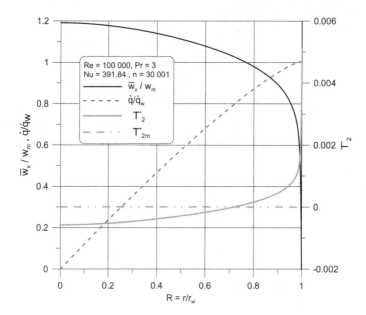

Fig. 6.13 Distribution of the fluid dimensionless velocity $\bar{w}_x/w_m$, heat flux $\dot{q}/\dot{q}_w$ and temperature T_2^* as a function of dimensionless radius R and mean temperature T_{2m}^* for the Reynolds number $Re = 100000$

6.3.4 Correlations for the Nusselt Number

The Nusselt number values are determined using the finite difference method
presented in Sect. 6.3.2. The fluid velocity distribution is calculated using the
Reichardt formula (4.114) determined experimentally. The Nusselt number
(Nu) values obtained as a function of the Reynolds number (Re) and the Prandtl
number (Pr) are listed in Table 6.3.

It should be noted that velocity u_m^+ in the u^+/u_m^+ ratio is calculated according to
the definition of the mean value in a cylindrical system of coordinates, i.e. using
formula (6.56).

The Nusselt numbers $Nu = Nu(Re, Pr)$ listed in Table 6.3 are approximated by
means of different functions (correlations) using the least squares method.
According to the least squares method [25, 261, 297–299, 310], the unknown
parameters vector $\mathbf{x} = (x_1, \dots, x_m)^{\mathrm{T}}$ is found from the following condition:

$$S(\mathbf{x}) = \sum_{i=1}^{n_{Re}} \sum_{j=1}^{n_{Pr}} \left(Nu_{ij}^m - Nu_{ij}^c \right)^2 = \min, \qquad (6.173)$$

where Nu_{ij}^m and Nu_{ij}^c, respectively, are values of the Nusselt number from Table 6.3
and those calculated according to the correlation in which the unknown coefficients
are determined, and n_{Re} and n_{Pr} denote the number of the Nusselt and the Prandtl
numbers based on which the correlation is found.

In this case, it is assumed that $n_{Re} = 10$ and $n_{Pr} = 10$, i.e. the total number of
points in Table 6.3 is 160. The Reynolds and the Prandtl numbers vary in the
following ranges: $3 \times 10^3 \le Re \le 10^6$, $0.1 \le Pr \le 1000$. It is assumed that for $3 \times
10^3 \le Re$ the fluid flow in tubes with a smooth surface is turbulent, which results
from the Moody chart presented in Fig. 6.2 and more recent studies carried out by
Ghajar [100, 101], Tam [333–335], Grote et al. [114] and Cheng Nian-Sheng [51].

The friction factors ξ calculated based on the flow velocity distribution found by
Reichardt experimentally (cf. Fig. 6.9) also indicate that for $3 \times 10^3 \le Re$ the form
of function $Nu = f(Re, Pr)$ can be adopted like for the turbulent flow. From the
research conducted by Ghajar and his team it follows that the Reynolds number
value at which the flow starts to change from laminar to transitional is affected by
the tube inlet type. For typical methods of connecting tubes to headers, the
Reynolds number value at which the flow starts to change from laminar to transitional is included in the range of $2157 \le Re \le 2524$ [334]. In heat exchangers, the
fluid flow at the tube inlet (on the tube-header interface) is characterized by a
complex distribution of velocity and temperature. It should be expected that the
change from the laminar to the turbulent flow may originate even at lower Reynolds
numbers smaller than 2300.

The unknown coefficients $x_1, \dots, x_m$ in the Nusselt number correlations are
found using the least squares method. The minimum of function (6.173) is found by
means of the Levenberg-Marquardt method [261, 297–299]. The uncertainty of the

Table 6.3 Values of the Nusselt number depending on the Reynolds and the Prandtl number for the fluid flow in a tube with a circular cross-section and determined by means of the finite difference method

Pr	Re									
	3×10^3	5×10^3	7.5×10^3	10^4	3×10^4	5×10^4	7.5×10^4	10^5	3×10^5	10^6
0.1	7.86	9.42	11.13	12.69	22.76	31.02	40.20	48.62	104.68	256.61
0.2	9.41	11.86	14.58	17.08	33.53	47.23	62.59	76.78	172.57	437.11
0.5	12.65	16.96	21.81	26.31	56.55	82.31	111.63	138.97	327.20	860.91
0.71	14.32	19.60	25.57	31.12	68.78	101.16	138.18	172.84	413.35	1102.29
1	16.22	22.61	29.86	36.61	82.91	123.04	169.15	212.50	515.41	1391.88
3	24.43	35.57	48.39	60.47	145.31	220.65	308.45	391.85	987.68	2764.63
5	29.56	43.64	59.94	75.36	184.68	282.64	397.43	506.89	1295.71	3677.45
7.5	34.34	51.15	70.68	89.21	221.38	340.53	480.68	614.66	1586.18	4544.73
10	38.16	57.14	79.24	100.25	250.64	386.73	547.14	700.78	1818.99	5242.89
12.5	41.40	62.20	86.47	109.56	275.31	425.71	603.25	773.51	2015.88	5834.61
15	44.23	66.62	92.78	117.70	296.85	459.72	652.22	836.98	2187.81	6352.24
30	56.73	86.09	120.53	153.40	391.25	608.73	866.75	1115.04	2941.80	8624.88
50	67.98	103.56	145.36	185.32	475.39	741.46	1057.70	1362.53	3612.61	10647.89
100	86.64	132.44	186.35	237.96	613.70	959.44	1371.11	1768.51	4711.28	13959.35
200	110.11	168.69	237.71	303.83	786.27	1231.01	1761.37	2273.65	6076.39	18067.10
1000	190.70	292.75	413.17	528.61	1373.01	2153.24	3085.03	3986.35	10692.06	31968.41

found coefficients is estimated using 95% confidence intervals determined using the formulae presented in [296–299, 325]. The mean-square error value is calculated from the following formula:

$$s_t = \sqrt{\frac{S_{min}}{n - m}},$$

where $n = n_{Re} n_{Pr}$ is the number of data (cf. Table 6.3) and m is the number of determined parameters.

6.3.4.1 Correlations for the Nusselt Number for the Turbulent Flow

Table 6.4 presents own formulae (correlations) (6.174)–(6.177) determined based on the data from Table 6.3 and the formulae frequently given in existing literature on the heat transfer phenomenon.

The formulae presented in Table 6.4 are valid for long ducts, when $L/d_w \to \infty$, where L is the duct length and d_w is the inner diameter with a circular cross-section or an equivalent hydraulic diameter for ducts with a cross-section other than circular. The temperature-related changes in the fluid viscosity can also be taken into account by calculating the Nusselt number Nu_L from the following formula [104–107]:

Table 6.4 Formulae for the Nusselt number as a function of the Reynolds and the Prandtl numbers

Curve number	Correlation	Validity range
1	$Nu = \dfrac{\frac{\xi}{8} Re\, Pr^{x_1}}{x_2 + x_3 \sqrt{\frac{\xi}{8}}\left(Pr^{2/3}-1\right)}$, (6.174) where: $\xi = \left(1.82 \log Re - 1.64\right)^{-2}$	$3 \times 10^3 \le Re \le 10^6$ $0.1 \le Pr \le 1000$ $S_{min} = 29555.9$ $s_t = 13.72$ $r^2 = 0.999984$ $x_1 = 1.0085 \pm 0.0050$ $x_2 = 1.0760 \pm 0.0089$ $x_3 = 12.4751 \pm 0.0080$
2	$Nu = \dfrac{\frac{\xi}{8} Re\, Pr}{x_1 + x_2 \sqrt{\frac{\xi}{8}}\left(Pr^{2/3}-1\right)}$, (6.175) where: $\xi = \left(1.82 \log Re - 1.64\right)^{-2}$	$3 \times 10^3 \le Re \le 10^6$ $0.1 \le Pr \le 1000$ $S_{min} = 99360.5$ $s_t = 25.08$ $r^2 = 0.999977$ $x_1 = 1.1252 \pm 0.0094$ $x_2 = 11.7826 \pm 0.0160$
3	$Nu = \dfrac{\frac{\xi}{8} Re\, Pr}{1 + x_1 \sqrt{\frac{\xi}{8}}\left(Pr^{2/3}-1\right)}$, (6.176) where: $\xi = \left(1.82 \log Re - 1.64\right)^{-2}$	$3 \times 10^3 \le Re \le 10^6$ $0.1 \le Pr \le 1000$ $S_{min} = 573698.4$ $s_t = 60.07$ $r^2 = 0.9996963$ $x_1 = 11.8927 \pm 0.0329$

(continued)

Table 6.4 (continued)

Curve number	Correlation	Validity range
4A	$\mathrm{Nu} = x_1 \mathrm{Re}^{x_2} \mathrm{Pr}^{x_3}$ (6.177)	$3 \times 10^3 \leq \mathrm{Re} \leq 10^6$ $3 \leq \mathrm{Pr} \leq 1000$ and $3 \times 10^3 \leq \mathrm{Re} \leq 2 \times 10^4$ $0.1 \leq \mathrm{Pr} \leq 3$ $S_{\min} = 6024118$ $s_t = 195.88$ $r^2 = 0.99683102$ $x_1 = 0.00881 \pm 0.0018$ $x_2 = 0.8991 \pm 0.0152$ $x_3 = 0.3911 \pm 0.0052$
4b	$\mathrm{Nu} = x_1 \mathrm{Re}^{x_2} \mathrm{Pr}^{x_3}$ (6.178)	$3 \times 10^3 \leq \mathrm{Re} \leq 10^6$ $0.1 \leq \mathrm{Pr} \leq 1$ $S_{\min} = 1704.98$ $s_t = 36.28$ $r^2 = 0.99955$ $x_1 = 0.02155 \pm 0.0024$ $x_2 = 0.8018 \pm 0.0083$ $x_3 = 0.7095 \pm 0.0115$
4c	$\mathrm{Nu} = x_1 \mathrm{Re}^{x_2} \mathrm{Pr}^{x_3}$ (6.179)	$3 \times 10^3 \leq \mathrm{Re} \leq 10^6$ $1 \leq \mathrm{Pr} \leq 3$ $S_{\min} = 1057.95$ $s_t = 62.23$ $r^2 = 0.99987$ $x_1 = 0.01253 \pm 0.0014$ $x_2 = 0.8413 \pm 0.0081$ $x_3 = 0.6179 \pm 0.0112$
5	$\mathrm{Nu} = 0.023 \mathrm{Re}^{0.8} \mathrm{Pr}^{0.3}$ (6.180)	Cooling $10^4 \leq \mathrm{Re} \leq 10^7$ $0.5 \leq \mathrm{Pr} \leq 120$
6	$\mathrm{Nu} = 0.023 \mathrm{Re}^{0.8} \mathrm{Pr}^{0.4}$ (6.181)	Heating $10^4 \leq \mathrm{Re} \leq 10^7$ $0.5 \leq \mathrm{Pr} \leq 120$
7	$\mathrm{Nu} = \dfrac{\frac{\xi}{8}\mathrm{Re}\,\mathrm{Pr}}{1.07 + 12.7\sqrt{\frac{\xi}{8}}\left(\mathrm{Pr}^{2/3} - 1\right)}$, (6.182) where: $\xi = \left(1.82 \log \mathrm{Re} - 1.64\right)^{-2}$	$10^4 \leq \mathrm{Re} \leq 5 \times 10^6$ $0.1 \leq \mathrm{Pr} \leq 2000$
8	$\mathrm{Nu} = \dfrac{\frac{\xi}{8}\mathrm{Re}\,\mathrm{Pr}}{1 + 12.7\sqrt{\frac{\xi}{8}}\left(\mathrm{Pr}^{2/3} - 1\right)}$, (6.183) where: $\xi = \left(1.8 \log \mathrm{Re} - 1.5\right)^{-2}$	$10^4 \leq \mathrm{Re} \leq 5 \times 10^6$ $0.1 \leq \mathrm{Pr} \leq 2000$
9	$\mathrm{Nu} = 0.1576 \frac{\xi}{8} \mathrm{Re}\,\mathrm{Pr}^{1.0685}$, (6.184) where: $\xi = 8\left(0.000695 + 0.054 \mathrm{Re}^{-0.308}\right)$	$13000 \leq \mathrm{Re} \leq 111000$ $\mathrm{Pr} = 8$
10	$\mathrm{Nu} = \dfrac{\frac{\xi}{8}\left(\mathrm{Re} - 1000\right)\mathrm{Pr}}{1 + 12.7\sqrt{\frac{\xi}{8}}\left(\mathrm{Pr}^{2/3} - 1\right)}$, (6.185) where: $\xi = \left(1.82 \log \mathrm{Re} - 1.64\right)^{-2}$	$4000 \leq \mathrm{Re} \leq 10^6$ $0.1 \leq \mathrm{Pr} \leq 1000$

$$\mathrm{Nu}_L = \mathrm{Nu}\left[1 + \left(\frac{d_w}{L}\right)^{2/3}\right]K, \qquad (6.186)$$

where Nu is calculated using one of the correlations listed in Table 6.4.

Factor K takes account of the changes in the fluid physical properties with temperature and is defined as:

$$K = \left(\frac{\mathrm{Pr}}{\mathrm{Pr}_w}\right)^{0.11} \text{ for liquids} \qquad (6.187)$$

and

$$K = \begin{cases} (T/T_w)^{0.45} & \text{for gas being heated} \\ 1 & \text{for gas being cooled,} \end{cases} \qquad (6.188)$$

where the fluid mass-averaged temperature T and the wall temperature T_w are expressed in Kelvins.

Pr_w stands for the Prandtl number in the wall temperature T_w. All the fluid physical properties while calculating Nu, Re, Pr and ξ in the formulae listed in Table 6.4 are found in the fluid mean temperature $T_m = (T_{m1} + T_{m2})/2$, where T_{m1} and T_{m2} are the fluid mass-averaged temperature at the duct inlet and outlet, respectively.

Figures 6.14 and 6.15, respectively, present changes in function 1 and function 2 from Table 6.4.

The red points in Figs. 6.14 and 6.15 mean that they are not approximated well by the adopted function Nu(Re, Pr). It can be seen that such points appear for Re = 1000000 and Pr = 0.1.

Table 6.4 also includes the Dittus-Boelter correlations defined by formulae (6.178) and (6.179) (functions 5 and 6). Function 7 is the known correlation

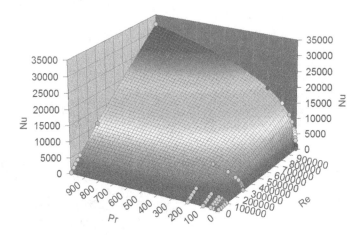

Fig. 6.14 Approximation of data from Table 6.3 by means of function 1, defined by formula (6.174); the points represent data from Table 6.3 and the surface presents function (6.174)

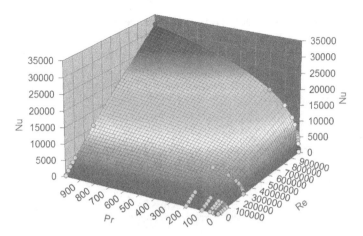

Fig. 6.15 Approximation of data from Table 6.3 by means of function 2, defined by formula (6.175); the points represent data from Table 6.3 and the surface presents function (6.175)

developed by Petukhov and Kirillov [23, 160, 226–230]. Petukhov and Popov [230] present modifications of the formula which are slightly more accurate but more complex. Correlation 8, where the number 1.07 in the Petukhov-Kirillov formula (6.180) is replaced by 1, is used by Gnielinski [104–109]. Formula (6.182) is proposed by Allen and Eckert for water based on experimental data if the Prandtl number is 8 [8]. The formula simplified by Gnielinski (6.181) was modified by him further in 1975 [105] to extend its validity for the transitional flow region (correlation 10 in Table 6.3).

It should be noted that the Petukhov-Kirillov formula (6.180) results from the Nusselt number approximations calculated from the Lyon integral [186].

Finding the Nusselt number by means of the Lyon integral, Petukhov and Kirillov adopted the following simplifications:

– the turbulent-to-molecular viscosity ratio is expressed by a single formula (4.104) in the entire range of y^+ under analysis:

$$\frac{\varepsilon_\tau}{\nu} = 0.4\left[y^+ - 11\tanh\left(\frac{y^+}{11}\right)\right],$$

– the mean velocity w_m-to-friction velocity u_τ ratio is defined by formula (6.46), i.e.:

$$u_m^+ = \frac{w_m}{u_\tau} = \sqrt{\frac{8}{\xi}},$$

where friction factor ξ is defined by the Filonenko formula (6.67).

The effect of the last assumption is that the Nu values determined using the Lyon integral differ from the Nusselt numbers found by solving the energy conservation equation for the turbulent flow. In order to obtain the same results in both cases, the mean velocity in the Lyon integral should be calculated from formula (6.54), i.e. according to the definition of the mean value in a cylindrical system of coordinates.

Table 6.5 presents a comparison between the Nusselt number values found herein by means of the finite difference method and the Nusselt numbers determined according to the procedure applied by Petukhov and Kirillov in [228].

The top values in individual rows present the Nusselt numbers determined from the Lyon integral and the bottom ones—the Nusselt numbers obtained by solving the energy conservation equation by means of the finite difference method.

Relative error e of the Nusselt number determination is found using the following formula:

$$e = \frac{\mathrm{Nu}_L - \mathrm{Nu}_d}{\mathrm{Nu}_d} 100\%, \tag{6.189}$$

where Nu_L and Nu_d are the Nusselt numbers determined using the Lyon integral [228] and the finite difference method, respectively.

An analysis of the results presented in Table 6.5 indicates that relative errors e range from -5.96 to 3.91%.

A comparison between correlations (6.182) and experimental testing [228] performed for the Reynolds numbers $4 \times 10^3 \leq \mathrm{Re} \leq 6 \times 10^5$ and for the Prandtl numbers $0.7 \leq \mathrm{Pr}$ points to very good agreement between the results obtained from correlation (6.182) and measurement data. Only for $\mathrm{Re} \leq 2 \times 10^4$ and $0.7 \leq \mathrm{Pr} \leq 1$ $0.7 \leq \mathrm{Pr} \leq 1$ are the obtained Nusselt numbers by 6–8% lower compared to the values found experimentally. For this reason, the following modification of for mula (6.182) is put forward in [227, 229]:

$$\mathrm{Nu} = \frac{\frac{\xi}{8}\mathrm{Re}\,\mathrm{Pr}}{1 + \frac{900}{\mathrm{Re}} + 12.7\sqrt{\frac{\xi}{8}}(\mathrm{Pr}^{2/3} - 1)}, \tag{6.190}$$

$$4 \times 10^3 \leq \mathrm{Re} \leq 5 \times 10^6, \quad 0.5 \leq \mathrm{Pr} \leq 2000,$$

where $\xi = (1.82 \log \mathrm{Re} - 1.64)^{-2}$.

Analysing the results presented in Fig. 6.16a, b, it may be concluded that the biggest differences between individual formulae occur for air. Especially the exponential expressions differ from the others. Compared to the other formulae, exponential correlation 4a gives lower Nusselt numbers for the Reynolds number values included in the range of $3000 \leq \mathrm{Re} \leq 20000$ (cf. Fig. 6.16b). However, in this range of the Reynolds number, formula 4a demonstrates very good agreement with experimental data. Analysing the results presented in Fig. 6.16a, b, it can be seen that for the Prandtl number $\mathrm{Pr} = 0.7$ the exponential correlation overestimates the Nusselt number values for $\mathrm{Re} > 20000$.

A comparison of own correlations 1–3 and 4a with the Petukhov and Kirillov correlation 7 and correlation 8 is presented in Table 6.6.

Table 6.5 Comparison between the Nusselt numbers determined using the finite difference method and the Lyon integral [228]

Pr	Re 3×10^3		10^4		10^5		10^6	
0.71	14.88	$e = 3.91\%$	31.32	$e = 0.64\%$	172.06	$e = -0.45\%$	1088.19	$e = -1.28\%$
	14.32		31.12		172.84		1102.29	
5	30.21	$e = 2.20\%$	72.97	$e = -3.17\%$	487.10	$e = -3.90\%$	3539.78	$e = -3.74\%$
	29.56		75.36		506.89		3677.45	
50	69.03	$e = 1.54\%$	176.40	$e = -4.81\%$	1286.20	$e = -5.60\%$	10096.92	$e = -5.17\%$
	67.98		185.32		1362.53		10647.89	
100	87.91	$e = 1.47\%$	226.05	$e = -5.01\%$	1665.65	$e = -5.82\%$	13208.82	$e = -5.38\%$
	86.64		237.96		1768.51		13959.35	
200	111.67	$e = 1.42\%$	288.25	$e = -5.13\%$	2138.16	$e = -5.96\%$	17072.88	$e = -5.50\%$
	110.11		303.83		2273.65		18067.10	

Fig. 6.16 Comparison of selected Nu = f(Re, Pr) correlations from Table 6.4 in a linear-logarithmic system of coordinates (**a**), and in a double logarithmic system (**b**); the curve numbers in the figures and in Table 6.4 are the same

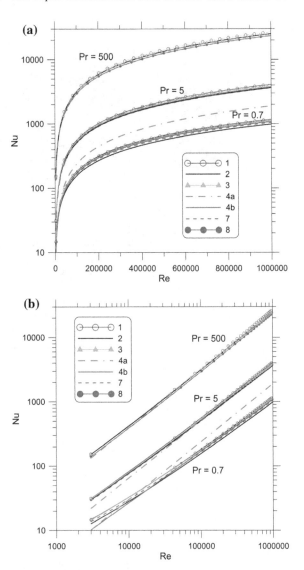

Symbol ε is the relative deviation from the mean value calculated from the following formula:

$$\varepsilon = \frac{Nu_i - Nu_m}{Nu_m} \cdot 100\%, \qquad (6.191)$$

where the Nusselt number mean value is found from the following expression:

Table 6.6 Comparison between the Nusselt numbers calculated using own formulae, the Petukhov-Kirillov formula and its modification applied by Gnielinski

Re	Correlation number in Table 6.4	Pr = 0.7		Pr = 5		Pr = 500	
		Nu	ε, %	Nu	ε, %	Nu	ε, %
3000	1	14.58	7.92	31.64	8.84	153.72	5.71
	2	12.74	−5.70	30.09	3.51	151.75	4.35
	3	14.74	9.10	31.30	7.67	150.71	3.64
	4a	10.25	−24.13	22.11	−23.94	133.91	−7.92
	7	13.77	1.92	29.29	0.76	141.12	−2.96
	8	14.98	10.88	30.01	3.23	141.29	−2.84
10000	1	32.67	4.51	82.34	7.72	428.75	5.48
	2	28.39	−9.18	77.16	0.94	418.79	3.03
	3	32.66	4.48	80.71	5.59	416.10	2.37
	4a	30.25	−3.23	65.27	−14.61	395.31	−2.74
	7	30.51	−2.40	75.52	−1.20	389.64	−4.14
	8	33.08	5.82	77.61	1.53	390.18	−4.01
100000	1	182.35	−0.88	559.52	6.55	3282.56	6.24
	2	156.13	−15.13	510.57	−2.77	3141.88	1.68
	3	178.52	−2.96	538.77	2.60	3124.47	1.12
	4a	239.82	30.36	517.42	−1.47	3133.60	1.41
	7	166.80	−9.33	504.05	−4.01	2925.77	−5.31
	8	180.18	−2.06	520.40	−0.90	2931.12	−5.14
1000000	1	1171.44	−4.56	4100.46	6.44	26723.50	8.08
	2	986.26	−19.65	3649.00	−5.28	25064.50	1.37
	3	1123.80	−8.44	3877.30	0.65	24947.70	0.90
	4a	1901.05	54.88	4101.50	6.47	24839.51	0.46
	7	1050.06	−14.45	3627.03	−5.85	23360.80	−5.52
	8	1131.95	−7.78	3758.51	−2.43	23413.60	−5.30

$$Nu_m = \frac{1}{6} \sum_{i=1}^{6} Nu_i.$$

Analysing the results presented in Table 6.6, it can be seen that the biggest deviation from the other correlations is demonstrated by formula **4a** for Pr = 0.7. For Re = 3000, deviation ε totals −24.13%, and for Re = 1000000 it is even bigger and totals 54.88%. In the case of liquids if Pr = 5, the agreement between correlation **4a** from Table 6.6 and the other correlations is very good. In the correlation proposed by Li and Xuan [179, 373]

$$Nu = 0.0059 Re^{0.9238} Pr^{0.4} \tag{6.192}$$

the exponent at the Reynolds number, like in correlation **4a**, also totals about 0.9.

The comparison between selected correlations from Table 6.4 is presented in Figs. 6.17, 6.18 and 6.19.

Analysing the results presented in Figs. 6.17, 6.18 and 6.19, it can be seen that the popular Dittus-Boelter correlations (6.180) and (6.181) [22, 160, 182] differ from the other correlations the most, showing deviation of the order of a few dozen per cent. For air (Pr = 0.71), the Dittus-Boelter correlations give higher values of the Nusselt number, and for liquids (Pr = 8 and Pr = 500) the obtained values are much lower. Especially big differences can be observed for higher values of the

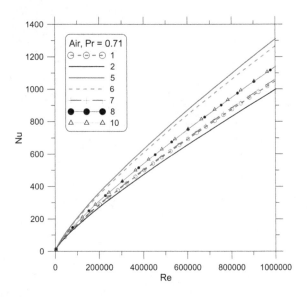

Fig. 6.17 Comparison of selected Nu = f(Re, Pr) correlations from Table 6.4 for air (Pr = 0.71)—the curve numbers in the figure and in Table 6.4 are the same

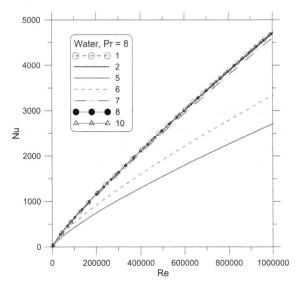

Fig. 6.18 Comparison of selected Nu = f(Re, Pr) correlations from Table 6.4 for water (Pr = 8)—the curve numbers in the figure and in Table 6.4 are the same

Fig. 6.19 Comparison of selected Nu = f(Re, Pr) correlations from Table 6.4 for oil (Pr = 500)—the curve numbers in the figure and in Table 6.4 are the same

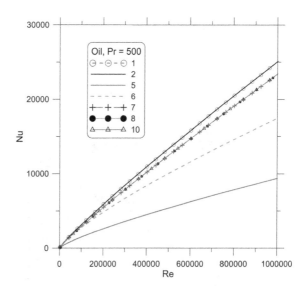

Reynolds number when Re $\geq$ 100000. For this reason, new and more accurate exponential correlations **4b** and **4c** are proposed (cf. Table 6.4). One essential advantage of exponential formulae is their simple form. Correlation **4b** defined by formula (6.178) is found using the least squares method based on 50 pieces of data from Table 6.3. The Prandtl and the Reynolds number values varied in the range of $0.1 \leq Pr \leq 1$ and $3000 \leq Re \leq 1000000$, respectively. Formula **4b** holds not only for air but for other gases as well. Figure 6.16a, b indicate that the correlation shows good agreement with correlations **1-3** and **7**.

A similar procedure is used to determine correlation **4c**, which holds for the following ranges of the Prandtl and the Reynolds numbers, respectively: $1 < Pr \leq 3$ and $3000 \leq Re \leq 1000000$. The coefficient and the exponents in formula **4c** are found based on 20 pieces of data from Table 6.3 for Pr = 1 and Pr = 3 for the Reynolds numbers included in the range of $3000 \leq Re \leq 1000000$. For the Prandtl numbers higher than 3, formula **4a** can be used because based on 100 pieces of data from Table 6.3 included in the range of $3 \leq Pr \leq 1000$ and $3000 \leq Re \leq 1000000$, an almost identical formula is obtained.

Figure 6.20 shows a comparison between selected correlations from Table 6.4 and the Allen and Eckert formula (6.184) based on experimental data and presented in [8].

The comparison presented in Fig. 6.20 shows very good agreement between correlation **9** (Allen and Eckert) and own correlations **1** and **2**. Also the Petukhov-Kirillov correlation **7** and the Gnielinski correlations **8** and **10** are close to the Nusselt number values found using the Allen and Eckert formula. Only the Dittus-Boelter correlations demonstrate considerable deviation from the others.

Figure 6.21 presents a comparison of correlations **2** and **4a** with the experimental data obtained by Li Xiao-Wei et al. [180–181] for a tube with inner diameter

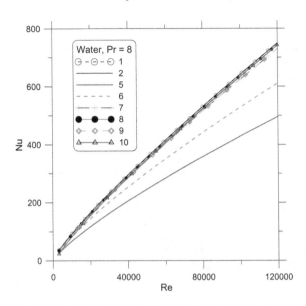

Fig. 6.20 Comparison of selected Nu = f(Re, Pr) correlations from Table 6.4 for water (Pr = 8); curve **9** represents the Allen and Eckert formula (6.184) determined based on experimental data; numbers in the figure and in Table 6.4 are the same

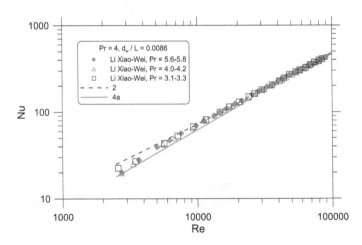

Fig. 6.21 Comparison between correlations **2** and **4a** from Table 6.4 for water and the experimental data obtained by Li Xiao-Wei et al. [180–181] for Pr = 4 and d_w/L = 0.0086

d_w = 17 mm and the tube inner diameter-to-length L ratio d_w/L = 0.0086. The experimental tests were performed for three different inlet temperatures of water flowing through the tube: 30, 45 and 60 °C. The Reynolds number varied from 2500 to 90000. Correlation **4a** provides a very good approximation of the experimental data, whereas correlation **2** slightly overestimates the Nusselt number values for small Reynolds numbers included in the range of $2500 \leq Re \leq 7000$ (cf. Fig. 6.21).

Figure 6.22 presents a comparison of correlation **4a** from Table 6.4 with the experimental data obtained by Olivier and Meyer [220]. The experiments were performed for tubes with the diameter of 15.88 mm (5/8″) and 19.02 mm (3/4″) and the length of about 5 m. The inner diameters of the tubes were 14.482 and 17.6 mm, respectively. The tested tubes, with hot distilled water flowing inside, were the inner tubes of a countercurrent tube-in-tube heat exchanger. Cold distilled water flowed through the ring channel between the inner and the outer tube. The tests were performed for four different conditions at the inlet of the inner tube under analysis. In the first case, marked FD (fully developed), the water flow was hydrodynamically developed, i.e. the velocity profile was fully developed. The other three cases differ in the structure of the inlet through which the fluid flows from the tank or chamber into the tube. In the most common connection, the tube is welded to the tank wall and the fluid flows into the tube through a sharp-edged hole with a diameter the same as the tube inner diameter. This inlet type is referred to as square-edged (SE) [100, 333–335].

Two other design solutions of the tube inlet were also investigated.

In one of them, the tube was inserted into a cylindrical chamber and its end was still in the region of the liquid. This inlet type is referred to as "re-entrant". In the other solution, referred to as "bellmouth", the liquid is introduced into the tube through a transitional section tapering gently. The results of the testing conducted by Ghajar et al. and Olivier and Meyer indicate that the onset of the change in the flow character from laminar to transitional for the Reynolds number values from the range of $2100 \leq Re \leq 12000$ depends on the type of the inlet. The transition from the laminar to the turbulent flow occurs the latest, i.e. for Re = 12000, for the

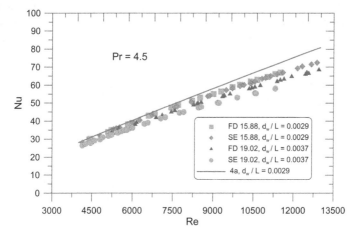

Fig. 6.22 Comparison of correlation 4a from Table 6.4 for water with the experimental data obtained by Olivier and Meyer [220] for Pr = 4.5 and $d_w/L = 0.0029$ and $d_w/L = 0.0037$

bellmouth inlet, characterized by the cross-section gentle taper. According to Ghajar and Tam, for the most common type of the tube/header interface (square-edged inlet), the transitional flow occurs for the Reynolds numbers included in the range from 2400 to 7300. Based on a survey of the existing literature results, it can be seen that the transitional flow occurs generally in the Reynolds number range from $Re = 2100$ to $Re = 20000$. This is also confirmed by the testing of finned-tube heat exchangers [68, 203, 204, 304, 306, 321]. Mirth and Ramadhyani [203, 204] found that for the water-side Reynolds numbers included in the range of $2300 \leq Re \leq 10000$, the real power of the finned-tube heat exchanger was by about 20% smaller compared to the value obtained from calculations where the heat transfer coefficient on the water side was determined using the Dittus-Boelter formula. It follows both from own studies [306, 307, 309] and from the testing performed by Cuevas et al. [68] that the heat transfer coefficients on the side of the liquid in the analysed car radiators were by 20–40% smaller compared to the values calculated from the Gnielinski formula (6.185).

Comparing correlation (6.177) with the experimental data presented by Olivier and Meyer [220], it can be seen that correlation (6.177) gives a good approximation of experimental data in the range of the Reynolds numbers from 4000 to 13000, despite the fact that compared to other correlations (cf. Fig. 6.16b), it underestimates the Nusselt number values in the range of the Reynolds number smaller values.

Attention should also be drawn to the experimental data uncertainty resulting from the method of the measurement data development. Thermal conductivity k was determined based on the testing of the countercurrent tube-in-tube heat exchanger and then, after the heat transfer coefficient on the outer surface of the inner tube was found, the heat transfer coefficient on the tube inner surface was calculated. Finding the heat transfer coefficient in this way is burdened with the error in the heat transfer coefficient determination on the tube outer surface on the ring channel side. The method of the heat transfer coefficient determination on one side of the wall based on the knowledge of the heat transfer coefficient on the wall other side is discussed in [299]. Although it is commonly used by many researchers, the heat transfer coefficient correlations found in this way can be burdened with rather big errors.

In this case, the heat transfer coefficient α_o on the inner tube outer surface was determined experimentally using the following formula:

$$\alpha_o = \frac{\dot{Q}_i}{A_o(T_o - T_{wo})}, \tag{6.193}$$

where: $\dot{Q}_i$—hot-to-cold water heat flow rate, W, A_o—inner tube outer surface area, m^2, T_o—mean temperature on the inner tube outer surface found based on the temperature measurement on the tube length and perimeter, T_{wo}—mean temperature of cold water in the ring channel.

The main cause of the uncertainty of determination of the heat transfer coefficient α_o is determination of water temperature T_{wo}, which, according to the heat transfer coefficient definition, should be the water temperature mass-averaged on the tube length.

In [220], the temperature is found as the mean of the cold water temperatures measured at the exchanger inlet and outlet and five temperatures indicated by thermocouples fixed on the outer tube outer surface along the tube length. Considering the changes in the water velocity and temperature in the radial direction in the ring channel, this is not the most correct method of finding temperature T_{wo}. Taking account of the fact that the heat transfer coefficient α_o is high due to the high velocity of the water flow in the ring channel, the errors in the coefficient determination are carried onto the error in the obtained values of the heat transfer coefficient and—consequently—of the Nusselt number on the inner tube inner surface.

Despite the uncertainties inherent in the measurement data and correlation **4a** (6.177), the agreement between experimental data and calculated values is very good (cf. Fig. 6.22).

Figure 6.23 presents a comparison between correlations **2** (6.175), **4a** (6.177) and **4b** (6.178) from Table 6.4 and the results obtained by American researchers [30, 150, 173, 360].

It can be seen that correlations **2** and **4b** approximate experimental data very well. The inclination angle of the line representing correlation 2 is slightly smaller compared to the line representing experimental data. Correlation **4a** provides a good approximation of experimental data only for the Reynolds number values smaller than 10000.

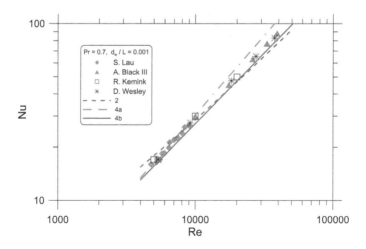

Fig. 6.23 Comparison between correlations **2**, **4a** and **4b** from Table 6.4 and the experimental data obtained by Lau [173], Black [30], Kemink [150] and Wesley [360]

Figure 6.24 presents a comparison of correlations **4a** and **4b** with the experimental data obtained by Huber and Walter [129] for the flow of air in a tube with diameter $d_w = 32.8$ mm and length 3000 mm in the range of the Reynolds numbers from 4095 to 10400. The tests were carried out for a cocurrent tube-in-tube heat exchanger. The air flowing through the inner tube is heated by water flowing in the ring channel between the inner and the outer tube. For a given measuring point, the heat transfer coefficient was first determined for the inner tube outer surface. Then, the air-side heat transfer coefficient was found assuming that the heat transfer coefficient on the water side is known. This method of experimental determination of the heat transfer coefficient is similar to the methodology adopted by Olivier and Meyer [220]. However, it is not specified whether the heat transfer coefficient on the inner tube outer surface was calculated from the formulae available in literature or whether it was determined experimentally. As shown in [299], the uncertainty of the water-side heat transfer coefficient determination has an impact on the uncertainty of the obtained heat transfer coefficient value on the air side. Much more accurate values of the heat transfer coefficients could be obtained if the heat transfer coefficient was determined simultaneously on the air and the water side according to the method presented in [296–299, 310, 313].

Correlations **4a** and **4b** from Table 6.4 are compared with the experimental data obtained by Huber and Walter [130] in Fig. 6.24. It can be seen that the correlations provide a good approximation of experimental data; correlation **4a** gives higher values of the Nusselt number compared to exponential correlation **4b** valid for gases.

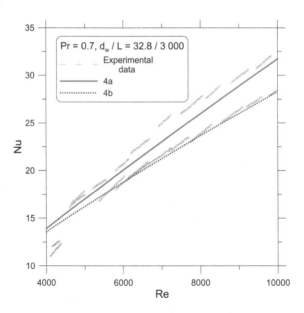

Fig. 6.24 Comparison of correlations **4a** and **4b** from Table 6.4 with the experimental data obtained by Huber and Walter [130]

Figure 6.25 presents a comparison of correlations **1, 2, 3** and **4b** from Table 6.4 with the experimental data obtained by Eiamsa-Ard et al. [86]. The measurements were performed for air flowing inside a copper tube with inner diameter $d_w = 47.5$ mm heated on the outer surface electrically. The tube heated section with length $L = 1500$ mm is preceded by a 2500 mm long unheated section. Therefore, it may be assumed that the air flow at the heated section inlet is fully developed hydrodynamically, i.e. the velocity profile is fully developed.

The tests were performed for the Reynolds number values ranging from 3500 to 20000. The mean heat transfer coefficient on the tube inner surface was found using the following formula:

$$\bar{\alpha} = \frac{\dot{m}_a c_{pa} \left(T_{a,out} - T_{a,in} \right)}{A_w \left(\bar{T}_w - \bar{T}_a \right)}, \tag{6.194}$$

where: $\dot{m}_a$—air mass flow rate, kg/s, c_{pa}—air specific heat at constant pressure, J/(kg · K), $T_{a,out}$, $T_{a,in}$—air mass-averaged temperature at the tube heated section outlet and inlet, respectively, K or °C, A_w—tube inner surface area, m^2, $\bar{T}_w$—tube wall mean temperature, K or °C, determined based on the wall temperature measurement in 15 equidistant points on the tube length, $\bar{T}_a$—air mean temperature on the tube length, K or °C, calculated from the following formula: $\bar{T}_a = \left(T_{a,out} + T_{a,in} \right)/2$.

The Nusselt number Nu is calculated from: $Nu = \bar{\alpha} d_w / \lambda_a$, where d_w is the tube inner diameter and λ_a—mean thermal conductivity for air.

Comparing the results presented in Fig. 6.25, it can be seen that correlations **2** and **4b** provide the best approximation of the experimental data presented in [86]. In this case, correlations **1** and **2** give results which are slightly higher compared to experimental testing data.

Fig. 6.25 Comparison of correlations **1, 2, 3** and **4b** from Table 6.4 with the experimental data obtained by Eiamsa-Ard et al. [86]

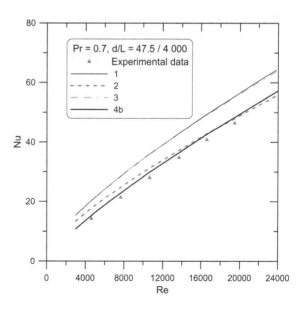

Figure 6.26 shows a comparison of correlations **1, 2, 3** and **4a** and **4b** from Table 6.4 with the experimental data presented by Buyukalaca et al. [41]. The measurements were performed for air flowing inside a copper tube with inner diameter d_w = 56 mm heated on the outer surface electrically. The tube heated section with length L = 3000 mm is preceded by a 6000 mm long unheated section. Therefore, it may be assumed that the air flow at the heated section inlet is fully developed hydrodynamically, i.e. the velocity profile is fully developed.

The tests were performed for the Reynolds number values ranging from 3115 to 24920. The local heat transfer coefficient on the tube inner surface was found using the following formula:

$$\alpha(x) = \frac{\dot{m}_a c_{pa}\left(T_{a,out} - T_{a,in}\right)}{A_w\left[\bar{T}_w(x) - T_a(x)\right]}, \tag{6.195}$$

where: $\dot{m}_a$—air mass flow rate, kg/s, c_{pa}—air specific heat at constant pressure, J/(kg·K), $T_{a,out}$, $T_{a,in}$—air mass-averaged temperature at the tube heated section outlet and inlet, respectively, K or °C, $\bar{T}_w(x)$—mean temperature of the tube inner surface on the perimeter at distance x from the tube inlet, K or °C, $T_a(x)$—air mass-averaged temperature at distance x from the tube inlet, K or °C, $A_w = \pi d_w L$—inner surface area of the tube heated section, m^2.

The tube inner surface mean temperature $\bar{T}_w(x)$ was found as the arithmetic mean of temperatures from 4 wall temperatures measured every 90° on the perimeter of the tubes. The air mass-averaged temperature $T_a(x)$ can be determined from linear interpolation of the air temperature at the tube inlet and outlet or from formula (6.99) because the air temperature changes linearly on the tube length due to the constant heat flux on the tube inner surface.

Fig. 6.26 Comparison of correlations **1, 2, 3** and **4a** and **4b** from Table 6.4 with the experimental data obtained by Buyukalaca et al. [41]

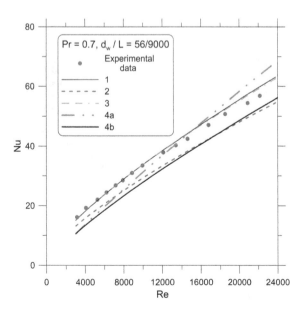

The local Nusselt number Nu(x) was calculated simultaneously in 24 points spaced uniformly on the tube length using the formula: $\text{Nu}(x) = \alpha(x)d_w/\lambda_a$, where d_w is the tube inner diameter and λ_a is the mean thermal conductivity for air. At a large distance from the tube inlet, the air velocity and temperature profiles do not change on the tube length. The Nusselt number was adopted as the arithmetic mean of the local Nusselt numbers determined in the last three points close to the tube outlet.

Analysing the results presented in Fig. 6.26, it can be seen that correlations **1** and **3** from Table 6.4 give results which are the closest to experimental data. In the case of the experimental data obtained by Eiamsa-Ard et al. [86], which are presented in Fig. 6.25, correlations **2** and **4b** provide their best approximation, whereas correlations **1** and **3** give results which are slightly overestimated compared to those obtained from experiments. Analysing the comparisons presented in Figs. 6.25 and 6.26, it can be seen that there are some differences both between the Nusselt number correlations and between the experimental data. In the case of the correlations, the differences result from the difficulties in approximating the data presented in Table 6.3 by means of one function only in the entire range of changes in the Prandtl and the Reynolds numbers. For this reason, the accuracy of a given correlation for Pr = 0.7 may be smaller than for other Prandtl number values. This can be seen clearly on the example of exponential correlations. Formula **4a** does not provide a very good approximation of the experimental data obtained by Buyukalaca et al. and presented in Fig. 6.26, and correlation **4b**, which is valid for gases only, approximates well the experimental data obtained by Eiamsa-Ard et al. and presented in Fig. 6.25.

In the case of the experimental data, the differences between the results are caused by systematic measuring errors which are common for the tube wall temperature measurement, and by the different structure of the testing stand, especially—different air conditions at the tube inlet.

Measuring the wall temperature, special care should be taken to ensure good contact between the jacket thermocouple and the tube, and to reduce measuring errors being the effect of heat conduction through the thermocouples from the temperature measuring point to the environment or of the heat flow from heaters to thermocouples. Good contact between thermocouples and the wall can be achieved by placing them in special grooves milled in the wall and soldering them to the wall, or filling the groove with the thermocouple with a ceramic-to-metal adhesive. The differences in experimental data can also be caused by natural convection occurring in the air flow.

In the transitional region, when the Reynolds numbers are small, natural convection manifesting itself in big temperature differences on the tube perimeter in a given cross-section can have a considerable impact.

Summing up the comparison of the correlations presented in Table 6.4 with the experimental data obtained by different authors, it can be seen that correlations **1** and **2** provide a very good approximation of the experimental data for water obtained by Allen and Eckert (cf. Fig. 6.20). Correlations **2** and **4a** are in good agreement with the experimental data presented by Li Xiao-Wei (cf. Fig. 6.21). Correlation **4a** also provides a very good approximation of the experimental data obtained by Olivier and Meyer (cf. Fig. 6.22).

Correlations **1-3** give similar results for different Reynolds and Prandtl numbers. Only for air, if Pr = 0.7, does correlation **2** give lower results, even compared to correlations **7** and **8**. It should be added that correlation **8** defined by formula (6.183) is a modification of the Petukhov-Kirillov formula (6.182) [226–228] used by Gnielinski [104–108]. Due to the replacement of the number 1.07 in the denominator of formula (6.182) with the number 1 used in formula (6.183), the formula modified by Gnielinski gives overestimated values of the Nusselt numbers, especially for air.

Correlations **2** and **4b** provide a very good approximation of experimental data for air for the Reynold number values included in the range of $3000 \leq Re \leq 40000$, which is confirmed by the results of experimental testing conducted for air by Petukhov, Kurganov and Gladuncov [229], Lau [173], Black [30], Kemink [150], Wesley [360] and Eiamsa-Ard et al. [86].

Correlations **1** and **3** approximate very well the experimental data obtained by Buyukalaca et al. (cf. Fig. 6.26), and correlations **2** and **4b** in this case give results which are slightly underestimated.

The comparisons made in this subsection indicate that the same correlation provides a different approximation of experimental data for the same value of the Prandtl number and the same range of changes in the Reynolds number. This is the effect of the uncertainty of experimental data obtained by means of different methods. In some works, the Nusselt number is determined as a function of the Reynolds number based on the testing of co- or countercurrent tube-in-tube heat exchangers, whereas in others—the analysed tube is heated electrically.

It also seems that some researchers tried to achieve very good agreement between their results and those obtained from calculations using formulae which in available literature are generally considered as standard procedures, e.g. the calculation results obtained from the Dittus-Boelter formula. One example of such an approach is the work of Liu and Liao [184], where very good agreement is achieved between experimental results and results of calculations performed using the Dittus-Boelter formula for water with the temperature of 22 °C (Pr = 6.62) flowing through a tube with the inner diameter of 25.6 mm. The extensive and detailed comparisons made in this subsection, such as the comparison presented in Fig. 6.18 for example, indicate that the Dittus-Boelter formula gives underestimated values of the Nusselt number for liquids.

6.3.4.2 Correlations for the Nusselt Number for the Transitional Turbulent Flow

The previous section presents a discussion of the experimental testing results obtained for the Reynolds numbers included in the range of $2300 \leq Re \leq 10000$, which is usually considered to be the transitional region. It follows from experimental testing of the heat transfer in heated tubes and from the testing of cross-flow heat exchangers that the transition from the laminar to the turbulent flow may occur at the Reynolds number values ranging from 2100 or 2300 to 12000 or to even higher Reynolds numbers. For the last dozens of years, the most frequently used correlations have been the Hausen formula [119–120]:

$$\mathrm{Nu} = 0.037\left(\mathrm{Re}^{0.75} - 180\right)\mathrm{Pr}^{0.42}\left[1 + (d_w/L)^{2/3}\right](\mu/\mu_w)^{0.14} \qquad (6.196)$$

and the Gnielinski formula:

$$\mathrm{Nu} = \frac{(\xi/8)(\mathrm{Re} - 1000)\,\mathrm{Pr}}{1 + 12.7\sqrt{\left(\frac{\xi}{8}\right)}\left(\mathrm{Pr}^{2/3} - 1\right)}\left[1 + \left(\frac{d_w}{L}\right)^{2/3}\right]\left(\frac{\mathrm{Pr}}{\mathrm{Pr}_w}\right)^{0.11}, \qquad (6.197)$$

where: d_w—the duct inner diameter, L—the duct length.

The Prandtl number Pr is calculated in the fluid mean temperature, and Pr_w—in the wall temperature. The friction factor in formula (6.197) is calculated from the Filonenko or the Konakov formula (6.67) or (6.66), respectively.

The Reynolds number lower limit at which formulae (6.196) and (6.197) can be applied is Re = 3000. If the Nusselt number values are found for Re = 2300 and Pr = 5, then for $d_w/L = 0$ and $\mu/\mu_w = 1$ or $\mathrm{Pr}/\mathrm{Pr}_w = 1$, formula (6.196) gives Nu = 11.07, and formulae (6.67) and (6.197) result in Nu = 13.83.

It can be seen that both these values of the Nusselt number are much higher than Nu = 3.66 or Nu = 4.36, i.e. than the Reynolds numbers for the fully developed laminar flow of a fluid with constant physical properties in ducts with a circular cross-section with the wall constant temperature or a constant heat flux value on the wall.

In order to ensure the Nusselt number continuity for Re = 2300, it is necessary to find appropriate correlations for the Nusselt number for the Reynolds number value Re $\geq$ 2300. Taborek [294] put forward linear interpolation of the Nusselt number in the range of $2000 \leq \mathrm{Re} \leq 8000$:

$$\mathrm{Nu} = \varepsilon \mathrm{Nu}_{m,1,2000} + (1 - \varepsilon)\mathrm{Nu}_{m,t,8000}, \qquad (6.198)$$

where

$$\varepsilon = \frac{8000 - \mathrm{Re}}{8000 - 2000} = 1.333 - \frac{\mathrm{Re}}{6000}. \qquad (6.199)$$

In formula (6.198), $\mathrm{Nu}_{m,l,2000}$ is the mean value of the Nusselt number on the duct entire length for the laminar flow for Re = 2000, and $\mathrm{Nu}_{m,t,8000}$ is the Nusselt number mean value on the entire length of the duct for the turbulent flow for Re = 8000.

Gnielinski proposed modification of the Taborek formula by linear interpolation of the Nusselt number in the range of $2300 \leq \text{Re} \leq 10000$ [105–107]:

$$\text{Nu}_m = (1 - \gamma)\text{Nu}_{m,l,2300} + \gamma\text{Nu}_{m,t,10000}, \tag{6.200}$$

where

$$\gamma = \frac{\text{Re} - 2300}{10000 - 2300}, \quad 2300 \leq \text{Re} \leq 10000. \tag{6.201}$$

The Nusselt number for the laminar flow $\text{Nu}_{m,l,2300}$ is calculated for a constant value of the heat flux on the tube surface or for a constant temperature of the tube wall.

Assuming that the heat flux is constant on the tube inner surface and the velocity and temperature profiles are still not fully developed, the Nusselt number for the laminar flow is calculated using the following formula [106–108]:

$$\text{Nu}_{m,q} = \left[\text{Nu}_{m,q,1}^3 + 0.6^3 + \left(\text{Nu}_{m,q,2} - 0.6\right)^3 + \text{Nu}_{m,q,3}^3\right]^{1/3}, \tag{6.202}$$

where

$$\text{Nu}_{m,q,1} = \frac{48}{11} = 4.364, \tag{6.203}$$

$$\text{Nu}_{m,q,2} = 1.953\left(\text{Re}\,\text{Pr}\frac{d_w}{L}\right)^{1/3}, \tag{6.204}$$

$$\text{Nu}_{m,q,3} = 0.924\text{Pr}^{1/3}\left(\text{Re}\frac{d_w}{L}\right)^{1/2}. \tag{6.205}$$

Substituting Re = 2300 in formulae (6.202)–(6.205), the Nusselt number $\text{Nu}_{m,l,2300}$ is defined as:

$$\text{Nu}_{m,l,2300} \equiv \text{Nu}_{m,q}(\text{Re} = 2300)$$
$$= \left[83.09 + 0.6^3 + \left(\text{Nu}_{m,q,2,2300} - 0.6\right)^3 + \text{Nu}_{m,q,3,2300}^3\right]^{1/3}, \tag{6.206}$$

where

$$\text{Nu}_{m,q,2,2300} \equiv \text{Nu}_{m,q,2}(\text{Re} = 2300) = 1.953\left(2300\,\text{Pr}\frac{d_w}{L}\right)^{1/3}, \tag{6.207}$$

$$\text{Nu}_{m,q,3,2300} \equiv \text{Nu}_{m,q,3} = 0.924\text{Pr}^{1/3}\left(2300\frac{d_w}{L}\right)^{1/2}. \tag{6.208}$$

Assuming that the tube inner surface temperature is constant and the velocity and temperature profiles are still not fully developed, the Nusselt number for the laminar flow is calculated using the following formula [106–107, 190]:

$$\mathrm{Nu}_{m,T} = \left[\mathrm{Nu}_{m,T,1}^3 + 0.7^3 + \left(\mathrm{Nu}_{m,T,2} - 0.7\right)^3 + \mathrm{Nu}_{m,T,3}^3\right]^{1/3}, \qquad (6.209)$$

where

$$\mathrm{Nu}_{m,T,1} = 3.658, \qquad (6.210)$$

$$\mathrm{Nu}_{m,T,2} = 1.615\left(\mathrm{Re}\,\mathrm{Pr}\,\frac{d_w}{L}\right)^{1/3}, \qquad (6.211)$$

$$\mathrm{Nu}_{m,T,3} = \left(\frac{2}{1+22\,\mathrm{Pr}}\right)^{1/6}\left(\mathrm{Re}\,\mathrm{Pr}\,\frac{d_w}{L}\right)^{1/2}. \qquad (6.212)$$

Substituting Re = 2300 in formulae (6.210)–(6.212), the Nusselt number $\mathrm{Nu}_{m,l,2300}$ for the tube inner surface constant temperature is defined as:

$$\mathrm{Nu}_{m,l,2300} \equiv \mathrm{Nu}_{m,T}(\mathrm{Re} = 2300)$$
$$= \left[48.95 + 0.7^3 + \left(\mathrm{Nu}_{m,T,2,2300} - 0.7\right)^3 + \mathrm{Nu}_{m,T,3,2300}^3\right]^{1/3}, \qquad (6.213)$$

where

$$\mathrm{Nu}_{m,T,2,2300} \equiv \mathrm{Nu}_{m,T,2}(\mathrm{Re} = 2300) = 1.615\left(2300\,\mathrm{Pr}\,\frac{d_w}{L}\right)^{1/3}, \qquad (6.214)$$

$$\mathrm{Nu}_{m,T,3,2300} \equiv \mathrm{Nu}_{m,T,3}(\mathrm{Re} = 2300) = \left(\frac{2}{1+22\,\mathrm{Pr}}\right)^{1/6}\left(2300\,\mathrm{Pr}\,\frac{d_w}{L}\right)^{1/2}. \qquad (6.215)$$

The value of $\mathrm{Nu}_{m,t,10000}$ is calculated from formula (6.183):

$$\mathrm{Nu}_{m,t,10000} = \frac{(0.0308/8)\cdot 10000\,\mathrm{Pr}}{1+12.7\sqrt{(0.0308/8)}\,(\mathrm{Pr}^{2/3}-1)}\left[1+\left(\frac{d_w}{L}\right)^{2/3}\right]\left(\frac{\mathrm{Pr}}{\mathrm{Pr}_w}\right)^{0,11}. \qquad (6.216)$$

Formulae (6.200)–(6.216) are valid in the following ranges of parameters:

$$0.7 \le \mathrm{Pr} \le 1000,$$
$$2300 \le \mathrm{Re} \le 10000, \qquad (6.217)$$
$$\frac{d_w}{L} \le 1.$$

The Nusselt number Nu, the Reynolds number Re and the Prandtl number Pr are defined by the following formulae:

$$\text{Nu} = \frac{\alpha d_w}{\lambda}, \text{Re} = \frac{w d_w}{\nu}, \text{Pr} = \frac{c_p \mu}{\lambda}. \tag{6.218}$$

The fluid physical properties are calculated in mean temperature T_m equal to the arithmetic mean of the inlet and outlet temperatures T_{in} and T_{out}, respectively, i.e. using the following formula:

$$T_m = (T_{in} + T_{out})/2. \tag{6.219}$$

Recently, formulae (6.200) and (6.201) were modified by Gnielinski [108] to give the following form:

$$\text{Nu}_m = (1 - \gamma)\text{Nu}_{m,l,2300} + \gamma \text{Nu}_{m,t,4000}, \tag{6.220}$$

where

$$\gamma = \frac{\text{Re} - 2300}{4000 - 2300}, \quad 2300 \le \text{Re} \le 4000. \tag{6.221}$$

The Nusselt number $\text{Nu}_{m,l,2300}$ for the laminar flow is calculated in the same manner as before. $\text{Nu}_{m,t,4000}$ in formula (6.220) is calculated from formula (6.197), and friction factor ξ is calculated from the Konakov formula (6.66). Substituting Re = 4000 in formulae (6.66) and (6.197), the following formula is obtained for $\text{Nu}_{m,t,4000}$:

$$
\begin{aligned}
\text{Nu}_{m,t,4000} &\equiv \frac{(\xi/8)(\text{Re} - 1000)\,\text{Pr}}{1 + 12.7\sqrt{\left(\frac{\xi}{8}\right)}\left(\text{Pr}^{2/3} - 1\right)} \left[1 + \left(\frac{d_w}{L}\right)^{2/3}\right]\left(\frac{\text{Pr}}{\text{Pr}_w}\right)^{0.11}\bigg|_{\text{Re}=2300} \\
&= \frac{(0.0403/8) \cdot 3000\,\text{Pr}}{1 + 12.7\sqrt{\left(\frac{0.0403}{8}\right)}\left(\text{Pr}^{2/3} - 1\right)} \left[1 + \left(\frac{d_w}{L}\right)^{2/3}\right]\left(\frac{\text{Pr}}{\text{Pr}_w}\right)^{0.11}.
\end{aligned}
\tag{6.222}
$$

The drawback of formulae (6.198), (6.200) and (6.220) is the adoption of the Reynolds number limit value of 8000, 10000 and 4000, respectively, which marks the end of the transitional flow region. Moreover, the first derivative dNu/dRe in the transition point is no longer continuous.

For this reason, two new correlations are proposed which are valid for the transitional and the turbulent flow, i.e. for Re $\ge$ 2300. The data from Table 6.3 are approximated using formulae (6.223) and (6.224), which are listed in Table 6.7 for comparison with other formulae. The formulae presented in the table are valid for very long ducts, where $d_w/L = 0$ and $\mu = \mu_w$.

Figures 6.27 and 6.28 present a comparison of the approximation results with the data listed in Table 6.3.

Table 6.7 Correlations valid for the laminar and the turbulent range

No.	Correlation	Validity range
1	$\mathrm{Nu} = 4.364 + \dfrac{\frac{\xi}{8}(\mathrm{Re}-2300)\,\mathrm{Pr}^{x_1}}{x_2 + x_3\sqrt{\frac{\xi}{8}}\left(\mathrm{Pr}^{2/3}-1\right)},$ (6.223) where: $\xi = (1.82\log\mathrm{Re} - 1.64)^{-2}$	$2300 \le \mathrm{Re} \le 10^6$ $0.1 \le \mathrm{Pr} \le 1000$ $S = 189933.7$ $s_t = 34.78$ $r^2 = 0.9999$ $x_1 = 1.007972 \pm 0.002143$ $x_2 = 1.079635 \pm 0.017748$ $x_3 = 12.3852 \pm 0.17911$
2	$\mathrm{Nu} = 4.364 + \dfrac{\frac{\xi}{8}\mathrm{Re}\,\mathrm{Pr}^{x_1}}{1 + \frac{x_2}{\mathrm{Re}-2300} + x_3\sqrt{\frac{\xi}{8}}\left(\mathrm{Pr}^{2/3}-1\right)},$ (6.224) where: $\xi = (1.82\log\mathrm{Re} - 1.64)^{-2}$	$2300 \le \mathrm{Re} \le 10^6$ $0.1 \le \mathrm{Pr} \le 1000$ $S = 116492.3$ $s_t = 27.24$ $r^2 = 0.999939$ $x_1 = 1.014225 \pm 0.001215$ $x_2 = 5885 \pm 2597.3$ $x_3 = 12.9953 \pm 0.10106$
3	$\mathrm{Nu} = 0.12\left(\mathrm{Re}^{2/3} - 125\right)\mathrm{Pr}^{1/3}$ (6.225)	$2300 \le \mathrm{Re} \le 10^6$ $0.6 \le \mathrm{Pr} \le 1000$ [120]
4	$\mathrm{Nu} = \dfrac{\frac{\xi}{8}(\mathrm{Re}-1000)\,\mathrm{Pr}}{1 + 12.7\sqrt{\frac{\xi}{8}}\left(\mathrm{Pr}^{2/3}-1\right)},$ (6.226) where: $\xi = (1.82\log\mathrm{Re} - 1.64)^{-2}$	$2300 \le \mathrm{Re} \le 10^6$ $0.1 \le \mathrm{Pr} \le 1000$
5	$\mathrm{Nu} = 4.364(1 - \gamma) + \gamma\,\mathrm{Nu}\vert_{\mathrm{Re}=10000}$ (6.227) where: $\gamma = \dfrac{\mathrm{Re} - 2300}{10000 - 2300},$ $\mathrm{Nu}\vert_{\mathrm{Re}=10000} = \dfrac{\frac{\xi\vert_{\mathrm{Re}=10000}}{8}\,10000\,\mathrm{Pr}}{1 + 12.7\sqrt{\frac{\xi\vert_{\mathrm{Re}=10000}}{8}}\left(\mathrm{Pr}^{2/3}-1\right)}$ $\xi\vert_{\mathrm{Re}=10000} = (1.8\log 10000 - 1.5)^{-2}$	$2300 \le \mathrm{Re} \le 10^4$ $0.6 \le \mathrm{Pr} \le 1000$
6	$\mathrm{Nu} = \dfrac{\frac{\xi}{8}(\mathrm{Re}-x_1)\,\mathrm{Pr}}{x_2 + x_3\sqrt{\frac{\xi}{8}}\left(\mathrm{Pr}^{2/3}-1\right)},$ (6.228) where: $\xi = (1.82\log\mathrm{Re} - 1.64)^{-2}$	$3000 \le \mathrm{Re} \le 12000$ $2 \le \mathrm{Pr} \le 7$ $S_{\min} = 0.5042\ \mathrm{K}^2$ $s_t = 0.0994\ \mathrm{K}$ $x_1 = 1179.4 \pm 0.1449$ $x_2 = 1.1785 \pm 0.2417$ $x_3 = 13.0574 \pm 0.2292$

It should be emphasized that function (6.223) and function (6.224) both give Nu = 4.364 for Re = 2300, thus ensuring the function continuity in the point that marks the change from the laminar to the transitional flow. Formulae **1**, **2**, **4** and **5** from Table 6.7 are compared in Fig. 6.29. It can be seen that for high Reynolds numbers, correlation **1** gives the smallest values of the Nusselt number.

The comparison presented in Fig. 6.30 indicates that correlations **1** and **2** from Table 6.7 give smaller values of the Nusselt number in the Reynolds number range from 3000 to 12000 than correlations **1** and **2** from Table 6.4. Correlation **2** from

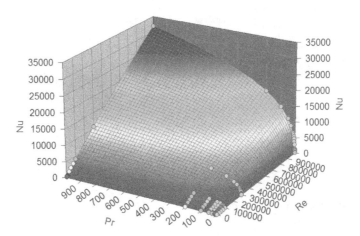

Fig. 6.27 Approximation of data from Table 6.3 by means of function **1**, defined by formula (6.223); the points represent data from Table 6.3 and the surface presents function (6.223)

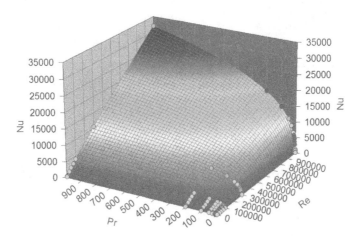

Fig. 6.28 Approximation of data from Table 6.3 by means of function **2**, defined by formula (6.224); the points represent data from Table 6.3 and the surface presents function (6.224)

Table 6.7, which ensures a continuous transition from the laminar to the turbulent flow, is highly sensitive to the Prandtl number. For Pr = 0.7, correlation **2** defined by formula (6.224) gives much lower values of the Nusselt number than correlation **1** defined by formula (6.223). But for Pr = 500, the opposite is the case, i.e. formula (6.223) gives lower values of the Nusselt number compared to formula (6.224).

It follows from Figs. 6.23, 6.24, 6.25 and 6.26 that for air for Re ≥ 4000 the correlations derived for the fully developed turbulent flow provide a good approximation of experimental data. For this reason, correlation **1** from Table 6.7 defined by formula (6.223) can be recommended for the Nusselt number calculations in the transitional and the turbulent range. Correlation (6.223) can be modified so that it takes account of the duct finite length. If the heat flux on the circular duct inner surface is constant, formula (6.223) takes the following form:

Fig. 6.29 Comparison of the
correlations from Table 6.7;
1—formula (6.223),
2—formula (6.224),
4—formula (6.226),
5—formula (6.227)

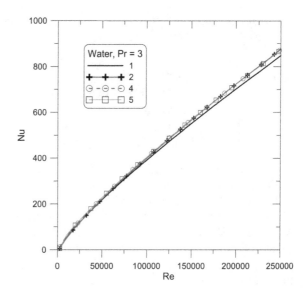

Fig. 6.30 Comparison of
correlations **1** (formula 6.223)
—curves 1, continuous line in
different colours, and **2**
(formula 6.224)—curves 2,
dashed lines in different
colours, from Table 6.7 with
correlations **1** (formula 6.174)
—curves 3 in different colours
and **2** (formula 6.175)—
curves 4 in different colours
from Table 6.4

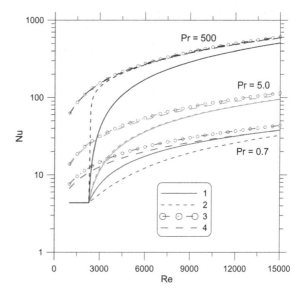

$$
\mathrm{Nu} = \mathrm{Nu}_{\mathrm{m,q}}(\mathrm{Re} = 2300) + \frac{\frac{\xi}{8}(\mathrm{Re} - 2300)\,\mathrm{Pr}^{1.008}}{1.08 + 12.39\sqrt{\frac{\xi}{8}}(\mathrm{Pr}^{2/3} - 1)}
$$
$$
\times \left[1 + \left(\frac{d_w}{L}\right)^{2/3}\right]\left(\frac{\mathrm{Pr}}{\mathrm{Pr}_w}\right)^{0.11}, \quad 2300 \leq \mathrm{Re} \leq 10^6, \quad 0.1 \leq \mathrm{Pr} \leq 1000. \tag{6.229}
$$

where $\mathrm{Nu}_{m,q}$ is defined by formula (6.202).

If constant temperature is set on the circular duct inner surface, formula (6.223) takes the following form:

$$\text{Nu} = \text{Nu}_{m,T}(\text{Re} = 2300) + \frac{\frac{\xi}{8}(\text{Re} - 2300)\,\text{Pr}^{1.008}}{1.08 + 12.39\sqrt{\frac{\xi}{8}}(\text{Pr}^{2/3} - 1)}$$

$$\times \left[1 + \left(\frac{d_w}{L}\right)^{2/3}\right]\left(\frac{\text{Pr}}{\text{Pr}_w}\right)^{0.11}, \quad 2300 \le \text{Re} \le 10^6, \quad 0.1 \le \text{Pr} \le 1000. \tag{6.230}$$

where $\text{Nu}_{m,T}$ is defined by formula (6.209).

Correlations (6.229) and (6.230) will be compared to experimental data.

Figure 6.31 presents a comparison of own correlations and the Gnielinski correlation for the transitional flow with the experimental data obtained by Li Xiao-Wei [180–181]. It can be seen that the experimental data indicate a fully developed turbulent flow as early as for the Reynolds numbers higher than 2500. Own formula (6.229) and the Gnielinski correlation (6.220), which are valid for the transitional and the turbulent flow, give underestimated values of the Nusselt number for small Reynolds numbers. Additionally, for Re > 7000, own formula (6.229) is a bit more accurate than the Gnielinski correlation.

Figure 6.32 presents a comparison of own correlation (6.229) and the Gnielinski correlation (6.220) with the experimental data obtained by Olivier and Meyer [220].

The comparison presented in Fig. 6.32 indicates that the agreement between own formula (6.229) and the experimental data obtained by Olivier and Meyer [220] is better compared to the Gnielinski correlation (6.220). Figure 6.33 presents a comparison of own correlations (6.175) and (6.229) and the Gnielinski correlation

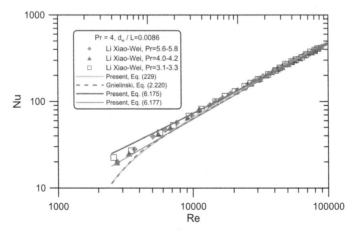

Fig. 6.31 Comparison of own correlation (6.229) for the transitional and the turbulent flow and own correlations (6.175) and (6.177) with the Gnielinski correlation (6.220) and experimental data obtained by Li Xiao-Wei [180–181]; all the correlations take account of the duct finite length

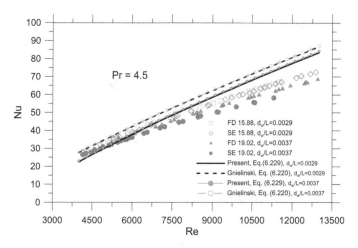

Fig. 6.32 Comparison of own correlation (6.229) for the transitional and the turbulent flow and the Gnielinski correlation (6.220) with experimental data obtained by Olivier and Meyer [220]; all the correlations take account of the duct finite length

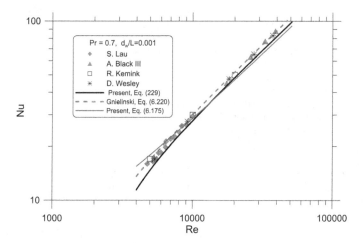

Fig. 6.33 Comparison of own correlation (6.175) for the fully developed turbulent flow and correlation (6.229) for the transitional and the turbulent flow and the Gnielinski correlation (6.220) with experimental data obtained by Lau [173], Black [30], Kemink [150] and Wesley [360]; all the correlations take account of the duct finite length

(6.220) with the experimental data obtained by Lau [173], Black [30], Kemink [150] and Wesley [360].

It can be seen that all three correlations give a good approximation of the experimental data, and the Gnielinski formula is the best.

Figure 6.34 illustrates a comparison of different formulae defining the Nusselt number with the experimental data obtained by Huber and Walter [130].

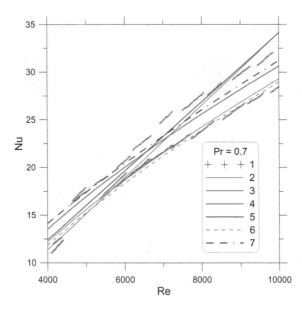

Fig. 6.34 A comparison between the experimental data obtained by Huber and Walter [130] and own correlations and the Gnielinski correlations taking account of the tube finite length; 1—Huber and Walter experimental data [130], 2—own formula (6.230) for the wall constant temperature, 3—the Gnielinski formula for the wall constant temperature—linear interpolation of the Nusselt number between Nusselt numbers for the Reynolds number values of 2300 and 10000 [105–107], 4—own formula (6.229) for a constant heat flux on the tube wall, 5—the Gnielinski formula for a constant heat flux on the tube wall—linear interpolation of the Nusselt number between the Nusselt numbers for the Reynolds number values of 2300 and 10000 [105–107], 6—formula (6.223), the formula right side multiplied by $[1 + (d_w/L)^{2/3}]$, 7—previous and present formula proposed by Gnielinski [104,108]

The experimental data were found based on the testing of a cocurrent tube-in-tube heat exchanger. Air flowed in the inner tube, and water flowed through the ring channel.

Analysing the results presented in Fig. 6.34, it can be seen that the linear interpolation of the Nusselt number in the transitional range of $2300 \leq \text{Re} \leq 10000$ proposed by Gnielinski (curves 3 and 5) [105–107] does not show good agreement with the experimental data. Much better agreement with the experimental data is demonstrated by the modification of the Petukhov-Kirillov formula (6.226) (curve 7) proposed by Gnielinski in 1975 [104]. Curve 7 represents formula (6.226), where the right side is multiplied by $\left[1 + (d_w/L)^{2/3}\right]$ to take account of the tube entrance length. The best agreement with experimental data is displayed by own correlation (6.229), which is valid for a constant value of the heat flux on the wall. Also own formulae (6.223) (curve 2) and (6.230) (curve 6) show satisfactory agreement with the experimental data obtained by Huber and Walter [130].

A comparison of own correlation (6.229) and the Gnielinski correlation (6.220) with the experimental data obtained by Buyukalaca et al. [41] and Eiamsa-Ard [86] is presented in Fig. 6.35.

The calculations were performed for the tube analysed by Buyukalaca et al. [41]. The tube entire length (9,000 mm) was adopted for the calculations, i.e. the analysis concerned both the unheated and the heated section of the tube (6,000 and 3,000 mm, respectively). Analysing the results presented in Fig. 6.35, it can be seen that both own formula (6.229) and the Gnielinski correlation (6.220) provide a good approximation of the experimental data.

Next, the Nusselt numbers were calculated using own formula (6.229) and the Gnielinski formula (6.220) taking account of the tube heated section only (cf. Fig. 6.36).

Analysing the results presented in Fig. 6.36, it can be seen that own formula (6.229) gives a bit lower values of the Nusselt number compared to the Gnielinski correlation (6.220). The Eiamsa-Ard et al. experimental data [86] differ from those obtained by Buyukalaca et al. [41]. Therefore, the agreement of the results obtained by means of formulae (6.220) and (6.229) with the experimental data should be considered as satisfactory.

Figure 6.37 presents a comparison of correlations **1-6** from Table 6.7. Own correlations **1-2** are derived for the transitional range. Similarly, correlation **3** proposed by Hausen [120] and correlation **5** put forward by Gnielinski [105–107] are valid only in the transitional range. Correlation **4** proposed by Gnielinski, which is a modification of the Petukhov-Kirillov formula, is recommended for Re higher than 3000, but, as indicated by Fig. 6.37, it can also be used with good effect for the Reynolds number lower values. Correlation **6** was determined based on

Fig. 6.35 A comparison of own correlation (6.229) and the Gnielinski correlation (6.220) with the experimental data obtained by Buyukalaca et al. [41] and Eiamsa-Ard [86]; the calculations were performed for a tube as presented by Buyukalaca et al., taking account of the tube entire length, i.e. the tube unheated and heated parts

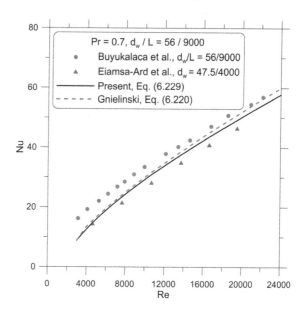

Fig. 6.36 Comparison of own correlation (6.229) and the Gnielinski correlation (6.220) with the experimental data obtained by Buyukalaca et al. [41] and Eiamsa-Ard [86]; the calculations were performed only for the tube heated part, i.e. in order to correct the Nusselt number values, the entrance length was assumed as $d_w/L = 56/3000$ for the tube tested by Buyukalaca et al. [41] and $d_w/L = 47.5/1500$ for the tube tested by Eiamsa-Ard [86]

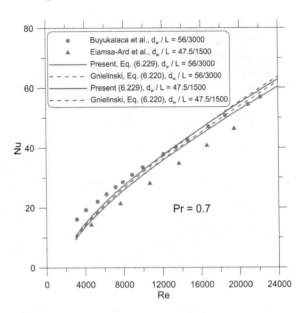

Fig. 6.37 Comparison of correlations **1–6** from Table 6.7; **1**—own correlation (6.223), **2**—own correlation (6.224), **3**—the Hausen correlation (6.225), **4**—the Gnielinski correlation (6.226), **5**—the Gnielinski correlation (6.227), **6**—own correlation (6.228) determined based on the results of the car radiator testing

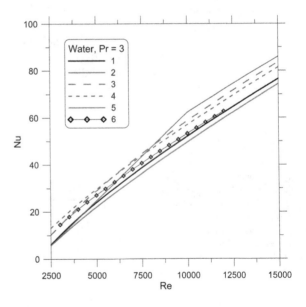

experimental data obtained for a car radiator [309]. It is a two-pass double-row cross-flow heat exchanger made of oval tubes. The hydraulic diameter on the water side is $d_h = 7.01$ mm, and the length of the exchanger tubes totals $L = 520$ mm. The Nusselt numbers are calculated using the correlations presented in Table 6.7 assuming that $d_w/L = 0$. It is also assumed that the heat flux on the tube inner surface is constant.

Analysing the results presented in Fig. 6.37, it can be seen that own correlations **1** and **2** display very good agreement with correlation **6** determined based on experimental testing of the radiator. The Nusselt numbers obtained from the Hausen formula and from the two formulae proposed by Gnielinski are overestimated.

Additional comparisons of correlations **1-6** were made to show how different assumptions affect the results of the Nusselt number calculations. Figure 6.38 presents a comparison of the correlations under the assumption that the tube wall inner surface temperature is constant. Moreover, the finite length of the exchanger tubes is taken into account assuming that $d_w/L = 7.01/520 = 1.057$.

Figure 6.39 presents a comparison of the Nusselt numbers found by means of the proposed formulae assuming that a constant heat flux is set on the tube wall inner surface.

Analysing Fig. 6.39, it can be seen that the agreement of the results obtained using the proposed correlations with the results given by correlation (6.228), which is based on experimental data [309], is worse compared to the case where the assumption is made of the tube wall inner surface constant temperature (cf. Fig. 6.38).

A comparison is also performed of the Nusselt numbers calculated using the correlations presented in Table 6.7 except correlation **5**. The right sides of correlations **1-4** and of correlation **6** from Table 6.7 are multiplied by a constant multiplier.

$$\left[1 + (d_w/L)^{2/3}\right] = \left[1 + (7.01/520)^{2/3}\right] = 1.057.$$

Fig. 6.38 Comparison of correlations **1–6** from Table 6.7 assuming the tube wall inner surface constant temperature and considering that $d_w/L = 7.01/520 = 1.057$; **1**—own correlation (6.230), **2**—own correlation (6.224) modified like correlation (6.230), **3**—the Hausen correlation (6.225), **4**—the Gnielinski correlation (6.226), **5**—the Gnielinski correlation (6.200), **6**—own correlation (6.228) determined based on the results of the car radiator testing

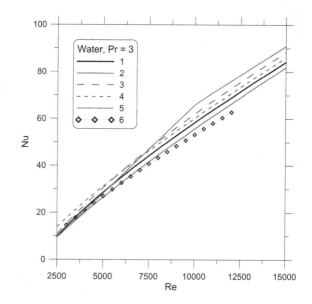

Fig. 6.39 Comparison of correlations **1–6** from Table 6.7 assuming a constant heat flux on the tube inner surface and considering that $d_w/L = 7.01/520 = 1.057$; **1**—own correlation (6.230), **2**—own correlation (6.224) modified like correlation (6.229), **3**—the Hausen correlation (6.225), **4**—the Gnielinski correlation (6.226), **5**—the Gnielinski correlation (6.200), **6**—own correlation (6.228) determined based on the results of the car radiator testing

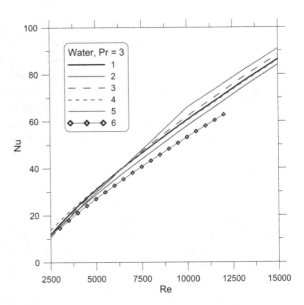

Curve **5** in Fig. 6.40 is plotted using formulae (6.200) and (6.209), i.e. it is assumed that the tube inner surface temperature is constant.

Analysing the results presented in Fig. 6.40, it can be seen that correlations **1** and **2** display good agreement with correlation **6** determined based on experimental data. Only for smaller values of the Reynolds number from the range of $2300 \le Re \le 4000$ are the differences between correlation **1** and **2** and correlation **6** a little bigger. The comparison of the results presented in Fig. 6.37 is very similar.

Fig. 6.40 Comparison of correlations **1–4** and correlation **6** from Table 6.7 and correlations (6.200) and (6.209) considering that $d_w/L = 7.01/520 = 1.057$; **1**—own correlation (6.223), **2**—own correlation (6.224), **3**—the Hausen correlation (6.225), **4**—the Gnielinski correlation (6.226), **5**—the Gnielinski correlation (6.200), **6**—own correlation (6.228) determined based on the results of the car radiator testing

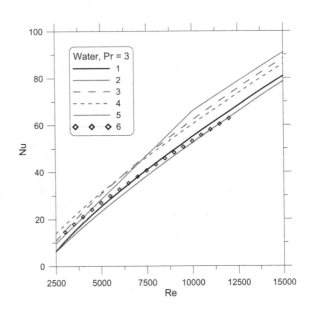

In the case of Fig. 6.37, correlations **1-4** or correlation **6** are not multiplied by factor 1.057 taking account of the entrance length.

It follows from the comparisons performed in Sects. 6.3.4.1 and 6.3.4.2 that the correlations derived in Sect. 6.3.4.1 for the turbulent flow can also be used in the transitional range for the Reynolds number values higher than Re = 3000. In such a case, a correction for the entrance length is taken into consideration like for the turbulent flow.

The correlations proposed in Sect. 6.3.4.2 can be used starting already from the Reynolds number Re = 2300 to Re = 1000000, ensuring the Nusselt number continuity for Re = 2300, i.e. in the point that marks the change from the laminar to the transitional flow

The method of the Nusselt number calculation in the transitional range as proposed by Gnielinski in 1995 [105], which consists in linear interpolation between the Nusselt numbers calculated for Re = 2300 and for Re = 10000, does not seem to be appropriate. Despite the fact that in 2013 Gnielinski proposed in [108] a modification of the formula (6.227), consisting in linear interpolation between the Nusselt number values calculated for Re = 2300 and Re = 4000 and in calculating the Nusselt number for Re = 4000 by means of his own formula proposed in 1975 for the turbulent range [104], the modification has the same drawback as before, i.e. a specific value of the Reynolds number is assumed that marks the end of the transitional region.

Figures 6.41 and 6.42 present a comparison between own formula (6.229) and the Gnielinski formulae (6.200) and (6.220).

It follows from the comparisons presented in Fig. 6.41a, b and c that for $L/d_w > 100$ the Gnielinski formula (6.220) gives higher Nusselt numbers for the Reynolds number values higher than Re = $\sim$7000. This results from calculating the Nusselt number from formula (6.216) for the Reynolds number values of Re $\geq$ 10000, which is a modification of the Petukhov-Kirillov formula. In the Petukhov-Kirillov original correlation, the denominator of formula (6.216) contains the number 1.07, and not the number 1 like in the formula modified by Gnielinski. The consequence of the change introduced by Gnielinski is that the Nusselt numbers calculated from formula (6.216) are overestimated. This finds confirmation in the comparison between formula (6.220) and the experimental data presented in Sects. 6.3.4.1 and 6.3.4.2.

Gnielinski modified formula (6.200) by introducing linear interpolation of the Nusselt number in the Reynolds number range from Re = 2300 to Re = 4000 (formula 6.220). Due to the fact that the Nusselt numbers are calculated from formula (6.226), the Nusselt number values for the Reynolds numbers Re $\geq$ 4000 are smaller. This improved the agreement of formula (6.220) with experimental data for air and water in the case of very long ducts. For short ducts, e.g. for $L/d_w = 5$, a decrease in the Nusselt number values occurs in the range of 2300 $\leq$ Re $\leq$ 4000 (cf. Fig. 6.42a–c). This is contrary to current experience indicating that the Nusselt number increases with a rise in the Reynolds number. New semi-empirical and empirical correlations for transitional and turbulent flow in tubes were developed by Taler [310, 325].

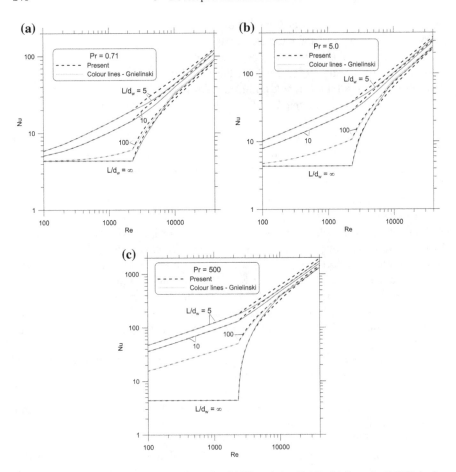

Fig. 6.41 Comparison between own formula (6.229) and the Gnielinski formula (6.200) in the Reynolds number range of $100 \leq \text{Re} \leq 40000$

6.3.4.3 Tubes with a Rough Surface

The tube surface roughness involves a rise in friction factor ξ and in the heat transfer coefficient α. The changes in the friction factor ξ as a function of the Reynolds number Re and relative roughness ε defined by formula (6.59) are illustrated by the Moody chart in Fig. 6.2. Considering that the Reynolds number Re_ε characterizing the tube inner surface roughness is defined as:

$$\text{Re}_\varepsilon \equiv \frac{u_\tau R_a}{\nu} = \text{Re}\varepsilon\sqrt{\frac{\xi}{8}} \qquad (6.231)$$

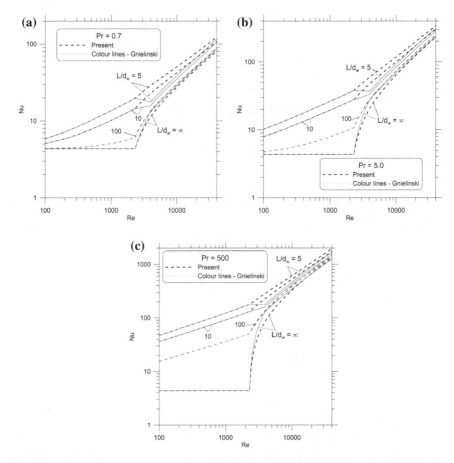

Fig. 6.42 Comparison between own formula (6.229) and the Gnielinski formula (6.220) in the Reynolds number range of $100 \le \text{Re} \le 40000$

three characteristic ranges of changes in friction factor ξ are distinguished, which depend on the following ranges of the Reynolds number values:

$$\begin{aligned} \text{Re}_\varepsilon &< 5, \\ 5 \le \text{Re}_\varepsilon &\le 70, \\ 70 &< \text{Re}_\varepsilon. \end{aligned} \qquad (6.232)$$

Bhatti and Shah [26, 182] put forward the following formula for tubes with a rough inner surface:

$$\text{Nu} = \frac{(\xi/8)\text{Re}\,\text{Pr}}{1 + \sqrt{\xi/8}\left(4.5\text{Re}_\varepsilon^{0.2}\,\text{Pr}^{0.5} - 8.48\right)} \qquad (6.233)$$

valid for:

$$10^4 \leq \mathrm{Re}, 0.5 \leq \mathrm{Pr} \leq 10, \quad 0.002 \leq \varepsilon \leq 0.05.$$

Friction factor ξ for rough tubes can be determined using one of the formulae given in Sect. 6.2.2.2.

For gases, when $\mathrm{Pr} \simeq 1$, Burck [40] proposed the following formula:

$$\frac{\mathrm{Nu}}{\mathrm{Nu}_s} = \sqrt{\frac{\xi}{\xi_s}}, \tag{6.234}$$

where Nu_s and ξ_s denote the Nusselt number and the friction factor for a smooth tube, respectively.

The heat transfer coefficient on a rough surface can be many times higher compared to a smooth surface. It should be noted, however, that the power needed to pump the fluid through rough ducts is also bigger. For internally rifled tubes or tubes with special protrusions on the inner surface, the Nusselt number correlations are determined experimentally. Many such correlations can be found in literature, e.g. in Webb's book [353].

6.3.4.4   Correlations for the Nusselt Number for Flows of Liquid Metals

The heat transfer in liquid metals sparks interest from many industries, e.g. in continuous steel casting, in the semiconductor industry in relation to crystal growing processes and in the construction of fast breeder reactors cooled by liquid metals [115]. Many research programmes focusing on breeder reactors were initiated in recent years, such as the European Sodium-cooled Fast Reactor (ESFR) project, or the Prototype Fast Breeder Reactor (PFBR) programme in India for the construction of a breeder reactor cooled with other heavy metals. Under the European and the Russian research programmes ELSY and BREST-OD-300, respectively, lead (Pb) is used as the breeder reactor coolant [115]. Due to the research projects now in progress, a need arises to find reliable Nusselt number correlations for the flow of liquid metals through ducts. There are considerable divergences between the testing results obtained so far, which necessitates further research on issues concerning the hydraulic and thermal processes occurring in liquid metals.

The Prandtl number for liquid metals is smaller than 0.1 (cf. Table 6.8).

For liquid metals, the assumption that the turbulent Prandtl number Pr_t is 1 is impossible. The Nusselt number Nu cannot be found numerically from formula (6.147) or (6.149) without the relation for the turbulent Prandtl number Pr_t. Many formulae have been worked out for the turbulent Prandtl number [40, 44–45, 140–141, 148]. However, they give different Nusselt number values for the same

Table 6.8 Prandtl number values for liquid metals [151]

Lead (Pb)	
$T = 400$ °C	Pr = 0.0216
$T = 500$ °C	Pr = 0.0178
$T = 600$ °C	Pr = 0.0144
Mercury (Hg)	
$T = 80$ °C	Pr = 0.0196
$T = 200$ °C	Pr = 0.0133
$T = 300$ °C	Pr = 0.0108
Sodium (Na)	
$T = 100$ °C	Pr = 0.0197
$T = 400$ °C	Pr = 0.0167
$T = 600$ °C	Pr = 0.0147
Sodium-potassium alloy (22% Na + 78% K)	
$T = 100$ °C	Pr = 0.018828
$T = 300$ °C	Pr = 0.008496
$T = 600$ °C	Pr = 0.005306
Sodium-potassium alloy (44% Na + 56% K)	
$T = 100$ °C	Pr = 0.02480
$T = 300$ °C	Pr = 0.01050
$T = 600$ °C	Pr = 0.00647

values of Re and Pr. One of the first formulae defining the turbulent Prandtl number is the Aoki relation [12]:

$$\Pr_t = \left\{ 0.014 \mathrm{Re}^{0.45} \mathrm{Pr}^{0.2} \left[1 - \exp\left(\frac{-1}{0.014 \mathrm{Re}^{0.45}\,\mathrm{Pr}^{0.2}} \right) \right] \right\}^{-1}. \qquad (6.235)$$

Using the form of relation (6.235), the turbulent Prandtl number will be approximated by means of the following two formulae:

$$\Pr_t = \left\{ a\mathrm{Re}^{0.45} \mathrm{Pr}^{0.2} \left[1 - \exp\left(\frac{-1}{a\mathrm{Re}^{0.45}\,\mathrm{Pr}^{0.2}} \right) \right] \right\}^{-1}, \qquad (6.236)$$

$$\Pr_t = \left\{ b\mathrm{Re}^{0.45} \mathrm{Pr}^{0.2} \left[1 - \exp\left(\frac{-1}{c\mathrm{Re}^{0.45}\,\mathrm{Pr}^{0.2}} \right) \right] \right\}^{-1}. \qquad (6.237)$$

The unknown parameter a in formula (6.236) and parameters b and c are found by means of the least squares method using the experimental data obtained by Sheriff and O'Kane [271] for liquid sodium. During the testing, the Reynolds number varied from 30000 to 120000. The Prandtl number was 0.0072 for the Reynolds number values: 30000, 40000, 60000 and 80000, and 0.0071 for Re = 100000 and Re = 120000. The experimental data [271] and the Prandtl numbers defined by formulae (6.236) and (6.237) for parameters $a = 0.01592$, $b = 0.01171$ and

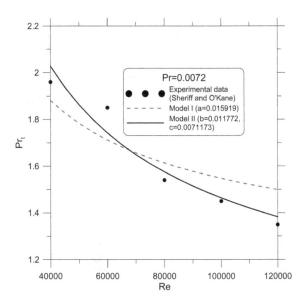

Fig. 6.43 Turbulent Prandtl number—Sheriff and O'Kane [271] experimental data approximation using functions (6.236) and (6.237)

$c = 0.00712$ determined by means of the least squares method are presented in Fig. 6.43.

The determination coefficients are rather low—they total $r^2 = 0.7736$ and $r^2 = 0.9332$ for approximation by means of function (6.236) and function (6.237), respectively, which is due to the rather considerable spread of the measurement data.

A different formula is proposed by Kays, Crawford and Weigand [148, 354–355]:

$$\mathrm{Pr}_t = \left\{ \frac{1}{2\,\mathrm{Pr}_{t\infty}} + d\mathrm{Pe}_t \sqrt{\frac{1}{\mathrm{Pr}_{t\infty}}} - (d\mathrm{Pe}_t)^2 \left[1 - \exp\left(-\frac{1}{d\mathrm{Pe}_t \sqrt{\mathrm{Pr}_{t\infty}}} \right) \right] \right\}^{-1}, \quad (6.238)$$

where the turbulent Peclet number is defined as:

$$\mathrm{Pe}_t = \mathrm{Pr}\,\frac{\varepsilon_\tau}{\nu}. \tag{6.239}$$

$\mathrm{Pr}_{t\infty}$ is the turbulent Prandtl number defined by the formula given by Jischa and Rieke [140–141]:

$$\mathrm{Pr}_{t\infty} = \mathrm{Pr}_{ts} + \frac{e}{\mathrm{Pr}\,\mathrm{Re}^{0.888}}. \tag{6.240}$$

Constants d, e and $\mathrm{Pr}_{t\infty}$ in formulae (6.239) and (6.240) are as follows: $d = 0.3$, $\mathrm{Pr}_{ts} = 0.85$, $e = 182.4$. $\mathrm{Pr}_{t\infty}$ denotes the turbulent Prandtl number for high values of the molecular Prandtl number Pr.

The turbulent Prandtl number formula proposed by Kays [147] has the following form:

$$Pr_t = 0.85 + \frac{f}{Pe_t} = 0.85 + \frac{f}{\frac{\varepsilon_\tau}{\nu} Pr}, \tag{6.241}$$

where $f = 1.46$.

One of the more popular formulae describing the Prandtl number is the Cebeci relation [44–46, 53], which is based on the mixing path concept proposed by Prandtl:

$$Pr_t = \frac{\varepsilon_\tau}{\varepsilon_q} = \frac{\kappa}{\kappa_q} \frac{1 - \exp(-y^+/A^+)}{1 - \exp(-y^+/B^+)}, \tag{6.242}$$

where $y^+ = yu_\tau/\nu$.

Constants A^+ and B^+ in formula (6.242) are as follows:

$$\kappa = 0.4, \quad \kappa_q = 0.44, \quad A^+ = 26,$$
$$B^+ = \frac{1}{\sqrt{Pr}} \sum_{i=1}^{5} C_i (\log_{10} Pr)^{i-1}. \tag{6.243}$$

Constants C_i in the formula are as follows:

$$C_1 = 34.96, C_2 = 28.79, C_3 = 33.95, C_4 = 6.3 \text{ and } C_5 = -1.186.$$

For small Prandtl numbers, which are typical of liquid metals and gases, the exponents at the Reynolds and the Prandtl numbers in the Nusselt number correlations are similar. The Sleicher and Rouse correlation [279] is an example of such a formula:

$$Nu = 6.3 + 0.0167 Re_d^{0.85} Pr^{0.93}. \tag{6.244}$$

The analysis of formula (6.178) from Table 6.4 also indicates that for small values of the Prandtl number included in the range of $0.1 \leq Pr \leq 1$, the exponents at the Reynolds and the Prandtl numbers are $x_2 = 0.8018$ and $x_3 = 0.7095$, respectively. It can be seen that the values of exponents x_2 and x_3 are close to each other. It may therefore be expected that for smaller values of the Prandtl number, which are typical of liquid metals, the values of the exponents will be similar. Assuming that $x_2 = x_3$, the exponential correlation for the Nusselt number can be reduced to the following form:

$$Nu = x_1 + x_2 Pe^{x_3}, \tag{6.245}$$

where the Peclet number is defined as:

$$Pe = Re\,Pr = \frac{wd}{\nu}\frac{c_p\mu}{\lambda} = \frac{wd\rho c_p}{\lambda} = \frac{wd}{a}. \tag{6.246}$$

and $a = \lambda/(\rho c_p)$ is thermal diffusivity. Using the Lyon integral, the Nusselt number values are determined as a function of the Reynolds and the Prandtl numbers (cf. Table 6.9).

Next, the data $Nu_{ij}^m = Nu(Re_i, Pr_j)$, $i = 1,\ldots,10$, $j = 1,\ldots,5$ from Table 6.9 were approximated using function (6.245) by means of the least squares method. In order to find the optimal values of parameters x_1, x_2 and x_3, at which the sum of squares (6.173) of the differences between the Nusselt numbers Nu_{ij}^c determined using formula (6.245) and the Nusselt number values Nu_{ij}^m found by means of the Lyon integral given in Table 6.9, the Levenberg-Marquardt method is used. In this case, n_{Re} equals 10, and n_{Pr} totals 5. Formula (6.245) takes the following for

$$Nu = 5.717 + 0.0184 Pe^{0.8205}, \quad 2300 \le Re \le 10^6, \\ 0.0001 \le Pr \le 0.1. \tag{6.247}$$

The 95% confidence intervals are:

$$x_1 = 5.7170 \pm 0.0015,$$
$$x_2 = 0.0184 \pm 0.0072,$$
$$x_3 = 0.8205 \pm 0.2239.$$

Similar calculations are performed for the modified Aoki formula (6.237) describing the turbulent Prandtl number Pr_t, where the two unknown parameters b and c are found based on the experimental data obtained by Sheriff and O'Kane. The Nusselt number values calculated by means of the Lyon integral for the turbulent Prandtl number Pr_t defined by formula (6.237) are listed in Table 6.10.

The data from Table 6.10 are approximated using the following function:

Table 6.9 Nusselt number (Nu) values as a function of the Reynolds (Re) and the Prandtl (Pr) numbers determined using the modified Aoki formula (6.236) defining the turbulent Prandtl number—$a = 0.01592$

Pr	Re									
	2300	3000	5000	7500	10000	30000	50000	100000	300000	1000000
0.0001	5.47	5.65	5.96	6.17	6.30	6.64	6.74	6.85	6.98	7.12
0.001	5.47	5.65	5.96	6.17	6.30	6.65	6.76	6.91	7.38	9.48
0.01	5.48	5.66	5.99	6.23	6.40	7.24	8.01	10.03	17.66	39.32
0.05	5.7	5.96	6.55	7.20	7.84	12.65	16.97	26.52	57.66	143.03
0.1	6.14	6.55	7.61	8.90	10.15	19.05	26.73	43.52	98.40	250.63

Table 6.10 Nusselt number (Nu) values as a function of the Reynolds (Re) and the Prandtl (Pr) numbers determined using the modified Aoki formula (6.237) defining the turbulent Prandtl number—$b = 0.01177$; $c = 0.00712$

Pr	Re									
	2300	3000	5000	7500	10000	30000	50000	100000	300000	1000000
0.0001	5.47	5.65	5.96	6.17	6.30	6.64	6.74	6.85	6.98	7.12
0.001	5.47	5.65	5.96	6.17	6.30	6.65	6.76	6.91	7.39	9.66
0.01	5.48	5.66	5.99	6.23	6.39	7.25	8.09	10.35	19.10	43.84
0.05	5.64	5.91	6.52	7.22	7.92	13.30	18.23	29.10	64.09	158.46
0.1	6.04	6.48	7.65	9.08	10.49	20.56	29.25	48.08	108.84	275.13

$$\mathrm{Nu} = 5.5123 + 0.015 \mathrm{Pe}^{0.865}, \quad 2300 \leq \mathrm{Re} \leq 10^6, \qquad 0.0001 \leq \mathrm{Pr} \leq 0.1. \tag{6.248}$$

The uncertainty of the found parameters is defined by the following 95% confidence intervals:

$$x_1 = 5.5123 \pm 0.0034,$$
$$x_2 = 0.0150 \pm 0.0201,$$
$$x_3 = 0.8650 \pm 0.4089.$$

Next, the turbulent Prandtl number is determined using the Kays and Crawford [148] formula modified by Weigand, Ferguson and Crawford [355]. The results obtained calculating the turbulent Prandtl number by means of formulae (6.238)-(6.240) are listed in Table 6.11.

The data from Table 6.11 are approximated using the following function:

$$\mathrm{Nu} = 5.5066 + 0.0179 \mathrm{Pe}^{0.8275}, \quad 2300 \leq \mathrm{Re} \leq 10^6, \qquad 0.0001 \leq \mathrm{Pr} \leq 0.1. \tag{6.249}$$

Table 6.11 Nusselt number (Nu) values as a function of the Reynolds (Re) and the Prandtl (Pr) numbers determined using the modified Kays-Crawford formula (6.238) defining the turbulent Prandtl number – $\mathrm{Pr}_{ts} = 0.85$; $d = 0.3$; $e = 182.4$

Pr	Re									
	2300	3000	5000	7500	10000	30000	50000	100000	300000	1000000
0.0001	5.47	5.65	5.96	6.17	6.30	6.65	6.76	6.88	7.09	7.50
0.001	5.47	5.65	5.97	6.18	6.32	6.73	6.92	7.25	8.29	11.38
0.01	5.52	5.71	6.09	6.38	6.61	7.85	8.88	11.21	19.10	40.74
0.05	5.77	6.07	6.74	7.45	8.12	12.75	16.83	25.87	55.59	137.51
0.1	6.11	6.53	7.58	8.78	9.93	18.04	25.14	40.86	92.84	237.90

The uncertainty of the found parameters is defined by the following 95% confidence intervals:

$$x_1 = 5.5066 \pm 0.3177,$$
$$x_2 = 0.0179 \pm 0.0020,$$
$$x_3 = 0.8275 \pm 0.0097.$$

The turbulent Prandtl number is also calculated using the modified Kays formula (6.241). Instead of the value of coefficient f proposed by Kays in formula (6.241) ($f = 2$) [147], $f = 1.46$ is adopted, as this value ensures better agreement of the calculation results with experimental data. The Nusselt number values calculated using the Lyon integral and the turbulent Prandtl number defined by formula (6.241) are listed in Table 6.12.

The data from Table 6.12 are approximated using the following function:

$$Nu = 5.3103 + 0.0221Pe^{0.8174}, \quad 2300 \leq Re \leq 10^6, \tag{6.250}$$
$$0.0001 \leq Pr \leq 0.1.$$

The uncertainty of the found parameters is defined by the following 95% confidence intervals:

$$x_1 = 5.3103 \pm 0.3675,$$
$$x_2 = 0.0221 \pm 0.0025,$$
$$x_3 = 0.8174 \pm 0.0100.$$

Figures 6.44, 6.45 and 6.46 present a comparison of the derived formulae (6.247), (6.248), (6.249) and (6.250) to the available experimental data and to the Seban and Shimazaki formula [151, 227, 277]:

$$Nu = 5.0 + 0.025Pe^{0.8}, \quad 10000 \leq Re \leq 500000, \tag{6.251}$$
$$100 \leq Pe \leq 10000.$$

Table 6.12 Nusselt number (Nu) values as a function of the Reynolds (Re) and the Prandtl (Pr) numbers determined using the modified Kays formula (6.241) defining the turbulent Prandtl number—$f = 1.46$

Pr	Re									
	2300	3000	5000	7500	10000	30000	50000	100000	300000	1000000
0.0001	5.47	5.65	5.96	6.17	6.30	6.64	6.75	6.88	7.08	7.55
0.001	5.47	5.65	5.97	6.18	6.31	6.71	6.89	7.22	8.41	12.42
0.01	5.50	5.70	6.06	6.34	6.55	7.75	8.85	11.55	21.60	51.25
0.05	5.70	5.98	6.59	7.26	7.90	12.83	17.59	28.79	68.29	183.81
0.1	5.98	6.36	7.33	8.49	9.63	18.59	27.09	46.91	116.65	321.95

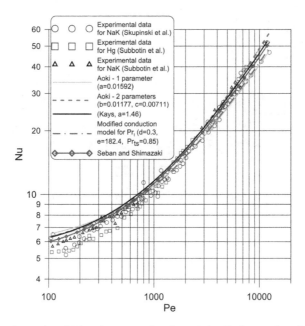

Fig. 6.44 Nusselt number (Nu) values as a function of the Peclet number (Pe) determined experimentally by Skupinski [277] and Subotin [227] and using different formulae defining the turbulent Prandtl number Pr_t

Fig. 6.45 Nusselt number (Nu) values as a function of the Reynolds (Re) and the Prandtl (Pr) numbers determined experimentally by Skupinski [277] and using different formulae defining the turbulent Prandtl number Pr_t

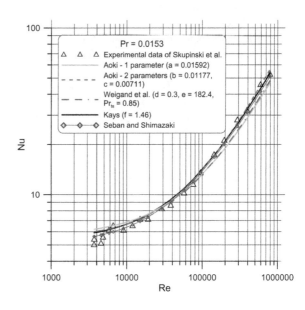

Fig. 6.46 Nusselt number
(Nu) values as a function of
the Reynolds (Re) and the
Prandtl (Pr) numbers
determined experimentally
by Fuchs [354] and using
different formulae defining
the turbulent Prandtl number
Pr_t

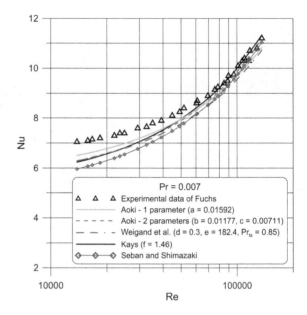

It can be seen from the comparisons presented in Figs. 6.44, 6.45 and 6.46 that the
formulae derived herein demonstrate very good agreement with experimental data.
Only for smaller values of the Reynolds and the Peclet numbers are the differences
between the calculated Nusselt numbers and those determined experimentally a
little bigger. Of all the correlations, formulae (6.247) and (6.248) based on the
turbulent Prandtl numbers defined by formulae (6.236) and (6.237) display the best
agreement with experimental data. Further information on heat transfer in liquid
metals can be found in the paper by Taler [327].

Part II
Methods of the Heat Exchanger Modelling

Chapter 7
Basics of the Heat Exchanger Modelling

This chapter presents the basic mass, momentum and energy conservation equations derived for the flowing medium, determination of the temperature distribution in the walls of tubes with a circular cross-section and with a complex shape, and determination of the heat transfer coefficient for non-finned and finned tubes.

7.1 Simplified Equations of Mass, Momentum and Energy Conservation

The temperature and pressure distributions of fluids in heat exchangers are usually found using one-dimensional equations describing the fluid plug flow through tubes. The mass conservation equation has the following form:

$$\frac{\partial(A\,\rho)}{\partial t} = -\frac{\partial(A\,\rho\,w)}{\partial s}. \tag{7.1}$$

In the case of the fluid steady-state flow $\partial(A\,\rho)/\partial t = 0$, and it follows from Eq. (7.1) that the fluid mass flow rate in the channel is constant, i.e. $\dot{m} = A\rho w = $ const. The fluid mass flow rate $\dot{m}$ is also constant in the case of an incompressible fluid flow ($\rho = $ const) through a channel with the cross-section constant surface area.

The momentum conservation equation used for the heat exchanger calculations has the following form:

$$\frac{\partial w}{\partial t} + w\frac{\partial w}{\partial s} = -\frac{1}{\rho}\frac{\partial p}{\partial s} - g\sin\varphi - \frac{\xi}{d_h}\frac{w|w|}{2}. \tag{7.2}$$

The last term on the right side of Eq. (7.2) takes account of the sense of the fluid velocity vector because the pressure value drops along the channel in the fluid flow

© Springer International Publishing AG, part of Springer Nature 2019
D. Taler, *Numerical Modelling and Experimental Testing of Heat Exchangers*,
Studies in Systems, Decision and Control 161,
https://doi.org/10.1007/978-3-319-91128-1_7

direction. Local pressure losses occurring on valves, bends and knees are added to pressure losses being the effect of friction. In the steady state, when $\partial w / \partial t = 0$, Eq. (7.2) can be written as follows:

$$\frac{\partial p}{\partial s} = -\rho w \frac{\partial w}{\partial s} - \rho g \sin \varphi - \rho \frac{\xi}{d_h} \frac{w|w|}{2}. \tag{7.3}$$

Equation (7.3) is commonly used to find the distribution of pressure $p(s)$ along the fluid flow path.

The analysis of the right side of the energy balance Eq. (2.187) indicates that for heat exchangers all terms on the equation right side can be omitted except $\dot{q} U / A$. The energy balance Eq. (2.189) then takes the following form:

$$\rho c_p \left(\frac{\partial T}{\partial t} + w \frac{\partial T}{\partial s} \right) = \frac{\dot{q} U}{A}. \tag{7.4}$$

It is assumed here that no volume heat sources occur in the fluid, i.e. $\dot{q}_v = 0$.

7.2 Determination of the Tube Wall Temperature Distribution

In many tube heat exchangers the drop in temperature on the wall thickness is slight. It may therefore be assumed that the temperature along the wall entire thickness is the same and depends on time only, which means that the wall is treated as an element with a lumped thermal capacity. Using this assumption, it is easier to model the heat transfer in both non-finned and finned tube exchangers.

7.2.1 Cylindrical Wall

Tubes with no fins fixed on the outer surface may have different cross-sections (cf. Fig. 7.1). The most common are tubes with a circular cross-section (cf. Fig. 7.1a) because they are easy to make [159, 169, 241]. Tubes in car radiators, hot water coolers using air and air-conditioning heat exchangers can also have oval or elliptical cross-sections (cf. Fig. 7.1b). Moreover, the tubes may be flattened (rectangular cross-section) to reduce flow resistance on the air side and lower the motor power needed to drive the fan. Compared to tubes with a circular cross-section, oval, elliptical and flattened (rectangular) tubes are also more efficient in terms of the heat transfer.

In such tubes, the dead zone occurring in the tube rear part due to the eddy flow of gas is relatively small. In the dead zone occurring in the rear stagnation point area

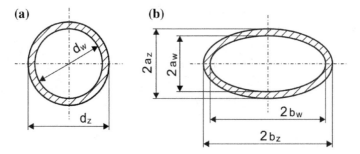

Fig. 7.1 Circular and oval cross-sections of a tube

for tubes located in the first row, and in the front and rear stagnation point areas for tubes in the second and subsequent rows, the heat transfer between the tube surface and the eddying gas is almost none.

The temperature of the gas circulating near the tube is close to the temperature of the tube wall surface. Due to the very small difference between the temperatures of the circulating gas and the tube surface in the dead zone, there is almost no heat exchange between the two fluids in that region, i.e. the heat flow rate is close to zero.

Figure 7.2 presents a diagram illustrating the energy balance equation for a wall with lumped thermal capacity.

Considering that the surface area of the wall cross-section is:

$$A_{wsc} = U_m \, \delta_w, \tag{7.5}$$

where

$$U_m = (U_w + U_z)/2 \tag{7.6}$$

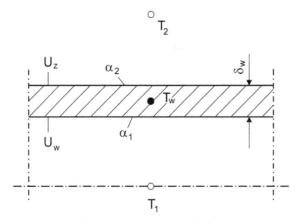

Fig. 7.2 Diagram illustrating the energy balance for a wall with a circular cross-section

is the tube mean perimeter, the energy balance for the wall of a 1-metre long tube takes the following form:

$$A_{wsc}\, \rho_w\, c_w\, \frac{\partial T_w}{\partial t} = \alpha_1\, U_w\, (T_1 - T_w) + \alpha_2\, U_z\, (T_2 - T_w). \tag{7.7}$$

Dividing Eq. (7.7) by A_{wsc}, the following is obtained:

$$c_w\, \rho_w\, \frac{\partial T_w}{\partial t} = \frac{\alpha_1\, U_w}{A_{wsc}}\, (T_1 - T_w) + \frac{\alpha_2\, U_z}{A_{wsc}}\, (T_2 - T_w). \tag{7.8}$$

In the steady state, when $\partial T_w / \partial t = 0$, Eq. (7.8) gives a simple formula for the wall temperature calculation:

$$T_w = \frac{\alpha_1\, U_w\, T_1 + \alpha_2\, U_z\, T_2}{\alpha_1\, U_w + \alpha_2\, U_z}. \tag{7.9}$$

For a cylindrical wall, the inner and outer perimeters are defined as $U_w = 2\pi r_w$ and $U_z = 2\pi r_z$, respectively. Formulae (7.8) and (7.9) can also be applied to tubes with a different shape of the cross-section, e.g. for rectangular tubes.

7.2.2   Wall with a Complex Cross-Section Shape

The heat exchanger tubes can have cross-sections with a complicated shape. This is the case with the *omega* tubes of the live steam second stage platen superheater of the OFZ-425 fluidized-bed boiler. The boiler live steam temperature, pressure and mass flow rate are 560 °C, 160 bar and 425×10^3 kg/h, respectively. The platen superheater is located in the furnace chamber upper part. The superheater tubes make up a panel (cf. Fig. 7.3a) with a smooth surface, which eliminates the for- mation of slag deposits on their surface. The *double omega* tube cross-section is presented in Fig. 7.3b. Due to the complex shape and considerable thickness of the *double omega* tubes, the temperature distribution in the cross-section has to be calculated more accurately to enable determination on the tube outer and inner surface, which in turn is essential for the correct determination of the heat flow rate transferred from flue gases to steam.

A similar situation occurs in the case of the condenser in compressor cooling devices with a modular structure as presented in Fig. 7.3c. Because in the con- denser the cooling fluid pressure is higher than in the evaporator, the walls of the condenser tubes must be thicker compared the evaporator tubes. In order to simplify the calculations, the temperature distribution may be determined in a single repetitive fragment of the tube (cf. Fig. 7.3d). The temperature distribution in tubes with a complex cross-section shape cannot be found without the application of numerical methods, such as the finite element method (FEM) or the combined finite

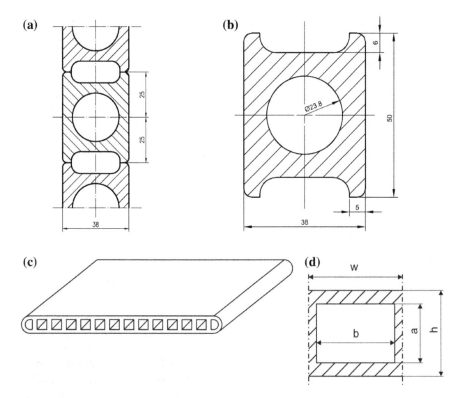

Fig. 7.3 Examples of tubes with a complex cross-section shape: **a** fluidized-bed furnace chamber platen made of *double omega* tubes, **b** single *double omega* tube, **c** condenser in a compressor cooling system, heat pump or air conditioner, **d** condenser repetitive fragment

volume-finite element method (FVM-FEM) [5, 14, 18, 19, 72, 177, 311, 316, 317]. The FVM-FEM method is more straightforward than the classical finite element approach based on the Galerkin method [72, 177, 263, 331]. In contrast to the traditional method, the heat flow rates transferred by the side common to two adjacent elements are identical, regardless of which element the heat flow rate is determined for. Even at a small number of finite elements, it is possible to achieve satisfactory accuracy needed to ensure a correct simulation of the entire exchanger. The FVM-FEM method can also be used to find the temperature field in rectangular or hexagonal fins and to determine their efficiency [311, 316].

7.2.2.1 Finite Volume Method—Finite Element Method (FVM-FEM)

In the finite volume method—finite element method (FVM-FEM) the analysed area is first divided into triangular and then into tetragonal finite elements. An example division of ¼ of the *double omega* tube into triangular finite elements is presented in Fig. 7.4.

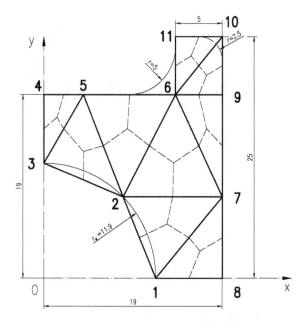

Fig. 7.4 Division of a quarter of the *double omega* tube cross-section into triangular finite elements and control areas (finite volumes)

Then, connecting the triangular elements centres of gravity, located on the intersection of lines connecting the triangle vertices to the midpoints of the opposite sides, to the sides midpoints, a control area (referred to as the control volume) is created around each mesh node (cf. Fig. 7.4). The boundaries of the control areas are marked with dashed lines. An energy conservation equation is written for each finite volume, which gives a system of first-order ordinary differential equations with respect to time for transient-state problems or a system of algebraic equations for steady-state ones. In both cases, the number of equations is equal to the number of nodes. If in nodes located on the boundary the temperatures are known, no energy conservation equations are written for them.

Figures 7.5 and 7.6, respectively, present the division of ¼ of the *double omega* tube cross-section into triangular elements and finite volumes using the STAR-CCM+ commercial program. The number of equations is equal to the number of the finite element mesh nodes.

This chapter presents a generalised FVM-FEM method taking account of the material temperature-dependent physical properties and the heat transfer on the sides of an elementary control area and on the surface of a two-dimensional control area with a constant thickness. By taking account of the heat transfer on the control area surface, it is possible to determine the unsteady-state temperature distribution in fins with complex shapes, where—due to their small thickness—the temperature drop on the fin thickness is omitted.

The energy conservation equations used to formulate the energy conservation equation for the control area will be derived using the following transient-state heat conduction equation:

Fig. 7.5 Division of a quarter of the *double omega* tube cross-section into triangular finite elements using the STAR-CCM+ program

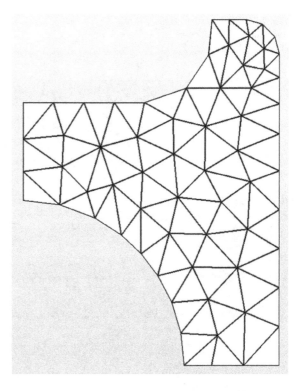

Fig. 7.6 Division of a quarter of the *double omega* tube cross-section into control areas (finite volumes) using the STAR-CCM+ program

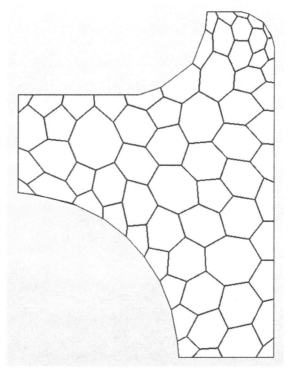

Fig. 7.7 Diagram illustrating transient heat conduction in area Ω with boundary Γ

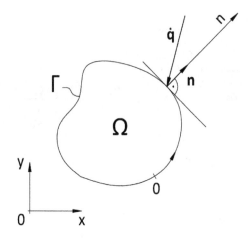

$$c(T)\,\rho(T)\,\frac{\partial T}{\partial t} = -\nabla \cdot \dot{\mathbf{q}}, \tag{7.10}$$

where $\dot{\mathbf{q}}$ is the heat flux vector defined by the Fourier law:

$$\dot{\mathbf{q}} = -\lambda(T)\,\nabla T. \tag{7.11}$$

Equation (7.10) will be volume-integrated in area Ω with boundary Γ (cf. Fig. 7.7).

$$\int_{\Omega} c(T)\,\rho(T)\,\frac{\partial T}{\partial t}\,dV = -\int_{\Omega} \nabla \cdot \dot{\mathbf{q}}\,dV. \tag{7.12}$$

Averaging the left side of Eq. (7.12) and applying the divergence theorem [112, 250] to the right side, the following is obtained:

$$V_{\Omega}\,c(\bar{T})\,\rho(\bar{T})\,\frac{d\bar{T}}{dt} = -\int_{\Gamma} \dot{\mathbf{q}} \cdot \mathbf{n}\,dS, \tag{7.13}$$

where the overscore denotes the relevant mean value in control area Ω, and $\mathbf{n}$ is the unit vector normal to the surface and directed outside area Ω (cf. Fig. 7.7).

Three different methods will be presented of deriving equations used in the combined finite volume-finite element method (FVM-FEM) to formulate the energy conservation equation for the control area.

Method I

Using Eq. (7.13) for control area Ω_{1aoc1} (cf. Fig. 7.8), the following is obtained:

Fig. 7.8 Diagram illustrating the energy conservation equation for control area 1-*a-o-c*-1, which is a fragment of the control area around point 1

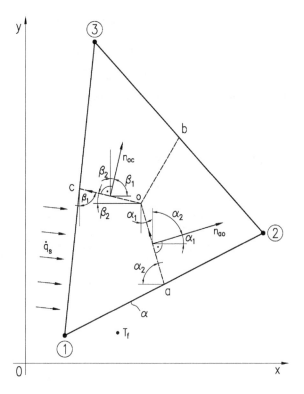

$$\frac{1}{3} A_{123}\, c(T_1)\, \rho(T_1)\, \frac{\mathrm{d}T_1}{\mathrm{d}t} = -\int_a^o \dot{\mathbf{q}} \cdot \mathbf{n}\,\mathrm{d}s - \int_o^c \dot{\mathbf{q}} \cdot \mathbf{n}\,\mathrm{d}s, \qquad (7.14)$$

where the surface area of the 1-2-3 triangle is defined as:

$$A_{123} = \frac{1}{2}\det \begin{vmatrix} 1 & x_1 & y_1 \\ 1 & x_2 & y_2 \\ 1 & x_3 & y_3 \end{vmatrix} = \frac{1}{2}[x_2(y_3 - y_1) + y_2(x_1 - x_3) + (x_3 y_1 - x_1 y_3)]. \quad (7.15)$$

The integrals on the right side of expression (7.14) are calculated anticlockwise, i.e. moving along the boundary, the area is on the left side.

Equation (7.14) will be formulated for an anisotropic body with temperature-dependent physical properties. In such a situation, the nonlinear transient heat conduction equation has the following form:

$$c(T)\,\rho(T)\,\frac{\partial T}{\partial t} = \frac{\partial}{\partial x}\left[\lambda_x(T)\frac{\partial T}{\partial x}\right] + \frac{\partial}{\partial y}\left[\lambda_y(T)\frac{\partial T}{\partial y}\right]. \qquad (7.16)$$

If the body is isotropic, the following equality is the case:

$$\lambda_x(T) = \lambda_y(T) = \lambda(T). \tag{7.17}$$

The temperature field inside the 1-2-3 triangle (cf. Fig. 7.8) will be approximated using the following linear function:

$$T = a_1 + a_2 x + a_3 y, \tag{7.18}$$

where constants: a_1, a_2 and a_3 determined from the following conditions:

$$T(x_1, y_1) = T_1, \quad T(x_2, y_2) = T_2, \quad T(x_3, y_3) = T_3, \tag{7.19}$$

are

$$a_1 = \frac{1}{2A_{123}}[(x_2 y_3 - x_3 y_2)T_1 + (x_3 y_1 - x_1 y_3)T_2 + (x_1 y_2 - x_2 y_1)T_3],$$

$$a_2 = \frac{1}{2A_{123}}[(y_2 - y_3)T_1 + (y_3 - y_1)T_2 + (y_1 - y_2)T_3], \tag{7.20}$$

$$a_3 = \frac{1}{2A_{123}}[(x_3 - x_2)T_1 + (x_1 - x_3)T_2 + (x_2 - x_1)T_3].$$

The first of the integrals on the right side of Eq. (7.14) will be determined considering that the following relations occur for the section a-o of boundary:

$$-dx = ds\,\sin\alpha_1 = ds\,\cos\alpha_2 = n_y\,ds,$$
$$dy = ds\,\sin\alpha_2 = ds\,\cos\alpha_1 = n_x\,ds, \tag{7.21}$$

$$\mathbf{n}_{ao} = \cos\alpha_1\,\mathbf{i} + \cos\alpha_2\,\mathbf{j} = n_{ao,x}\mathbf{i} + n_{ao,y}\mathbf{j}$$
$$= \frac{y_o - y_a}{\sqrt{(x_a - x_o)^2 + (y_a - y_o)^2}}\,\mathbf{i} + \frac{x_a - x_o}{\sqrt{(x_a - x_o)^2 + (y_a - y_o)^2}}\,\mathbf{j} \tag{7.22}$$

the unit vector normal to the section o-c of the boundary can be found similarly:

$$\mathbf{n}_{oc} = \cos\beta_1\,\mathbf{i} + \cos\beta_2\,\mathbf{j} = n_{oc,x}\mathbf{i} + n_{oc,y}\mathbf{j}$$
$$= \frac{y_c - y_o}{\sqrt{(x_o - x_c)^2 + (y_o - y_c)^2}}\,\mathbf{i} + \frac{x_o - x_c}{\sqrt{(x_o - x_c)^2 + (y_o - y_c)^2}}\,\mathbf{j}. \tag{7.23}$$

Assuming that the heat transfer coefficients are calculated at temperature T_o, the heat flux vector in the entire area 1-2-3 is defined as:

$$\dot{\mathbf{q}} = \dot{q}_x\mathbf{i} + \dot{q}_y\mathbf{j} = -\lambda_x(T_o)\frac{\partial T}{\partial x}\mathbf{i} - \lambda_y(T_o)\frac{\partial T}{\partial y}\mathbf{j}, \tag{7.24}$$

where:

$$\dot{q}_x = -\lambda_x (T_o) \frac{\partial T}{\partial x} = -\lambda_x(T_o) \frac{1}{2A_{123}} [(y_2 - y_3) T_1 + (y_3 - y_1) T_2 \\ + (y_1 - y_2) T_3], \tag{7.25}$$

$$\dot{q}_y = -\lambda_y(T_o) \frac{\partial T_1}{\partial y} = -\lambda_y(T_o) \frac{1}{2A_{123}} [(x_3 - x_2) T_1 + (x_1 - x_3) T_2 \\ + (x_2 - x_1) T_3]. \tag{7.26}$$

The first integral in Eq. (7.14) can be calculated as follows:

$$-\int_a^o \dot{\mathbf{q}} \cdot \mathbf{n}\, ds = -\int_a^o \dot{\mathbf{q}} \cdot \mathbf{n}_{ao}\, ds = -\int_a^o \left(\dot{q}_x n_{ao,x} + \dot{q}_y n_{ao,y}\right) ds \\ = -\left(\dot{q}_x n_{ao,x} + \dot{q}_y n_{ao,y}\right) \int_a^o ds = -\left(\dot{q}_x n_{ao,x} + \dot{q}_y n_{ao,y}\right) s_{ao}. \tag{7.27}$$

Considering that the length of section s_{ao} is defined by the following formula:

$$s_{ao} = \sqrt{(x_a - x_o)^2 + (y_a - y_o)^2} \tag{7.28}$$

formula (7.27) can be written as:

$$-\int_a^o \dot{\mathbf{q}} \cdot \mathbf{n}\, ds = -\lambda_x(T_o) \frac{y_a - y_o}{2A_{123}} [(y_2 - y_3) T_1 + (y_3 - y_1) T_2 + (y_1 - y_2) T_3] \\ + \lambda_y(T_o) \frac{x_a - x_o}{2A_{123}} [(x_3 - x_2) T_1 + (x_1 - x_3) T_2 + (x_2 - x_1) T_3]. \tag{7.29}$$

A similar procedure can be used to calculate the following integral:

$$-\int_0^c \dot{\mathbf{q}} \cdot \mathbf{n}\, ds = -\left(\dot{q}_x n_{oc,x} + \dot{q}_y n_{oc,y}\right) s_{oc}. \tag{7.30}$$

Substitution of relevant components results in:

$$-\int_0^c \dot{\mathbf{q}} \cdot \mathbf{n}\, ds = \lambda_x(T_o) \frac{y_c - y_o}{2A_{123}} [(y_2 - y_3) T_1 + (y_3 - y_1) T_2 + (y_1 - y_2) T_3] \\ - \lambda_y(T_o) \frac{x_c - x_o}{2A_{123}} [(x_3 - x_2) T_1 + (x_1 - x_3) T_2 + (x_2 - x_1) T_3]. \tag{7.31}$$

Substituting integrals (7.29) and (7.31) in (7.14), after more transformations the following ordinary differential equation is obtained:

$$c(T_1)\rho(T_1)\frac{A_{123}}{3}\frac{dT_1}{dt} = \lambda_x(T_o)\frac{y_c - y_a}{2A_{123}}[(y_2 - y_3)T_1 + (y_3 - y_1)T_2 + (y_1 - y_2)T_3 \\ + \lambda_y(T_o)\frac{x_c - x_a}{2A_{123}}[(x_2 - x_3)T_1 + (x_3 - x_1)T_2 + (x_1 - x_2)T_3].$$

$$(7.32)$$

Equation (7.32) does not take account of the boundary conditions on sides 1-a and 1-c, i.e. it is assumed that sides 1-a and 1-c are thermally insulated.

If side 1-3 is heated with heat flux $\dot{q}_s$ and if the convective heat transfer, defined by the heat transfer coefficient α and the fluid temperature T_f, occurs on side 1-2, additional integrals should be taken into account in Eq. (7.14):

$$-\int_1^c \dot{\mathbf{q}} \cdot \mathbf{n} ds = \dot{q}_s s_{1c},$$

$$(7.33)$$

where side length s_{1c} is defined as follows:

$$s_{1c} = \sqrt{(x_c - x_1)^2 + (y_c - y_1)^2}$$

$$(7.34)$$

and

$$-\int_1^a \dot{\mathbf{q}} \cdot \mathbf{n} ds = \alpha\left(T_f - \frac{T_1 + T_a}{2}\right)s_{1a} = \alpha\left(T_f - \frac{T_1 + \frac{T_1 + T_2}{2}}{2}\right)s_{1a} \\ = \alpha\left[T_f - \left(\frac{3T_1}{4} + \frac{T_2}{4}\right)\right]s_{1a},$$

$$(7.35)$$

where the following formula defines side length s_{1a}:

$$s_{1a} = \sqrt{(x_a - x_1)^2 + (y_a - y_1)^2}.$$

$$(7.36)$$

Taking account of integrals (7.33) and (7.35) in Eqs. (7.14), (7.32) takes the following form:

$$c(T_1)\rho(T_1)\frac{A_{123}}{3}\frac{dT_1}{dt} = \lambda_x(T_o)\frac{y_c - y_a}{2A_{123}}[(y_2 - y_3)T_1 + (y_3 - y_1)T_2 \\ + (y_1 - y_2)T_3] + \lambda_y(T_o)\frac{x_c - x_a}{2A_{123}}[(x_2 - x_3)T_1 + (x_3 - x_1)T_2 \\ + (x_1 - x_2)T_3] + \dot{q}_s s_{1c} + \alpha\left[T_f - \left(\frac{3T_1}{4} + \frac{T_2}{4}\right)\right]s_{1a}.$$

$$(7.37)$$

In Eqs. (7.32) and (7.37), specific heat c and density ρ are calculated in node 1, around which a control area is constructed, and thermal conductivity λ is calculated in the temperature of the centre of gravity of the 1-2-3 triangle. The temperature in the centre of gravity of triangle 1-2-3 is:

$$T_o = \frac{T_1 + T_2 + T_3}{3}. \qquad (7.38)$$

If on the triangle surface a convective heat transfer occurs to the environment with temperature T_f, a term that takes account of this heat exchange should be added to the right side of Eq. (7.38). If the heat transfer is analysed in a fin with thickness δ_f exchanging heat on both surfaces, the energy balance is performed for a triangle with surface area A_{123} and half of the thickness: $\delta_f/2$. Equation (7.37) is then as follows:

$$
\begin{aligned}
c(T_1)\rho(T_1)\frac{A_{123}}{3}\frac{\delta_f}{2}\frac{dT_1}{dt} &= \lambda_x(T_o)\frac{y_c - y_a}{2A_{123}}\frac{\delta_f}{2}[(y_2 - y_3)T_1 + (y_3 - y_1)T_2 \\
&\quad + (y_1 - y_2)T_3] + \lambda_y(T_o)\frac{x_c - x_a}{2A_{123}}\frac{\delta_f}{2}[(x_2 - x_3)T_1 \\
&\quad + (x_3 - x_1)T_2 + (x_1 - x_2)T_3] + \dot{q}_s s_{1c}\frac{\delta_f}{2} \\
&\quad + \left[T_f - \left(\frac{3T_1}{4} + \frac{T_2}{4}\right)\right]s_{1a}\frac{\delta_f}{2} + \alpha A_{123}(T_f - T_o).
\end{aligned}
\qquad (7.39)
$$

Equation (7.39) is the basis for the formulation of the energy conservation equation for the control area. If node 1 is common to n triangles forming the control area (control volume), Eq. (7.30) is written for each triangle, and then all the equations are added by sides. In this way, a first-order differential equation with respect to time is obtained for the temperature in node 1.

The FVM-FEM method advantage is higher accuracy compared to the traditional finite element approach based on the Galerkin method [177, 263]. Absolute values of the heat flow rate on the control area boundary are identical irrespective of whether they are calculated for the control area under analysis or an adjacent control area sharing a common boundary part. It results from heat flow rate continuity through the common part of the boundary between two adjacent control areas. This ensures stability of the space marching methods used to solve inverse heat conduction problems [82, 330]. In the case of the finite element method based on the Galerkin approach, the marching method used to solve the inverse problem gives unstable results.

Method II

An analysis will be conducted of the two-dimensional temperature field described by Eq. (7.16) under the following assumptions: $\lambda_x = \lambda_x(T_o)$, $\lambda_y = \lambda_y(T_o)$, $c = c(T_1)$ and $\rho = \rho(T_1)$ (cf. Fig. 7.8). Equation (7.12) will be transformed using the Green formula [112, 250]. Integration is performed anticlockwise along the boundary of the domain 1-*a*-*o*-*c*-1. Considering that the heat flux vector is given by:

$$\dot{q} = -\left[\lambda_x(T_o)\frac{\partial T}{\partial x}\mathbf{i} + \lambda_y(T_o)\frac{\partial T}{\partial y}\mathbf{j}\right] = -G\mathbf{i} + F\mathbf{j}. \tag{7.40}$$

Introducing the following symbols:

$$F = -\lambda_y(T_o)\frac{\partial T}{\partial y} \text{ and } G = \lambda_x(T_o)\frac{\partial T}{\partial x} \tag{7.41}$$

and using the Green theorem, the following is obtained:

$$\int_\Omega \left(\frac{\partial G}{\partial x} - \frac{\partial F}{\partial y}\right) dx\,dy = \int_\Gamma (F\,dx + G\,dy), \tag{7.42}$$

$$\int_\Omega \left\{\frac{\partial}{\partial x}\left[\lambda_x(T_o)\frac{\partial T}{\partial x}\right] + \frac{\partial}{\partial y}\left[\lambda_y(T_o)\frac{\partial T}{\partial y}\right]\right\} dx\,dy$$

$$= \int_\Gamma \left[-\lambda_y(T_o)\frac{\partial T}{\partial y}\,dx + \lambda_x(T_o)\frac{\partial T}{\partial x}\,dy\right]. \tag{7.43}$$

Integrals on sides *a*-*o* and *o*-*c* will be determined using Eq. (7.43) for the 1-*a*-*o*-*c*-1 domain, considering that $\lambda_y(T_o)\frac{\partial T}{\partial y}$ and $\lambda_x(T_o)\frac{\partial T}{\partial x}$ are constant in the 1-2-3 triangle. Taking account of formulae (7.25) and (7.26), the result is:

$$\int_a^o -\lambda_y(T_o)\frac{\partial T}{\partial y}\,dx = -\lambda_y(T_o)\frac{\partial T}{\partial y}\int_a^o dx$$

$$= -\lambda_y(T_o)(x_o - x_a)\left\{\frac{1}{2A_{123}}\left[(x_3 - x_2)T_1 + (x_1 - x_3)T_2 + (x_2 - x_1)T_3\right]\right\}, \tag{7.44}$$

$$\int_a^o \lambda_x(T_o)\frac{\partial T}{\partial x}\,dy = \lambda_x(T_o)\frac{\partial T}{\partial x}\int_a^o dy$$

$$= \lambda_x(T_o)(y_o - y_a)\left\{\frac{1}{2A_{123}}\left[(y_2 - y_3)T_1 + (y_3 - y_1)T_2 + (y_1 - y_2)T_3\right]\right\}. \tag{7.45}$$

The right side of Eq. (7.43) for side a-o is then of the following form:

$$\int\limits_a^o \left[-\lambda_y(T_o)\frac{\partial T}{\partial y}\,dx + \lambda_x(T_o)\frac{\partial T}{\partial x}\,dy \right] = \frac{\lambda_y(T_o)}{2A_{123}}[(x_3 - x_2)T_1$$

$$+ (x_1 - x_3)T_2 + (x_2 - x_1)T_3](x_a - x_o)$$

$$- \frac{\lambda_x(T_o)}{2A_{123}}[(y_2 - y_3)T_1$$

$$+ (y_3 - y_1)T_2 + (y_1 - y_2)T_3](y_a - y_o).$$

$$(7.46)$$

The next to be calculated is integral on side o-c:

$$\int\limits_o^c (F\,dx + G\,dy) = \int\limits_o^c \left[-\lambda_y(T_o)\frac{\partial T}{\partial y}\,dx + \lambda_x(T_o)\frac{\partial T}{\partial x}\,dy \right]. \qquad (7.47)$$

Considering that

$$\int\limits_o^c -\lambda_y(T_o)\frac{\partial T}{\partial y}\,dx = -\lambda_y(T_o)\,(x_c - x_o)\frac{1}{2A_{123}}[(x_3 - x_2)\,T_1$$

$$+ (x_1 - x_3)\,T_2 + (x_2 - x_1)\,T_3],$$

$$\int\limits_o^c \lambda_x(T_o)\frac{\partial T}{\partial x}\,dy = \lambda_x(T_o)\,(y_c - y_o)\frac{1}{2A_{123}}[(y_2 - y_3)\,T_1$$

$$+ (y_3 - y_1)\,T_2 + (y_1 - y_2)\,T_3]$$

$$(7.48)$$

the following is obtained:

$$\int\limits_o^c \left[-\lambda_y(T_o)\frac{\partial T}{\partial y}\,dx + \lambda_x(T_o)\frac{\partial T}{\partial x}\,dy \right]$$

$$= -\lambda_y(T_o)\,(x_c - x_o)\frac{1}{2A_{123}}[(x_3 - x_2)\,T_1 + (x_1 - x_3)\,T_2 + (x_2 - x_1)\,T_3] \qquad (7.49)$$

$$+ \lambda_x(T_o)\,(y_c - y_o)\frac{1}{2A_{123}}[(y_2 - y_3)\,T_1 + (y_3 - y_1)\,T_2 + (y_1 - y_2)\,T_3].$$

Consequently, the surface integral on the right side of Eq. (7.43) can be written as:

$$
\int_a^o \left[-\lambda_y(T_o) \frac{\partial T}{\partial y} \, dx + \lambda_x(T_o) \frac{\partial T}{\partial x} \, dy \right] + \int_0^c \left[-\lambda_y(T_o) \frac{\partial T}{\partial y} \, dx + \lambda_x(T_o) \frac{\partial T}{\partial x} \, dy \right]
$$

$$
= -\lambda_x(T_o) \frac{y_a - y_o}{2A_{123}} \left[(y_2 - y_3) T_1 + (y_3 - y_1) T_2 + (y_1 - y_2) T_3 \right]
$$

$$
+ \lambda_y(T_o) \frac{x_a - x_o}{2A_{123}} \left[(x_3 - x_2) T_1 + (x_1 - x_3) T_2 + (x_2 - x_1) T_3 \right]
$$

$$
+ \lambda_x(T_o) \frac{y_c - y_o}{2A_{123}} \left[(y_2 - y_3) T_1 + (y_3 - y_1) T_2 + (y_1 - y_2) T_3 \right]
$$

$$
- \lambda_y(T_o) \frac{x_c - x_a}{2A_{123}} \left[(x_3 - x_2) T_1 + (x_1 - x_3) T_2 + (x_2 - x_1) T_3 \right].
$$

$$(7.50)$$

After the terms are grouped, expression (7.50) can be written as:

$$
\int_a^o \left[-\lambda_y(T_o) \frac{\partial T}{\partial y} \, dx + \lambda_x(T_o) \frac{\partial T}{\partial x} \, dy \right] + \int_0^c \left[-\lambda_y(T_o) \frac{\partial T}{\partial y} \, dx + \lambda_x(T_o) \frac{\partial T}{\partial x} \, dy \right]
$$

$$
= \lambda_x(T_o) \frac{y_c - y_a}{2A_{123}} \left[(y_2 - y_3) T_1 + (y_3 - y_1) T_2 + (y_1 - y_2) T_3 \right]
$$

$$
- \lambda_y(T_o) \frac{x_c - x_a}{2A_{123}} \left[(x_3 - x_2) T_1 + (x_1 - x_3) T_2 + (x_2 - x_1) T_3 \right].
$$

$$(7.51)$$

The energy conservation equation for the 1-a-o-c-1 control area around node 1, where sides 1-a and c-1 are thermally insulated and where the heat transfer occurs through sides a-o and o-c is as follows:

$$
c(T_1)\rho(T_1) \frac{A_{123}}{3} \frac{dT_1}{dt} = \lambda_x(T_o) \frac{y_c - y_a}{2A_{123}} \left[(y_2 - y_3) T_1 + (y_3 - y_1) T_2 + (y_1 - y_2) T_3 \right]
$$

$$
- \lambda_y(T_o) \frac{x_c - x_a}{2A_{123}} \left[(x_3 - x_2) T_1 + (x_1 - x_3) T_2 + (x_2 - x_1) T_3 \right].
$$

$$(7.52)$$

This is the basic equation for making energy balance equations for a finite area (volume). Equations (7.32) and (7.52) are identical. Terms taking account of the boundary conditions should be added to the right side of Eq. (7.52) so that the conditions can be taken into consideration on sides 1-a and 1-c and on the perimeter of the 1-a-o-c-1 area. The effect of the procedure is Eq. (7.39).

Method III

Application of the divergence theorem to the transient heat conduction equation results in Eq. (7.13), where surface integral $-\int_\Gamma \dot{\mathbf{q}} \cdot \mathbf{n}\,\mathrm{d}S$ appears on the right side. In the case of a two-dimensional temperature field, a curvilinear integral occurs, which can be calculated as described below.

Considering that integration along the domain boundary is performed anti-clockwise, the vector differential element of a curvilinear boundary $(-\mathbf{n}\mathrm{d}s)$ is found from the following formula (cf. Fig. 7.9)

$$
\begin{aligned}
-\int_\Gamma \dot{\mathbf{q}} \cdot \mathbf{n}\,\mathrm{d}s &= -\int_\Gamma \left(\dot{q}_x \mathbf{i} + \dot{q}_y \mathbf{j} \right) \cdot \left(-\mathbf{j}\mathrm{d}x + \mathbf{i}\mathrm{d}y \right) \\
&= \int_\Gamma \dot{q}_y \mathrm{d}x - \int_\Gamma \dot{q}_x \mathrm{d}y = \int_\Gamma \left(-\lambda_y \frac{\partial T}{\partial y}\mathrm{d}x + \lambda_x \frac{\partial T}{\partial x}\mathrm{d}y \right).
\end{aligned}
\tag{7.53}
$$

Assuming in (7.53) that $\lambda_x = \lambda_x(T_o)$ and $\lambda_y = \lambda_y(T_o)$, an expression is obtained which is the same as in Method II and which is defined by formula (7.43).

Methods II and III enable a straightforward calculation of the surface integral appearing in formula (7.13) for any shape of the finite element and any function interpolating the temperature distribution inside the element. If boundary 1–2 of area Ω (cf. Fig. 7.10) is adjacent to a tetragonal finite element, the temperature distribution inside the element can be approximated by means of the following function:

$$
T(x, y) = a_1 + a_2 x + a_3 y + a_4 xy.
\tag{7.54}
$$

Derivatives $\partial T/\partial x$ and $\partial T/\partial y$ are then defined as:

$$
\frac{\partial T}{\partial x} = a_2 + a_4 y,
\tag{7.55}
$$

Fig. 7.9 Diagram illustrating determination of the differential vector element of curvilinear coordinate s

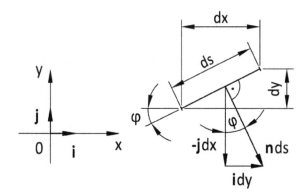

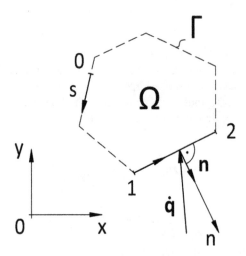

Fig. 7.10 Diagram illustrating determination of the heat flow rate through boundary 1-2 of area Ω

$$\frac{\partial T}{\partial y} = a_3 + a_4 x. \tag{7.56}$$

Integral (7.53) on boundary 1–2 (cf. Fig. 7.10) can, in this case, be calculated from the following formula:

$$
\begin{aligned}
-\int_{x_1}^{x_2} \dot{\mathbf{q}} \cdot \mathbf{n} ds &= \int_{\Gamma_{1-2}} \left(-\lambda_y \frac{\partial T}{\partial y} dx + \lambda_x \frac{\partial T}{\partial x} dy \right) \\
&= -\int_{x_1}^{x_2} \lambda_y (a_3 + a_4 x) dx + \int_{y_1}^{y_2} \lambda_x (a_2 + a_4 y) dy.
\end{aligned}
\tag{7.57}
$$

If thermal conductivities λ_x and λ_y are constant, the integrals in expression (7.57) can be calculated easily. The temperature-dependent coefficients λ_x and λ_y can be approximated using their mean values expressed as follows:

$$\bar{\lambda}_x = \frac{1}{2}[\lambda_x(T_1) + \lambda_x(T_2)], \quad \bar{\lambda}_y = \frac{1}{2}[\lambda_y(T_1) + \lambda_y(T_2)] \tag{7.58}$$

and the integrals from expression (7.57) can be calculated like for constant values of coefficients λ_x and λ_y. The surface integral defined by formula (7.53) can be determined for other forms of interpolating functions in the same manner.

The calculations of the heat flowing through the control area boundary can be simplified considering that the sum of heat flow rates $\dot{Q}_{ao}$ and $\dot{Q}_{oc}$ through the boundary straight sections is equal to the heat flow rate $\dot{Q}_{ac}$ through a straight

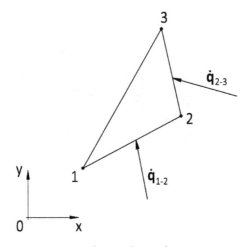

Fig. 7.11 Diagram illustrating equality $\dot{Q}_{1-3} = \dot{Q}_{1-2} + \dot{Q}_{2-3}$

section connecting points a and c. This can also be seen analysing the right side of Eq. (7.32), expressing the heat flow rate reaching control area $1\text{-}a\text{-}o\text{-}c\text{-}1$ through the straight section connecting points a and c. In the general case (cf. Fig. 7.11), this can be expressed as follows:

$$\dot{Q}_{1-3} = \dot{Q}_{1-2} + \dot{Q}_{2-3}, \qquad (7.59)$$

where

$$\dot{Q}_{1-3} = \int_{s_1}^{s_3} \dot{\mathbf{q}}_{1-3} \cdot \mathbf{n}_{1-3} \, ds, \quad \dot{Q}_{1-2} = \int_{s_1}^{s_2} \dot{\mathbf{q}}_{1-2} \cdot \mathbf{n}_{1-2} \, ds, \quad \dot{Q}_{2-3} = \int_{s_2}^{s_3} \dot{\mathbf{q}}_{2-3} \cdot \mathbf{n}_{2-3} \, ds.$$

$$(7.60)$$

In the case of the entire control area (cf. Fig. 7.12), heat flow rate $\dot{Q}_{1-2-3-4-5-6-1}$ into the control volume through the sides of the 1-2-3-4-5-6-1 polygon marked by a continuous line is equal to heat flow rate $\dot{Q}_{A-B-C-D-E-F-A}$ through the sides of the $A\text{-}B\text{-}C\text{-}D\text{-}E\text{-}F\text{-}A$ polygon marked in Fig. 7.12 by a dotted line.

Equation (7.32) is the basis for the temperature field calculation. If around point n a control area is built with surface area $A_{cv,n} = \sum_{i=1}^{n} \frac{1}{3} A_i$, where A_i is the surface area of a triangle with a vertex located at point n, the energy balance equation can be written as:

$$c(T_n) \, \rho(T_n) \, A_{cv,n} \frac{dT_n}{dt} = \sum_{i=1}^{N} \dot{Q}_i, \qquad (7.61)$$

where N is the number of triangles with a common point n.

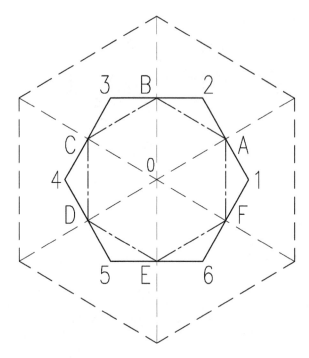

Fig. 7.12 Diagram illustrating the equality of heat flow rates $\dot{Q}_{1-2-3-4-5-6-1}$ and $\dot{Q}_{A-B-C-D-E-F-A}$

Equation (7.61) or a system of equations for the entire area under analysis can be solved using the explicit Euler method [311, 331]:

$$c(T_n^n)\,\rho(T_n^n)\,\frac{T_n^{n+1} - T_n^n}{\Delta t} = \sum_{i=1}^{N} \dot{Q}_i^n. \tag{7.62}$$

The solution of Eq. (7.62) is then of the following form:

$$T_n^{n+1} = \frac{\Delta t}{c(T_n^n)\,\rho(T_n^n)} \sum_{i=1}^{N} \dot{Q}_i^n + T_n^n. \tag{7.63}$$

In order to ensure the stability of the calculations, an appropriately small time step must be used [98, 111, 127, 152, 160, 182, 203]. This, however, does not hinder practical calculations, considering the high computing power of state-of-the-art personal computers. Although implicit methods enable integration of Eq. (7.63) at a bigger time step, only an appropriately small time step makes it possible to achieve a satisfactory accuracy of the calculations. The calculations of the transient temperature field in rectangular fins fixed on oval tubes performed by means of the explicit Euler method and the fifth-order Runge-Kutta-Verner method [134, 239] prove that both procedures give practically identical results [158, 311, 316].

7.2.2.2 Example Application of the FVM-FEM for Determination of the Temperature Distribution in a Tube with a Complex Cross-Section Shape

This subsection presents an example application of the FVM-FEM method for determination of the steady-state temperature field in the cross-section of a *double omega* tube (cf. Fig. 7.3a). The calculations are performed for three different passes of the platen superheater in the OFZ-425 fluidized-bed boiler. The passes are made of the 15Mo3, 13CrMo44 and 10CrMo910 steel grades. Temperature-dependent changes in thermal conductivity are shown in Fig. 7.13. In all three passes the outer dimensions of the tubes are identical. However, the tubes differ in the inner surface radius, which is: $r_w = 13.4$ mm, $r_w = 12.7$ mm, and $r_w = 11.9$ mm (cf. Fig. 7.1) for pass I, II and III, respectively. The steady-state heat conduction equation describing the temperature distribution in the tube cross-section has the following form:

$$\frac{\partial}{\partial x}\left[\lambda_w(T) \frac{\partial T_w}{\partial x}\right] + \frac{\partial}{\partial y}\left[\lambda_w(T) \frac{\partial T_w}{\partial y}\right] = 0. \tag{7.64}$$

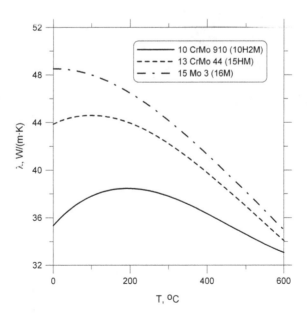

Fig. 7.13 Temperature-dependent changes in thermal conductivity

The boundary conditions set on the tube surface (cf. Fig. 7.14) are as follows:

$$\lambda_w \frac{\partial T_w}{\partial n}\Big|_{A-B} = \alpha_1 \left(T_1 - T_w\big|_{A-B}\right), \tag{7.65}$$

$$\lambda_w \frac{\partial T_w}{\partial n}\Big|_{B-C-D-E} = 0, \tag{7.66}$$

$$\lambda_w \frac{\partial T_w}{\partial n}\Big|_{E-F} = \alpha_2 \left(T_2 - T_w\big|_{E-F}\right), \tag{7.67}$$

$$\lambda_w \frac{\partial T_w}{\partial n}\Big|_{A-F} = 0. \tag{7.68}$$

The calculations are performed using the following data: furnace chamber fluidized-bed temperature $T_2 = 850\ °C$, steam-side heat transfer coefficient $\alpha_1 = 3500\ W/(m^2K)$, flue-gas side heat transfer coefficient $\alpha_2 = 200\ W/(m^2K)$ [317]. The division of the cross-section quarter into finite elements is presented in

Fig. 7.14 A quarter of the *double omega* tube cross-section with set boundary conditions

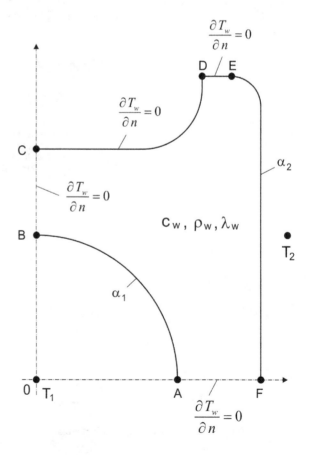

Table 7.1 Fluid temperature and temperatures in nodes

Steel	T_1, °C	T_{w1}, °C	T_{w2}, °C	T_{w3}, °C	T_{w4}, °C	T_{w5}, °C
15Mo3	425.00	455.28	452.56	443.70	449.04	452.86
13CrMo44	460.00	488.66	486.21	478.63	484.79	487.91
10CrMo910	495.00	522.10	519.94	513.50	520.70	523.23

Fig. 7.4. The fluid temperature inside the tubes and the results of the steady-state temperature field calculations are listed in Table 7.1.

In the case of pass III made of German 10CrMo910 steel, calculations were also performed for a constant thermal conductivity value λ_w = 33.8 W/(mK).

The following steady-state values of the wall temperature were obtained in individual nodes: T_{w1} = 522.10 °C, T_{w3} = 513.50 °C, T_{w4} = 520.70 °C, T_{w8} = 537.79 °C, T_{w9} = 550.71 °C, T_{w10} = 555.39 °C, T_{w11} = 549.89 °C.

The determined temperatures demonstrate very good agreement with the temperature values obtained from the ANSYS program at the tube quarter division into 21101 finite elements: T_{w1} = 519.14 °C, T_{w3} = 514.71 °C, T_{w4} = 523.44 °C, T_{w8} = 536.64 °C, T_{w9} = 554.14 °C, T_{w10} = 558.84 °C, T_{w11} = 552.14 °C.

An application of the FVM-FEM method to determine the unsteady-state temperature field in ceramic elements of a heat accumulator is presented in [253, 254]. The heat accumulator computations performed using the Ansys-CFX program confirm the high accuracy of the accumulator in-house mathematical model, where the calculations of the unsteady-state temperature distribution in the ceramic elements are made using the FVM-FEM method.

7.3 Overall Heat Transfer Coefficient

Overall heat transfer coefficient k is the inverse of the equivalent heat resistance which takes account of the wall heat resistance and heat resistance of the convective heat transfer on both sides of the wall.

This section discusses determination of overall heat transfer coefficient k for smooth non-finned tubes and walls and also for finned surfaces.

7.3.1 Bare (Non-finned) Tubes with Circular, Oval and Elliptical Cross-Sections

In the case of circular tubes with the outer and the inner surface radius r_z and r_w, respectively, overall heat transfer coefficient k is defined as:

$$\dot{Q} = kA_{zrg}\left(T_{cz} - T_p\right), \tag{7.69}$$

where $\dot{Q}$—liquid-to-air heat flow rate (W), A_{zrg}—non-finned tube outer surface area (m^2), T_{cz}—the liquid temperature inside the tube (°C), T_p—ambient air temperature (°C).

The temperature difference $(T_{cz} - T_p)$ obtained from formula (7.69) can be expressed as follows:

$$T_{cz} - T_p = \frac{\dot{Q}}{A_{zrg}k} = \frac{\dot{q}_z}{k}, \tag{7.70}$$

where $\dot{q}_z$ is the heat flux on surface area A_{zrg}. The temperature difference $(T_c - T_p)$ is made up of three temperature drops (cf. Fig. 7.15):

$$T_{cz} - T_p = \Delta T_w + \Delta T_{\acute{s}c} + \Delta T_z, \tag{7.71}$$

where the drops in temperature are defined by the following formulae:

$$\Delta T_w = \frac{\dot{q}_z\, r_z}{r_w} \frac{1}{\alpha_c} = \frac{\dot{q}_w}{\alpha_c}, \tag{7.72}$$

$$\Delta T_{\acute{s}c} = \frac{\dot{q}_z\, r_z}{\lambda} \ln\!\left(\frac{r_z}{r_w}\right), \tag{7.73}$$

$$\Delta T_z = \frac{\dot{q}_z}{\alpha_{zr}}. \tag{7.74}$$

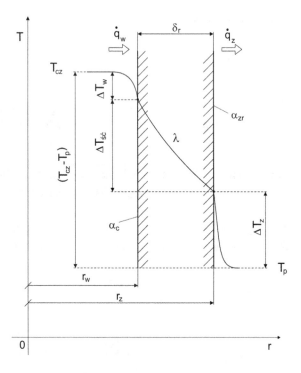

Fig. 7.15 Heat transfer through a cylindrical wall from a hot liquid with temperature T_{cz} to air with temperature $T_p < T_{cz}$

The following symbols are adopted in formulae (7.72)–(7.74): $\dot{q}_w$—heat flux on the tube inner surface (W/m^2), α_c—liquid-side (the tube inner surface) heat transfer coefficient (W/(m^{2}K)), α_{zr}—gas-side (the tube outer surface) weighted heat transfer coefficient (W/(m^{2}K)).

For a smooth non-finned tube it should be assumed that $\alpha_{zr} = \alpha_p$, where α_p is the gas-side heat transfer coefficient. The other symbols are shown in Fig. 7.15.

Substituting (7.72), (7.73), (7.74) and (7.70) in (7.71), the following is obtained:

$$\frac{\dot{q}_z}{k} = \frac{\dot{q}_z\, r_z}{r_w\, \alpha_c} + \frac{\dot{q}_z\, r_z}{\lambda}\ln\left(\frac{r_z}{r_w}\right) + \frac{\dot{q}_z}{\alpha_{zr}}, \tag{7.75}$$

from which it follows that

$$\frac{1}{k} = \frac{r_z}{r_w}\frac{1}{\alpha_c} + \frac{r_z}{\lambda}\ln\left(\frac{r_z}{r_w}\right) + \frac{1}{\alpha_{zr}}. \tag{7.76}$$

After overall heat transfer coefficient k is found from formula (7.76), heat flow rate $\dot{Q}$ can be determined using formula (7.69). For a flat tube, the overall heat transfer coefficient k is defined by the following formula:

$$\frac{1}{k} = \frac{1}{\alpha_c} + \frac{\delta_r}{\lambda} + \frac{1}{\alpha_{zr}}, \tag{7.77}$$

where δ_r is the wall thickness (m).

Formula (7.76) can also be applied to tubes with cylindrical, elliptical and other shapes, for channels with a very small wall thickness (δ_r) or characterized by high thermal conductivity λ. Using the symbols A_{zrg} and A_{wrg} to denote the non-finned channel outer and inner surface area, respectively, the overall heat transfer coefficient related to the channel outer surface area A_{zrg} is defined as:

$$\frac{1}{k} = \frac{A_{zrg}}{A_{wrg}}\frac{1}{\alpha_c} + \frac{A_{zrg}}{A_m}\frac{\delta_r}{\lambda} + \frac{1}{\alpha_{zr}}, \tag{7.78}$$

where $A_m = \left(A_{zrg} + A_{wrg}\right)/2$.

The heat flow rate exchanged between the fluids is calculated from formula (7.69), where the overall heat transfer coefficient is defined by formula (7.78).

7.3.2 Finned Tubes

Calculations of the overall heat transfer coefficient in heat exchangers with individual fins or with fins which are continuous (lamella exchangers) are performed in a very similar manner. The continuous fin (lamella) is divided into conventional rectangular fins for the in-line tube arrangement or into hexagonal fins—for the staggered configuration. In order to simplify the calculations of finned surfaces, weighted heat transfer coefficient α_{zr} is introduced on the gas side, which is an equivalent heat transfer coefficient taking account of the presence of fins. The coefficient will be related to the outer surface area of a non-finned tube. The heat transferred from the fin and the tube surfaces to the flowing gas can be expressed as follows (cf. Fig. 7.16):

$$\alpha_{zr} A_{zrg} \left(T_{rg} - T_{cz}\right) = \alpha_p A_{mz} \left(T_{rg} - T_{cz}\right) + \alpha_p A_z \eta_z \left(T_{rg} - T_{cz}\right), \qquad (7.79)$$

which gives the following formula:

$$\alpha_{zr} = \alpha_p \left(\frac{A_{mz}}{A_{zrg}} + \frac{A_z}{A_{zrg}} \eta_z \right), \qquad (7.80)$$

where α_{zr}—gas-side weighted heat transfer coefficient related to the non-finned tube outer surface area (W/(m^2 K)), α_p—gas-side heat transfer coefficient, (W/(m^{2}K)), A_{mz}—non-finned tube surface area in between the fins (m^2), A_{zrg}—non-finned tube (outer) surface area (m^2), A_z—surface area of the fins (m^2), η_z—fin efficiency defined as the ratio between heat flow rate transferred by a real fin to the heat flow rate transferred by an isothermal fin with the temperature of the fin base T_b.

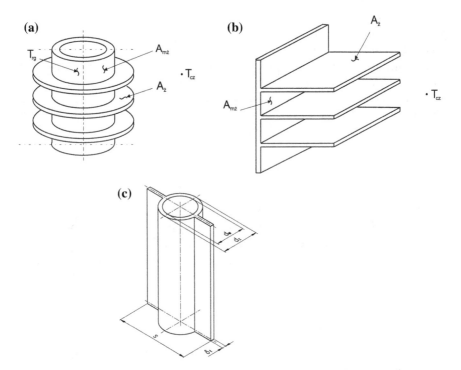

Fig. 7.16 Diagram of finned surfaces: **a** tube with round fins, **b** flat surface with straight fins, **c** longitudinally finned tube

It is possible to determine the overall heat transfer coefficient k related to the non-finned tube outer surface A_{zrg} using formula (7.78) knowing the weighted heat transfer coefficient α_{zr}.

Figure 7.17 shows a circular cross-section tube with round fins (Fig. 7.17a), a finned tube with straight fins (Fig. 7.17b), longitudinally finned tubes (Fig. 7.17c) and what is referred to as the membrane heating surface (Fig. 17.7d).

Further below formulae are given for the calculation of surface areas A_z, A_{zrg} and A_{mz}, which are necessary to determine the weighted heat transfer coefficient using formula (7.80). Assuming that the pitch of both round and straight fins is s, the respective surface areas are defined by the following expressions:

– round fins fixed on a tube with a circular cross-section (cf. 7.17a)

$$A_z = 2\pi\left(r_2^2 - r_1^2\right) n\, m_t, \quad A_{zrg} = 2\pi r_1 s\, n\, m_t, \quad A_{mz} = 2\pi r_1 \left(s - \delta_f\right) n\, m_t, \quad (7.81)$$

Fig. 7.17 Selected types of finned surfaces: **a** circular cross-section tube with individual round fins, **b** flat wall with fins, **c** longitudinally finned tube, **d** membrane heating surface

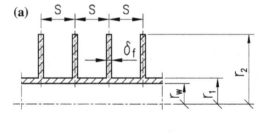

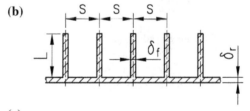

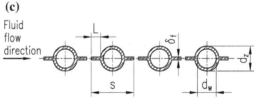

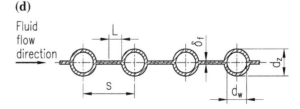

– straight fins on a flat wall (cf. Fig. 7.17b)

$$A_z = 2\,Lw\,n\,m_t, \quad A_{zrg} = \left(s - \delta_f\right)w\,n\,m_t, \quad A_{mz} = \left(s - \delta_f\right)w\,n\,m_t, \tag{7.82}$$

where w is the straight fin length, n—the number of fins, and m_f—the number of tubes.

In order to take account of the heat transfer on the fin tip, the round fin radius or the flat fin height can be increased by $\delta_f/2$, i.e. the round fin bigger radius and the straight fin larger height are $(r_2 + \delta_f/2)$ and $(L + \delta_f/2)$, respectively.

Respective surface areas of an exchanger made of m_t longitudinally finned tubes (cf. Fig. 7.17c) or of an exchanger made of membrane heating surfaces where the number of tubes is m_t (cf. Fig. 7.17d) are defined by the following formulae:

$$A_z = 2 \cdot 2\,L\,s\,m_t, \quad A_{zrg} = 2\,\pi\,r_z\,w\,m_t, \quad A_{mz} = 2\,(\pi - 2\,\varphi)\,r_z\,w\,m_t, \tag{7.83}$$

where

$$L = \frac{s}{2} - r_z \cos\varphi, \quad \varphi = \arcsin\left(\frac{\delta_f}{2\,r_z}\right). \tag{7.84}$$

Formulae (7.81)–(7.84) enable calculations of the weighted heat transfer coefficient using formula (7.80) for the finned surfaces presented in Fig. 7.17.

7.4   Fin Efficiency

Fins are fixed on tubes on the gas side to enhance the heat transfer from the hot to the cold medium. In order to simplify the finned-tube heat exchanger calculations, the notion of the fin efficiency is introduced. The fin efficiency is defined as the ratio between the heat flow rates of a real fin and an isothermal fin with the temperature of the fin base T_b. It is assumed that the fin base temperature is equal to the temperature of the surface that the fin is fixed to. For fins with simple shapes, such as straight or round fins with a constant thickness, analytical formulae can be derived to determine their efficiency [21, 23, 34, 65, 123,125, 159, 312, 331]. In the case of fins with a complicated shape, the fin efficiency can only be found using approximate methods, mainly numerical [159, 177, 311, 316, 323, 331, 347, 355].

7.4.1   Fins with Simple Shapes

This subsection presents determination of the temperature field and efficiency of straight and rectangular fins with a constant thickness, which are often used in practice.

Straight fin with a constant thickness

The straight fin diagram is presented in Fig. 7.18 [312, 331].

It is assumed that the fin is very wide, i.e. $w \gg L$ and the heat transfer on the two side walls with the surface area of $L \times \delta_f$ each is omitted. The temperature drop on the fin thickness is also omitted. Moreover, the considerations do not take account of the heat transfer between the fin tip and the environment.

The heat transfer on the fin tip can be taken into account by increasing the fin height to $L_c = L + \delta_f/2$. The temperature field in the fin is described by the heat conduction Eq. (7.85):

$$\frac{d^2 T}{dx^2} - m^2 \left(T - T_f \right) = 0 \tag{7.85}$$

and boundary conditions (7.86)–(7.87):

$$T\big|_{x=0} = T_b, \tag{7.86}$$

$$\frac{dT}{dx}\bigg|_{x=L} = 0, \tag{7.87}$$

where parameter m^2 is defined as:

$$m^2 = \frac{\alpha P}{\lambda A}. \tag{7.88}$$

Considering that the fin perimeter P and the cross-section surface area A are $P = 2(w + \delta_f)$ and $A = w\delta_f$, respectively, parameter m^2 is expressed as:

$$m^2 = \frac{2\alpha \left(w + \delta_f \right)}{\lambda w \delta_f} \approx \frac{2\alpha}{\lambda \delta_f}, \tag{7.89}$$

Fig. 7.18 Diagram of a straight fin with height L, thickness δ_f and width w

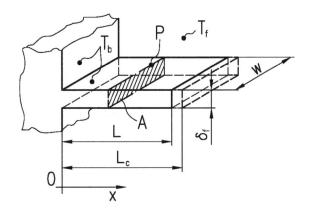

because $\delta_f \ll w$. Introducing the following new variable:

$$\theta = T - T_f \tag{7.90}$$

Equation (7.85) and boundary conditions (7.86)–(7.87) can be written in the following form:

$$\frac{d^2\theta}{dx^2} - m^2\theta = 0, \tag{7.91}$$

$$\theta|_{x=0} = \theta_b, \quad \theta_b = T_b - T_f, \tag{7.92}$$

$$\frac{d\theta}{dx}\bigg|_{x=L} = 0. \tag{7.93}$$

The solution of the boundary value problem (7.91)–(7.92) takes the following form:

$$\frac{\theta(x)}{\theta_b} = \frac{T(x) - T_f}{T_b - T_f} = \frac{\cosh m(L - x)}{\cosh mL}. \tag{7.94}$$

Heat flow rates $\dot{Q}$ and $\dot{Q}_{max}$ are defined as:

$$\dot{Q} = -\lambda A \frac{dT}{dx}\bigg|_{x=0} = -\left[\lambda w \delta_f (T_b - T_f) \frac{-m \sinh m(L - x)}{\cosh mL}\right]\bigg|_{x=0}$$

$$= \lambda w \delta_f \sqrt{\frac{2\alpha}{\lambda \delta_f}}(T_b - T_f) \operatorname{tgh} mL, \tag{7.95}$$

$$\dot{Q}_{max} = 2\alpha wL(T_b - T_f). \tag{7.96}$$

The following simple formula defines the fin efficiency:

$$\eta = \frac{\dot{Q}}{\dot{Q}_{max}} = \frac{\lambda w \delta_f \sqrt{\frac{2\alpha}{\lambda \delta_f}}(T_b - T_f) \operatorname{tgh} mL}{2\alpha wL(T_b - T_f)} = \frac{\operatorname{tgh} mL}{mL}. \tag{7.97}$$

where hyperbolic function $\operatorname{tgh} x$ is expressed as:

$$\operatorname{tgh} x = \frac{\sinh x}{\cosh x} = \frac{e^x - e^{-x}}{e^x + e^{-x}}, \quad \sinh x = \frac{e^x - e^{-x}}{2}, \quad \cosh x = \frac{e^x + e^{-x}}{2}. \tag{7.98}$$

In order to take account of the heat transfer on the fin tip, L in formulae (7.94)–(7.97) should be substituted with $L_c = L + \delta_f/2$. The temperature distribution and the fin efficiency are then defined as:

$$\frac{\theta(x)}{\theta_b} = \frac{T(x) - T_f}{T_b - T_f} = \frac{\cosh\ m(L_c - x)}{\cosh\ mL_c}, \qquad (7.99)$$

$$\eta = \frac{\text{tgh}\ mL_c}{mL_c}, \qquad (7.100)$$

where $m = \sqrt{2\alpha/\lambda\delta_f}$.

Round fin with a constant thickness

The round fin diagram is presented in Fig. 7.19.

Considering that the fin cross-section surface area that heat is transferred through is $A = 2\pi r\delta_f$ and the perimeter on which the convective heat transfer occurs is $P = 4\pi r$, parameter m is defined as:

$$m = \sqrt{\frac{\alpha P}{\lambda A}} = \sqrt{\frac{4\alpha\pi r}{2\lambda\pi r\delta_f}} = \sqrt{\frac{2\alpha}{\lambda\delta_f}}. \qquad (7.101)$$

If written in a differential form, the fin energy conservation equation is as follows:

$$\frac{1}{r}\frac{d}{dr}\left(r\frac{dT}{dr}\right) - \frac{2\alpha}{\lambda\delta_f}(T - T_f) = 0. \qquad (7.102)$$

Introducing the temperature surplus $\theta = T - T_f$, Eq. (7.102) can be written as:

$$\frac{d^2\theta}{dr^2} + \frac{1}{r}\frac{d\theta}{dr} - m^2\theta = 0. \qquad (7.103)$$

Fig. 7.19 Diagram of a round fin with height L and constant thickness δ_f

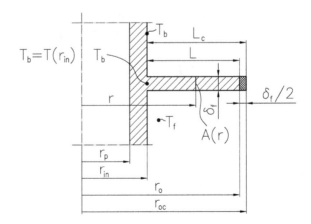

The modified Bessel Eq. (7.103) has the following solution:

$$\theta(r) = C_1 I_0(mr) + C_2 K_0(mr), \tag{7.104}$$

where constants C_1 and C_2 will be found using the following boundary conditions:

$$\theta|_{r=r_{in}} = \theta_b, \tag{7.105}$$

$$\frac{d\theta}{dr}\bigg|_{r=r_o} = 0, \tag{7.106}$$

where $\theta_b = T_b - T_f$.

After the constants are determined from conditions (7.105) and (7.106) and substituted in (7.104), an expression is obtained for the temperature distribution in the fin:

$$\frac{\theta}{\theta_b} = \frac{T(r) - T_f}{T_b - T_f} = \frac{K_0(mr)I_1(mr_o) + I_0(mr)K_1(mr_o)}{I_0(mr_{in})K_1(mr_o) + K_0(mr_{in})I_1(mr_o)}. \tag{7.107}$$

Heat flow rate $\dot{Q}$ exchanged by the fin with the environment and the maximum heat flow rate $\dot{Q}_{max}$ that the fin would exchange with the environment if the fin temperature was uniform and equal to T_b are defined by the following respective formulae:

$$\dot{Q} = -\lambda A_b \frac{dT}{dr}\bigg|_{r=r_{in}} = 2\pi\lambda \, r_{in}\delta_f\theta_b m \frac{K_1(mr_{in})I_1(mr_o) - I_1(mr_{in})K_1(mr_o)}{K_0(mr_{in})I_1(mr_o) + I_0(mr_{in})K_1(mr_o)}, \tag{7.108}$$

$$\dot{Q}_{max} = \alpha A_{fin}(T_b - T_f) = 2\pi\alpha(r_o^2 - r_{in}^2)\theta_b, \tag{7.109}$$

$$\eta = \frac{\dot{Q}}{\dot{Q}_{max}} = \frac{2r_{in}}{m(r_o^2 - r_{in}^2)} \frac{K_1(mr_{in})I_1(mr_o) - I_1(mr_{in})K_1(mr_o)}{K_0(mr_{in})I_1(mr_o) + I_0(mr_{in})K_1(mr_o)}$$

$$= \frac{2}{mL(1+u)} \frac{K_1\left(\frac{mL}{u-1}\right)I_1\left(\frac{mLu}{u-1}\right) - I_1\left(\frac{mL}{u-1}\right)K_1\left(\frac{mLu}{u-1}\right)}{K_0\left(\frac{mL}{u-1}\right)I_1\left(\frac{mLu}{u-1}\right) + I_0\left(\frac{mL}{u-1}\right)K_1\left(\frac{mLu}{u-1}\right)}, \tag{7.110}$$

where: $L = r_o - r_{in}$, $u = r_o/r_{in}$.

The heat transfer on the fin tip can be taken into account by increasing the fin tip radius by half of the fin thickness. Substituting radius $r_{oc} = r_o + \delta_f/2$ for radius r_o and increased fin height $L_c = L + \delta_f/2$ for L in formulae (7.107) and (7.110), it is possible to take account of the heat transfer between the fin tip surface and the environment. Figures 7.20 and 7.21 present the straight fin efficiency calculated from formula (7.97) and the round fin efficiency determined using formula (7.110).

Due to the complex form of expression (7.110), literature offers approximation formulae for the round fin efficiency calculation [34, 109, 258, 259, 311, 331]. The round fin efficiency is usually approximated using a function that is used for the straight fin efficiency.

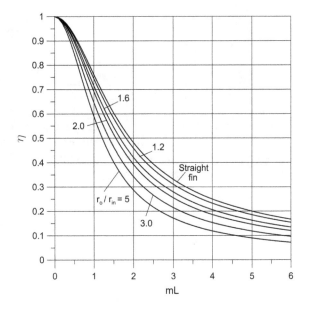

Fig. 7.20 Efficiency η of the straight and round fin with a constant thickness depending on parameter mL and as a function of the r_o/r_{in} ratio; the numbers: 1.2; 1.6; 2.0; 3.0 and 5.0 are the values of the outer-to-inner radius ratio (r_o/r_{in}) for the round fin

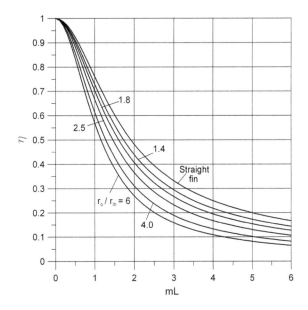

Fig. 7.21 Efficiency η of the straight and round fin with a constant thickness depending on parameter mL and as a function of the r_o/r_{in} ratio; the numbers: 1.4; 1.8; 2.5; 4.0 and 6.0 are the values of the outer-to-inner radius ratio (r_o/r_{in}) for the round fin

Round fin with a constant thickness—approximation formulae

The following approximation formulae for the calculation of the round fin efficiency can be found in literature:

- the Schmidt formula [258, 259]:

$$\eta_S = \frac{\text{tgh}\, mL_c \varphi}{mL_c \varphi}, \tag{7.111}$$

where

$$\varphi = 1 + 0.35 \ln\left(1 + \frac{L_c}{r_{in}}\right), \quad L_c = r_{oc} - r_{in}, \quad m = \sqrt{\frac{2\alpha}{\lambda \delta_f}}. \tag{7.112}$$

If the fin efficiency is higher than 0.5 ($\eta > 0.5$), the difference between efficiency η calculated using formula (7.111) and the exact value does not exceed $\pm 1\%$. Rabas and Taborek [241] do not recommend using the modifications of formula (7.111) that are available in the literature.

- the Brandt formula [34]

$$\eta_B = \frac{2r_{in}}{2r_{in} + L_c}\frac{\text{tgh}\, mL_c}{mL_c}\left[1 + \frac{\text{tgh}\, mL_c}{2mr_{in}} - C\frac{(\text{tgh}\, mL_c)^p}{(mr_{in})^n}\right], \tag{7.113}$$

where

$$C = 0.071882, p = 3.7482, n = 1.4810. \tag{7.114}$$

The maximum error in efficiency determination by means of formula (7.113) is smaller than 0.6% compared to the error in efficiency determination using formula (7.111).

- the formula presented in [109]:

$$\eta_H = \frac{1}{1 + \frac{1}{3}(mL_c)^2\sqrt{r_{oc}/r_{in}}}. \tag{7.115}$$

Formula (7.115) gives good results if the fin efficiency is higher than 0.75 ($\eta > 0.75$).

7.4.2 Fins with Complex Shapes

Tube heat exchangers with continuous (plate) fins are commonly used in water cooling systems in power plants, car radiators and in the heating and air-conditioning of buildings. Continuous fins (lamellae) are fixed on circular or oval tubes. Calculating the heat transfer in such exchangers, efficiency of rectangular or hexagonal fins has to be determined. At present, there are still no simple and accurate methods of calculating fins with complex shapes, either in steady- or unsteady-states [159, 285, 323]. Continuous fins, also known as lamellae or plate fins, can be divided into conventional rectangular fins (cf. Fig. 7.22b) in the case of the in-line configuration of the heat exchanger tubes (cf. Fig. 7.22a) or into hexagonal fins (cf. Fig. 7.22d) in the case of the tubes staggered configuration (cf. Fig. 7.22c) [158, 159, 162, 285, 311, 316].

A square fin efficiency can be calculated in approximation like the efficiency of a round fin with the same surface area [159, 169]. The accuracy of the equivalent-round-fin method may be too low, especially in the case of rectangular fins with shapes substantially different from the round fin. The comparison of the fin efficiency calculation using different approximation formulae [249, 331] indicates that the efficiency calculation error arising from the equivalent-round-fin method can be as high as 6%.

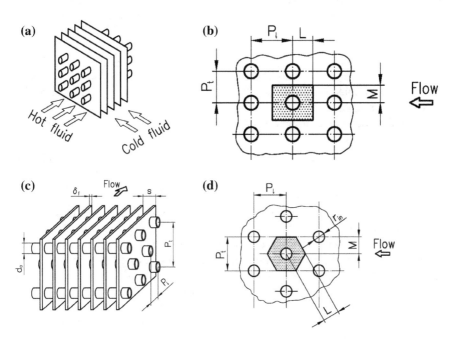

Fig. 7.22 Cross-flow tube heat exchangers with plate fins: **a** in-line arrangement of tubes, **b** conventional rectangular fin for the tubes in-line configuration, **c** staggered arrangement of tubes, **d** conventional hexagonal fin for the tubes staggered configuration

Efficiency can also be calculated in approximation using the sector method [14, 249, 270, 331], which is more accurate compared to the method based on the equivalent round fin. Schmidt [259–260] proposes approximation formulae to determine the efficiency of rectangular and hexagonal fins. Their accuracy is sufficient for engineering applications. The Schmidt method is based on determining an equivalent outer radius ($r_{o,e}$) of a circular fin with an efficiency equal to the efficiency of a rectangular or hexagonal fin. The round fin efficiency η is calculated from the following modified formula used to find the efficiency of straight fins with a constant thickness:

$$\eta = \frac{\tanh(mr_{in}\phi)}{mr_{in}\phi},\tag{7.116}$$

where parameters m and ϕ are defined as follows:

$$m = \sqrt{\frac{2\alpha}{\lambda\,\delta_f}}, \quad \phi = \left(\frac{r_{o,e}}{r_{in}} - 1\right)\left[1 + 0.35 \ln\left(\frac{r_{o,e}}{r_{in}}\right)\right].\tag{7.117}$$

The following symbols are used in formulae (7.116) and (7.117): α—heat transfer coefficient, W/(m^2 K); λ—fin material thermal conductivity, W/(m K); r_{in}, $r_{o,e}$—round fin inner and outer radius, respectively, m; δ_f—fin thickness, m.

In the case of a rectangular fin, Schmidt proposes the following correlations:

$$\frac{r_{o,e}}{r_{in}} = 1.28\,\psi(\beta - 0.2)^{1/2}, \quad \psi = \frac{M}{r_{in}}, \quad \beta = \frac{L}{M}.\tag{7.118}$$

Dimensions L and M are marked in Fig. 7.22b. They should be selected so that L is bigger or equal to M.

A similar equation is proposed for the hexagonal fin:

$$\frac{r_{o,e}}{r_{in}} = 1.27\,\psi(\beta - 0.3)^{1/2}.\tag{7.119}$$

where L and M are shown in Fig. 7.22d, and $L \geq M$.

Zeller and Grewe [376] improved the Schmidt formula (7.116), but they made it more complicated.

Another way of calculating the efficiency of fins with complex shapes is the sector method, which, although quite accurate, is somewhat laborious [164-166, 169, 249, 270, 331]. It was developed in the 1940s and is still in use today. D. G. Rich developed charts for the calculation of the fin efficiency using the sector method, which are included in the ASHRAE guidebook [14]. The smallest repetitive fragment of the rectangular fin (Fig. 7.23b) or of the hexagonal fin is divided into n sectors (segments).

Each sector with surface area A_i is treated as a circular sector with outer radius $r_{o,\,i}$ with the same surface area. Radius $r_{o,i}$ is determined from the condition of equality

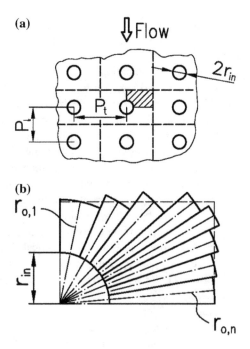

Fig. 7.23 Diagram illustrating the rectangular fin efficiency determination using the sector method: **a** division of the continuous fin into conventional rectangular fins; **b** division of the conventional fin quarter into segments

between the surface area of a given segment and the surface area of an equivalent circular sector. Efficiency η_i of the i-th circular sector is then found from formula (7.110) describing the circular fin efficiency.

The entire fin efficiency η is the area-weighted average of efficiencies η_i of all n circular sectors:

$$\eta = \frac{\sum_{i=1}^{n} \eta_i A_i}{\sum_{i=1}^{n} A_i}. \tag{7.120}$$

The fin efficiency determined from formula (7.120) is smaller than the fin actual efficiency because in the sector method the efficiencies of circular sectors are calculated from the formula describing the circular fin efficiency, which is derived assuming heat conduction in the radial direction only.

An example application of the equivalent-circular-fin method and the sector method for the hexagonal fin efficiency calculation is presented in [331].

Approximate temperature distributions and efficiencies of rectangular and hexagonal fins fixed on tubes with a circular cross-section can be determined by means of the collocation method [164, 285], which can easily be generalised to elliptical tubes [166]. First, using the variables separation method, a solution of the fin steady-state heat conduction equation is found. Then, selecting collocation points on the analysed area boundary, it is demanded that the boundary conditions in these points are satisfied.

As a result, a system of algebraic equations is obtained that is used to find the unknown coefficients in the analytical solution that exactly satisfies the heat conduction equation for the fin. The number of the algebraic equations is equal to the number of the collocation points. After the temperature distribution is determined, the heat flow rate from the fin base is calculated and then divided by the maximum heat flow rate, which gives the fin efficiency. Determination of the efficiency of rectangular and hexagonal fins using the collocation method is very complex and laborious. The solution of the two-dimensional steady-state heat conduction equation for the fin is determined in cylindrical coordinates, which means that setting the second- or the third-type boundary conditions on the fin outer boundary is more difficult because the radial coordinate is not normal to the fin boundary. The fin efficiency can be determined more easily using numerical methods, such as the finite element method or the finite volume method, using professional ANSYS-CFX, ANSYS-FLUENT, STAR-CCM+ software packages, or other programs available on the market.

If the fin shape is complicated, professional computer programs based on the finite volume method or the finite element method can be used successfully to calculate the temperature distribution in the fin, the heat transferred by the fin and the fin efficiency [296, 311, 323].

The fin efficiency is the ratio between heat flow rate $\dot{Q}$ transferred from the fin to the environment and the maximum heat flow rate $\dot{Q}_{max}$ that would be transferred from the fin to the environment if the temperature of the entire fin was equal to the temperature of the fin base T_b:

$$\eta = \frac{\dot{Q}}{\dot{Q}_{max}}. \tag{7.121}$$

Heat flow rate $\dot{Q}$ is defined as follows:

$$\dot{Q} = \int_{A_l} \alpha(T - T_f)\, dA, \tag{7.122}$$

where α is the heat transfer coefficient on the fin surface, and A_l is the fin surface area on which the heat transfer occurs. If the heat transfer coefficient α and ambient temperature T_f are constant, expression (7.122) can be written in the following form:

$$\dot{Q} = \alpha A_l(\bar{T} - T_f), \tag{7.123}$$

where $\bar{T}$ is the fin surface mean temperature calculated from the following formula:

$$\bar{T} = \frac{\int_{A_l} T\,dA}{\sum_{i=1}^{N_e} A_{e,i}} = \frac{\sum_{i=1}^{N_e} \bar{T}_{e,i}\, A_{e,i}}{A_l}, \tag{7.124}$$

where T is the temperature of the fin surface in contact with the fluid, and $\bar{T}_{e,i}$ denotes the mean temperature of the surface of the finite element. Symbol N_e denotes the number of finite elements on the surface of which the heat transfer occurs.

If the temperature field is determined using the finite volume method, symbol $\bar{T}_{e,i}$ denotes the mean temperature of the control area surface around a node, and N_e is the number of control areas. The maximum heat flow rate $\dot{Q}_{max}$ is defined as follows:

$$\dot{Q}_{max} = \alpha A_l \left(T_b - T_f \right). \tag{7.125}$$

Substitution of (7.123) and (7.125) in (7.121) gives a simple formula for the fin efficiency calculation:

$$\eta = \frac{\bar{T} - T_f}{T_b - T_f}. \tag{7.126}$$

In the calculations of the temperature distribution in fins by means of the finite element method or the finite volume method, it is easy to take account of the heat transfer coefficient dependence on location, time or the fin surface temperature.

An application of the FLUENT program will be illustrated on the example of determination of the efficiency of a rectangular fin fixed on a round tube with the diameter of 38.1 mm, as presented in Fig. 7.24.

The fin base and the flue gas temperatures are $T_b = 250\ °C$ and $T_f = 650\ °C$, respectively. The fin is made of 15Mo3 steel with a temperature-dependent thermal conductivity λ defined as:

$$\lambda = 48.4947 + 1.2455 \times 10^{-3} T - 6.4604 \times 10^{-5} T^2 + 4.1751 \times 10^{-8} T^3. \tag{7.127}$$

where λ is expressed in W/(m K), and T in °C.

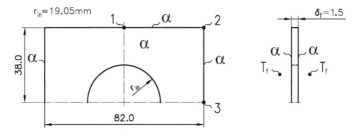

Fig. 7.24 Rectangular fin used in the boiler finned water heaters

The three-dimensional temperature field in the fin is found dividing the fin quarter into 20248 finite volumes. Then $\bar{T}$ (the mean temperature of the fin surface on which the heat transfer to the environment occurs) and the fin efficiency are determined. Figure 7.25 presents curves illustrating changes in the fin temperatures T_1, T_2 and T_3 in the points marked in Fig. 7.24, the fin mean temperature $\bar{T}$ and the fin efficiency η as a function of the heat transfer coefficient α.

In the second example, the temperature distribution in a rectangular fin fixed on an oval tube will be determined. Then, the fin efficiency will be found using formula (7.126). The analysed fin dimensions, expressed in mm, are presented in Fig. 7.26. The fin is divided into 19 finite volumes (cf. Fig. 7.27). Side areas 4–19, 1-13, 2-3 are thermally insulated, whereas the convective heat transfer occurs on areas 4-6-3 and 1-5-2. The calculations are performed for the following data: $c = 896$ J/(kg K), $\rho = 2707$ kg/m^3, $\lambda = 207$ W/(m K), $\delta_f = 0.08$ mm, $T_f = 0$ °C, $T_b = 100$ °C, $T_0 = 0$ °C, $\alpha = 50$ W/(m^2 K). The temperature distribution is also calculated using the ANSYS v.15.0. package. The fin division into finite elements used in the calculations by means of the ANSYS program is presented in Fig. 7.28 [158]. The analysis of the temperature distribution obtained by means of the ANSYS program (cf. Fig. 7.29) indicates that the greatest temperature gradient occurs in the area adjacent to the fin base (Fig. 7.27).

The steady-state temperatures calculated using the FVM-FEM method: $T_1 = 96.355$ °C, $T_2 = 86.359$ °C, $T_3 = 85.337$ °C, $T_4 = 93.328$ °C (cf. Fig. 7.29) in nodes 1, 2, 3 and 4 agree very well with the results obtained by means of the ANSYS program: $T_1 = 96.403$ °C, $T_2 = 86.903$ °C, $T_3 = 84.479$ °C, $T_4 = 93.021$ °C.

The unsteady-state history of the fin temperature in points 1 and 2 is shown in Fig. 7.30, and in points 3 and 4—in Fig. 7.31. The results obtained by means of the developed method and of the ANSYS program show good agreement, despite the

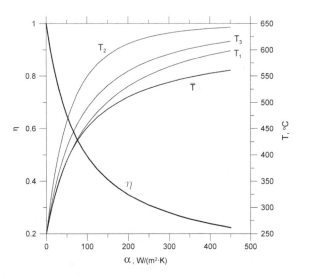

Fig. 7.25 Fin efficiency η, temperatures T_1, T_2 and T_3 in points 1, 2 and 3 marked in Fig. 7.24, and the fin mean temperature $\bar{T}$ as a function of the heat transfer coefficient α

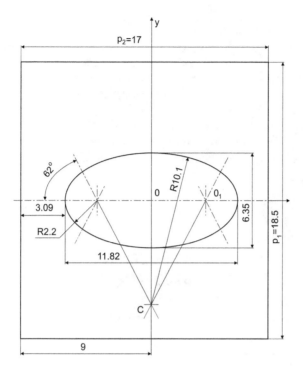

Fig. 7.26 Rectangular fin fixed on an oval tube

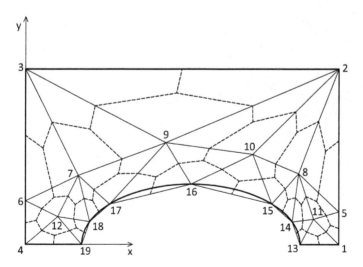

Fig. 7.27 Division of the fin model into finite volumes

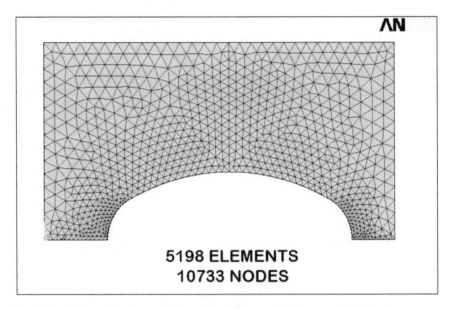

Fig. 7.28 Division of the fin model into finite elements, 5198 elements, 10,733 nodes

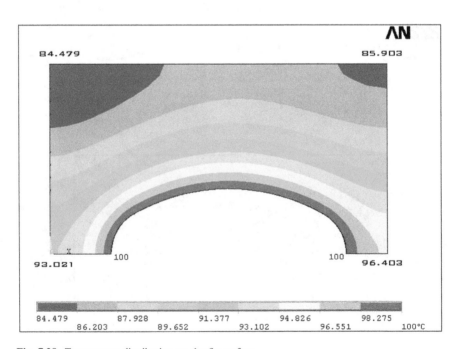

Fig. 7.29 Temperature distribution on the fin surface

Fig. 7.30 The fin
temperature at nodes 1 and 2
as a function of time

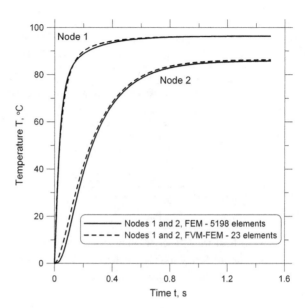

Fig. 7.31 The fin
temperature at nodes 3 and 4
as a function of time

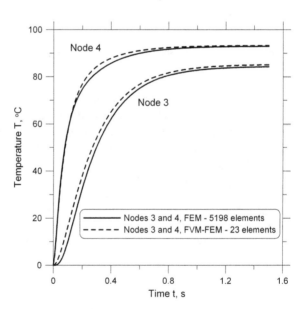

small number of finite elements and, consequently, the small number of control
areas (finite volumes) (cf. Fig. 7.27) applied in the FVM-FEM method. Using
calculations performed using the classical FEM based on the Galerkin method, the
following formula for the fin efficiency is found:

Table 7.2 Comparison of the fin efficiency determined using the finite element method based on the Galerkin approach [formula (7.128)] and using the FVM-FEM method

α, W/(m^2 K)	0	25	50	75	100	125	150	175
η_{eb}, FVM-FEM	1	0.9491	0.9081	0.8712	0.8376	0.8070	0.7789	0.7531
η_{eg}, classical FEM, Eq. (7.128)	1	0.9502	0.9060	0.8664	0.8308	0.7986	0.7692	0.7424
$e = \frac{\eta_{eg}-\eta_{eb}}{\eta_{eg}} 100$, %	0	0.12	−0.23	−0.55	−0.82	−1.05	−1.26	−1.44

$$\eta_{eg} = \frac{1+6.45184 \cdot 10^{-4}\,\alpha}{1+2.761444 \cdot 10^{-3}\,\alpha + 5.16617 \cdot 10^{-7}\,\alpha^2}, \quad 0 \le \alpha \le 250 \text{ W/(m}^2 \text{ K)},$$

$$(7.128)$$

where the heat transfer coefficient α is expressed in W/(m^2 K).

The fin efficiency is also determined based on the temperature distribution obtained from the FVM-FEM method:

$$\eta_{eb} = \frac{\sum_{i=1}^{N} A_{123,\,i}\left(T_{o,i} - T_f\right) + \sum_{j=1}^{N_l} A_{l,\,j}\left(\bar{T}_{l,j} - T_f\right)}{A_l\left(T_b - T_f\right)}, \qquad (7.129)$$

where: $A_{123,i}$—surface area of the i-th triangle, $T_{o,i}$—fin temperature in the centre of gravity of the i-th triangle, N—number of triangles, $A_{l,j}$—surface area of the j-th side area with a width equal to the fin thickness δ_f, $\bar{T}_{l,j}$—mean temperature of the j-th side area with width δ_f, N_l—number of side areas with width δ_f.

The fin efficiency values obtained from expressions (7.128) and (7.129) are compared in Table 7.2. The accuracy of the presented method is perfect despite the small number of finite volumes in the mesh presented in Fig. 7.27. The results obtained using the two methods demonstrate excellent agreement. It can be noticed that the relative difference (cf. Table 7.2) increases with a rise in the heat transfer coefficient α. For higher values of α, the differences in the fin temperature get bigger, which at the small number of finite volumes involves a rise in the relative difference e. It should be added, however, that the heat transfer coefficient values on the surface of fins are smaller than 100 W/(m^2 K).

Chapter 8
Engineering Methods for Thermal Calculations of Heat Exchangers

There is a significant number of books and publications on the heat exchanger design and calculation [97, 121–123, 125–126, 143, 162, 169, 222, 267, 270, 306]. Due to the wide variety of design options, a need arises to develop new thermal and flow calculation methods.

This chapter presents two fundamental methods of thermal calculations which can be applied to various types of heat exchangers. The first to be discussed is the method based on the logarithmic mean temperature difference (LMTD), which is commonly used in Europe, and the second is the ε-NTU method (ε—heat exchanger effectiveness, NTU—the number of heat transfer units), which is mainly popular in the USA.

In steady-state conditions, the hot and the cold fluid mass flow rates ($\dot{m}_h$ and $\dot{m}_c$, respectively) are constant along the exchanger entire length. Considering perfect thermal insulation of the exchanger outer surface, the heat losses from the exchanger to the environment may be omitted. The heat flow rate from the hot fluid is therefore equal to the heat flow rate absorbed by the cold medium $\dot{Q}_c$, i.e.

$$\dot{m}_h = \text{const},$$
$$\dot{m}_c = \text{const},$$
$$\dot{Q}_h = \dot{Q}_c = \dot{Q} = \text{const}, \tag{8.1}$$

where

$$\dot{Q}_h = \dot{m}_h c_h (T_{h1} - T_{h2}), \tag{8.2}$$

$$\dot{Q}_c = \dot{m}_c c_c (T_{c2} - T_{c1}). \tag{8.3}$$

The energy balance equations for the cocurrent and the countercurrent heat exchanger (cf. Fig. 8.1), written in a differential form, are as follows:

© Springer International Publishing AG, part of Springer Nature 2019
D. Taler, *Numerical Modelling and Experimental Testing of Heat Exchangers*,
Studies in Systems, Decision and Control 161,
https://doi.org/10.1007/978-3-319-91128-1_8

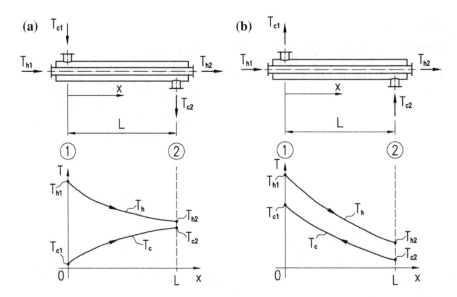

Fig. 8.1 Diagram of the tube-in-tube heat exchanger: **a** cocurrent heat exchanger, **b** countercurrent heat exchanger

$$\dot{m}_h c_h \frac{dT_h}{dx} = -k_A\, U_A (T_h - T_c), \qquad (8.4)$$

$$\dot{m}_c c_c \frac{dT_c}{dx} = \pm k_A\, U_A (T_h - T_c), \qquad (8.5)$$

where the positive sign "+" on the right side of Eq. (8.5) is taken for the cocurrent heat exchanger and the negative sign "−" for the countercurrent one.

Relations (8.1)–(8.5) are the basis for the heat exchanger thermal calculations.

Two methods of the heat exchanger calculations will be discussed: the method based on the logarithmic mean temperature difference between the fluids and the ε-NTU method.

8.1 Method Based on the Logarithmic Mean Temperature Difference

The formula for the logarithmic mean temperature difference between the mediums will be derived on the example of the countercurrent exchanger.

The following simplifications are assumed:

- the overall heat transfer coefficient k_A is constant and depends neither on coordinate x nor on the temperature of the fluids,
- the values of the hot and the cold fluid specific heat c_h and c_c, respectively, are constant and independent of coordinate x and of the temperature of the fluids.

From the first assumption, it follows that also the heat transfer coefficients on both the hot and the cold fluid side and thermal conductivity of the wall material are independent of location and temperature. Should the cold fluid be heated along the exchanger length to saturation temperature and—further on—it boils, the assumptions presented above will not be satisfied because the heat transfer coefficient values in the single- and in the two-phase flow area differ considerably. In such cases, the entire exchanger is divided into several parts where the heat transfer coefficients are constant, and the logarithmic mean temperature difference between the fluids is calculated for each of them separately.

Dividing Eq. (8.4) by $\dot{m}_h c_h$ and Eq. (8.5) by $\dot{m}_c c_c$, and then subtracting Eq. (8.5) from Eq. (8.4), the following expression is obtained for the countercurrent heat exchanger:

$$\frac{d(T_h - T_c)}{(T_h - T_c)} = -k_A U_A \left(\frac{1}{\dot{m}_h c_h} - \frac{1}{\dot{m}_c c_c} \right) dx. \tag{8.6}$$

Integrating both sides of Eq. (8.6), the following is obtained:

$$\ln(T_h - T_c) = -k_A U_A \left(\frac{1}{\dot{m}_h c_h} - \frac{1}{\dot{m}_c c_c} \right) x + C, \tag{8.7}$$

which can be transformed into:

$$T_h - T_c = e^{-k_A U_A \left(\frac{1}{\dot{m}_h c_h} - \frac{1}{\dot{m}_c c_c} \right) x + C} = C_1 \exp\left[-k_A U_A \left(\frac{1}{\dot{m}_h c_h} - \frac{1}{\dot{m}_c c_c} \right) x \right], \tag{8.8}$$

where $C_1 = e^C$.

Constant C_1 is determined from the boundary condition for $x = 0$ (cf. Fig. 8.1b)

$$(T_h - T_c)|_{x=0} = (T_{h1} - T_{c1}). \tag{8.9}$$

Substituting solution (8.8) in boundary condition (8.9), the following is obtained:

$$C_1 = T_{h1} - T_{c1}. \tag{8.10}$$

Next, substitution of C_1 into expression (8.8) gives:

$$T_h - T_c = (T_{h1} - T_{c1}) \exp\left[-k_A U_A \left(\frac{1}{\dot{m}_h c_h} - \frac{1}{\dot{m}_c c_c} \right) x \right]. \tag{8.11}$$

Considering in expression (8.11) the temperature of the mediums on boundary $x = L$:

$$T_{h2} - T_{c2} = (T_{h1} - T_{c1}) \exp\left[-k_A U_A \left(\frac{1}{\dot{m}_h c_h} - \frac{1}{\dot{m}_c c_c} \right) L \right], \qquad (8.12)$$

and then dividing both sides of Eq. (8.12) by $(T_{h1} - T_{c1})$ and taking the logarithm thereof, the following is obtained:

$$\ln\left(\frac{T_{h2} - T_{c2}}{T_{h1} - T_{c1}} \right) = -k_A U_A \left(\frac{1}{\dot{m}_h c_h} - \frac{1}{\dot{m}_c c_c} \right) L. \qquad (8.13)$$

The heat flow rate transferred in the exchanger (the exchanger thermal power) is defined as:

$$\dot{Q} = \dot{m}_c c_c (T_{c1} - T_{c2}) = \dot{m}_h c_h (T_{h1} - T_{h2}) \qquad (8.14)$$

After transformations, the following is obtained:

$$\dot{m}_c c_c = \frac{\dot{Q}}{T_{c1} - T_{c2}} \quad \text{and} \quad \dot{m}_h c_h = \frac{\dot{Q}}{T_{h1} - T_{h2}}. \qquad (8.15)$$

Next, substitution of (8.15) into (8.13) gives:

$$\ln\left(\frac{T_{h2} - T_{c2}}{T_{h1} - T_{c1}} \right) = \frac{k_A U_A L}{\dot{Q}} [(T_{h2} - T_{h1}) - (T_{c2} - T_{c1})], \qquad (8.16)$$

which can be transformed into:

$$\ln\left(\frac{T_{h2} - T_{c2}}{T_{h1} - T_{c1}} \right) = \frac{k_A U_A L}{\dot{Q}} [(T_{h2} - T_{c2}) - (T_{h1} - T_{c1})]. \qquad (8.17)$$

Introducing the following relations:

$$A = U_A L, \qquad (8.18)$$

$$\Delta T_2 = T_{h2} - T_{c2},$$
$$\Delta T_1 = T_{h1} - T_{c1}, \qquad (8.19)$$

Eq. (8.16) can be written as follows:

$$\ln\left(\frac{\Delta T_2}{\Delta T_1} \right) = \frac{k_A A}{\dot{Q}} (\Delta T_2 - \Delta T_1). \qquad (8.20)$$

Equation (8.20) leads to the formula for the heat flow rate $\dot{Q}$ exchanged by the two mediums:

$$\dot{Q} = k_A A \frac{\Delta T_2 - \Delta T_1}{\ln\left(\frac{\Delta T_2}{\Delta T_1}\right)}. \tag{8.21}$$

Introducing the logarithmic mean temperature difference between the mediums:

$$\Delta T_m = \frac{\Delta T_2 - \Delta T_1}{\ln\left(\frac{\Delta T_2}{\Delta T_1}\right)}, \tag{8.22}$$

the following formula expresses the heat flow rate transferred from the hot to the cold fluid (the exchanger thermal power):

$$\dot{Q} = k_A A \, \Delta T_m. \tag{8.23}$$

This formula is valid for co- and countercurrent heat exchangers, and is typically used to calculate the heat transfer surface area (the surface area of tubes or plates). It is also valid if the temperature of one of the mediums is constant.

In the case of cross-flow or other heat exchangers with a complex structure, the $\dot{Q}$ formula needs the following modification:

$$\dot{Q} = F k_A A \, \Delta T_m, \tag{8.24}$$

where correction factor F depends on the heat exchanger design. In the case of tube heat exchangers, it depends on the number of tube rows and passes. The logarithmic mean temperature difference between the mediums ΔT_m in formula (8.24) is calculated like for a countercurrent heat exchanger. Correction factor F is thus the ratio between the thermal power of the analyzed heat exchanger and the thermal power of a countercurrent one. Figure 8.3 presents a chart which enables determination of correction factor F for the exchanger shown in Fig. 8.2.

Correction factor F for a cross-flow plate heat exchanger with unmixed fluids (cf. Fig. 8.4) is presented in Fig. 8.5 [344].

In the case of heat exchangers with a complex flow structure, the temperature distribution can be determined using the finite volume method. The calculation accuracy is then higher, even for mixed cross-cocurrent and cross-countercurrent flows. It is also easy to take account of the dependence of thermophysical properties and the heat transfer coefficients on temperature and location. This is especially important for small-size heat exchangers, where the heat transfer on the entrance lengths has an essential impact on the heat flow rate transferred from the hot to the cold medium.

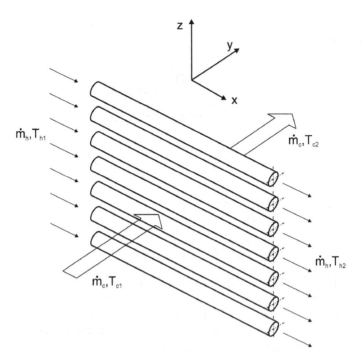

Fig. 8.2 Diagram of a single-row, cross-flow heat exchanger

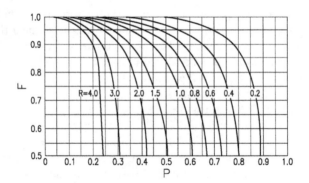

Fig. 8.3 Correction factor F for a single-row tube cross-flow heat exchanger (cf. Fig. 8.2), $P = (T_{c2} - T_{c1})/(T_{h1} - T_{c1})$, $R = \dot{m}_c c_c/(\dot{m}_h c_h) = (T_{h1} - T_{h2})/(T_{c2} - T_{c1})$

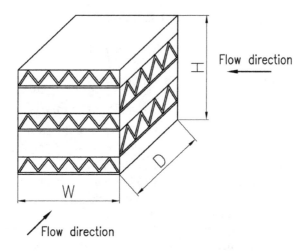

Fig. 8.4 Diagram of a cross-flow heat exchanger with unmixed fluids

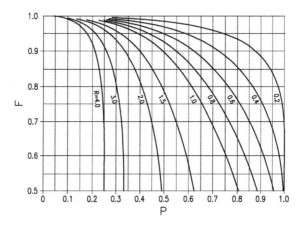

Fig. 8.5 Correction factor F [344] for a cross-flow plate heat exchanger with unmixed fluids (cf. Fig. 8.4), $P = (T_{c2} - T_{c1})/(T_{h1} - T_{c1})$, $R = \dot{m}_c c_c/(\dot{m}_h c_h) = (T_{h1} - T_{h2})/(T_{c2} - T_{c1})$

8.2 ε-NTU Method

The method based on the logarithmic mean temperature difference between the fluids requires the knowledge of the inlet and outlet temperatures of fluids. At least three temperatures must be set beforehand, and the fourth can be determined using the formulae for the exchanger thermal power:

- for the cocurrent flow of fluids (cf. Fig. 8.1a)

$$\dot{Q} = \dot{m}_h \, c_h (T_{h1} - T_{h2}) = \dot{m}_c \, c_c (T_{c2} - T_{c1}), \qquad (8.25)$$

- for the countercurrent flow of fluids (cf. Fig. 8.1b)

$$\dot{Q} = \dot{m}_h \, c_h (T_{h1} - T_{h2}) = \dot{m}_c \, c_c (T_{c1} - T_{c2}). \qquad (8.26)$$

If the following quantities are known for the exchanger: inlet temperatures T_{h1} and T_{c1} for the cocurrent flow or T_{h1} and T_{c2} for the countercurrent flow, the heat transfer surface area A and overall heat transfer coefficient k_A, it is difficult to calculate the exchanger thermal power based on the logarithmic mean temperature difference because the outlet temperatures of the fluids: T_{h2} and T_{c2} for the cocurrent flow or T_{h2} and T_{c1} for the countercurrent flow are unknown (cf. Fig. 8.1). To determine the outlet temperatures of the two fluids, the heat flow rate transferred between them $(\dot{Q})$ is determined first. To determine $\dot{Q}$ for the cocurrent heat exchanger, Eq. (8.25) is used to find temperatures T_{h2} and T_{c2}:

$$T_{h2} = T_{h1} - \frac{\dot{Q}}{\dot{m}_h \, c_h}, \qquad (8.27)$$

$$T_{c2} = T_{c1} + \frac{\dot{Q}}{\dot{m}_c \, c_c}. \qquad (8.28)$$

Then, formulae (8.27) and (8.28) are substituted into the formula for the heat exchanger thermal power:

$$\dot{Q} = k_A \, A \frac{(T_{h1} - T_{c1}) - (T_{h2} - T_{c2})}{\ln\left(\dfrac{T_{h1} - T_{c1}}{T_{h2} - T_{c2}}\right)}, \qquad (8.29)$$

which gives:

$$\dot{Q} = k_A \, A \frac{(T_{h1} - T_{c1}) - \left[T_{h1} - \dfrac{\dot{Q}}{\dot{m}_h \, c_h} - \left(T_{c1} + \dfrac{\dot{Q}}{\dot{m}_c \, c_c} \right) \right]}{\ln \dfrac{(T_{h1} - T_{c1})}{T_{h1} - \dfrac{\dot{Q}}{\dot{m}_h \, c_h} - \left(T_{c1} + \dfrac{\dot{Q}}{\dot{m}_c \, c_c} \right)}}$$

$$= k_A \, A \frac{\left(\dfrac{\dot{Q}}{\dot{m}_h \, c_h} + \dfrac{\dot{Q}}{\dot{m}_c \, c_c} \right)}{\ln \dfrac{(T_{h1} - T_{c1})}{(T_{h1} - T_{c1}) - \left(\dfrac{\dot{Q}}{\dot{m}_h \, c_h} + \dfrac{\dot{Q}}{\dot{m}_c \, c_c} \right)}}. \qquad (8.30)$$

Transposing the right side of Eq. (8.30) to the left gives the nonlinear algebraic equation $f(\dot{Q}) = 0$:

$$f(\dot{Q}) = \dot{Q} - k_A A \frac{\dot{Q}\left(\dfrac{1}{\dot{m}_h c_h} + \dfrac{1}{\dot{m}_c c_c}\right)}{\ln \dfrac{(T_{h1} - T_{c1})}{(T_{h1} - T_{c1}) - \dot{Q}\left(\dfrac{1}{\dot{m}_h c_h} + \dfrac{1}{\dot{m}_c c_c}\right)}} = 0. \qquad (8.31)$$

The same procedure can be applied to derive $f(\dot{Q}) = 0$ for the countercurrent flow:

$$f(\dot{Q}) = \dot{Q} - k_A A \frac{\dot{Q}\left(\dfrac{1}{\dot{m}_h c_h} - \dfrac{1}{\dot{m}_c c_c}\right)}{\ln \left[\dfrac{(T_{h1} - T_{c2}) - \dfrac{\dot{Q}}{\dot{m}_c c_c}}{(T_{h1} - T_{c2}) - \dfrac{\dot{Q}}{\dot{m}_h c_h}}\right]} = 0. \qquad (8.32)$$

The above are algebraic equations which are nonlinear with respect to $\dot{Q}$, and they can be solved using one of the known iterative methods, e.g., the interval search method, the bisection (interval halving) method or the secant method. The easiest way is to determine $\dot{Q}$ from Eqs. (8.31) to (8.32) using the interval search method, e.g., calculating in the Excel program the value of the function $f(\dot{Q})$ for $\dot{Q}$ values rising from zero with a constant step, or making a graph of the function $f(\dot{Q})$. The coordinate of $\dot{Q}^*$ on the x-axis for which $f(\dot{Q}^*) = 0$ is the sought thermal power of the heat exchanger. Substituting $\dot{Q}^*$ in formulae (8.25) the hot and the cold fluid outlet temperatures T_{h2} and T_{c2} are determined for the cocurrent flow. For the countercurrent flow, outlet temperatures T_{h2} and T_{c1} are found from Eq. (8.26).

In the ε-NTU method, determination of temperatures T_{h2} and T_{c2} is much more comfortable. It should be added, however, that finding these temperatures based on the logarithmic mean temperature difference as described above does not present significant difficulties, either. The ε-NTU method will be discussed in detail because it is often referred to in both European and American literature [169, 270, 308, 353, 357–358].

The symbol NTU denotes the number of (heat) transfer units, and it is expressed as:

$$NTU = \frac{k_A A}{\dot{C}_{min}}. \qquad (8.33)$$

The heat exchanger effectiveness ε is defined as the ratio between the exchanger real thermal power $\dot{Q}_{rz}$ (heat flow rate from the hot to the cold medium) and the exchanger maximum power $\dot{Q}_{max}$:

$$\varepsilon = \frac{\dot{Q}_{rz}}{\dot{Q}_{max}} = \frac{\dot{Q}_{rz}}{\dot{C}_{min}\left(T_{h,\,wlot} - T_{c,\,wlot}\right)}, \qquad (8.34)$$

where $\dot{C}_{min}$ is the smaller value of two heat capacity rates:

$$\dot{C}_{min} = min\left(\dot{C}_h, \dot{C}_c\right), \qquad (8.35)$$

where $\dot{C}_h = \dot{m}_h\,c_h$, $\dot{C}_c = \dot{m}_c\,c_c$.

Heat flow rate $\dot{Q}_{max}$ is the exchanger thermal power calculated under the assumption that the heat capacity rate is $\dot{C}_{min}$, and the change in the fluid temperature is equal to the maximum temperature difference in the exchanger $(T_{h,wlot} - T_{c,wlot})$. Definition (8.34) is valid both for the cocurrent and the countercurrent flow. The following relation occurs for cocurrent heat exchangers (cf. Fig. 8.1a):

$$T_{h,wlot} - T_{c,wlot} = T_{h1} - T_{c1}, \qquad (8.36)$$

whereas for countercurrent exchangers (cf. Fig. 8.1b), the following is the case:

$$T_{h,wlot} - T_{c,wlot} = T_{h1} - T_{c2}. \qquad (8.37)$$

The definition of $\dot{Q}_{max}$ is based on the premise that a more significant change in the fluid temperature between the exchanger inlet and outlet occurs for the fluid for which the heat capacity rate is $\dot{C}_{min}$. In the extreme case, the temperature change of fluid with $\dot{C}_{min}$ would reach the value of $(T_{h,wlot} - T_{c,wlot})$. The formulae for thermal effectiveness ε will be derived for the following selected types of heat exchangers:

- a cocurrent tube-in-tube heat exchanger,
- a countercurrent tube-in-tube heat exchanger,
- a single-row tube cross-flow heat exchanger,
- a cross-flow heat exchanger.

8.2.1   Cocurrent Heat Exchanger

Formula (8.34) takes the following form for the cocurrent heat exchanger:

$$\varepsilon = \frac{\dot{C}_h(T_{h1} - T_{h2})}{\dot{C}_{min}(T_{h1} - T_{c1})} = \frac{\dot{C}_c(T_{c2} - T_{c1})}{\dot{C}_{min}(T_{h1} - T_{c1})}. \qquad (8.38)$$

In the following considerations, it is assumed that:

$$\dot{C}_h = \dot{C}_{min}, \quad \dot{C}_c = \dot{C}_{max}. \tag{8.39}$$

Considering assumption (8.39), formula (8.38) is written as:

$$\varepsilon = \frac{T_{h1} - T_{h2}}{T_{h1} - T_{c1}}. \tag{8.40}$$

An identical form of the ε formula is obtained using a different assumption: $\dot{C}_c = \dot{C}_{min}$, $\dot{C}_h = \dot{C}_{max}$. Considering in (8.16) that:

$$\dot{Q} = \dot{C}_{min}(T_{h1} - T_{h2}) \tag{8.41}$$

and after simple rearrangements of the right side of the expression in the square bracket, the following is obtained:

$$\ln\left(\frac{T_{h2} - T_{c2}}{T_{h1} - T_{c1}}\right) = \frac{-k_A A_A}{\dot{C}_{min}(T_{h1} - T_{h2})}[(T_{h1} - T_{h2}) + (T_{c2} - T_{c1})]. \tag{8.42}$$

Considering the definition of *NTU* expressed in (8.33) as $NTU = k_A A_A / \dot{C}_{min}$, Eq. (8.42) can be written in the following form:

$$\ln\left(\frac{T_{h2} - T_{c2}}{T_{h1} - T_{c1}}\right) = -NTU\left(1 + \frac{T_{c2} - T_{c1}}{T_{h1} - T_{h2}}\right). \tag{8.43}$$

Equation (8.25) gives:

$$\frac{\dot{C}_{min}}{\dot{C}_{max}} = \frac{\dot{C}_h}{\dot{C}_c} = \frac{\dot{m}_h c_h}{\dot{m}_c c_c} = \frac{T_{c2} - T_{c1}}{T_{h1} - T_{h2}}. \tag{8.44}$$

Taking account of Eq. (8.44) in formula (8.43), the following expression is obtained:

$$\ln\left(\frac{T_{h2} - T_{c2}}{T_{h1} - T_{c1}}\right) = -NTU\left(1 + \frac{\dot{C}_{min}}{\dot{C}_{max}}\right), \tag{8.45}$$

which results in the following formula:

$$\frac{T_{h2} - T_{c2}}{T_{h1} - T_{c1}} = \exp\left[-NTU\left(1 + \frac{\dot{C}_{min}}{\dot{C}_{max}}\right)\right]. \tag{8.46}$$

After temperature T_{c2} is found from formula (8.44):

$$T_{c2} = T_{c1} + \frac{\dot{C}_{min}}{\dot{C}_{max}}(T_{h1} - T_{h2}) \tag{8.47}$$

and after substitution of expression (8.47), with respect to formula (8.40) for the exchanger effectiveness ε, the left side of formula (8.46) can be transformed into the following form:

$$
\begin{aligned}
\frac{T_{h2} - T_{c2}}{T_{h1} - T_{c1}} &= \frac{T_{h2} - T_{c1} - \frac{\dot{C}_{min}}{\dot{C}_{max}}(T_{h1} - T_{h2})}{T_{h1} - T_{c1}} \\
&= \frac{-(T_{h1} - T_{h2}) + (T_{h1} - T_{c1}) - \frac{\dot{C}_{min}}{\dot{C}_{max}}(T_{h1} - T_{h2})}{T_{h1} - T_{c1}} \\
&= -\varepsilon + 1 - \frac{\dot{C}_{min}}{\dot{C}_{max}}\varepsilon = 1 - \varepsilon\left(1 + \frac{\dot{C}_{min}}{\dot{C}_{max}}\right).
\end{aligned} \tag{8.48}
$$

Substitution the left side of (8.46) by the expression (8.46) results in:

$$1 - \varepsilon\left(1 + \frac{\dot{C}_{min}}{\dot{C}_{max}}\right) = \exp\left[-NTU\left(1 + \frac{\dot{C}_{min}}{\dot{C}_{max}}\right)\right], \tag{8.49}$$

which, after more transformations, gives the formula for the cocurrent heat exchanger effectiveness:

$$\varepsilon = \frac{1 - \exp\left[-NTU\left(1 + \frac{\dot{C}_{min}}{\dot{C}_{max}}\right)\right]}{1 + \frac{\dot{C}_{min}}{\dot{C}_{max}}}. \tag{8.50}$$

Figure 8.6 presents changes in the cocurrent heat exchanger effectiveness ε as a function of NTU determined using formula (8.50).

8.2.2 Countercurrent Heat Exchanger

For the countercurrent heat exchanger, thermal effectiveness is defined as follows:

$$\varepsilon = \frac{\dot{C}_h(T_{h1} - T_{h2})}{\dot{C}_{min}(T_{h1} - T_{c2})} = \frac{\dot{C}_c(T_{c1} - T_{c2})}{\dot{C}_{min}(T_{h1} - T_{c2})}. \tag{8.51}$$

Performing transformations similar to the cocurrent exchanger, the following formula for the countercurrent heat exchanger effectiveness is obtained:

Fig. 8.6 Cocurrent heat exchanger effectiveness, $C^* = \dot{C}_{min}/\dot{C}_{max}$

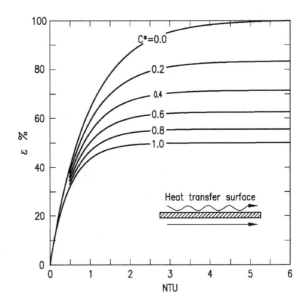

$$\varepsilon = \frac{1 - \exp\left[-NTU\left(1 - \dfrac{\dot{C}_{min}}{\dot{C}_{max}}\right)\right]}{1 - \dfrac{\dot{C}_{min}}{\dot{C}_{max}}\exp\left[-NTU\left(1 - \dfrac{\dot{C}_{min}}{\dot{C}_{max}}\right)\right]}. \tag{8.52}$$

Figure 8.7 presents changes in the countercurrent heat exchanger effectiveness ε as a function of *NTU* determined using formula (8.52).

8.2.3 Single-Row Cross-Flow Tube Heat Exchanger

The exchanger is presented in the diagram in Fig. 8.2. The hot fluid (subscript h) flowing inside the tubes is mixed, whereas the cold fluid (subscript c) is unmixed and flows perpendicularly to the axis of the tubes. The fact that the hot fluid is mixed means that all the tubes are fed from a single collector, where water is very well mixed and has a uniform temperature. This also means that the temperature of the fluid at the inlet of the tubes is identical. Furthermore, at the same distance from the inlet of the tube, the fluid temperature is the same. It is assumed that the medium flowing perpendicularly to the tubes is unmixed, which is the effect of the application of individually finned tubes or tubes with continuous fins. Gas or air flows through narrow channels created by two adjacent fins and does not mix in the exchanger cross-section. The gas or air mixing process usually occurs only after the medium leaves the exchanger. It is assumed that $\dot{C}_{min} = \dot{m}_c c_c$ and $\dot{C}_{max} = \dot{m}_h c_h$. The heat exchanger thermal effectiveness is defined as:

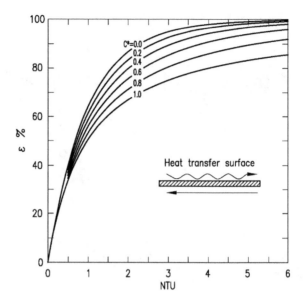

Fig. 8.7 Countercurrent heat exchanger effectiveness, $C^* = \dot{C}_{min}/\dot{C}_{max}$

$$\varepsilon = \frac{\dot{C}_{min}(T_{c2m} - T_{c1})}{\dot{C}_{min}(T_{h1} - T_{c1})} = \frac{(T_{c2m} - T_{c1})}{(T_{h1} - T_{c1})}, \tag{8.53}$$

where the air mean temperature T_{c2m} is defined by the following formula [308]:

$$T_{c2m} = T_{c1} + \frac{N_c}{N_h}(T_{h1} - T_{c1})[1 - \exp(-B)], \tag{8.54}$$

where

$$B = \frac{N_h}{N_c}[1 - \exp(-N_c)], \tag{8.55}$$

$$N_c = \frac{k_A A}{\dot{m}_c\, c_c}, \quad N_h = \frac{k_A A}{\dot{m}_h\, c_h}. \tag{8.56}$$

Substituting (8.54) into (8.53), the following is obtained after transformations:

$$\varepsilon = \frac{N_c}{N_h}\left\{1 - \exp\left[\frac{N_h}{N_c}\left(e^{-NTU} - 1\right)\right]\right\} = \frac{1}{C^*}\left\{1 - \exp\left[C^*\left(e^{-NTU} - 1\right)\right]\right\}, \tag{8.57}$$

where

$$NTU = N_c = \frac{k_A A}{\dot{C}_{min}} = \frac{k_A A}{\dot{m}_c\, c_c}, \quad C^* = \frac{\dot{C}_{min}}{\dot{C}_{max}}. \tag{8.58}$$

If $\dot{C}_{\min} = \dot{m}_h\,c_h$ and $\dot{C}_{\max} = \dot{m}_c\,c_c$, the heat exchanger thermal effectiveness is expressed as:

$$\varepsilon = \frac{\dot{C}_{\min}(T_{h1} - T_{h2})}{\dot{C}_{\min}(T_{h1} - T_{c1})} = \frac{(T_{h1} - T_{h2})}{(T_{h1} - T_{c1})}, \tag{8.59}$$

where outlet temperature T_{h2} is the hot fluid temperature for $x^+ = 1$ and is defined by the following formula [308]:

$$T_{h2} = T_{c1} + (T_{h1} - T_{c1})\exp\left[-\frac{N_h}{N_c}\left(1 - e^{-N_c}\right)\right]. \tag{8.60}$$

Substituting (8.60) into (8.59), the following is obtained after transformations:

$$\varepsilon = 1 - \exp\left[\frac{N_h}{N_c}\left(e^{-N_c} - 1\right)\right] = 1 - \exp\left[\frac{1}{C^*}\left(e^{-C^* \cdot NTU} - 1\right)\right], \tag{8.61}$$

where

$$NTU = N_h = \frac{k_A A}{\dot{C}_{\min}} = \frac{k_A A}{\dot{m}_h\,c_h}. \tag{8.62}$$

The changes in the exchanger thermal effectiveness ε defined by formula (8.61) as a function of parameters C^* and NTU are shown in Fig. 8.8.

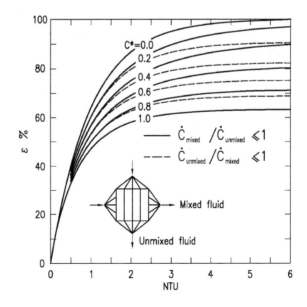

Fig. 8.8 Thermal effectiveness of the cross-flow heat exchanger where one of the fluids is mixed whereas the other remains unmixed (in the heat exchanger presented in Fig. 8.2 the medium flowing inside the tubes is mixed, whereas the fluid flowing perpendicularly to the tubes axis is unmixed); solid line—formula (8.61), dashed line—formula (8.57)

8.2.4   Cross-Flow Heat Exchanger

In the cross-flow plate heat exchanger presented in Fig. 8.4 the fluids are not mixed because the corrugated sheets of steel acting as fins make the two fluids flow through channels with a small-cross section, which prevents the fluids from getting mixed in the channel cross-section created by two adjacent plates and side areas. The effectiveness of the cross-flow heat exchanger in which the fluids are unmixed (cf. Fig. 8.4) can be calculated, assuming that $\dot{C}_{min} = \dot{C}_h = \dot{m}_h \, c_h$, from the following formula:

$$\varepsilon = \frac{T_{h,\text{wlot}} - T_{hm,\text{wylot}}}{T_{h,\text{wlot}} - T_{c,\text{wlot}}} = \frac{T_{h1} - T_{h2m}}{T_{h1} - T_{c1}}, \tag{8.63}$$

where $T_{hm,\text{wylot}} = T_{h2m}$ is the hot fluid mean outlet temperature.

A similar formula defining the exchanger thermal effectiveness is obtained for $\dot{C}_{min} = \dot{C}_c = \dot{m}_c \, c_c$. In this case, the exchanger effectiveness is expressed as:

$$\varepsilon = \frac{T_{cm,\text{wylot}} - T_{c,\text{wlot}}}{T_{h,\text{wlot}} - T_{c,\text{wlot}}} = \frac{T_{c1m} - T_{c1}}{T_{h1} - T_{c1}}, \tag{8.64}$$

where $T_{cm,\text{wylot}} = T_{c1m}$ is the cold fluid mean outlet temperature.

The following formula for the calculation of thermal effectiveness ε is derived in [15, 27, 159]:

$$\varepsilon = 1 - \exp[-(1 + C^*)NTU] \left[I_0\left(2NTU\sqrt{C^*}\right) \right.$$
$$\left. + C^* I_1\left(2NTU\sqrt{C^*}\right) - \frac{1 - C^*}{C^*} \sum_{n=2}^{\infty} (C^*)^{n/2} I_n\left(2NTU\sqrt{C^*}\right) \right], \tag{8.65}$$

where

$$C^* = \frac{(\dot{m}c_p)_{min}}{(\dot{m}c_p)_{max}}, \quad NTU = \frac{k_A A}{(\dot{m}c_p)_{min}}. \tag{8.66}$$

In the case of a balanced flow of both fluids, when $C^* \to 1$, formula (8.65) is reduced to the following form:

$$\varepsilon = 1 - e^{-2NTU}[I_0(2NTU) + I_1(2NTU)]. \tag{8.67}$$

A different form of the formula describing the thermal effectiveness of the cross-flow heat exchanger with unmixed fluids is given in [178, 217]:

$$\varepsilon = \frac{1}{C^*NTU} \sum_{k=0}^{\infty} \left\{ \left[1 - \exp(-NTU) \sum_{m=0}^{k} \frac{(NTU)^m}{m!} \right] \right.$$
$$\left. \times \left[1 - \exp(-C^*NTU) \sum_{m=0}^{k} \frac{(C^*NTU)^m}{m!} \right] \right\}. \tag{8.68}$$

The thermal effectiveness of a cross-flow plate heat exchanger with unmixed flows as a function of NTU and C^* is depicted in Fig. 8.9.

Exact formulae (8.65) and (8.68) can be approximated using the following simple expression [126, 169, 210, 271]:

$$\varepsilon = 1 - \exp\left\{ \frac{NTU^{0.22}}{C^*} \left[\exp\left(-C^*NTU^{0.78}\right) - 1 \right] \right\}. \tag{8.69}$$

Figure 8.10 presents a comparison of the effectiveness of different types of heat exchangers for $\dot{C}_{min}/\dot{C}_{max} = 1$.

Analysing the calculation results presented in Fig. 8.10, it can be seen that the countercurrent heat exchanger is characterized by the highest thermal effectiveness, whereas for the cocurrent one—effectiveness reaches the lowest values.

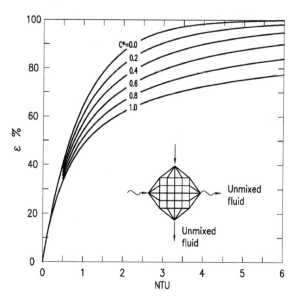

Fig. 8.9 Thermal effectiveness of a cross-flow plate heat exchanger with unmixed mediums

Fig. 8.10 Comparison of thermal effectiveness values for different types of heat exchangers if $\dot{C}_{min}/\dot{C}_{max} = 1$

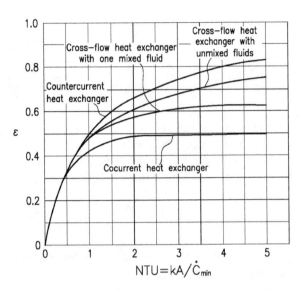

The effectiveness of the cross-flow plate heat exchanger with unmixed fluids is also high. For higher *NTU* values, the effectiveness of the cross-flow tube heat exchanger with one unmixed medium is lower compared to either the counter-current heat exchanger or the cross-flow plate exchanger with unmixed fluids.

Chapter 9
Mathematical Models of Heat Exchangers

In this chapter, formulae will be derived that define the temperature of fluids in four basic types of the heat exchanger. If in a given cross-section the hot and the cold medium temperatures (T_h and T_c, respectively) and the heat transfer coefficients on the hot and the cold medium side (α_h and α_c, respectively) are known, it is possible to find the wall temperature T_w.

Neglecting the temperature drop on the wall thickness and considering that the heat flow rates on the wall two sides are identical, the temperature of the wall can be determined from the following equation:

$$\alpha_h U_h (T_h - T_w) = \alpha_c U_c (T_w - T_c), \tag{9.1}$$

where U_h and U_c denote the tube (channel) perimeter on the side of the hot and the cold medium, respectively.

If the hot fluid flows inside a circular tube with the inner and outer diameters d_w and d_z, respectively, and the cold fluid flows outside, then $U_h = \pi d_w$ and $U_c = \pi d_z$.

Solving Eq. (9.1) with respect to T_w, the following is obtained:

$$T_w = \frac{\alpha_h U_h T_h + \alpha_c U_c T_c}{\alpha_h U_h + \alpha_c U_c}. \tag{9.2}$$

The knowledge of the wall temperature is essential in high-temperature heat exchangers because it enables correct selection of the steel grade or other materials needed to make the device. For considerable changes in the fluid viscosity, the correction factor in the Nusselt number formula includes the ratio between the fluid and the wall temperatures. Also, in this case, the knowledge of the wall temperature is indispensable.

The distributions of the temperatures of fluids will be determined for the heat exchanger following types [308]:

© Springer International Publishing AG, part of Springer Nature 2019
D. Taler, *Numerical Modelling and Experimental Testing of Heat Exchangers*,
Studies in Systems, Decision and Control 161,
https://doi.org/10.1007/978-3-319-91128-1_9

- a cocurrent tube-in-tube heat exchanger,
- a countercurrent tube-in-tube heat exchanger,
- a single-row tube cross-flow heat exchanger,
- a cross-flow plate heat exchanger with unmixed fluids.

9.1   Tube-in-Tube Cocurrent Heat Exchanger

The cocurrent tube-in-tube heat exchanger is presented in the diagram in Fig. 8.1a. The energy conservation equations for the cocurrent exchanger have the following form:

$$\dot{m}_h c_h \frac{dT_h}{dx} = -k_A U_A (T_h - T_c), \tag{9.3}$$

$$\dot{m}_c c_c \frac{dT_c}{dx} = k_A U_A (T_h - T_c). \tag{9.4}$$

The hot and the cold fluid inlet temperatures (T_{h1} and T_{c1}, respectively) are set for the same coordinate $x = 0$, i.e., the boundary conditions take the following form:

$$T_h|_{x=0} = T_{h1}, \tag{9.5}$$

$$T_c|_{x=0} = T_{c1}. \tag{9.6}$$

Subtracting Eq. (9.4) from Eq. (9.3), the following is obtained after simple transformations:

$$\frac{d(T_h - T_c)}{(T_h - T_c)} = -k_A U_A \left(\frac{1}{\dot{m}_h c_h} + \frac{1}{\dot{m}_c c_c} \right) dx. \tag{9.7}$$

Integration of both sides of Eq. (9.7):

$$\int \frac{d(T_h - T_c)}{(T_h - T_c)} = - \int k_A U_A \left(\frac{1}{\dot{m}_h c_h} + \frac{1}{\dot{m}_c c_c} \right) dx + C, \tag{9.8}$$

gives:

$$\ln(T_h - T_c) = -k_A U_A \left(\frac{1}{\dot{m}_h c_h} + \frac{1}{\dot{m}_c c_c} \right) x + C. \tag{9.9}$$

After simple transformations, the following is obtained from Eq. (9.9):

$$T_h - T_c = \exp\left[-\left(k_A U_A \left(\frac{1}{\dot{m}_h c_h} + \frac{1}{\dot{m}_c c_c}\right) x + C\right)\right]$$
$$= C_1 \exp\left[-k_A U_A \left(\frac{1}{\dot{m}_h c_h} + \frac{1}{\dot{m}_c c_c}\right) x\right], \tag{9.10}$$

where $C_1 = \exp(C)$.

Subtracting by sides boundary conditions (9.5) and (9.6), the following is obtained:

$$(T_h - T_c)|_{x=0} = T_{h1} - T_{c1}. \tag{9.11}$$

After (9.10) is substituted in condition (9.11) and constant C_1 is determined, formula (9.10) defining the temperature difference between the mediums is expressed as follows:

$$T_h - T_c = (T_{h1} - T_{c1}) \exp\left[-k_A U_A \left(\frac{1}{\dot{m}_h c_h} + \frac{1}{\dot{m}_c c_c}\right) x\right], \tag{9.12}$$

Substituting expression (9.12) on the right side of Eq. (9.3), the following is obtained after transformations:

$$\dot{m}_h c_h \frac{dT_h}{dx} = -k_A U_A (T_{h1} - T_{c1}) \exp\left[-k_A U_A \left(\frac{1}{\dot{m}_h c_h} + \frac{1}{\dot{m}_c c_c}\right) x\right]. \tag{9.13}$$

Integration of Eq. (9.13) using the variables separation method under boundary condition (9.5) gives the following expression for the hot medium temperature distribution:

$$T_h = T_{h1} - \frac{(T_{h1} - T_{c1})}{1 + \frac{\dot{m}_h c_h}{\dot{m}_c c_c}} \left\{1 - \exp\left[-k_A U_A \left(\frac{1}{\dot{m}_h c_h} + \frac{1}{\dot{m}_c c_c}\right) x\right]\right\}, \quad 0 \le x \le L. \tag{9.14}$$

The formula for the cold medium temperature is determined similarly. Considering expression (9.12) in Eq. (9.4), the following differential equation is obtained:

$$\dot{m}_c c_c \frac{dT_c}{dx} = -k_A U_A (T_{h1} - T_{c1}) \exp\left[-k_A U_A \left(\frac{1}{\dot{m}_h c_h} + \frac{1}{\dot{m}_c c_c}\right) x\right], \tag{9.15}$$

which, after integration using the variables separation method under boundary condition (9.6), results in:

$$T_c = T_{c1} + \frac{(T_{h1} - T_{c1})}{1 + \frac{\dot{m}_c c_c}{\dot{m}_h c_h}} \left\{ 1 - \exp\left[-k_A U_A \left(\frac{1}{\dot{m}_h c_h} + \frac{1}{\dot{m}_c c_c} \right) x \right] \right\}, \quad 0 \leq x \leq L.$$

$$(9.16)$$

Substitution of expressions (9.14) and (9.16) in formula (9.2) makes it possible to determine the inner tube wall temperature in the exchanger presented in Fig. 8.1a.

9.2   Tube-in-Tube Countercurrent Heat Exchanger

The countercurrent tube-in-tube heat exchanger is presented in the diagram in Fig. 8.1b. For this exchanger type, the negative sign appears on the right side of the energy balance Eq. (9.4) for the medium with the lower temperature:

$$\dot{m}_h c_h \frac{dT_h}{dx} = -k_A U_A (T_h - T_c), \tag{9.17}$$

$$\dot{m}_c c_c \frac{dT_c}{dx} = -k_A U_A (T_h - T_c). \tag{9.18}$$

Dividing both sides of Eq. (9.17) by $\dot{m}_h c_h$ and both sides of Eq. (9.18) by $\dot{m}_c c_c$, and then subtracting Eq. (9.18) from Eq. (9.17), the following is obtained:

$$\frac{d(T_h - T_c)}{(T_h - T_c)} = -k_A U_A \left(\frac{1}{\dot{m}_h c_h} - \frac{1}{\dot{m}_c c_c} \right) dx. \tag{9.19}$$

Integration of Eq. (9.19) under boundary condition (9.11) gives the following formula defining the temperature difference between the two mediums:

$$T_h - T_c = (T_{h1} - T_{c1}) \exp\left[-k_A U_A \left(\frac{1}{\dot{m}_h c_h} - \frac{1}{\dot{m}_c c_c} \right) x \right]. \tag{9.20}$$

Substituting (9.20) in (9.17) and integrating the obtained differential equation under boundary condition (9.5), the following is obtained:

$$T_h = T_{h1} - \frac{(T_{h1} - T_{c1})}{1 - \frac{\dot{m}_h c_h}{\dot{m}_c c_c}} \left\{ 1 - \exp\left[-k_A U_A \left(\frac{1}{\dot{m}_h c_h} - \frac{1}{\dot{m}_c c_c} \right) x \right] \right\}, \quad 0 \leq x \leq L.$$

$$(9.21)$$

Equation (9.21) includes temperature T_{c1}, which is unknown and which will be eliminated from Eq. (9.21) using Eq. (9.20) for $x = L$ and Eq. (8.14):

$$\frac{T_{h2} - T_{c2}}{(T_{h1} - T_{c1})} = \exp\left[-k_A\,U_A\left(\frac{1}{\dot{m}_h c_h} - \frac{1}{\dot{m}_c c_c}\right)L\right], \tag{9.22}$$

$$\dot{m}_h c_h(T_{h1} - T_{h2}) = \dot{m}_c c_c(T_{c1} - T_{c2}), \tag{9.23}$$

where $A = U_A\,L$.

Equation (9.23) is first used to determine temperature T_{h2}:

$$T_{h2} = T_{h1} - \frac{\dot{m}_c c_c}{\dot{m}_h c_h}(T_{c1} - T_{c2}). \tag{9.24}$$

Then, substituting (9.24) in (9.22), temperature T_{c1} is found:

$$T_{c1} = \frac{T_{h1}\left\{1 - \exp\left[-k_A\,A\left(\frac{1}{\dot{m}_h c_h} - \frac{1}{\dot{m}_c c_c}\right)\right]\right\} + T_{c2}\left(\frac{\dot{m}_c c_c}{\dot{m}_h c_h} - 1\right)}{\frac{\dot{m}_c c_c}{\dot{m}_h c_h} - \exp\left[-k_A\,A\left(\frac{1}{\dot{m}_h c_h} - \frac{1}{\dot{m}_c c_c}\right)\right]}. \tag{9.25}$$

Considering (9.25) in (9.21), the following equation for the hot fluid temperature is obtained after transformations:

$$T_h = T_{h1} - \frac{(T_{h1} - T_{c2})\left\{1 - \exp\left[-k_A\,U_A\left(\frac{1}{\dot{m}_h c_h} - \frac{1}{\dot{m}_c c_c}\right)x\right]\right\}}{1 - \frac{\dot{m}_h c_h}{\dot{m}_c c_c}\exp\left[-k_A\,A\left(\frac{1}{\dot{m}_h c_h} - \frac{1}{\dot{m}_c c_c}\right)\right]}. \tag{9.26}$$

The cold medium temperature can be calculated using the following formula:

$$T_c = T_h - (T_h - T_c), \tag{9.27}$$

which, considering expressions (9.20) and (9.25), gives:

$$T_c = T_{h1} - \frac{(T_{h1} - T_{c2})\left\{1 - \frac{\dot{m}_h c_h}{\dot{m}_c c_c}\exp\left[-k_A\,U_A\left(\frac{1}{\dot{m}_h c_h} - \frac{1}{\dot{m}_c c_c}\right)x\right]\right\}}{1 - \frac{\dot{m}_h c_h}{\dot{m}_c c_c}\exp\left[-k_A\,A\left(\frac{1}{\dot{m}_h c_h} - \frac{1}{\dot{m}_c c_c}\right)\right]}. \tag{9.28}$$

Formulae (9.26) and (9.28) make it possible to determine the temperature distribution of the medium as a function of coordinate x.

9.3 Single-Row Cross-Flow Tube Heat Exchanger

The single-row cross-flow heat exchanger is presented in the diagram in Fig. 8.2. In this case, it is impossible to apply the energy balance equations derived in Sect. 3.1 for the fluid flow inside a duct.

The tubes are fed from a single header, so the cooling fluid temperature at the inlet is the same and equals T_{h1}. The fluid temperature is a function of axis coordinate x only because for a given value of x the fluid temperature in all tubes is the same, i.e., it does not change in the direction of axis z. The cooling fluid outlet temperature is identical in all tubes and equals T_{h2}. The hot fluid can be treated as an entirely mixed medium. The stream of the air (gas) with a uniform inlet temperature T_{c1} is transverse to the tube row. The air temperature T_c is a function of coordinates x and y, i.e., it varies on the exchanger width and along the flow path in the direction of axis y. For set values of coordinates x and y, the air temperature on the exchanger height remains constant. The cooling fluid and air mass flow rates are $\dot{m}_h$ and $\dot{m}_c$, respectively. A case will be considered where the medium flowing perpendicularly to the axis of the tubes is unmixed. The gas-side tubes usually have individual or continuous fins (the latter are referred to as lamellae). The gas streams flowing through narrow channels in between the fins do not mix, and the gas temperature is a function of coordinate x. In order to derive differential equations describing the heat transfer, the real heat exchanger presented in Fig. 8.2 will be replaced with a generalized cross-flow heat exchanger shown in Fig. 9.1.

The tube surface area A, on which the heat transfer occurs between the hot fluid with temperature T_h and the cold one with temperature T_c, is replaced with a rectangular area characterized by dimensions L_x and L_y. The cooler gas flows through the exchanger upper part, whereas the fluid with the higher temperature flows through the exchanger lower part. The changes in the temperatures of the fluids in the generalized heat exchanger are illustrated in the diagram in Fig. 9.2.

The considerations presented below neglect the fluids kinetic and potential energy because their values are small compared to enthalpy i.

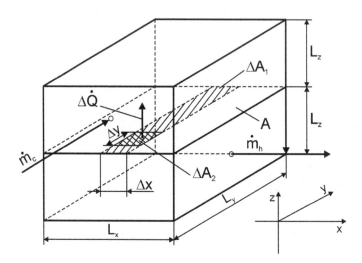

Fig. 9.1 Generalized diagram of a cross-flow heat exchanger—the hot medium is mixed, whereas the cold medium does not get mixed in the cross-section perpendicular to its flow direction

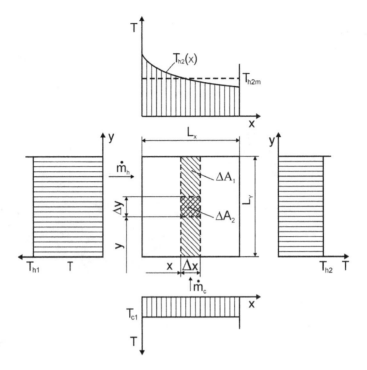

Fig. 9.2 Computational scheme of the generalized cross-flow heat exchanger; hot fluid—mixed, cold fluid—unmixed

The heat flow rate transferred from the fluid to the air through elementary surface area $\Delta A_2 = \Delta x \Delta y$ $(\Delta \dot{Q})$ is expressed as:

$$\Delta \dot{Q} = k_A \, \Delta A_2 \, [T_h(x) - T_c(x, y)]. \tag{9.29}$$

The mass flow rate of air flowing through the cross-section with dimensions Δx and height L_z equal to the height of the entire tube row is $\dot{m}_c \, \Delta x / L_x$ (cf. Figs. 8.2, 9.1 and 9.2). The energy balance for the control volume presented in Figs. 9.2 and 9.3 is expressed as:

$$\dot{m}_c \, \frac{\Delta x}{L_x} \, i_c\big|_y + k_A \, \Delta A_2 \, [T_h(x) - T_c(x, y)] = \dot{m}_c \, \frac{\Delta x}{L_x} \, i_c\big|_{y \, + \, \Delta y}. \tag{9.30}$$

Considering that enthalpy $i_c = c_{pc}\big|_0^{T_c} T_c$ and $\Delta A_2 = \Delta x \Delta y$, Eq. (9.30) can be written as:

$$\frac{\dot{m}_c \, c_{pc} \, \Delta x}{L_x} \, \big[T_c(x, y)\big|_{y \, + \, \Delta y} - T_c(x, y)\big|_y\big] = k_A \, \Delta x \, \Delta y \, [T_h(x) - T_c(x, y)]. \tag{9.31}$$

Fig. 9.3 Diagram illustrating
the heat transfer through
elementary surface area
$\Delta A = \Delta A_2$ (cf. Fig. 9.1)

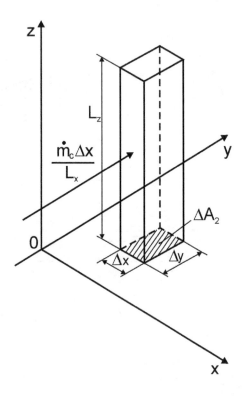

Dividing Eq. (9.31) by ($\Delta x \Delta y$), the following is obtained:

$$\frac{\dot{m}_c \, c_{pc}}{L_x} \frac{T_c(x,y)|_{y+\Delta y} - T_c(x,y)|_y}{\Delta y} = k_A \, [T_h(x) - T_c(x,y)]. \tag{9.32}$$

If $\Delta y \rightarrow 0$, Eq. (9.32) takes the following form:

$$\frac{\dot{m}_c \, c_{pc}}{L_x} \frac{\partial T_c(x,y)}{\partial y} = k_A \, [T_h(x) - T_c(x,y)]. \tag{9.33}$$

Introducing dimensionless coordinates $x^+ = x/L_x$, $y^+ = y/L_y$ and considering that:

$$\frac{\partial T_c}{\partial y} = \frac{\partial T_c}{\partial y^+} \frac{\partial y^+}{\partial y} = \frac{1}{L_y} \frac{\partial T_c}{\partial y^+}$$

Equation (9.33) gives:

$$\frac{\dot{m}_c \, c_{pc}}{k_A \, A} \frac{\partial T_c(x^+,y^+)}{\partial y^+} = T_h\,(x^+) - T_c(x^+,y^+), \tag{9.34}$$

where: $A = L_x L_y$.

Introducing the following relation:

$$N_c = \frac{k_A A}{\dot{m}_c c_{pc}}, \tag{9.35}$$

Eq. (9.34) can be written as follows:

$$\frac{\partial T_c(x^+, y^+)}{\partial y^+} = N_c \left[T_h(x^+) - T_c(x^+, y^+) \right]. \tag{9.36}$$

The energy balance for the medium characterized by the higher temperature (cf. Figs. 9.1 and 9.2) gives:

$$\dot{m}_h i_h|_x = \dot{m}_h i_h|_{x+\Delta x} + k_A \Delta A_1 \left[T_h(x^+) - T_{mc}(x^+) \right], \tag{9.37}$$

where

$$T_{mc}(x^+) = \int_0^1 T_c(x^+, y^+)\, dy^+, \quad \Delta A_1 = L_y \Delta x. \tag{9.38}$$

The symbol $T_{mc}(x^+)$ denotes the air mean temperature on length L_y. Considering that $i_h = c_{ph} T_h$ and dividing the equation by $k_A L_y \Delta x$, the following is obtained:

$$\frac{\dot{m}_h c_{ph}}{k_A L_y} \frac{T_h|_{x+\Delta x} - T_h|_x}{\Delta x} = -[T_h(x^+) - T_{mc}(x^+)]. \tag{9.39}$$

If $\Delta x \to 0$, Eq. (9.39) gives the following differential equation:

$$\frac{\dot{m}_h c_{ph}}{k_A L_y} \frac{dT_h}{dx} = -[T_h(x^+) - T_{mc}(x^+)]. \tag{9.40}$$

Considering that $A = L_x L_y$ and introducing the following dimensionless quantities:

$$x^+ = \frac{x}{L_x} \text{ and } N_h = \frac{k_A A}{\dot{m}_h c_{ph}}, \tag{9.41}$$

Eq. (9.40) can be written as:

$$\frac{dT_h}{dx^+} = -N_h[T_h(x^+) - T_{mc}(x^+)]. \tag{9.42}$$

Equations (9.36) and (9.42) were solved under the following boundary conditions:

$$T_h|_{x^+=0} = T_{h1},\tag{9.43}$$

$$T_c|_{y^+=0} = T_{c1}.\tag{9.44}$$

The first to be solved is Eq. (9.36), which can be written in the following form:

$$\frac{\partial\,[T_h(x^+) - T_c(x^+,y^+)]}{\partial y^+} = -N_c[T_h(x^+) - T_c(x^+,y^+)].\tag{9.45}$$

Separating the variables:

$$\frac{\partial\,[T_h(x^+) - T_c(x^+,y^+)]}{T_h(x^+) - T_c(x^+,y^+)} = -N_c\,\partial y,\tag{9.46}$$

and performing integration, the following is obtained:

$$\ln\,[T_h(x^+) - T_c(x^+,y^+)] = -N_c\,y^+ + C,$$

which gives:

$$T_h(x^+) - T_c(x^+,y^+) = C_1 e^{-N_c y^+}, \quad \text{where: } C_1 = e^C.\tag{9.47}$$

After constant C_1 is determined from boundary condition (9.44), the following is obtained:

$$T_c(x^+,y^+) = T_h(x^+) - [T_h(x^+) - T_{c1}]\,e^{-N_c y^+}.\tag{9.48}$$

Using the expression (9.48), it is possible to calculate the air temperature along its flow path. Substituting (9.48) in (9.38), an expression is obtained that defines the air mean temperature on the thickness of a single tube row:

$$T_{mc}(x^+) = \int_0^1 T_c(x^+,y^+)dy^+ = \int_0^1 \left\{T_h(x^+) - [T_h(x^+) - T_{c1}]\,e^{-N_c y^+}\right\}dy^+$$

$$= T_h(x^+) + [T_h(x^+) - T_{c1}]\frac{1}{N_c}e^{-N_c y^+}\Big|_0^1,$$

from which it follows that:

$$T_{mc}(x^+) = T_h(x^+) - \frac{1}{N_c}[T_h(x^+) - T_{c1}]\big(1 - e^{-N_c}\big).\tag{9.49}$$

Substitution of $T_{mc}(x^+)$ defined by formula (9.49) in Eq. (9.42) gives:

$$\frac{\mathrm{d}T_h}{\mathrm{d}x^+} = -N_h \left\{ T_h(x^+) - T_h(x^+) + \frac{1}{N_c} \left[T_h(x^+) - T_{c1} \right] \left(1 - \mathrm{e}^{-N_c} \right) \right\},$$

from which the following differential equation is obtained:

$$\frac{\mathrm{d}T_h}{\mathrm{d}x^+} = -\frac{N_h}{N_c} \left[T_h(x^+) - T_{c1} \right] \left(1 - \mathrm{e}^{-N_c} \right). \qquad (9.50)$$

Transforming this equation into:

$$\frac{\mathrm{d}(T_h - T_{c1})}{\left[T_h(x^+) - T_{c1} \right]} = -\frac{N_h}{N_c} \left(1 - \mathrm{e}^{-N_c} \right) \mathrm{d}x^+,$$

and performing integration, the following is obtained:

$$\ln \left[T_h(x^+) - T_{c1} \right] = -\frac{N_h}{N_c} \left(1 - \mathrm{e}^{-N_c} \right) x^+ + C_2,$$

from which it follows that:

$$T_h(x^+) - T_{c1} = \mathrm{e}^{\left[-\frac{N_h}{N_c} \left(1 - \mathrm{e}^{-N_c} \right) x^+ + C_2 \right]} = C_3 \mathrm{e}^{-\frac{N_1}{N_2} \left(1 - \mathrm{e}^{-N_2} \right) x^+}, \qquad (9.51)$$

where: $C_3 = \mathrm{e}^{C_2}$.

After constant C_3 was determined from the boundary condition (9.43), a formula that describes changes in the temperature of the cooling medium $T_h(x^+)$ was obtained:

$$T_h(x^+) = T_{c1} + (T_{h1} - T_{c1}) \exp \left[-\frac{N_h}{N_c} \left(1 - \mathrm{e}^{-N_c} \right) x^+ \right]. \qquad (9.52)$$

The air temperature downstream the tube row, found from formula (9.48) considering (9.52), is given by the following formula:

$$T_{c2}(x^+) = T_c(x^+, y^+) \big|_{y^+ = 1}$$
$$= T_{c1} + (T_{h1} - T_{c1}) \exp \left[-\frac{N_h}{N_c} \left(1 - \mathrm{e}^{-N_c} \right) x^+ \right] \left(1 - \mathrm{e}^{-N_c} \right). \qquad (9.53)$$

To calculate the tube row thermal power, i.e., to determine the heat flow rate transferred from the hot liquid to the stream of cooling air, the air mean temperature on the exchanger width L_x is needed:

$$T_{c2m} = \int_0^1 T_{c2}(x^+)\,dx^+$$

$$= \int_0^1 \left\{ T_{c1} + (T_{c1} - T_{c2}) \, \exp\left[-\frac{N_h}{N_c}\left(1 - e^{-N_c}\right)x^+\right]\left(1 - e^{-N_c}\right)\right\} dx^+$$

$$= T_{c1} + \frac{N_c}{N_h}(T_{h1} - T_{c1})\left(1 - e^{-B}\right),$$

$$\tag{9.54}$$

where

$$B = \frac{N_h}{N_c}\left(1 - e^{-N_c}\right). \tag{9.55}$$

The derived formulae make it possible to determine the hot and the cold medium temperatures T_h and T_c, respectively, as well as the mean temperature of the cold medium T_{c2m} downstream the exchanger.

9.4   Plate-Fin Cross-Flow Heat Exchanger

The plate cross-flow heat exchanger is presented in the diagram in Fig. 8.2. The corrugated sheets placed between the plates act as fins. Excellent contact should be ensured between the tips of the corrugated sheets and the plates so that the temperature difference between the fins and the flowing gas should be high, which substantially increases the heat transfer between the fluids.

The corrugated sheets may be soldered to the surface of the plates along the tips of individual elements, or they may be joined to the plate using a different technique, e.g., spot welding (cf. Fig. 9.4).

The plates (cf. Fig. 9.5) are to increase the heat transfer between (usually gaseous) fluids. The hot or the cold medium stream flowing into the channel formed by two adjacent plates and side areas is separated into several or a few dozen streams which do not mix with each other, either on the hot or on the cold medium side. This is why the device is referred to as a cross-flow heat exchanger with unmixed fluids.

Because the exchanger width is much bigger compared to the distance between two plates, it is assumed that the medium is not mixed in the cross-section of the channel created by two adjacent plates and two sides of the exchanger. This assumption is also adopted if there is no corrugated sheet in the channel. Figure 9.6 presents a diagram illustrating the control volume energy balance equation. The energy balance equations for the medium characterized by the higher and the lower temperature (subscript h and c, respectively, cf. Fig. 9.6) have the following form:

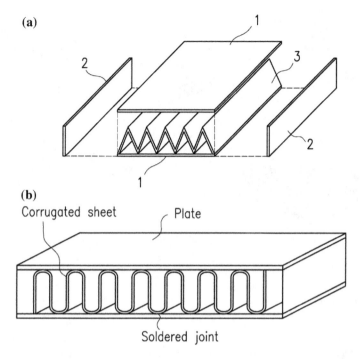

Fig. 9.4 Diagrams of fins made of corrugated sheets placed in between plates in cross-flow heat exchangers: **a** components of a single channel: 1—plate, 2—sides, 3—corrugated sheet with a series of triangular elements; **b** space in between two plates filled with a corrugated sheet with a series of rectangular elements with rounded edges—the tips of individual elements in the series are soldered to the plates

$$\dot{m}_h \frac{\Delta y}{L_y} i_h|_x = \dot{m}_h \frac{\Delta y}{L_y} i_h|_{x+\Delta x} + k_A \, \Delta A \, [T_h(x,y) - T_c(x,y)], \qquad (9.56)$$

$$\dot{m}_c \frac{\Delta x}{L_x} i_c|_y + k_A \, \Delta A \, [T_h(x,y) - T_c(x,y)] = \dot{m}_c \frac{\Delta x}{L_x} i_c|_{y+\Delta y}. \qquad (9.57)$$

Considering that $A = L_x \, L_y$ and $\Delta A = \Delta x \Delta y$, Eqs. (9.56) and (9.57) can be transformed into:

$$\dot{m}_h \frac{i_h|_{x+\Delta x} - i_h|_x}{\frac{\Delta x}{L_x}} = -k_A \, L_x \, L_y \, [T_h(x,y) - T_c(x,y)], \qquad (9.58)$$

$$\dot{m}_c \frac{i_c|_{y+\Delta y} - i_c|_y}{\frac{\Delta y}{L_y}} = k_A \, L_x \, L_y \, [T_h(x,y) - T_c(x,y)]. \qquad (9.59)$$

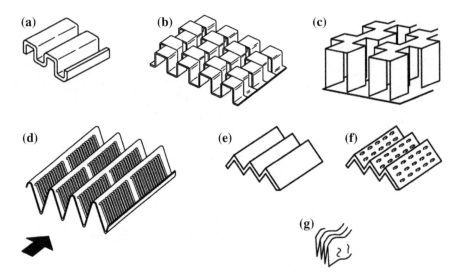

Fig. 9.5 Typical channel configurations in between the plates: **a** fins made of a corrugated sheet with a series of rectangular elements, **b** staggered strips of fins made of a corrugated sheet with a series of rectangular elements, **c** fins made of a corrugated sheet with a series of rectangular elements—gas flows perpendicularly to the rectangular elements, **d** grille-type fins made of a corrugated sheet with a series of triangular elements, **e** fins made of a corrugated sheet with a series of triangular elements, **f** fins made of a corrugated sheet with a series of perforated triangular elements

Introducing dimensionless coordinates $x^+ = x/L_x$ and $y^+ = y/L_y$ and assuming that $\Delta x \to 0$ and $\Delta y \to 0$, Eqs. (9.58) and (9.59) can be written as:

$$\frac{\partial T_h}{\partial x^+} = -N_h \left[T_h(x, y) - T_c(x, y)\right], \tag{9.60}$$

$$\frac{\partial T_c}{\partial y^+} = N_c \left[T_h(x, y) - T_c(x, y)\right], \tag{9.61}$$

where the numbers of the heat transfer units N_h and N_c are defined by the following formulae:

$$N_h = \frac{k_A A}{\dot{m}_h c_h}, \quad N_c = \frac{k_A A}{\dot{m}_c c_c}. \tag{9.62}$$

Equations (9.61) and (9.62) will be solved under the following boundary conditions:

$$T_h(x^+, y^+)\big|_{x^+=0} = T_{h1}, \tag{9.63}$$

$$T_c(x^+, y^+)\big|_{y^+=0} = 0. \tag{9.64}$$

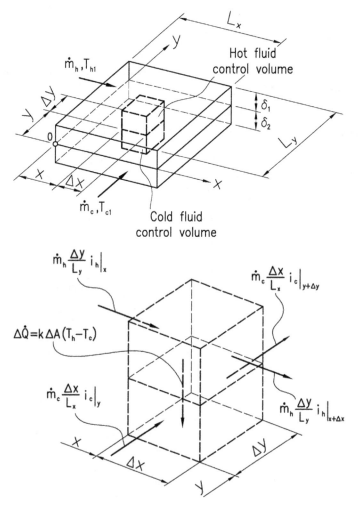

Fig. 9.6 Diagram illustrating the energy balance equation for an elementary control area in a cross-flow heat exchanger

Analysing boundary conditions (9.63) and (9.64), it can be seen that T_{h1} is the surplus of the hot fluid inlet temperature over the inlet temperature of the cold fluid. Introducing dimensionless values:

$$\theta = \frac{T}{T_{h,wlot}},\qquad(9.65)$$

$$\xi = N_h x^+ = \frac{k_A L_x L_y}{\dot{m}_h c_h}\frac{x}{L_x} = \frac{k_A L_y}{\dot{m}_h c_h}x,\qquad(9.66)$$

$$\eta = N_c y^+ = \frac{k_A L_x L_y}{\dot{m}_c c_c} \frac{y}{L_y} = \frac{k_A L_x}{\dot{m}_c c_c} y, \tag{9.67}$$

differential Eqs. (9.60) and (9.61) and boundary conditions (9.63) and (9.64) can be written in the following form:

$$\frac{\partial \theta_h}{\partial \xi} = -(\theta_h - \theta_c), \tag{9.68}$$

$$\frac{\partial \theta_c}{\partial \eta} = (\theta_h - \theta_c), \tag{9.69}$$

$$\theta_h(\xi, \eta)|_{\xi=0} = 1, \tag{9.70}$$

$$\theta_c(\xi, \eta)|_{\eta=0} = 0. \tag{9.71}$$

The solution of the problem (9.68)–(9.71) takes the following form [217, 266]:

$$\theta_h(\xi, \eta) = e^{-\xi}\left[e^{-\eta}I_0\left(2\sqrt{\xi\eta}\right) + \int_0^\eta e^{-u}I_0\left(2\sqrt{\xi u}\right)du\right], \tag{9.72}$$

$$\theta_c(\xi, \eta) = e^{-\xi}\int_0^\eta e^{-u}I_0\left(2\sqrt{\xi u}\right)du. \tag{9.73}$$

Alternative solutions of the problem expressed in (9.68)–(9.71) are found which are easier to use in practical calculations [266]:

$$\theta_h(\xi, \eta) = e^{-(\xi+\eta)}U(\xi, \eta), \tag{9.74}$$

$$\theta_c(\xi, \eta) = e^{-(\xi+\eta)}\left[U(\xi, \eta) - I_0\left(2\sqrt{\xi\eta}\right)\right]. \tag{9.75}$$

The function can be calculated using the following formula [266]:

$$U(\xi, \eta) = e^{(\xi+\eta)} - \sum_{n=1}^\infty \left(\frac{\xi}{\eta}\right)^{\frac{n}{2}}I_n\left(2\sqrt{\xi\eta}\right) = \sum_{n=0}^\infty \left(\frac{\eta}{\xi}\right)^{\frac{n}{2}}I_n\left(2\sqrt{\xi\eta}\right), \tag{9.76}$$

alternatively, using a different expression:

$$U(\xi, \eta) = 1 + \frac{\eta}{1!}\left(1 + \frac{\xi}{1!}\right) + \frac{\eta^2}{2!}\left(1 + \frac{\xi}{1!} + \frac{\xi^2}{2!}\right) + \cdots$$

$$= \sum_{n=0}^{\infty}\sum_{k=0}^{n} \frac{\eta^n \xi^k}{n!k!}, \quad n > 400.$$

(9.77)

Formulae (9.74) and (9.75), where the function $U(\xi, \eta)$ is calculated using formula (9.76), are the most convenient for numerical computations. Functions I_0 and I_n in formulae (9.72)–(9.76) denote modified first-kind zero- and nth-order Bessel functions, respectively.

Chapter 10
Mathematical Modelling of Tube Cross-Flow Heat Exchangers Operating in Steady-State Conditions

This chapter presents two numerical models and one exact analytical model of a two-pass, two-row heat exchanger being a cross-flow plate fin-and-tube heat exchanger made of oval tubes. The computation results obtained using the two numerical models will be compared to those obtained using the exact analytical model.

10.1 Energy Balance Equations Describing the Heat Transfer in Tube Heat Exchangers with the Perpendicular Direction of the Flow of Mediums

Formulation of the initial-boundary problem for cross-flow tube heat exchangers presents some difficulties. This is due to the exchanger discontinuous structure—subsequent tube rows are separated by a fluid flowing transversely to the tubes. Fins or lamellae, i.e. continuous fins, are usually fixed on the outer surface of the tubes because the medium flowing perpendicularly to the axis of the tubes is air or another gas. In such heat exchangers, the mass, momentum and energy conservation equations derived in Chap. 2 cannot be applied directly to describe the flow and the heat transfer phenomena on the gaseous medium side. The modelling of the unsteady-state operation of tube heat exchangers where a liquid flows inside the tubes and gas flows outside transversely to the axis of the tubes is based only on the energy conservation equations formulated for the two mediums. The energy conservation equations will be written in a form suitable for modelling the heat transfer in tube cross-flow heat exchangers, i.e. they will take account of the tube heat exchanger discrete structure.

Equation (2.189) for medium l flowing inside a tube (cf. Fig. 10.1) takes the following form:

© Springer International Publishing AG, part of Springer Nature 2019
D. Taler, *Numerical Modelling and Experimental Testing of Heat Exchangers*,
Studies in Systems, Decision and Control 161,
https://doi.org/10.1007/978-3-319-91128-1_10

Fig. 10.1 Diagram of a
single tube where water
(*l*) flows inside the tube and
air (*g*) flows perpendicularly
to the tube axis

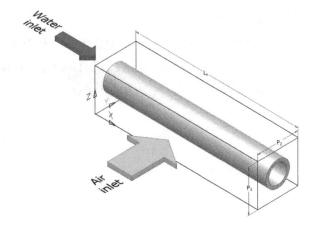

$$\rho_l c_{pl}\left(\frac{\partial T_l}{\partial t} + w_l \frac{\partial T_l}{\partial x}\right) = \frac{\dot{q}_l U_w}{A_l}, \tag{10.1}$$

where U_w is the tube inner perimeter, A_l is the tube cross-section surface area and $\dot{q}_l$
is the heat flux on the tube inner surface. For a tube with a circular cross-section
with the inner surface area radius $r_l = r_w$, quantities U_w and A_l are defined as
follows:

$$U_w = 2\pi r_w, \ A_l = \pi r_w^2. \tag{10.2}$$

Considering that the heat flux on the tube inner surface is defined as:

$$\dot{q}_l = -\alpha_l(T_l - T_w) \tag{10.3}$$

and introducing dimensionless coordinate $x^+ = x/L_x$ and multiplying Eq. (10.1) by
$A_1 Lx$, the equation can be written in the following form:

$$\tau_l \frac{\partial T_l}{\partial t} + \frac{1}{N_l}\frac{\partial T_l}{\partial x^+} = -(T_l - T_w), \tag{10.4}$$

where time constant τ_l and the number of the heat transfer units N_l are defined as
follows:

$$\tau_l = \frac{m_l c_{pl}}{\alpha_l A_{wrg}}, \ N_l = \frac{\alpha_l A_{wrg}}{\dot{m}_l c_{pl}}, \tag{10.5}$$

where $A_{wrg} = U_w L_x = \pi d_w L_x$ is the tube inner surface area.
   Symbol m_l stands for the mass of medium *l* contained in a tube with length L_x
and $\dot{m}_l$ is the mass flow rate.

$$m_l = A_l L \rho_l, \quad \dot{m}_l = (w_l A_l \rho_l)|_{x=0} = (w_l A_l \rho_l)|_{x=L}. \tag{10.6}$$

Considering that the temperature drop on the tube wall is small, a lumped thermal capacity wall model may be adopted, which is described by Eq. (7.7). In this case, Eq. (7.7) takes the following form:

$$m_w c_w \frac{\partial T_w}{\partial t} = \alpha_l A_{wrg} (T_l - T_w) + \alpha_g A_{zrg} (T_{mg} - T_w), \tag{10.7}$$

where the wall mass m_w and the outer and the inner surface areas A_{zrg} and A_{wrg}, respectively, are defined as follows:

$$m_w = A_{wr} L \rho_w, \quad A_{zrg} = U_z L_x, \quad A_{wrg} = U_w L_x. \tag{10.8}$$

The following is obtained after more transformations:

$$\tau_w \frac{\partial T_w}{\partial t} + T_w = \frac{\alpha_l A_{wrg} T_l + \alpha_g A_{zrg} T_{mg}}{\alpha_l A_{wrg} + \alpha_g A_{zrg}}, \tag{10.9}$$

where time constant τ_w is expressed as:

$$\tau_w = \frac{m_w c_w}{\alpha_l A_{wrg} + \alpha_g A_{zrg}}. \tag{10.10}$$

The gas mean temperature T_{mg} on length p_2 (cf. Fig. 10.1) is defined by the following formula:

$$T_{mg} = \frac{1}{p_2} \int_0^{p_2} T_g(x, y, t) dy. \tag{10.11}$$

The right side of Eq. (10.9) expresses the weighted average of temperatures of the two fluids: liquid and gas. In the case of a tube cross-flow heat exchanger, the energy conservation equation for the tube wall differs from the equation written for co- and countercurrent exchangers.

Additionally, the change in the temperature of medium g on the length equal to pitch p_2 (cf. Fig. 10.2) is replaced by a continuous temperature change despite the fact that the medium absorbs heat only from the tube surface. In the heat exchanger discrete model, the gas temperature $T_{g,i}''$ at the control area outlet is determined from the energy balance equation for the control volume presented in Fig. 10.2. It is assumed that in this control volume the liquid temperature on the width of longitudinal pitch p_2 is constant, i.e. the tube occupies the entire width of pitch p_2. The liquid mean temperature $\bar{T}_{l,i}$ is defined by the following formula:

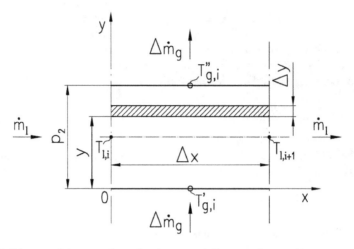

Fig. 10.2 Diagram of the control area for the energy balance on the gas side

$$\bar{T}_{l,i} = \frac{T_{l,i} + T_{l,i+1}}{2}. \tag{10.12}$$

$\bar{T}_{l,i}$ is the liquid temperature in the entire control volume with dimensions $\Delta x p_2$. The changes in the gas temperature along the gas flow path, inside the control area, will be determined using an analytical formula. For this purpose, an unsteady-state energy conservation equation will be written for a smaller control area with dimensions $\Delta x \cdot \Delta y$, located inside a bigger control area with dimensions $\Delta x p_2$.

$$\Delta \dot{m}_g i_g \big|_y + \alpha_g \Delta x \Delta y \left(T_w \big|_{y + \Delta y/2} - T_g \big|_y \right)$$
$$= \Delta \dot{m}_g i_g \big|_{y + \Delta y} + \Delta V (\rho c_{pg} \frac{\partial T_g}{\partial t}) \Big|_{y + \Delta y/2}, \tag{10.13}$$

where $\Delta \dot{m}_g = \dot{m}_g \frac{\Delta x}{L_x}$, $\Delta V = (p_1 p_2 - A_{zr}) \Delta x \frac{\Delta y}{p_2}$.

Considering that enthalpy i_g of gas with temperature T_g expressed in °C is defined as:

$$i_g = c_{pg} \big|_0^{T_g} T_g, \tag{10.14}$$

Eq. (10.13) can be written in the following form:

$$\Delta \dot{m}_g c_{pg,i} \big|_0^{T_g |_y} T_g \big|_y + \alpha_g \Delta x \Delta y \left(T_w \big|_{y + \Delta y/2} - T_g \big|_y \right)$$
$$= \Delta \dot{m}_g c_{pg,i} \big|_0^{T_g |_{y + \Delta y}} T_g \big|_{y + \Delta y} + \Delta V (\rho c_{pg} \frac{\partial T_g}{\partial t}) \Big|_{y + \Delta y/2}. \tag{10.15}$$

Introducing the air mean specific heat $\bar{c}_{pg,i}$ defined as:

$$\bar{c}_{pg,i} = \frac{c_{pg,i}\Big|_0^{T_g|_y} T_g\Big|_y - c_{pg,i}\Big|_0^{T_g|_{y+\Delta y}} T_g\Big|_{y+\Delta y}}{T_g\Big|_y - T_g\Big|_{y+\Delta y}}, \tag{10.16}$$

Eq. (10.15) can be written as:

$$\Delta \dot{m}_g \bar{c}_{pg,i}\left(T_g\Big|_y - T_g\Big|_{y+\Delta y}\right) + \alpha_g \Delta x \Delta y\left(T_w\Big|_{y+\Delta y/2} - T_g\Big|_y\right)$$
$$= (p_1 p_2 - A_{zr})\Delta x \frac{\Delta y}{p_2}\left(\rho c_{pg}\frac{\partial T_g}{\partial t}\right)\Bigg|_{y+\Delta y/2}. \tag{10.17}$$

Transforming Eq. (10.17) into:

$$\Delta \dot{m}_g \bar{c}_{pg,i}\frac{T_g\Big|_{y+\Delta y} - T_g\Big|_y}{\Delta y} + \alpha_g \Delta x\left(T_g\Big|_y - T_{w,i}\right)$$
$$= -(p_1 p_2 - A_{zr})\Delta x \frac{1}{p_2}\left(\rho c_{pg}\frac{\partial T_g}{\partial t}\right)\Bigg|_{y+\Delta y/2} \tag{10.18}$$

and assuming that $\Delta y \to 0$, the following is obtained:

$$-\Delta \dot{m}_g c_{pg}\left(T_g\right)\frac{\partial T_g}{\partial y} - \alpha_g \Delta x\left(T_g - T_w\right) = (p_1 p_2 - A_{zr})\frac{\Delta x}{p_2}\rho c_{pg}\frac{\partial T_g}{\partial t}. \tag{10.19}$$

Introducing the following quantities:

$$y^+ = \frac{y}{p_2}, N_g = \frac{\alpha_g p_2 \Delta x}{\Delta \dot{m}_g c_{pg}} = \frac{\alpha_g N p_2 \Delta x}{N \Delta \dot{m}_g c_{pg}} = \frac{\alpha_g A_{zrg}}{\dot{m}_g c_{pg}},$$
$$\tau_g = \frac{(p_1 p_2 - A_{zr})N \Delta x \rho c_{pg}}{\alpha_g N \Delta x p_2} = \frac{m_g c_{pg}}{\alpha_g A_{zrg}}, \tag{10.20}$$

and noting that

$$\frac{\partial T_g}{\partial y} = \frac{\partial T_g}{\partial y^+}\frac{\partial y^+}{\partial y} = \frac{1}{p_2}\frac{\partial T_g}{\partial y^+} \tag{10.21}$$

Eq. (10.19) can be written as:

$$\tau_g \frac{\partial T_g}{\partial t} + \frac{1}{N_g}\frac{\partial T_g}{\partial y^+} = -\left(T_g - T_w\right), \tag{10.22}$$

where the time constant τ_g and the number of the heat transfer units N_g for a tube heat exchanger are defined by formulae (10.20), where the gas mass m_g and the tube outer surface area A_{zrg} are defined by the following formulae (cf. Fig. 10.3)

Fig. 10.3 Diagram of a cross-flow heat exchanger control area; medium l flows in the direction of axis x, and medium g—in the direction of axis y; **a** control area, **b** cross-section

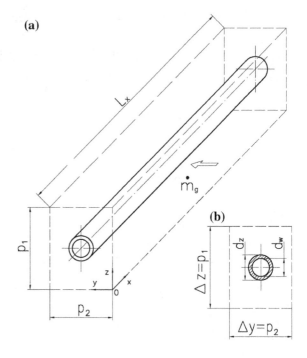

$$A_{zrg} = \pi d_z L_x, m_g = (p_1 p_2 - A_{zr}) L_x \rho, A_{zr} = \frac{\pi}{4} d_z^2. \qquad (10.23)$$

The gas temperature upstream the tube row is known and totals T_g'. Therefore, the boundary condition for Eq. (10.22) has the following form:

$$T_g\big|_{y^+ = 0} = T_g'. \qquad (10.24)$$

Equation (10.22) with boundary condition (10.24) will be used in the heat exchanger numerical model to calculate the gas temperature downstream a given tube row.

Differential equations will be derived next that describe the temperature distribution of the medium flowing transversely to the axis of the tubes and the wall temperature in finned exchangers with individual or continuous fins.

The energy balance equation for medium g, flowing perpendicularly to the axis of the tubes, will be written first. The exchanger length in the direction of axis x is L_x. The analysis will cover one horizontal tube of the exchanger. The derived equation can be used for any tube of the heat exchanger, both for heat exchangers with the in-line and staggered configuration of tubes (cf. Fig. 7.22). The mass flow rate of fluid g through the front surface area $p_1 L_x$ is $\dot{m}_g$ (cf. Fig. 10.3).

To derive a differential equation describing changes in the temperature of medium g, the heat exchanger presented in Fig. 10.3, made of a single tube, with air flowing transversely to it, will be replaced by the exchanger presented in Fig. 10.4 and characterized by the following dimensions: length L_x, width p_2, height p_1. Medium g flows in the direction of axis y along the tube surface with temperature $T_w(x)$. The tube wall temperature in the direction of axis y is constant for a given coordinate x, and the temperature of medium g in the direction of axis z does not change. On the exchanger each elementary area with volume $\Delta x \Delta y p_1$ (cf. Fig. 10.4) fall the tube surface area ΔA_{zrg} and the volume of medium g equal to ΔV_g in the real exchanger:

$$\Delta A_{zrg} = A_{zrg} \frac{\Delta x}{L_x} \cdot \frac{\Delta y}{p_2}, \Delta V_g = V_g \frac{\Delta x}{L_x} \cdot \frac{\Delta y}{p_2}. \tag{10.25}$$

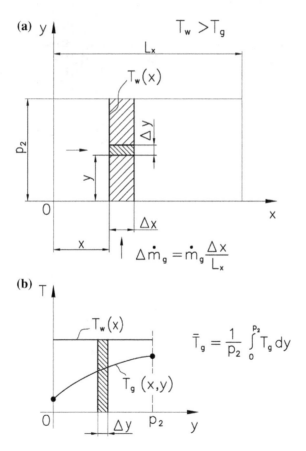

Fig. 10.4 Diagram illustrating the energy balance for medium g flowing perpendicularly to the axis of tubes which contain medium l flowing inside; **a** control area, **b** changes in the temperature of the wall and medium g

Volume V_g depends on the type of tubes the exchanger is made of and is defined as follows:

– for exchangers with continuous fins (lamellae) (cf. Fig. 10.5a):

$$V_g = (p_1 p_2 - A_{zr})(s - \delta_{\dot{z}})n_{\dot{z}}, \qquad (10.26)$$

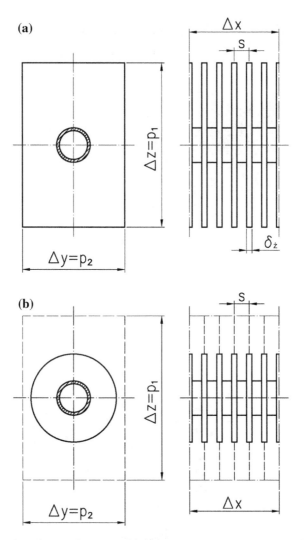

Fig. 10.5 Diagram illustrating calculation of volume ΔV_g of the control volume for plate-fin (lamella) heat exchangers and for exchangers made of finned tubes with round fins; p_1—tube pitch perpendicular to the flow direction of medium g, p_2—tube pitch parallel with the flow direction of medium g; **a** lamella exchanger, **b** finned tube

- for finned tubes (cf. Fig. 10.5b):

$$V_g = \left[(p_1 p_2 - A_{zr})s - \left(\frac{\pi D_{\dot{z}}^2}{4} - A_{otw} \right) \delta_{\dot{z}} \right] n_{\dot{z}},$$
(10.27)

where A_{zr}—surface area of the fin opening where the tube is located, p_1 and p_2—tube pitches for the in-line arrangement, s—fin pitch, $D_{\dot{z}}$—fin outer diameter, $\delta_{\dot{z}}$—fin thickness, $n_{\dot{z}}$—the number of fins on a single tube with length L_x.

For tubes with a circular cross-section, the surface area of the opening in a fin or lamella is $A_{zr} = \pi d_{\dot{z}}^2 / 4$: The energy conservation equation will be written for the control area presented in Fig. 10.4. The control area has the following dimensions: length—Δx, width—Δy, height—p_1.

The energy conservation equation for the control area with volume $\Delta x \Delta y p_1$ has the following form:

$$\Delta \dot{m}_g c_{pg} T_g|_{x+\Delta x/2, y} + \alpha_{zr} \Delta A_{zrg} (T_w - T_g)|_{x+\Delta x/2, y+\Delta y/2}$$
$$= \Delta \dot{m}_g c_{pg} T_g\Big|_{x+\Delta x/2, y+\Delta y} + \Delta V_g \rho_g c_{pg} \frac{\partial T_g}{\partial t},$$
(10.28)

which, considering expressions (10.25) and performing relevant transformations, can be written as:

$$V_g \frac{\Delta x \Delta y}{L_x \, p_2} \rho_g c_{pg} \frac{\partial T_g}{\partial t} + \dot{m}_g \frac{\Delta x}{L_x} c_{pg} \left(T_g|_{x+\Delta x/2, y+\Delta y} - T_g|_{x+\Delta x/2, y} \right)$$
$$= \alpha_{zr} A_{zrg} \frac{\Delta x \Delta y}{L_x \, p_2} (T_w - T_g)|_{x+\Delta x/2, y+\Delta y/2}.$$
(10.29)

Multiplying both sides of Eq. (10.29) by $L_x \, p_2/(\Delta x \, \Delta y)$, the following is obtained:

$$V_g \rho_g c_{pg} \frac{\partial T_g}{\partial t} + \dot{m}_g c_{pg} p_2 \frac{\left(T_g|_{x+\Delta x/2, y+\Delta y} - T_g|_{x+\Delta x/2, y} \right)}{\Delta y}$$
$$= \alpha_{zr} A_{zrg} (T_w - T_g)|_{x+\Delta x/2, y+\Delta y/2}.$$
(10.30)

Assuming that $\Delta x \to 0$ and $\Delta y \to 0$, Eq. (10.30) gives:

$$V_g \rho_g c_{pg} \frac{\partial T_g}{\partial t} + \dot{m}_g c_{pg} p_2 \frac{\partial T_g}{\partial y} = \alpha_{zr} A_{zrg} (T_w - T_g).$$
(10.31)

Introducing dimensionless coordinate $y^+ = y/p_2$, time constant τ_g and the number of the heat transfer units N_g:

$$\tau_g = \frac{V_g \rho_g c_{pg}}{\alpha_{zr} A_{zrg}} = \frac{m_g c_{pg}}{\alpha_{zr} A_{zrg}}, N_2 = \frac{\alpha_{zr} A_{zrg}}{\dot{m}_g c_{pg}}. \tag{10.32}$$

Equation (10.22) is obtained, which is identical to the equation derived for an exchanger made of unfinned tubes.

Next, the energy conservation equation will be derived for a tube wall to which the medium flows transversely. The tube contains medium *l* flowing inside, while medium *g* flows outside the tube, along its surface. For a given coordinate x and time t, the tube outer surface temperature T_w is constant on the perimeter. The temperature of medium *g* changes along the flow direction, i.e. it is not constant on the width of a single pitch p_2 (cf. Fig. 10.4). The heat flow rate $\Delta \dot{Q}_g (x + \Delta x/2, y + \Delta y/2, t)$ absorbed by medium *g* in the control area with width Δx and length Δy, as presented in Fig. 10.4, is:

$$\Delta \dot{Q}_g \left(x + \frac{\Delta x}{2}, y + \frac{\Delta y}{2}, t \right)$$
$$= \alpha_g \Delta A_{zrg} \left[T_w \left(x + \frac{\Delta x}{2}, t \right) - T_g \left(x + \frac{\Delta x}{2}, y + \frac{\Delta y}{2}, t \right) \right]. \tag{10.33}$$

Considering that $A_{zrg} = U_z L_x$ and $\Delta A_{zrg} = U_z L_x \frac{\Delta x}{L_x} \frac{\Delta y}{p_2}$, and assuming that $\Delta x \to 0$ and $\Delta y \to 0$, Eq. (10.33) results in the following relation:

$$\frac{d \dot{Q}_g (x, y, t)}{dxdy} = \frac{\alpha_g U_z}{p_2} \left[T_w(x, t) - T_g(x, y, t) \right]. \tag{10.34}$$

Noting that the local heat flux is defined as:

$$\dot{q}_g (x, y, t) = \frac{d \dot{Q}_g (x, y, t)}{dxdy} \tag{10.35}$$

and considering that the mean heat flux on the width of a single pitch p_2 is:

$$\bar{\dot{q}}_g (x, t) = \frac{1}{U_z} \int_0^{p_2} \dot{q}_g (x, y, t) dy, \tag{10.36}$$

the following is obtained from formula (10.34):

$$\bar{\dot{q}}_g (x, t) = \int_0^{p_2} \frac{\alpha_g}{p_2} \left[T_w(x, t) - T_g(x, y, t) \right] dy = \alpha_g \left[T_w(x, t) - T_{mg}(x, t) \right], \tag{10.37}$$

where

$$T_{mg}(x,t) = \frac{1}{p_2} \int_0^{p_2} T_g(x,y,t) \, dy.$$ (10.38)

It follows from formula (10.37), describing the mean heat flux on the tube perimeter $\bar{q}_g(x,t)$, that the heat flow rate transferred by the tube outer surface is proportional to the difference between the tube outer surface temperature and the mean temperature of the medium on the width of a single pitch p_2. The energy conservation equation written for the tube wall in finned, tube cross-flow heat exchangers has the following form [306]:

$$\left(U_m \delta_w \rho_w c_w + \frac{m_{\dot{z}} c_{\dot{z}} \eta_{\dot{z}}}{s} \right) \frac{\partial T_w}{\partial t} + \frac{m_{\dot{z}} c_{\dot{z}} (1 - \eta_{\dot{z}})}{s} \frac{\partial T_{mg}}{\partial t}$$
$$= \alpha_l U_w (T_l - T_w) + \alpha_{zr} U_z (T_{mg} - T_w).$$ (10.39)

Temperature T_{mg} in Eq. (10.39) is defined by formula (10.38). Equation (10.39) can be transformed into:

$$\tau_w \frac{\partial T_w}{\partial t} + \tau_{\dot{z}} \frac{\partial T_{mg}}{\partial t} + T_w = \frac{\alpha_l U_w}{\alpha_l U_w + \alpha_{zr} U_z} T_l + \frac{\alpha_{zr} U_z}{\alpha_l U_w + \alpha_{zr} U_z} T_{mg},$$ (10.40)

where

$$\tau_w = \frac{m_w c_w + \eta_{\dot{z}} m_{\dot{z}} c_{\dot{z}}}{\alpha_l A_{wrg} + \alpha_{zr} A_{zrg}}, \tau_{\dot{z}} = \frac{(1 - \eta_{\dot{z}}) m_{\dot{z}} c_{\dot{z}}}{\alpha_l A_{wrg} + \alpha_{zr} A_{zrg}}.$$ (10.41)

Considering that the fin efficiency $\eta_{\dot{z}}$ is high and close to one, time constant $\tau_{\dot{z}}$ is much smaller compared to time constant τ_w. For exchangers made of unfinned tubes, $m_{\dot{z}} = 0$ and $\alpha_{zr} = \alpha_g$ are assumed in Eq. (10.41).

The differential equations derived in Sect. (5.3) for a tube heat exchanger with a perpendicular flow of mediums can be applied because both mediums are in the liquid or the gaseous state, or one of the mediums is liquid and the other—gaseous.

10.2 Numerical Modelling of the Heat Transfer in Tube Cross-Flow Heat Exchangers

In the heat transfer numerical modelling the entire exchanger is divided into smaller cells, also referred to as finite volumes or control areas, (cf. Fig. 10.6).

Typically, the liquid medium flows inside the tubes, whereas the gaseous one flows outside, perpendicularly to their axis.

An energy balance equation is written for each control area. This gives a system of balance equations used to determine the temperature of the two mediums in

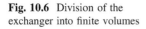
Fig. 10.6 Division of the
exchanger into finite volumes

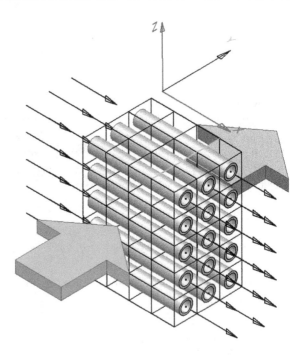

points, referred to as nodes, which are usually located in the centre of the control
area or at the control area inlet and outlet.

Two methods of numerical modelling of the tube cross-flow heat exchanger
steady-state operation will be presented. In all the energy conservation equations
derived in Sect. 10.1 for the two mediums and the wall, the time derivatives will be
equal to zero. The wall temperature can be found from formula (10.9) or (10.40).
The heat exchanger steady-state calculations can be simplified by eliminating from
the energy conservation Eq. (10.4), for the medium flowing inside the tube, and
from Eq. (10.22), for the medium flowing outside and transversely to the tube axis,
the wall temperature defined by formula (10.9) for unfinned tubes or by formula
(10.40) for finned ones. By introducing into the energy conservation equations
written for the two fluids overall heat transfer coefficient k, defined by formula (7.76)
for unfinned tubes with a circular cross-section or by formula (7.78) for finned tubes,
the heat exchanger mathematical modelling can be simplified.

Two methods of numerical modelling of the heat transfer in cross-flow tube heat
exchangers will be presented. The methods differ in the way the gas temperature is
averaged on a single tube row thickness. The gas mean temperature T_{mg} appears in
differential Eqs. (10.9) and (10.40), which describe changes in the wall temperature. In the first method, the mean temperature of the medium flowing perpendicularly to the axis of the tubes is the arithmetic mean of the temperature values at a
given tube row inlet and outlet. In the second—the medium mean temperature is
calculated from formula (10.38), which means that this is the integral-mean
temperature.

10.2.1 *Arithmetic Averaging of the Gas Temperature on the Thickness of a Single Tube Row*

An analytical model of a single- and a two-row tube heat exchanger is developed in [296, 298, 306]. This subsection presents the first numerical model of the exchanger, further below referred to as numerical Model I. The mediums thermo-physical properties may depend on temperature. A discrete mathematical model describing the heat transfer phenomenon will be derived using the control volume method. The temperature of the liquid flowing through the first tube row is a function of coordinate x only. The air mass flow rate $\Delta \dot{m}_g$ per a single control volume can be calculated by the following formula:

$$\Delta \dot{m}_g = \frac{\Delta x}{L_x} \dot{m}_g, \qquad (10.42)$$

where L_x is the radiator width and $\dot{m}_g$ is the air mass flow per a single tube. If there are N control volumes (cells) on the exchanger width, then $\Delta x = L_x/N$ and formula (10.42) takes the following form:

$$\Delta \dot{m}_g = \dot{m}_g/N. \qquad (10.43)$$

Next, formulae will be derived to calculate the temperature of the two mediums.

10.2.1.1 Liquid Energy Conservation Equation

Figure 10.7 presents a diagram of the control area for which the energy balance equation will be written.

The energy balance equation for the liquid flowing inside the tubes, written for the control volume presented in Fig. 10.7, has the following form [296, 303, 306, 307]:

$$\dot{m}_{l,i} i_{l,i} = k \, \Delta A \left(\frac{T_{l,i} + T_{l,i+1}}{2} - \frac{T'_{g,i} + T''_{g,i}}{2} \right) + \dot{m}_{l,i} i_{l,i+1}, \qquad (10.44)$$

which gives the following formula:

$$\dot{m}_{l,i} \left(i_{l,i+1} - i_{l,i} \right) = -k \, \Delta A \left(\frac{T_{l,i} + T_{l,i+1}}{2} - \frac{T'_{g,i} + T''_{g,i}}{2} \right). \qquad (10.45)$$

Considering that

$$i_{l,i+1} - i_{l,i} = \bar{c}_{l,i} \left(T_{l,i+1} - T_{l,i} \right), \qquad (10.46)$$

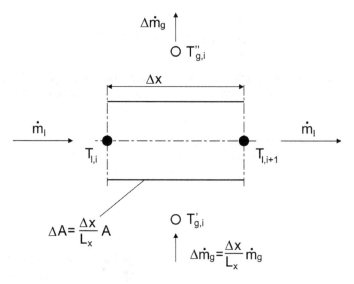

Fig. 10.7 Diagram illustrating the energy conservation equation for a finite volume in the liquid region

where

$$\bar{c}_{l,i} = \frac{\int_{T_{l,i+1}}^{T_{l,i}} c_l(T) \mathrm{d}T}{T_{l,i} - T_{l,i+1}} = c_l \Big|_{T_{l,i+1}}^{T_{l,i}}, \tag{10.47}$$

and then substituting (10.46) in (10.45), the following is obtained:

$$\dot{m}_{l,i} \bar{c}_{l,i} (T_{l,i+1} - T_{l,i}) = -k\,\Delta A \left(\frac{T_{l,i} + T_{l,i+1}}{2} - \frac{T'_{g,i} + T''_{g,i}}{2} \right). \tag{10.48}$$

Assuming that mean specific heat can be expressed as the arithmetic mean:

$$\bar{c}_{l,i} = \frac{c_l(T_{l,i}) + c_l(T_{l,i+1})}{2}$$

and introducing the following quantity:

$$\Delta N_{l,i} = \frac{k\,\Delta A}{\dot{m}_l \bar{c}_{l,i}} = \frac{2k\,\Delta A}{\dot{m}_l [c_l(T_{l,i}) + c_l(T_{l,i+1})]} \tag{10.49}$$

Eq. (10.48) gives:

$$T_{l,i+1} = \frac{1}{1 + \frac{\Delta N_{l,i}}{2}} \left[T_{l,i}\left(1 - \frac{\Delta N_{l,i}}{2}\right) + \frac{\Delta N_{l,i}}{2}\left(T'_{g,i} + T''_{g,i}\right) \right],$$

$$i = 1, \ldots, N.$$

(10.50)

The system of Eq. (10.50) can be solved efficiently using the Gauss-Seidel iterative method. This facilitates the calculation of $\Delta N_{l,i}$, which is updated in every iterative step with respect to temperature-dependent specific heat. Calculating $T_{l,i+1}$ at iterative step $(k + 1)$, i.e. calculating $T_{l,i+1}^{(k+1)}$, parameter $\Delta N_{l,i}$ is calculated at the k-th iterative step.

$$\Delta N_{l,i} = \frac{2 \, k\Delta A}{\dot{m}_l \left[c_l\left(T_{l,i}\right) + c_l\left(T_{l,i+1}^{(k)}\right) \right]}.$$

(10.51)

Usually, the dependence of the liquid specific heat on temperature is not very strong, and it may, therefore, be assumed that $\bar{c}_{l,i}$ is constant along the entire length of the exchanger tube. However, if the medium flowing inside tubes is water vapour, temperature and pressure have a substantial impact on its specific heat. In such a situation, $\bar{c}_{l,i}$ is calculated separately for each cell.

10.2.1.2 Gas Energy Conservation Equation

In the case of the gas flow, the energy conservation equation for the control volume presented in Fig. 10.7 has the following form:

$$\Delta \dot{m}_g i'_{g,i} + k\Delta A \left(\frac{T_{l,i} + T_{l,i+1}}{2} - \frac{T'_{g,i} + T''_{g,i}}{2} \right) = \Delta \dot{m}_g i''_{g,i},$$

(10.52)

where

$$T'_{g,i} = T'_{gm}, \quad i = 1, \ldots, N.$$

Equation (10.52) can be written as:

$$\Delta \dot{m}_g c_{pg} \Big|_0^{T'_{g,i}} T'_{g,i} + k\Delta A \left(\frac{T_{l,i} + T_{l,i+1}}{2} - \frac{T'_{g,i} + T''_{g,i}}{2} \right) = \Delta \dot{m}_g \, c_{pg} \Big|_0^{T''_{g,i}} T''_{g,i},$$

(10.53)

Introducing mean specific heat defined as:

$$\bar{c}_{pg,i} = \frac{c_{pg}\big|_0^{T''_{g,i}} T''_{g,i} - c_{pg}\big|_0^{T'_{g,i}} T'_{g,i}}{T''_{g,i} - T'_{g,i}} \cong c_{pg,i}\left(\frac{T'_{gm} + T''_{gm}}{2}\right) \qquad (10.54)$$

and dividing both sides of Eq. (10.53) by $\Delta \dot{m}_g \bar{c}_{pg,i}$, the following is obtained:

$$T''_{g,i} = \frac{1}{1 + \frac{\Delta N_{g,i}}{2}}\left[\left(1 - \frac{\Delta N_{g,i}}{2}\right)T'_{g,i} + \Delta N_{g,i}\frac{T_{l,i} + T_{l,i+1}}{2}\right], \quad i = 1,\ldots,N, \quad (10.55)$$

where

$$\Delta N_{g,i} = \frac{k\,\Delta A}{\Delta \dot{m}_g \bar{c}_{pg,i}}. \qquad (10.56)$$

In the case of the first tube row, it may be assumed that the inflowing gas temperature is constant and totals T'_{gm}. The boundary condition for gas at the exchanger inlet is then of the following form:

$$T'_{g,i} = T'_{gm}, \quad i = 1,\ldots,N. \qquad (10.57)$$

Temperatures T'_{gm} and T''_{gm} denote the gas (air) mean temperature at the single-row heat exchanger inlet and outlet, respectively. The derived difference equations can be used for the second and—if the device is a multi-row exchanger—also for the subsequent tube rows.

10.2.2 Integral Averaging of the Gas Temperature on the Thickness of a Single Tube Row

The modelling of multi-row cross-flow heat exchangers is based on the analysis of the heat transfer in a single tube row.

Depending on the working medium, the energy balance equations for the heat exchanger single tube have the following form [296, 306, 307]:

- for water:

$$\frac{dT_l}{dx^+} = -N_l\left[T_l(x^+) - T_{mg}(x^+)\right], \qquad (10.58)$$

- for gas (air):

$$\frac{\partial T_g}{\partial y^+} = N_g\left[T_l(x^+) - T_g(x^+, y^+)\right], \qquad (10.59)$$

where T_{mg} denotes the gas mean temperature on the thickness of a single tube row. Equations (10.58) and (10.59) will be solved under the following boundary conditions:

$$T_l|_{x^+=0} = T_l', \qquad (10.60)$$

$$T_g|_{y^+=0} = T_g'. \qquad (10.61)$$

The following notation is adopted in formulae (10.58)–(10.61): T_l—liquid temperature, °C; T_g—gas (air) temperature, °C; T_l', T_g'—temperature of the liquid at the tube inlet and of the gas upstream the heat exchanger, °C; $N_g = \frac{kA_{zrg}}{\dot{m}_g \bar{c}_g} = \frac{k(U_z L_x)}{\dot{m}_g \bar{c}_g}$ number of the heat transfer units, gas side; $N_l = \frac{kA_{rg}}{\dot{m}_l \bar{c}_l} = \frac{k(U_z L_x)}{\dot{m}_l \bar{c}_l}$ number of the heat transfer units, liquid side; $x^+ = x/L_x$—dimensionless coordinate; $y^+ = y/p_2$—dimensionless coordinate; $\dot{m}_l, \dot{m}_g$—liquid and gas mass flow rate per tube, respectively, kg/s.

The boundary problem expressed in (10.58)–(10.61) can be solved analytically assuming that physical properties of the mediums are independent of temperature. To take account of the dependence of the fluid physical properties on temperature and pressure, a numerical model of the exchanger was developed using the finite volume method. The exchanger division into control areas (finite volumes) is presented in Fig. 10.6. Figure 10.7 presents the control area for which the gas- and the liquid-side energy balance equations will be formulated. The height of a single cell is p_1, the width (depth) is p_2 (cf. Fig. 10.3) and the length is $\Delta x = L_{ch}/N = L_x/N$, where N is the number of control volumes with length Δx on the tube entire length.

10.2.2.1   Gas Energy Conservation Equation

In the heat exchanger discrete model, the gas temperature at the control area outlet $T_{g,i}''$ is determined from the energy balance equation for the control volume presented in Fig. 10.7. It is assumed that in this control volume the liquid temperature on the width of longitudinal pitch p_2 is constant, i.e. the tube occupies the entire width of pitch p_2. The liquid mean temperature $\bar{T}_{l,i}$ is defined by the following formula:

$$\bar{T}_{l,i} = \frac{T_{l,i} + T_{l,i+1}}{2}. \qquad (10.62)$$

$\bar{T}_{l,i}$ is the liquid temperature in the entire control volume with dimensions $\Delta x p_2$. The changes in the gas temperature along the gas flow path, inside the control area, will be determined using an analytical formula. For this purpose, an energy balance equation will be written for a smaller control area (dimensions: $\Delta x \Delta y$) located inside a bigger control area (dimensions: $\Delta x p_2$) (cf. Fig. 10.4).

$$\Delta \dot{m}_g i_g \big|_y + k \, \Delta x \Delta y \big(\bar{T}_{l,i} - T_g \big|_y \big) = \Delta \dot{m}_g i_g \big|_{y + \Delta y}. \tag{10.63}$$

Considering that enthalpy i_g of the gas with temperature T_g expressed in °C is defined as:

$$i_g = c_{pg} \big|_0^{T_g} T_g, \tag{10.64}$$

Eq. (10.63) can be written as follows:

$$\Delta \dot{m}_g c_{pg,i} \big|_0^{T_g |_y} T_g \big|_y + k \, \Delta x \, \Delta y \big(\bar{T}_{l,i} - T_g \big|_y \big) = \Delta \dot{m}_g c_{pg,i} \big|_0^{T_g |_{y + \Delta y}} T_g \big|_{y + \Delta y}. \tag{10.65}$$

Introducing the air mean specific heat $\bar{c}_{pg,i}$ defined as:

$$\bar{c}_{pg,i} = \frac{c_{pg,i} \big|_0^{T_g |_y} T_g \big|_y - c_{pg,i} \big|_0^{T_g |_{y + \Delta y}} T_g \big|_{y + \Delta y}}{T_g \big|_y - T_g \big|_{y + \Delta y}}, \tag{10.66}$$

Eq. (10.65) can be written as follows:

$$\Delta \dot{m}_g \, \bar{c}_{pg,i} \big(T_g \big|_y - T_g \big|_{y + \Delta y} \big) + k \, \Delta x \, \Delta y \big(\bar{T}_{l,i} - T_g \big|_y \big) = 0. \tag{10.67}$$

Transforming Eq. (10.67) into:

$$\Delta \dot{m}_g \bar{c}_{pg,i} \frac{T_g \big|_{y + \Delta y} - T_g \big|_y}{\Delta y} + k \, \Delta x \big(T_g \big|_y - \bar{T}_{l,i} \big) = 0 \tag{10.68}$$

and assuming that $\Delta y \to 0$, the following is obtained:

$$\Delta \dot{m}_g c_{pg} (T_g) \frac{\partial T_g}{\partial y} + k \, \Delta x \big(T_g - \bar{T}_{l,i} \big) = 0. \tag{10.69}$$

Equation (10.69) will be integrated for the i-th control volume (cf. Fig. 10.2) using the following boundary condition:

$$T_g \big|_{y=0} = T'_{g,i}. \tag{10.70}$$

Assuming that the gas specific heat is defined as Eq. (10.69) will be approximated for cell i presented in Fig. 10.7 using the following difference equation:

$$\frac{\partial\left(\bar{T}_{l,i} - T_{g,i}\right)}{\partial y} = -\frac{k\,\Delta x}{\Delta\dot{m}_g\bar{c}_{pg,i}}\left(\bar{T}_{l,i} - T_{g,i}\right). \tag{10.71}$$

Integrating both sides of Eq. (10.71), the following is obtained:

$$\int\frac{\mathrm{d}\left(\bar{T}_{l,i} - T_{g,i}\right)}{\bar{T}_{l,i} - T_{g,i}} = -\frac{k\Delta x}{\Delta\dot{m}_g\bar{c}_{pg,i}}\int\mathrm{d}y + C, \tag{10.72}$$

from which it follows that:

$$\ln\left(\bar{T}_{l,i} - T_{g,i}\right) = -\frac{k\,\Delta x}{\Delta\dot{m}_g\bar{c}_{pg,i}}y + C. \tag{10.73}$$

Expression (10.73) gives:

$$\bar{T}_{l,i} - T_{g,i} = e^C\exp\left(-\frac{k\,\Delta x}{\Delta\dot{m}_g\bar{c}_{pg,i}}y\right). \tag{10.74}$$

Considering that $C_1 = e^C$, Eq. (10.74) can be written in the following form:

$$T_{g,i} = \bar{T}_{l,i} - C_1\exp\left(-\frac{k\cdot\Delta x}{\Delta\dot{m}_g\bar{c}_{pg,i}}y\right). \tag{10.75}$$

Substitution of solution (10.75) in boundary condition (10.70) gives the following equation: $T'_{g,i} = \bar{T}_{l,i} - C_1\cdot 1$, from which constant C_1 is determined:

$$C_1 = \bar{T}_{l,i} - T'_{g,i}. \tag{10.76}$$

Substituting (10.76) in (10.75), an expression is obtained for the gas temperature distribution along the gas flow path:

$$T_{g,i} = \bar{T}_{l,i} - \left(\bar{T}_{l,i} - T'_{g,i}\right)\exp\left(-\frac{k\,\Delta x}{\Delta\dot{m}_g\bar{c}_{pg,i}}y\right), \quad i = 1,\dots,N. \tag{10.77}$$

The temperature at the control area outlet is found substituting $y = p_2$ in formula (10.77):

$$T''_{g,i} = \bar{T}_{l,i} - \left(\bar{T}_{l,i} - T'_{g,i}\right)\exp\left(-\frac{k\,\Delta x\,p_2}{\Delta\dot{m}_g\bar{c}_{pg,i}}\right). \tag{10.78}$$

The heat transfer surface area ΔA in a single control volume for a single tube is defined as $\Delta A = \Delta x p_2$. In a real heat exchanger, the heat transfer surface area in a single control volume is:

$$\Delta A = U_z \Delta x. \tag{10.79}$$

In the case of an exchanger made of tubes with a circular cross-section, ΔA is $\Delta A = \pi d_z \Delta x$. If $\Delta x p_2$ in formula (10.78) is replaced by expression (10.79), the following is obtained:

$$T''_{g,i} = \bar{T}_{l,i} - \left(\bar{T}_{l,i} - T'_{g,i}\right) \exp\left(-\Delta N_{g,i}\right), \tag{10.80}$$

where

$$\Delta N_{g,i} = \frac{k \, \Delta A}{\Delta \dot{m}_g \bar{c}_{pg,i}}, \quad A = U_z L_x, \quad \Delta A = \frac{A}{N}, \quad \bar{T}_{l,i} = \frac{T_{l,i} + T_{l,i+1}}{2}. \tag{10.81}$$

If the mass flow rate $\dot{m}_g$ is constant on the exchanger width, i.e. if it is not a function of coordinate x, then $\Delta \dot{m}_g = \dot{m}_g/N$, and formula (10.81) results in:

$$\Delta N_{g,i} = \frac{k \, \Delta A}{\Delta \dot{m}_g \bar{c}_{pg,i}} = \frac{k \, N \Delta A}{N \Delta \dot{m}_g \, \bar{c}_{pg,i}} = \frac{k \, A}{\dot{m}_g \, \bar{c}_{pg,i}}. \tag{10.82}$$

It can easily be seen that formula (10.80) is valid for overall heat transfer coefficient k and mass flow rate $\dot{m}_g$ both dependent on coordinate x. In this case, $\Delta N_{g,i}$ in formula (10.80) is defined by the following expression:

$$\Delta N_{g,i} = \frac{k_i \, \Delta A}{\Delta \dot{m}_{g,i} \bar{c}_{pg,i}}, \tag{10.83}$$

where k_i and $\Delta \dot{m}_{g,i}$, respectively, are overall heat transfer coefficient in the i-th finite volume and the mass flow rate through the i-th cell with width Δx and height p_1.

In the case of the entire tube row, it is easy to take account of the changes in $\dot{m}_g$ on the exchanger height, i.e. the changes in $\dot{m}_g$ as a function of coordinate z. Therefore, in the exchanger mathematical model, it is easy to take account of changes in overall heat transfer coefficient on the entrance length or the non-uniformity of the velocity field in the cross-section of the channel where the heat exchanger is located. Equation (10.80) is used to calculate the gas temperature downstream a given tube row in the numerical model of an exchanger with any number of tube rows and passes.

The gas integral-mean temperature on the thickness of a single tube row is determined from the following formula:

$$\bar{T}_{g,i} = \frac{1}{p_2} \int_0^{p_2} T_g(y)dy = \bar{T}_{l,i} - \left(\bar{T}_{l,i} - T'_{g,i}\right) \frac{1}{p_2} \int_0^{p_2} \exp\left(-\frac{k_i \Delta x}{\Delta \dot{m}_{g,i} \bar{c}_{pg,i}} y\right) dy$$

$$= \bar{T}_{l,i} - \left(\bar{T}_{l,i} - T'_{g,i}\right) \left(-\frac{\Delta \dot{m}_g \bar{c}_{pg,i}}{k_i \, \Delta x \, p_2}\right) \left[\exp\left(-\frac{k_i \Delta x p_2}{\Delta \dot{m}_g \, \bar{c}_{pg,i}}\right) - 1\right]. \tag{10.84}$$

Formula (10.84) can also be written in a different form:

$$\bar{T}_{g,i} = \bar{T}_{l,i} - \frac{1}{\Delta N_{g,i}} \left(\bar{T}_{l,i} - T'_{g,i} \right) \left[1 - \exp\left(-\Delta N_{g,i}\right) \right], \tag{10.85}$$

where $\Delta N_{g,i}$ is defined by formula (10.82).

10.2.2.2 Liquid Energy Conservation Equation

The energy conservation equation for the flowing liquid, written for the finite volume presented in Fig. 10.7, can be expressed as:

$$\dot{m}_l \, \bar{c}_{l,i} \left(T_{l,i+1} - T_{l,i} \right) = -k \, \Delta A \left(\frac{T_{l,i} + T_{l,i+1}}{2} - \bar{T}_{g,i} \right). \tag{10.86}$$

Assuming next that specific heat can be approximated by the arithmetic mean:

$$\bar{c}_{l,i} = \frac{c_l\left(T_{l,i}\right) + c_l\left(T_{l,i+1}\right)}{2}, \tag{10.87}$$

and introducing:

$$\Delta N_{l,i} = \frac{k \, \Delta A}{\dot{m}_l \, \bar{c}_{l,i}} = \frac{2 \, k \, \Delta A}{\dot{m}_l \left[c_l\left(T_{l,i}\right) + c_l\left(T_{l,i+1}\right) \right]}, \tag{10.88}$$

Eq. (10.86) gives:

$$T_{l,i+1} = \frac{1}{1 + \frac{\Delta N_{l,i}}{2}} \left[\left(1 - \frac{\Delta N_{l,i}}{2} \right) T_{l,i} + \Delta N_{l,i} \bar{T}_{g,i} \right], \quad i = 1, \ldots, N, \tag{10.89}$$

where the temperature $\bar{T}_{g,i}$ is defined by formula (10.85).

The system of Eq. (10.89) is used to calculate the liquid temperature on the tube length along the liquid flow path, considering that temperature $T_{l,1}$ in the first node $i = 1$ is known from boundary condition (10.60), i.e. the boundary condition takes the following form:

$$T_{l,1} = T'_l. \tag{10.90}$$

The system of nonlinear algebraic equations made of Eqs. (10.77) and (10.89) is solved using the Gauss-Seidel method, which converges for any initial approximation. It may be assumed that the initial values of node temperatures in the Gauss-Seidel iterative procedure are equal to the inlet temperatures of the liquid or gas, i.e.:

$$T_{l,i}^{(0)} = T_l', \quad i = 1, \ldots, N+1, \tag{10.91}$$

$$T_{g,i}^{(0)} = T_g', \quad i = 1, \ldots, N. \tag{10.92}$$

After the liquid and the gas temperatures are found, it is possible to determine the temperature of the wall.

10.2.3 Tube Wall Temperature

Knowing the liquid and the gas temperatures, it is possible to calculate the temperature of the wall. Assuming that the temperature of the liquid flowing inside the tube is higher than the temperature of the medium flowing perpendicularly to the tube axis, heat flux $\dot{q}_{z,i}$ on the tube outer surface ($r = r_z$) can be calculated from the following formula:

$$\dot{q}_{z,i} = k\left(\frac{T_{1,i} + T_{1,i+1}}{2} - \bar{T}_{g,i}\right), \tag{10.93}$$

where the overall heat transfer coefficient k related to the outer surface area of a bare (unfinned) tube (i.e. without taking account of the surface area of the fins) is defined by Eq. (7.78). The tube inner and outer surface temperatures $T_{w,i}|_{r=r_w}$ and $T_{w,i}|_{r=r_z}$, respectively, can be determined using the following formulae:

$$T_{w,i}|_{r=r_w} = \frac{T_{1,i} + T_{1,i+1}}{2} - \frac{\dot{q}_{z,i}\frac{U_z}{U_w}}{\alpha_w}, \tag{10.94}$$

$$T_{w,i}|_{r=r_z} = \bar{T}_{g,i} + \frac{\dot{q}_{z,i}}{\alpha_{zr}}, \tag{10.95}$$

where U_w and U_z, respectively, represent the tube inner and outer perimeter.

In the case of tubes with a circular cross-section, $U_w = \pi d_w$ and $U_z = \pi d_z$. The temperature drop in the tube wall is found by subtracting the outer surface temperature (10.95) from the inner surface temperature (10.94).

$$\begin{aligned}
\Delta T_{w,i} = T_{w,i}|_{r=r_w} - T_{w,i}|_{r=r_z} &= \frac{T_{1,i} + T_{1,i+1}}{2} - \frac{\dot{q}_{z,i}}{\alpha_w}\frac{U_z}{U_w} - \bar{T}_{g,i} - \frac{\dot{q}_{z,i}}{\alpha_{zr}} \\
&= \left(\frac{T_{1,i} + T_{1,i+1}}{2} - \bar{T}_{g,i}\right) - \dot{q}_{z,i}\left(\frac{U_z}{U_w}\frac{1}{\alpha_w} + \frac{1}{\alpha_{zr}}\right).
\end{aligned} \tag{10.96}$$

The temperature drop can also be found treating the tube wall as a flat surface through which heat flows with the heat flux mean value defined by the following formula:

$$\dot{q}_{m,i} = \frac{\dot{q}_{w,i} + \dot{q}_{z,i}}{2}. \tag{10.97}$$

The temperature drop in the wall is expressed as:

$$\Delta T_{w,i} = \frac{\dot{q}_{m,i}\delta_r}{\lambda_r} = \frac{(\dot{q}_{w,i} + \dot{q}_{z,i})\delta_r}{2\lambda_r} = \frac{\dot{q}_{z,i}\delta_r}{2\lambda_r}\left(\frac{U_z}{U_w} + 1\right), \tag{10.98}$$

where δ_r—tube wall thickness, λ_r—tube material thermal conductivity.

In the case of tubes with a circular cross-section, especially if their wall is thicker, the overall heat transfer coefficient k can also be calculated using formula (7.76). The temperature of the wall can then be found by the following formula:

$$T_{w,i}(r) = \frac{T_{l,i} + T_{l,i+1}}{2} - \frac{\dot{q}_{z,i}}{\alpha_l}\frac{r_z}{r_w} - \frac{\dot{q}_{z,i}r_z}{\lambda_w}\ln\frac{r}{r_w}. \tag{10.99}$$

The formulae derived in Sect. 10.2 can be applied to build mathematical models of cross-flow tube heat exchangers made of bare (unfinned) and finned tubes.

10.3 Mathematical Modelling of Multi-pass Heat Exchangers with Multiple Tube Rows

The presented principles of constructing mathematical models of heat exchangers with a complex structure will be applied to develop a numerical model of a car radiator being a two-pass heat exchanger with two tube rows. The model will be used to simulate the exchanger operation under steady-state conditions. It is a plate fin-and-tube heat exchanger, with an in-line tube arrangement, made of $(n_u + n_l) = 38$ tubes with an oval cross-section, $n_u = 20$ of which are placed in the upper pass, 10 tubes per row (cf. Fig. 10.8).

In the lower pass, there are $n_l = 18$ tubes, 9 tubes per row. The tube pitches are as follows: transverse to the air flow $p_1 = 18.5$ mm, longitudinal to the flow $p_2 = 17$ mm. The oval tube outer diameters are: $d_{min} = 6.35$ mm, $d_{max} = 11.82$ mm. The wall thickness is $\delta_w = 0.4$ mm. The plate fins (lamellae) are strips of thin aluminium sheets which are 34 mm wide, 359 mm long and 0.08 mm thick. There are 520 lamellae, and they are spaced with pitch $s = 1$ mm. Both the tubes and the lamellae are made of aluminium. The exchanger width is 520 mm. The fluid flows along a U-shaped path. The heat exchanger under consideration is the radiator of a 1600 cm^3 spark-ignition combustion engine. A more detailed description of the exchanger can be found in [296, 322].

The medium flowing inside the exchanger tubes is water, which flows into the inlet header feeding the first and the second tube row in the exchanger upper (first) pass (cf. Fig. 10.8). The first pass ends with a mixing manifold with the outlets of tubes from the upper pass first and second tube row. From the lower part of this

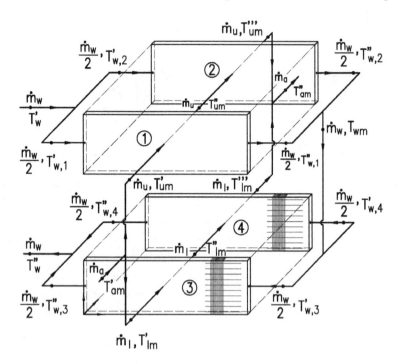

Fig. 10.8 Diagram of the mediums flow in the heat exchanger under consideration: 1—first tube row in the upper pass, 2—second tube row in the upper pass, 3—first tube row in the lower pass, 4 —second tube row in the lower pass

mixing manifold, water flows simultaneously into the first and the second tube row in the exchanger lower (second) pass. Air flows transversely to the axis of the exchanger horizontal tubes (Fig. 10.9).

The heat exchanger numerical models are constructed according to the procedures discussed in detail in subsections 10.2.1 and 10.2.2. The division of the exchanger upper and lower pass into control areas is presented in Figs. 10.9 and 10.10, respectively. The input data for the exchanger calculations are as follows (cf. Fig. 10.8):

- water volume flow rate at the exchanger inlet $\dot{V}_w(t)$,
- water temperature at the exchanger inlet $T'_w(t)$,
- air volume flow rate at the exchanger inlet $\dot{V}_a(t)$,
- air temperature at the exchanger inlet $T'_w(t)$.

The water mass flow rate in the radiator is calculated based on measured values of $\dot{V}_w(t)$ and $T'_w(t)$:

$$\dot{m}_w(t) = \dot{V}_w(t)\rho_w\left[T'_w(t)\right]. \tag{10.100}$$

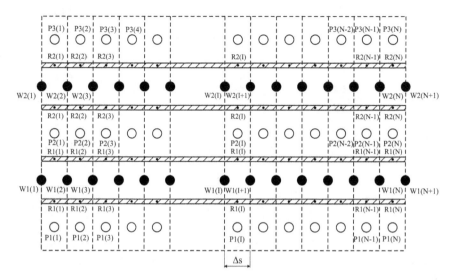

Fig. 10.9 Division of a repetitive fragment of the exchanger upper pass into control areas filled circle—water, open circle—air, dot—tube wall, $I = 1,..., N + 1$ for the liquid inside the tubes, $I = 1,..., N$ for the tube wall and the medium flowing outside the tubes, N—number of control volumes on a single tube length

Considering that the total numbers of tubes in the upper and in the lower pass are n_u and n_l, respectively, the water mass flow rates in a single tube in the upper pass $\dot{m}_{1u}(t)$ and the lower pass $\dot{m}_{1l}(t)$ are (Fig. 10.10):

$$\dot{m}_{1u}(t) = \frac{\dot{m}_w(t)}{n_u}, \quad \dot{m}_{1l}(t) = \frac{\dot{m}_w(t)}{n_l}. \tag{10.101}$$

Both in the exchanger upper and lower pass, the mass flow rates in the first and the second tube row are equal and total $\dot{m}_w/2$ (cf. Fig. 10.8). The air temperature $T'_{am}(t)$ and mean velocity w_m are measured in a circular duct with the outer surface insulated thermally and with inner diameter D_w. The air passing through the duct is sucked in from the outside of the building. The air mass flow rate in the radiator $\dot{m}_a$ is determined by the following formula:

$$\dot{m}_a(t) = \dot{V}_a(t)\rho\left[T'_{am}(t)\right], \tag{10.102}$$

where the air volume flow rate upstream the radiator $\dot{V}_a(t)$ is defined as:

$$\dot{V}_a(t) = w_m \frac{\pi D_w^2}{4} = w_0 L_x L_z. \tag{10.103}$$

Symbols L_x and L_z, respectively, denote the radiator effective cross-section width and height, which are equal to the width and height of the rectangular cross-section

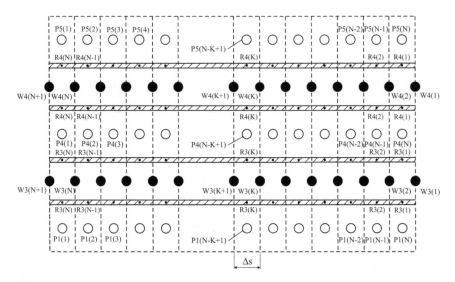

Fig. 10.10 Division of a repetitive fragment of the exchanger lower pass into control areas filled circle—water, open circle—air, dot—tube wall, $K = 1,…, N + 1$ for the liquid inside the tubes, $K = 1,…, N$ for the tube wall and the medium flowing outside the tubes, N—the number of control volumes on a single tube length

of the channel where the exchanger is located. The air flow velocity upstream the exchanger in the channel with a rectangular cross-section can be determined using the following formula:

$$w_0 = w_m \frac{\pi D_w^2}{4 L_x L_z}. \tag{10.104}$$

Considering that $D_w = 0.313$ m, $L_x = 0.52$ m and $L_z = 0.359$ m, formula (10.104) gives the following result:

$$w_0 = 0.4122\, w_m. \tag{10.105}$$

The air mass flow rate per one tube in the first or in the second tube row, both in the exchanger first and second pass, is determined by the following expression:

$$\dot{m}_{2u} = \dot{m}_{2l} = \dot{m}_2 = \frac{2\dot{m}_a}{n_u + n_l} = w_0 p_1 L_x \rho \left[T'_{am}(t) \right]. \tag{10.106}$$

Cooled water with a temperature $T''_{w,1}$ from the first tube row in the upper pass is mixed in the outlet chamber with water with a temperature $T''_{w,2}$ from the second tube row. The mixed water temperature T_{wm} is found from the energy balance equation for the upper pass outlet chamber:

$$\frac{\dot{m}_w}{2} c_w \Big|_0^{T''_{w,1}} T''_{w,1} + \frac{\dot{m}_w}{2} c_w \Big|_0^{T''_{w,2}} T''_{w,2} = \dot{m}_w c_w \Big|_0^{T_{wm}} T_{wm}, \qquad (10.107)$$

which, after simple transformations, gives:

$$T_{wm} = \frac{c_w \Big|_0^{T''_{w,1}} T''_{w,1} + c_w \Big|_0^{T''_{w,2}} T''_{w,2}}{2 c_w \Big|_0^{T_{wm}}}. \qquad (10.108)$$

Due to the small differences, usually of less than 10 K, between temperatures $T''_{w,1}$ and $T''_{w,2}$, it may be assumed that:

$$c_w \Big|_0^{T''_{w,1}} = c_w \Big|_0^{T''_{w,2}} = c_w \Big|_0^{T_{wm}}. \qquad (10.109)$$

Considering equality (10.109) in formula (10.108), the water temperature downstream the first pass reduces to the arithmetic mean:

$$T_{wm} = \frac{T''_{w,1} + T''_{w,2}}{2}. \qquad (10.110)$$

It is possible first to calculate temperature T_{wm} from formula (10.110), and then determine $c_w \Big|_0^{T_{wm}}$ and find T_{wm} again using formula (10.108).

The water temperature at the inlet of all tubes in the lower pass in the first and the second tube row is T_{wm}. A similar procedure can be applied to calculate the water temperature at the exchanger outlet T''_w:

$$T''_w = \frac{c_w \Big|_0^{T''_{w,3}} T''_{w,3} + c_w \Big|_0^{T''_{w,4}} T''_{w,4}}{2 c_w \Big|_0^{T''_w}}, \qquad (10.111)$$

which, assuming equal values of the water specific heat, gives:

$$T''_w = \frac{T''_{w,3} + T''_{w,4}}{2}. \qquad (10.112)$$

Formulae defining temperatures of the mediums and the tube wall in the upper pass will be determined first. Then, the formulae will be written for the lower pass. Because the operating conditions are identical, two tubes will be analysed in each pass, one of which is located in the first tube row and the other—in the second (cf. Fig. 10.8). The heat transfer coefficients on the air and the water side are calculated using in-house correlations for the Nusselt number determined experimentally, based on 57 measuring series [309]:

$$\mathrm{Nu}_a = x_1 \mathrm{Re}_a^{x_2} \, \mathrm{Pr}_a^{1/3}, \quad 150 \le \mathrm{Re}_a \le 350, \tag{10.113}$$

$$\mathrm{Nu}_w = \frac{\frac{\xi}{8}(\mathrm{Re}_w - x_3)\,\mathrm{Pr}_w}{1 + x_4\sqrt{\frac{\xi}{8}(\mathrm{Pr}_w^{2/3} - 1)}}\left[1 + \left(\frac{d_{h,w}}{L_{ch}}\right)^{2/3}\right], \quad 4000 \le \mathrm{Re}_w \le 12{,}000, \tag{10.114}$$

where Nu_a, Re_a and Pr_a, and Nu_w, Re_w and Pr_w denote the Nusselt, Reynolds and Prandtl numbers for air and water, respectively. Using the least squares method, the following values of coefficients x_1, x_2, x_3 and x_4 are obtained: $x_1 = 0.08993$, $x_2 = 0.699$, $x_3 = 1079$, $x_4 = 16.38$. Numbers Nu_a and Re_a are calculated for the equivalent hydraulic diameter $d_{h,a}$ on the air side. The Reynolds number Re_a takes account of the air flow maximum velocity that occurs in the smallest cross-section in between the tubes. Symbol d_z denotes the oval tube equivalent hydraulic diameter on the water side calculated from the equation $d_{h,w} = 4A_{\mathrm{wew}}/U_w$, where: A_{wew}—surface area of the oval tube water flow cross-section, U_w—the oval tube inner perimeter. Symbol L_{ch} denotes the exchanger tube length. Friction factor ξ related to pressure losses due to local resistance to the water flow inside an oval tube is calculated in the same manner as for tubes with a circular cross-section with diameter $d_{h,w} = 7.06$ mm by means of the following formula:

$$\xi = \frac{1}{(1.82 \log \mathrm{Re}_w - 1.64)^2} = \frac{1}{(0.79 \ln \mathrm{Re}_w - 1.64)^2}. \tag{10.115}$$

Friction factor ξ can also be calculated using other formulae presented in Chap. 4. The Reynolds number on the air side is defined as:

$$\mathrm{Re}_a = \frac{w_{\max} d_{h,a}}{\nu_h},$$

where $w_{\max}$ denotes the air flow maximum velocity in the exchanger that occurs in the narrowest cross-section in between two adjacent tubes.

The equivalent hydraulic diameter $d_{h,a}$ is calculated according to the procedure proposed by Kays and London [149]:

$$d_{h,a} = \frac{8A_{\min}p_2}{A_z' + A_{mz}}, \tag{10.116}$$

where $A_{\min}$—an surface area of the narrowest cross-section with the air flow maximum velocity, A_z'—an surface area of the sides of a fin with the dimensions $2p_2p_1$, A_{mz}—an outer surface area of two tubes in between two fins.

In the case of the heat exchanger under analysis, diameter d_h is 1.41 mm. The surface area of the fin sides (the surface area of the lateral surfaces making up the channel that the air flows through in between the fins) is defined by the following formula:

$$A'_z = 2 \cdot 2(p_1 p_2 - A_{\text{owal}}) = 4(p_1 p_2 - A_{\text{owal}}), \qquad (10.117)$$

and the tube surface area in between the fins is expressed as:

$$A'_{m\dot{z}} = 2A_{m\dot{z}} = 2U_z(s - \delta_{\dot{z}}), \qquad (10.118)$$

where A_{owal}—surface area of the oval opening in the fin, U_z—oval tube outer surface perimeter, s—fin pitch, $\delta_{\dot{z}}$—fin thickness.

Due to the tubes in-line arrangement in the exchanger, the maximum air flow velocity $w_{\max}$ occurs in the vertical plane passing through longitudinal axes of the tubes. The velocity of air flowing through the exchanger rises as the medium heats up from inlet temperature T'_{am} to temperature T''_{am} at the exchanger outlet.

Assuming that the air mean temperature on the exchanger thickness is $\bar{T}_{am} = \left(T'_{am} + T''_{am}\right)/2$, the air maximum velocity is defined by the following formula:

$$w_{\max} = \frac{s \cdot p_1}{(s - \delta_{\dot{z}})(p_1 - d_{\min})} \frac{\bar{T}_{am} + 273}{T'_{am} + 273} w_0, \qquad (10.119)$$

where $d_{\min}$—oval tube outer surface minor diameter, w_0—air velocity at the inlet (upstream the exchanger).

Two numerical models of the car radiator were developed based on the equations derived in subsections (10.2.1) and (10.2.2) for the exchanger single tube to evaluate the effectiveness and accuracy of the presented method of the cross-flow tube heat exchanger modelling. The models were then compared to the analytical model presented in [296, 297]. The exchanger tube was divided into $N = 27$ control volumes. A comparison was performed of the results obtained for the following two sets of data (cf. Table 10.1): $\dot{V}_w = 1000$ l/h, $T'_{am} = 15$ °C, $T'_w = 80$ °C and $\dot{V}_w = 2000$ l/h, $T'_{am} = 15$ °C, $T'_w = 80$ °C. Air velocity w_0 upstream the radiator varies from 0.5 to 2.5 m/s. The accuracy of the obtained results was assessed calculating the relative error from the following formula:

$$e = 100 \cdot \left(\dot{Q}_{ch} - \dot{Q}^d_{ch}\right)/\dot{Q}^d_{ch} \qquad (10.120)$$

where $\dot{Q}^d_{ch}$ is the heat flow rate determined using the exact analytical model presented in [296, 297], and $\dot{Q}_{ch}$ is the heat flow rate obtained from the numerical model. The radiator two numerical models based on the methods presented in subsections (10.2.2) and (10.2.1) will be referred to as Model I and Model II, respectively. In Model I, the gas mean temperature on the thickness of a single tube row is the integral-mean temperature, whereas in Model II, the gas mean temperature is the arithmetic mean of the gas temperatures at the tube row inlet and outlet.

The calculation results of the heat flow rate transferred in the car radiator from hot water to air flowing with a different velocity are presented in Table 10.1. The calculations were performed using the exact analytical model and the two numerical models I and II.

Table 10.1 Comparison of the heat exchanger calculation results obtained using the exact analytical method and the developed numerical models

$\dot{V}_w = 1000$ l/h, $T'_{am} = 15$ °C, $T'_w = 80$ °C

w_0 (m/s)	Analytical model			Numerical model I			Numerical model II		
	T''_w (°C)	T''_{am} (°C)	$\dot{Q}^e_{ch}$ (W)	T''_w (°C)	T''_{am} (°C)	$\dot{Q}^{I}_{ch}$ (W)	T''_w (°C)	T''_{am} (°C)	$\dot{Q}^{II}_{ch}$ (W)
0.5	74.37	71.10	6380.38	74.37	71.10	6380.38	74.23	72.57	6540.27
1.0	70.31	63.24	10969.85	70.31	63.24	10969.85	70.13	64.23	11184.92
1.5	67.26	57.28	14419.76	67.26	57.28	14419.76	67.04	58.10	14687.18
2.0	64.87	52.64	17117.00	64.87	52.64	17117.05	64.60	53.41	17453.07
2.5	62.95	48.94	19290.50	62.95	48.94	19290.53	62.61	49.70	19708.42

$\dot{V}_w = 2000$ l/h, $T'_{am} = 15$ °C, $T'_w = 80$ °C

w_0 (m/s)	Analytical model			Numerical model I			Numerical model II		
	T''_w (°C)	T''_{am} (°C)	$\dot{Q}^e_{ch}$ (W)	T''_w (°C)	T''_{am} (°C)	$\dot{Q}^{I}_{ch}$ (W)	T''_c (°C)	T''_{am} (°C)	$\dot{Q}^{II}_{ch}$ (W)
0.5	77.06	73.51	6655.18	77.06	73.51	6655.19	76.99	75.07	6823.80
1.0	74.71	67.65	11975.28	74.71	67.65	11975.31	74.62	68.71	12202.55
1.5	72.77	62.97	16365.14	72.77	62.97	16365.14	72.67	63.74	16611.88
2.0	71.13	59.14	20075.77	71.13	59.14	20075.78	71.03	59.76	20337.12
2.5	69.72	55.93	23269.19	69.72	55.93	23269.18	69.61	56.46	23549.89

Table 10.2 Comparison of relative errors $e_I = 100 \cdot (\dot{Q}^I_{ch} - \dot{Q}^d_{ch})/\dot{Q}^d_{ch}$ and $e_{II} = 100 \cdot (\dot{Q}^{II}_{ch} - \dot{Q}^d_{ch})/\dot{Q}^d_{ch}$ of the heat exchanger power calculation by means of the developed numerical models

w_0 (m/s)	$\dot{V}_w = 1000$ l/h, $T'_{am} = 15$ °C, $T'_w = 80$ °C		$\dot{V}_w = 2000$ l/h, $T'_{am} = 15$ °C, $T'_w = 80$ °C	
	Numerical model I e_I, (%)	Numerical model II e_{II}, (%)	Numerical model I e_I, (%)	Numerical model II e_{II}, (%)
0.5	0.00	2.51	0.00	2.53
1.0	0.00	1.96	0.00	1.90
1.5	0.00	1.85	0.00	1.51
2.0	0.00	1.96	0.00	1.30
2.5	0.00	2.17	0.00	1.21

The relative errors in heat flow rate values calculated by the two proposed numerical models are presented in Table 10.2.

The analysis of the results listed in Tables 10.1 and 10.2 indicates that the accuracy of the heat exchanger numerical models presented herein and developed based on the finite volume method is very good. The results obtained by means of Model I are practically identical to those given by the analytical model. The advantage of the numerical model is the ease of modelling heat exchangers with complex flow systems characterized by a significant number of passes and tube rows. The dependence of the fluid physical properties on temperature can also be taken into consideration easily. In the case of modelling steam superheaters, steam specific heat is a function of temperature and pressure have to be taken into account. The developed numerical methods are advantageous in this respect.

Part III
Experimental Testing of Heat Exchangers

Chapter 11
Assessment of the Indirect Measurement Uncertainty

The heat exchanger experimental testing usually draws on indirect methods of determining quantities appearing in thermal and flow calculations. One example of such indirect measurement is determination of the mean heat transfer coefficient values on the sides of the two mediums flowing in the heat exchanger. The indirect measurement uncertainty assessment is often a complex task, which is presented in detail in this chapter.

11.1 Characteristics of Basic Terms

A lot of quantities are measured in laboratory testing, industry and everyday life in general. The rapid development of the computer technology enables precise monitoring of complex technological processes, which involves the need to perform measurements of even up to a few hundred quantities at a time. This chapter discusses the basic concepts related to measurement defined as assigning a specific value to a physical variable which thus becomes a measured quantity. Three basic subsystems can be distinguished in a measuring system (cf. Fig. 11.1):

(a) the sensor (detector)—transducer subsystem, which uses a certain physical phenomenon to measure a quantity and converts the information obtained in this manner into an easily measurable mechanical, electrical or optical signal. The converted signal can be analog or digital. The digital signal advantage is that it can easily be stored in memory, which in turn enables easy processing of the signal by means of appropriate computer programs;

(b) the signal processing (conditioning) subsystem, where the signal can be amplified or filtered;

(c) the output subsystem, where the signal can be indicated or recorded; the output subsystem can be either analog or digital.

© Springer International Publishing AG, part of Springer Nature 2019
D. Taler, *Numerical Modelling and Experimental Testing of Heat Exchangers*,
Studies in Systems, Decision and Control 161,
https://doi.org/10.1007/978-3-319-91128-1_11

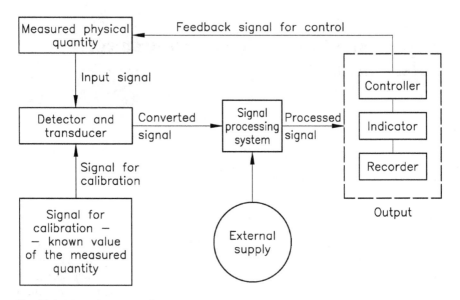

Fig. 11.1 Measuring system flowchart

If the measuring system is a part of the control system, the signal measured value is compared to the set value and the feedback system introduces relevant alterations into the process to bring the measured and the set values closer to each other. The relation between the measuring system input and output values is found based on calibration. Calibration consists in feeding the measuring system with a measured signal with a known value and observing the output signal. The known value of the input signal is referred to as the standard. Errors that occur during calibration, or during any measurement, can be divided into systematic and random. The measurement error is the difference between the measurement result and the measured quantity real value. The systematic error is the difference between the mean value of an infinite number of measurements of the same measured quantity performed in conditions of repeatability and the measured quantity true value.

The random error is defined as the difference between the measurement result and the mean value of an infinite number of results of measurements of the same measured quantity that are performed in conditions of repeatability [35, 91, 116, 127, 135]. The instrument measuring range r_0 is equal to the difference between the output signal maximum and minimum values: y_{max} and y_{min}, respectively. The instrument static sensitivity is defined by the following formula:

$$K(x) = \left(\frac{dy}{dx}\right)\Big|_x. \tag{11.1}$$

The relation between output signal y and input signal x is determined based on the instrument calibration, after measuring data $(x_i, y_i), i = 1, \ldots, n$ are approximated by means of function $y = f(x)$ using the least squares method (cf. Fig. 11.2).

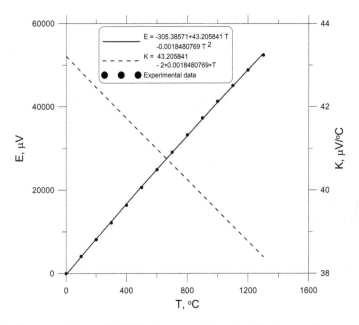

Fig. 11.2 The sensor (K-type (Ni–NiCr) thermocouple) static calibration result

If function $f(x)$ is not linear, the instrument sensitivity, also referred to as the calibration curve, is not a constant value.

The measuring system error can be estimated during calibration. The measure of the system inaccuracy is the absolute error ε, defined as the difference between the true value y_p and the measuring instrument indication y_m.

$$\varepsilon = y_p - y_m. \qquad (11.2)$$

The relative error is the ratio between absolute error ε and true value y_p. If the maximum relative error ε_{max} is expressed in percentages, $y_{p,\,max}$ is the value of this relative error defined as the instrument accuracy (or inaccuracy) class.

$$k_d = \left| \frac{\varepsilon_{max}}{y_{p,\,max}} \right| \cdot 100\%. \qquad (11.3)$$

The accuracy class makes it possible to assess the instrument suitability for planned laboratory or industrial testing.

If very high measuring accuracy is required, the measurement accuracy class should not be smaller than 0.5. In industrial measurements, the accuracy class may be higher and reach 2.5, i.e. higher values of maximum error ε_{max} are allowed. Many manufacturers also give the value $\left| \varepsilon_{max}/y_{p,i} \right| \cdot 100\%$ to characterize the relative error occurring during the measurement of quantity $y_{p,i}$, which is smaller than the measuring range upper value y_{max}. This is justified considering that the instrument accuracy is different in different measuring ranges. It often happens that the measurement accuracy of quantity y is much higher compared to what the instrument accuracy class might suggest. Usually, the measurement lowest accuracy, i.e. the biggest ε error, occurs close to the lower and upper values of the instrument measuring range.

In laboratory or industrial measurements, the measuring error usually cannot be known.

11.2   Measurements of Physical Quantities

Physical measurements can be divided into direct and indirect ones. Direct measurements are performed by means of measuring instruments, e.g. thermometers, manometers, etc. In indirect measurements, the sought quantity z is defined using a mathematical formula.

$$z = f(x_1, x_2, \ldots, x_n), \tag{11.4}$$

where quantities $x_1, \ldots, x_n$ are measured directly. It is also assumed that relation (11.4) is exact. Consequently, if quantities $x_1, \ldots, x_n$ are measured without an error, the determined quantity y is also error-free. Sometimes, however, the situation is more complex, e.g. in transient inverse heat conduction problems. In inverse problems, relation (11.4) is usually implicit because it is most often impossible to find an explicit relation between measured transient-state temperature histories inside a body and the determined (indirectly measured) heat flux on the body boundary. Moreover, due to the damping and delay of the signal measured inside a body compared to the changes taking place on the body surface, it is impossible to reconstruct the real conditions on the body boundary based on the temperature measurements at points located inside. And this means that even if accurate direct measurements are performed of quantities $x_1, \ldots, x_n$ the obtained result y is inaccurate. Neither direct nor indirect measurements are accurate. The measuring error is usually defined as the difference between the true and the measured value. Simple as it is, the definition is difficult to apply in practice because the measured quantity true value is unknown.

For this reason, the measurement is used to establish an approximate value of x_i^z and interval $[x_i^z - \Delta x_i, x_i^z + \Delta x_i]$ which includes the true (actual) value x_i, i.e.

$$x_i = x_i^z \pm \Delta x_i, \tag{11.5}$$

where x_i^z is the measured value and Δx_i is the measurement uncertainty. The term "uncertainty" denotes the value which the measuring error can take [33, 35, 116, 135, 168, 292, 300]. At a bigger number of measurements n, equality $x_i^z = \bar{x}_{i,n}$ occurs, i.e. the measured value $\bar{x}_{i,n}$ is assumed to be the mean value.

There are two kinds of the measurement uncertainty: systematic and random.

The systematic uncertainty value is identical in all measurements performed in the same way using the same measuring instruments.

The random uncertainty value differs even for measurements performed in the same way. Subsequent measurements do not give identical results, i.e. $x_1 \neq x_2 \neq x_3 \dots$ Random uncertainty is the effect of the measured object deviation from the assumed model, of the applied measuring methods and instruments, and also of the observer's senses.

11.2.1   Calculation of the Direct Measurement Uncertainty

The considerations are limited to direct measurements where systematic uncertainty values are very small compared to random uncertainty. The sample, i.e. a series of testing results $x_1, x_2, \dots, x_i, \dots, x_n$ burdened with random uncertainty, can be approximated by the Gaussian curve or by the probability density function for the normal distribution given by equation

$$p(x) = \frac{1}{\sigma\sqrt{2\pi}} \exp\left[-\frac{(x-\mu)^2}{2\sigma^2}\right], \quad -\infty < x < \infty, \tag{11.6}$$

where μ is the expected value and the number that determines the location of the curve maximum, and σ is standard deviation characterizing the curve width, i.e. the results deviation from μ. Quantity $p(x)$ is the probability density of measurement results. The estimator of expected value μ is the arithmetic mean.

$$\bar{x}_n = \frac{1}{n}\sum_{i=1}^{n} x_i. \tag{11.7}$$

The estimator of standard deviation σ is the standard deviation of the sample, which is expressed as:

$$s_x = \sqrt{\frac{1}{n-1}\sum_{i=1}^{n}(x_i - \bar{x})^2}. \tag{11.8}$$

The standard deviation of the sample is characterized by a spread of individual measurement results compared to the mean value, and it is the best assessment of the mean square deviation. Quantity s_x is also referred to as the mean square error.

If the number of observations n is very high, e.g. $n \geq 30$, the randomly varying quantity s_x approaches σ:

$$E(s_x^2) = \sigma^2, \tag{11.9}$$

where E denotes the random variable expected value. Marking the real mean value of the measured quantity as $\bar{x} = E(x)$ and the probability that the measurement result x differs from the real value by less than Δx as P:

$$P[\%] = P(\bar{x} - \Delta x \leq x \leq \bar{x} + \Delta x), \tag{11.10}$$

confidence interval $[\bar{x} - \Delta x, \bar{x} + \Delta x]$ can be determined for a set value of Δx using the probability density curve $\rho(x)$ for the normal distribution.

Probability P that the measurement is included in interval $[x_1, x_2]$ is calculated from the following formula:

$$P(x_1 \leq x \leq x_2) = \int_{x_1}^{x_2} \rho(x)\,dx. \tag{11.11}$$

The graphical interpretation of formula (11.11) is the surface area under the $\rho(x)$ $\rho(x)$ curve. If the number of measurements n is large, then there is a relationship

$$\mu = E(x) \cong \bar{x}_n. \tag{11.12}$$

The value of probability P expressed in percentages is referred to as the confidence coefficient (confidence level). The probability that random variable x with the normal distribution is included in interval $[x_1, x_2]$ is found from formula (11.11). Substituting (11.6) in (11.11) and performing integration, the following is obtained:

$$P(x_1 \leq x \leq x_2) = F(x_2) - F(x_1) = \varphi\left(\frac{x_2 - \mu}{\sigma}\right) - \varphi\left(\frac{x_1 - \mu}{\sigma}\right)$$
$$= \frac{1}{2}\left[erf\left(\frac{x_2 - \mu}{\sqrt{2}\cdot\sigma}\right) - erf\left(\frac{x_1 - \mu}{\sqrt{2}\cdot\sigma}\right)\right], \tag{11.13}$$

where the Gauss error function is defined as:

$$erf(u) = \frac{2}{\sqrt{\pi}}\int_{0}^{u}\exp\left(-t^2\right)dt = 2\varphi\left(\frac{u}{\sqrt{2}}\right) - 1. \tag{11.14}$$

Table 11.1 and Fig. 11.3 present confidence intervals for set probability P, calculated according to formula (11.13).

Table 11.1 and Fig. 11.3 indicate that for the commonly adopted confidence level of P = 95%, the real measurement values are approximately included in interval $\mu - 2\sigma \leq x \leq \mu + 2\sigma$ (or, more precisely, $\mu - 1.96\sigma \leq x \leq \mu + 1.96\sigma$).

Table 11.1 Confidence intervals $\mu - \Delta x \leq x \leq \mu + \Delta x$ for a set probability value

Confidence interval	Probability P (%)
$\mu - \sigma \leq x \leq \mu + \sigma$	63.8
$\mu - 1.96\sigma \leq x \leq \mu + 1.96\sigma$	95
$\mu - 2\sigma \leq x \leq \mu + 2\sigma$	95.5
$\mu - 2.58\sigma \leq x \leq \mu + 2.58\sigma$	99
$\mu - 3\sigma \leq x \leq \mu + 3\sigma$	99.7

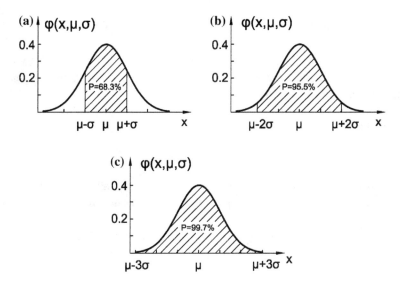

Fig. 11.3 Confidence coefficient P [%] for set confidence intervals; **a** for 68.3%, **b** for 95.5%, **c** for 99.7%

Figure 11.3 presents probability P for confidence intervals with width: 2σ, 4σ and 6σ.

Expected value μ is calculated from formula (11.12). If the number of measurements is small ($n < 30$), the confidence interval is established using the t-Student distribution. Marking the confidence level as $(1 - \alpha)$

$$\frac{P[\%]}{100} = 1 - \alpha \qquad (11.15)$$

the confidence interval is defined as:

$$\bar{x}_n - t_{n-1,\alpha} s_x \le x \le \bar{x}_n + t_{n-1,\alpha} s_x, \qquad (11.16)$$

where $t_{n-1,\alpha}$ is the t-Student coefficient. The values of coefficients $t_{r,\alpha}$ are listed in Table 11.1, where $r = n - 1$ is the number of freedom degrees. Assuming that the confidence level is P = 95% ($1 - \alpha = 0.95$), it can be seen that already for $n = 20$ measuring points, the value $t_{n-1,\alpha} = 2.086$ is close to the number 2 occurring in the normal distribution (cf. Fig. 11.3b). Therefore, the real measured value x is included in the following interval:

$$\bar{x}_n - 2.096\, s_x \le x \le \bar{x}_n + 2.096\, s_x, \qquad (11.17)$$

where $\bar{x}_n$ and s_x are defined by formula (11.7) and formula (11.8), respectively.

11.2.2   Calculation of the Indirect Measurement Uncertainty

In many cases the physical quantity z_i cannot be measured directly and it has to be calculated based on other measured quantities $x_1, x_2, \ldots, x_N$ which are independent of each other.

In order to simplify the considerations, it is assumed that quantity z depends on two measurable quantities x_1 and x_2:

$$z = f(x_1, x_2). \tag{11.18}$$

The mathematical means of quantities x_1 and x_2 are $\bar{x}_1$ and $\bar{x}_2$, respectively. The value of z measured at point: $x_{1,i}, x_{2,k}$ is z_{ik}, and the mean value is expressed as:

$$\bar{z} = \frac{1}{n^2} \sum_{i=1}^{n} \sum_{k=1}^{n} z(x_{1,i}, x_{2,k}) = \frac{1}{n^2} \sum_{i=1}^{n} \sum_{k=1}^{n} z_{ik}, \tag{11.19}$$

where n is the number of measurements of x_1 and x_2.

The x_1 and x_2 measurement errors defined by standard deviations s_{x_1} and s_{x_2} are also known. If the measurement errors are relatively small, the $z(x_{1,i}, x_{2,k})$ value can be found by expanding z_{ik} into the Taylor series close to point $\bar{x}(\bar{x}_1, \bar{x}_2)$:

$$z_{ik} = z(\bar{x}_1, \bar{x}_2) + \left(\frac{\partial z}{\partial x_1}\right)\Bigg|_{x=\bar{x}} (x_{1,i} - \bar{x}_1) + \cdots + \left(\frac{\partial z}{\partial x_2}\right)\Bigg|_{x=\bar{x}} (x_{2,i} - \bar{x}_2) + \cdots. \tag{11.20}$$

Omitting higher-order terms in (11.20) and substituting the formula in (11.19), the following is obtained:

$$
\begin{aligned}
\bar{z} &= \frac{1}{n^2} \sum_{i=1}^{n} \sum_{k=1}^{n} \left[z(\bar{x}_1, \bar{x}_2) + \left(\frac{\partial z}{\partial x_1}\right)\Bigg|_{x=\bar{x}} (x_{1,i} - \bar{x}_1) + \left(\frac{\partial z}{\partial x_2}\right)\Bigg|_{x=\bar{x}} (x_{2,i} - \bar{x}_2) + \cdots \right] \\
&= \frac{1}{n^2} \left[n^2 z(\bar{x}_1, \bar{x}_2) + n \left(\frac{\partial z}{\partial x_1}\right)\Bigg|_{x=\bar{x}} \sum_{i=1}^{n} (x_{1,i} - \bar{x}_1) + n \left(\frac{\partial z}{\partial x_2}\right)\Bigg|_{x=\bar{x}} \sum_{i=1}^{n} (x_{2,i} - \bar{x}_2) \right].
\end{aligned}
\tag{11.21}
$$

Noting that:

$$\sum_{i=1}^{n} (x_{1,i} - \bar{x}_1) = 0 \quad \text{and} \quad \sum_{k=1}^{n} (x_{2,i} - \bar{x}_2) = 0 \tag{11.22}$$

(11.21) gives:

$$\bar{z} = z(\bar{x}_1, \bar{x}_2). \tag{11.23}$$

The mean value of quantity z is thus equal (with accuracy up to the second-order terms) to value z calculated for mean values $\bar{x}_1$ and $\bar{x}_2$:

$$\bar{x}_1 = \frac{1}{n}\sum_{i=1}^{n} x_{1,i} \quad \text{and} \quad \bar{x}_2 = \frac{1}{n}\sum_{k=1}^{n} x_{2,k}. \tag{11.24}$$

Standard deviation s_z is:

$$s_z^2 = \frac{1}{n^2-1}\sum_{i=1}^{n}\sum_{k=1}^{n}(z_{ik}-\bar{z})^2$$

$$= \frac{1}{n^2-1}\sum_{i=1}^{n}\sum_{k=1}^{n}\left[\left(\frac{\partial z}{\partial x_1}\right)\Big|_{\mathbf{x}=\bar{\mathbf{x}}}(x_{1,i}-\bar{x}_1) + \left(\frac{\partial z}{\partial x_2}\right)\Big|_{\mathbf{x}=\bar{\mathbf{x}}}(x_{2,i}-\bar{x}_2)\right]^2.$$

Considering (11.8) and (11.22), the following is obtained:

$$s_z^2 = \frac{n(n-1)}{n^2-1}\left\{\left[\left(\frac{\partial z}{\partial x_1}\right)\Big|_{\mathbf{x}=\bar{\mathbf{x}}}\right]^2 s_{x_1}^2 + \left[\left(\frac{\partial z}{\partial x_2}\right)\Big|_{\mathbf{x}=\bar{\mathbf{x}}}\right]^2 s_{x_2}^2\right\}. \tag{11.25}$$

Considering that for high values of n: $\frac{n(n-1)}{(n^2-1)} \cong 1$, standard deviation s_z is:

$$s_z^2 = \left[\left(\frac{\partial z}{\partial x_1}\right)\Big|_{\mathbf{x}=\bar{\mathbf{x}}}\right]^2 s_{x_1}^2 + \left[\left(\frac{\partial z}{\partial x_2}\right)\Big|_{\mathbf{x}=\bar{\mathbf{x}}}\right]^2 s_{x_2}^2. \tag{11.26}$$

Formula (11.26) expresses the law of summing up errors formulated by Gauss.

The considerations may be generalized to cover the case where indirectly measured quantities $\mathbf{z} = (z_1, z_2, \ldots, z_r)$ depend linearly on n $\mathbf{x} = (x_1, x_2, \ldots, x_N)$ quantities. In this situation, the mean values are calculated from the following formula:

$$\bar{z}_i = z_i(\bar{x}_1, \bar{x}_2, \ldots, \bar{x}_N) = z_i(\bar{\mathbf{x}}), \quad i = 1, \ldots, r.$$

Assuming that variables x are independent, standard deviation takes the following form:

$$s_{z_i}^2 = \sum_{j=1}^{n}\left(\frac{\partial z_i}{\partial x_j}\right)^2\Big|_{\mathbf{x}=\bar{\mathbf{x}}} s_{x_j}^2 \tag{11.27}$$

known as the error propagation rule [32, 33, 35, 168, 292, 300]. If the number of measurements is high ($n > 30$), formula (11.27) can be written as:

$$\sigma_{z_i}^2 = \sum_{j=1}^{n} \left(\frac{\partial z_i}{\partial x_j} \right)^2 \bigg|_{x=\bar{x}} \sigma_{x_j}^2. \tag{11.28}$$

The error propagation rule can also be written in a matrix form.
The covariance matrix for quantity z is defined by the following formula:

$$\mathbf{C_z} = E\left\{ (\mathbf{z} - \bar{\mathbf{z}})(\mathbf{z} - \bar{\mathbf{z}})^{\mathrm{T}} \right\}. \tag{11.29}$$

Assuming that errors are relatively small, the value of function z_i can be calculated expanding z_i into the Taylor series with respect to point $\bar{\mathbf{x}}$:

$$z_i = z_i(\bar{\mathbf{x}}) + \left(\frac{\partial z_i}{\partial x_1} \right)\bigg|_{\mathbf{x}=\bar{x}} (x_1 - \bar{x}_1) + \cdots + \left(\frac{\partial z_i}{\partial x_N} \right)\bigg|_{\mathbf{x}=\bar{x}} (x_n - \bar{x}_N) + \cdots, \tag{11.30}$$

$$i = 1, \ldots, r.$$

Expression (11.30) can be written in the following matrix form:

$$\mathbf{z} = \mathbf{z}(\bar{\mathbf{x}}) + \mathbf{J}(\mathbf{x} - \bar{\mathbf{x}}), \tag{11.31}$$

where $\mathbf{J}$ is the Jacobian matrix:

$$\mathbf{J} = \begin{bmatrix} \left(\frac{\partial z_1}{\partial x_1} \right) & \cdots & \left(\frac{\partial z_1}{\partial x_N} \right) \\ \vdots & & \vdots \\ \left(\frac{\partial z_r}{\partial x_1} \right) & \cdots & \left(\frac{\partial z_r}{\partial x_N} \right) \end{bmatrix}_{\mathbf{x}=\bar{x}}. \tag{11.32}$$

Substituting (11.31) in (11.29), the following is obtained:

$$\mathbf{C_z} = E\left\{ (\mathbf{z}(\bar{\mathbf{x}}) + \mathbf{Jx} - \mathbf{J}\bar{\mathbf{x}} - \bar{\mathbf{z}})(\mathbf{z}(\bar{\mathbf{x}}) + \mathbf{Jx} - \mathbf{J}\bar{\mathbf{x}} - \bar{\mathbf{z}})^{\mathrm{T}} \right\}$$

$$= E\left\{ \mathbf{J}(\mathbf{x} - \bar{\mathbf{x}})(\mathbf{x} - \bar{\mathbf{x}})^{\mathrm{T}} \mathbf{J}^{\mathrm{T}} \right\} = \mathbf{J}E\left\{ (\mathbf{x} - \bar{\mathbf{x}})(\mathbf{x} - \bar{\mathbf{x}})^{\mathrm{T}} \right\} \mathbf{J}^{\mathrm{T}}, \tag{11.33}$$

$$\mathbf{C_z} = \mathbf{J}\mathbf{C_x}\mathbf{J}^{\mathrm{T}}.$$

Covariance matrix $\mathbf{C_x}$ has the following form:

$$\mathbf{C_x} = E\left\{ (\mathbf{x} - \bar{\mathbf{x}})(\mathbf{x} - \bar{\mathbf{x}})^{\mathrm{T}} \right\} = \begin{bmatrix} \sigma^2(x_1) & & & 0 \\ & \sigma^2(x_2) & & \\ & & \ddots & \\ 0 & & & \sigma^2(x_N) \end{bmatrix}. \tag{11.34}$$

Partial derivatives $\partial z_i / \partial x_j$, also known as sensitivity coefficients, are calculated analytically if the relation $z = z(\mathbf{x})$ is proportional, or using a difference quotient, e.g. the central one as presented below:

$$\frac{\partial z_i}{\partial x_j} \cong \frac{z_i(x_1, \ldots, x_j + \delta, \ldots, x_N) - z_i(x_1, \ldots, x_j - \delta, \ldots, x_N)}{2\delta}, \tag{11.35}$$

where δ is a small positive number.

The diagonal elements of matrix $\mathbf{C_z}$ are defined as follows:

$$c_{ii} = \sigma^2(z_i) = \sum_{j=1}^{N} \left(\frac{\partial z_i}{\partial x_j}\right)_{x=\bar{x}}^2 \sigma^2(x_j) \tag{11.36}$$

and express the error propagation rule.

It can be noticed that for a high number of measurements n, formulae (11.28) and (11.36) become identical because:

$$E\left(s_{x_j}^2\right) = \sigma_{x_j}^2. \tag{11.37}$$

Covariance matrix $\mathbf{C_z}$ is symmetric because the following relation holds for the matrix elements c_{ij}:

$$c_{ij} = c_{ji}.$$

Assuming that the confidence level is $P = 95\%$ and the number of measurements n is high, the uncertainty of quantity z measurement due to random errors with a normal distribution and with the mean equal to zero is $\pm 2\sigma(z_i)$. Moreover, if the measurement systematic uncertainty $\Delta z_{i,s}$ is known, total uncertainty Δz_i^c can be calculated from the following formula [232, 292, 300]:

$$\Delta z_i^c = \sqrt{\left(\Delta z_{i,s}\right)^2 + [2\sigma(z_i)]^2}. \tag{11.38}$$

An example calculation of the measurement total uncertainty by means of formula (11.38) is presented in [232].

Appendix D presents the method of determining the measurement 95% uncertainty levels recommended by the ASME [232].

11.2.2.1   Calculation of the Maximum Uncertainty

Quantity y is found by means of an exact formula using quantities x_i measured directly. Each direct quantity x_i is measured with accuracy Δx_i. Quantity z found indirectly is burdened with a random and a systematic error. Measurement uncertainty $\pm \Delta z$ caused by the measurement random uncertainties Δx_i is discussed below. For the adopted confidence level of $P = 95\%$, it is usually assumed that $\Delta x_i = \pm 2\sigma_{x_i}$ or $\Delta x_i = \pm t_{n-1,\alpha} \cdot s_{x_i}$. For $\alpha = 0.05$ ($P = 95\%$), the value of $t_{n-1;0.05}$ can be read from Table 11.2. Quantity σ_{x_i} is the sample standard deviation, $t_{n-1,\alpha}$ is the quantile of the t-Student distribution for $n - 1$ degrees of freedom and confidence level $P = (1 - \alpha) \cdot 100\%$. In order to assess the Δz maximum error, formula (11.4) can be written as:

$$z \pm \Delta z = f(x_1 \pm \Delta x_1, x_2 \pm \Delta x_2, \ldots, x_N \pm \Delta x_N). \tag{11.39}$$

Expanding function f into the Taylor series and omitting terms of an order higher than the first, the following is obtained:

$$z \pm \Delta z = f(\hat{\mathbf{x}}) \pm \frac{\partial f(\hat{\mathbf{x}})}{\partial x_1} \Delta x_1 \pm \frac{\partial f(\hat{\mathbf{x}})}{\partial x_2} \Delta x_2 \pm \cdots \pm \frac{\partial f(\hat{\mathbf{x}})}{\partial x_N} \Delta x_N, \tag{11.40}$$

where $\hat{\mathbf{x}} = (x_1, x_2, \ldots, x_N)^{\mathrm{T}}$.

Considering that $z = f(\hat{\mathbf{x}})$, the following is obtained from (11.40):

$$\Delta z = \sum_{i=1}^{N} \pm \frac{\partial f(\hat{\mathbf{x}})}{\partial x_i} \Delta x_i. \tag{11.41}$$

In the least favourable case, all the terms in expression (11.40) have the same sign. The maximum uncertainty formula thus takes the following form:

$$\Delta z = \sum_{i=1}^{N} \left| \frac{\partial f(\hat{\mathbf{x}})}{\partial x_i} \Delta x_i \right|. \tag{11.42}$$

Formula (11.42) usually gives an overestimated value of the z measurement uncertainty because it is very seldom for all Δx_i values to have the same sign and take the highest values at the same time. Uncertainties Δx_i usually have a random character because systematic errors can be detected during the measuring instrument calibration or by measuring the same quantity x_i by means of a few or a dozen or so sensors with an identical structure. If one of the sensors indicates values different from those indicated by the other sensors, the systematic error that the sensor indications are burdened with can be corrected. Therefore, there is no need to take account of this error in formula (11.42). The maximum values of the error in quantity Δx_i are often difficult to establish because they are the effect of not only the measuring instrument accuracy class but also the adopted simplifications concerning observed phenomena. For example, if the temperature is measured of a liquid with a high volume, it is usually incorrect to assume that the temperature is identical in the entire area under analysis. Due to natural convection in the first place, big differences caused by the liquid heating or cooling by the vessel walls often occur between the fluid temperatures measured in different points.

Table 11.2 t-Student coefficients (distribution quantiles) as a function of probability level $\frac{P[\%]}{100} = 1 - \alpha$ and number of freedom degrees r; $r = n - N$, n—number of measuring points, N—number of parameters to be determined

$1-\alpha$	0.5	0.6826	0.7	0.8	0.9	0.95[a]	0.98	0.99	0.995	0.998	0.999
α	0.5	0.3174	0.3	0.2	0.1	0.05	0.02	0.01	0.005	0.002	0.001
r											
1	1	1.8367	1.9626	3.0777	6.3138	12.7062	31.8205	63.6567	127.3213	318.3088	636.6192
2	0.8165	1.321	1.3862	1.8856	2.92	4.3027	6.9646	9.9248	14.089	22.3271	31.5991
3	0.7649	1.1966	1.2498	1.6377	2.3534	3.1824	4.5407	5.8409	7.4533	10.2145	12.924
4	0.7404	1.1414	1.1896	1.5332	2.1318	2.7764	3.7469	4.6041	5.5976	7.1732	8.6103
5	0.7267	1.1103	1.1558	1.4759	20150	2.5706	3.3649	4.0321	4.7733	5.8934	6.8688
6	0.7176	1.0903	1.1342	1.4398	1.9432	2.4469	3.1427	3.7074	4.3268	5.2076	5.9588
7	0.7111	1.0765	1.1192	1.4149	1.8946	2.3646	2.998	3.4995	4.0293	4.7853	5.4079
8	0.7064	1.0663	1.1082	1.3968	1.8595	2.306	2.8965	3.3554	3.8325	4.5008	5.0413
9	0.7024	1.0585	1.0997	1.383	1.8331	2.2622	2.8214	3.2498	36897	4.2968	4.7809
10	0.6998	1.0524	1.0931	1.3722	1.8125	2.2281	2.7638	3.1693	3.5814	4.1437	4.5869
11	0.6974	1.0474	1.0877	1.3634	1.7959	2.201	2.7181	3.1058	3.4966	4.0247	4.437
12	0.6955	1.0432	1.0832	1.3562	1.7823	2.1788	2.681	3.0545	3.4284	3.9296	4.3178
13	0.6938	1.0398	1.0795	1.3502	1.7709	2.1604	2.6503	3.0123	3.3725	3.852	4.2208
14	0.6924	1.0368	1.0763	1.345	1.7613	2.1448	2.6245	2.9768	3.3257	3.7874	4.1405
15	0.6012	1.0343	1.0735	1.3406	1.7531	2.1314	2.6025	2.9467	3.286	3.7228	4.0728
16	0.6901	1.032	1.0711	1.3368	1.7459	2.1199	2.5835	2.9208	3.252	3.6862	4.015
17	0.6892	1.0301	1.069	1.3334	1.7396	2.1098	2.5669	2.8982	3.2224	3.6458	3.9651
18	0.6884	1.0284	1.0672	1.3304	1.7341	2.1009	2.5524	2.8784	3.1966	3.6105	3.9216
19	0.6876	1.0268	1.0655	1.3277	1.7291	2.093	2.5395	2.8609	3.1737	3.5794	3.8834
20	0.687	1.0245	1.064	1.3253	1.7247	2.086	2.528	2.8453	3.1534	3.5518	3.8495

(continued)

Table 11.2 (continued)

$1-\alpha$	0.5	0.6826	0.7	0.8	0.9	0.95[a]	0.98	0.99	0.995	0.998	0.999
α	0.5	0.3174	0.3	0.2	0.1	0.05	0.02	0.01	0.005	0.002	0.001
r											
21	0.6864	1.0242	1.0627	1.3232	1.7207	2.0796	2.5176	2.8314	3.1352	3.5272	3.8193
22	0.6858	1.0231	1.0614	1.3212	1.7171	2.0739	2.5083	2.8188	3.1188	3.505	3.7921
23	0.6853	1.022	1.0603	1.3195	1.7139	2.0687	2.4999	2.8073	3.104	3.485	3.7676
24	0.6848	1.0211	1.0593	1.3178	1.7109	2.0639	2.4922	2.7969	3.0905	3.4668	3.7454
25	0.6844	1.0202	1.0584	1.3163	1.7081	2.0596	2.4851	2.7874	3.0782	3.4502	3.7251
26	0.684	1.0194	1.0575	1.315	1.7056	2.0555	2.4786	2.7787	3.0669	3.435	3.7066
27	0.6837	1.0187	1.0567	1.3137	1.7033	2.0518	2.4727	2.7707	3.0565	3.421	3.6896
28	0.6834	1.018	1.056	1.3125	1.7011	2.0484	2.4671	2.7633	3.0469	3.4082	3.6739
29	0.683	1.0173	1.0553	1.3114	1.6991	2.0452	2.462	2.7564	3.038	3.3962	3.6594
30	0.6828	1.0168	1.0547	1.3104	1.6973	2.0423	2.4573	2.75	3.0298	3.3852	3.646
40	0.6807	1.0125	1.05	1.3031	1.6839	2.0211	2.4233	2.7045	2.9712	3.3069	3.551
50	0.6794	1.0099	1.0473	1.2987	1.6759	2.0086	2.4033	2.6778	2.937	3.2614	3.496
10^2	0.677	1.0048	1.0418	1.2901	1.6602	1.984	2.3642	2.6559	2.8707	3.1737	3.3905
10^3	0.6747	1.0003	1.037	1.2824	1.6464	1.9623	2.3301	2.5808	2.8133	3.0984	3.3003
∞	0.6745	1	1.0364	1.2816	1.6449	1.96	2.3263	2.5808	2.8133	3.0902	3.2905

[a] $t_{r,\alpha}$ quantiles for the 95% confidence interval

For this reason, it is better to apply the measurement uncertainty random assessment according to the error propagation rule expressed by formula (11.28). Instead of the mean square error s_{x_i} or standard deviation $\sigma(x_i)$ in formulae (11.27) or (11.28), measurement uncertainty Δx_i is assumed. Consequently, the uncertainty of quantity z measured indirectly can be determined from the following formula [32, 33, 35, 91, 116, 126, 127, 135, 168, 292, 293, 300]:

$$\Delta z = \left[\sum_{i=1}^{N} \left(\frac{\partial f(\hat{\mathbf{x}})}{\partial x_i} \Delta x_i \right)^2 \right]^{\frac{1}{2}}. \tag{11.43}$$

If function f has a simple form, an analytical relation describing the measurement uncertainty Δz can be found. In the case of function

$$z = x_1^a x_2^b x_3^c \ldots x_N^m \tag{11.44}$$

formula (11.43) enables determination of a relation describing relative uncertainty:

$$\frac{\Delta z}{z} = \left[\left(a \frac{\Delta x_1}{x_1} \right)^2 + \left(b \frac{\Delta x_2}{x_2} \right)^2 + \cdots + \left(m \frac{\Delta x_N}{x_N} \right)^2 \right]^{\frac{1}{2}}. \tag{11.45}$$

If relation $z = f(x_1, \ldots, x_N)$ is implicit, e.g. if numerical algorithms are used to find the value of y based on directly measured quantities $x_1, \ldots, x_N$, derivatives $\frac{\partial f(\hat{\mathbf{x}})}{\partial x_i}$ are approximated by means of difference quotients.

11.2.2.2   Indirect Measurement Uncertainty Assessment Based on System Simulations with Input Data Burdened by Pseudorandom Errors

In inverse heat conduction problems [20, 118, 136, 223, 239, 329], generators of pseudorandom numbers are commonly used to assess the impact of the temperature measurement random errors on the heat flux determination on the body surface.

First, the so-called "exact measuring data" are found from the solution of the boundary or the initial-boundary value problem using exact methods. Then, pseudorandom numbers, generated by means of available library programs, are added to so-obtained exact measuring data to simulate the temperature measurement random errors.

The generated pseudorandom numbers may have a uniform or a normal distribution. In indirect measurements, other quantities on which the indirectly determined quantity depends can be taken into consideration as well, e.g. the temperature sensors location coordinates, thermal conductivity, the medium temperature, the heater power, the medium flow velocity, etc. To all these quantities small pseudorandom numbers simulating the measurement random errors should be added.

After a few hundred calculations of the quantity the measurements of which are disturbed with input data random errors, values are obtained of quantity z_i measured indirectly for a few hundred sets of measuring data. Based on that, mean square error s_z can be determined, which in this case is a very good approximation of mean standard deviation σ_z. Then the 68.3% confidence interval $\bar{z} - \sigma_z < z < \bar{z} + \sigma_z$ can be calculated, where $\bar{z}$ is the mean value, and z is the true value of the quantity measured indirectly. A similar procedure can be used to calculate the 95% confidence interval $\bar{z} - 2\sigma_z < z < \bar{z} + 2\sigma_z$.

This method of the measurement uncertainty calculations has a number of advantages, the most important of which is the identical procedure adopted both in linear and nonlinear problems. There is no need to calculate sensitivity coefficients, which is rather troublesome at a high number of quantities measured directly.

The pseudorandom numbers simulating the measurement random errors can be generated using easily available library programs.

Due to the great number of calculations, this method of the measurement uncertainty assessment requires a computer.

11.2.2.3   Example Measurement Uncertainty Calculations

The heat flux in boiler furnace chambers, industrial kiln furnaces and space-dividing elements in buildings is often measured using flat conductivity sensors (cf. Fig. 11.4) [329].

Fig. 11.4 Determination of heat flux $\dot{q}$ by means of a flat sensor with thickness L based on temperature measurement at two points: d_1 and d_2

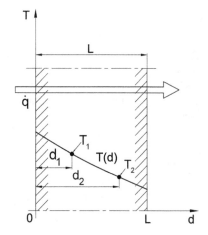

Heat flux $\dot{q}$ is also found from the following relation:

$$\dot{q} = \frac{\lambda(T_1) + \lambda(T_2)}{2} \frac{T_1 - T_2}{d_2 - d_1}. \tag{11.46}$$

If thermal conductivity $\lambda(T)$ is constant and temperature-independent, formula (11.46) is reduced to the following form:

$$\dot{q} = \frac{\lambda(T_1 - T_2)}{d_2 - d_1}, \tag{11.47}$$

where: λ—sensor material thermal conductivity [W/(m K)], d_1 and d_2—distance between the thermocouples fixing points and the sensor face surface, T_1 and T_2—temperatures measured at points d_1 and d_2, respectively.

Noting that $z = \dot{q}$ and $x_1 = T_1$ and $x_2 = T_2$, formula (11.47) can be written as:

$$z = \frac{\lambda(x_1 - x_2)}{d_2 - d_1}. \tag{11.48}$$

Assume that standard deviations $\sigma(x_1)$ and $\sigma(x_2)$ of temperatures x_1 and x_2 measured in points d_1 and d_2 are known. According to the error propagation rule (11.28), the following is obtained:

$$\sigma_z^2 = \left(\frac{\partial z}{\partial x_1}\right)^2 \sigma^2(x_1) + \left(\frac{\partial z}{\partial x_2}\right)^2 \sigma^2(x_2), \tag{11.49}$$

where:

$$\frac{\partial z}{\partial x_1} = \frac{\lambda}{d_2 - d_1}, \frac{\partial z}{\partial x_2} = -\frac{\lambda}{d_2 - d_1}. \tag{11.50}$$

The same is obtained by determining the covariance matrix. Considering that the Jacobian matrix

$$\mathbf{J} = \left(\frac{\partial z}{\partial x_1}, \frac{\partial z}{\partial x_2}\right) \tag{11.51}$$

is in this case a row vector and covariance matrix $\mathbf{C_x}$ has the following form:

$$\mathbf{C_x} = \begin{bmatrix} \sigma^2(x_1) & 0 \\ 0 & \sigma^2(x_2) \end{bmatrix}, \tag{11.52}$$

covariance matrix $\mathbf{C}_z$ takes the following form:

$$
\begin{aligned}
\mathbf{C}_z = \mathbf{J}\mathbf{C}_x\mathbf{J}^T &= \left(\frac{\partial z}{\partial x_1}, \frac{\partial z}{\partial x_2}\right) \cdot \begin{bmatrix} \sigma^2(x_1) & 0 \\ 0 & \sigma^2(x_2) \end{bmatrix} \cdot \left\{\begin{matrix} \frac{\partial z}{\partial x_1} \\ \frac{\partial z}{\partial x_2} \end{matrix}\right\} \\
&= \left(\frac{\partial z}{\partial x_1}\sigma^2(x_1), \frac{\partial z}{\partial x_2}\sigma^2(x_2)\right) \cdot \left\{\begin{matrix} \frac{\partial z}{\partial x_1} \\ \frac{\partial z}{\partial x_2} \end{matrix}\right\} \\
&= \left(\frac{\partial z}{\partial x_1}\right)^2 \sigma^2(x_1) + \left(\frac{\partial z}{\partial x_2}\right)^2 \sigma^2(x_2) = \sigma_z^2.
\end{aligned}
\tag{11.53}
$$

$\mathbf{C}_z$ is here a scalar equal to σ_z^2. Comparing the results of (11.49) and (11.53), it can be seen that formulae (11.28) and (11.33) give identical results.

Maximum uncertainty Δz at the assumed confidence level $P = 95\%$ can be calculated using formula (11.42), which gives:

$$
\Delta z = \left|\frac{\partial z}{\partial x_1}2\sigma(x_1)\right| + \left|\frac{\partial z}{\partial x_2}2\sigma(x_2)\right|.
\tag{11.54}
$$

It is also possible to calculate the 95% confidence interval $z - 2\sigma_z \le z \le z + 2\sigma_z$, where $2\sigma_z$ is found from the following formula:

$$
2\sigma_z = \left\{\left[\frac{\partial z}{\partial x_1}2\sigma(x_1)\right]^2 + \left[\frac{\partial z}{\partial x_2}2\sigma(x_2)\right]^2\right\}^{\frac{1}{2}}.
\tag{11.55}
$$

The maximum uncertainty value calculated according to formula (11.54) is higher compared to $2\sigma_z$ found from formula (11.55).

Example 11.1 Calculate the measurement uncertainty of the difference in temperature determined indirectly by measuring two temperatures. The measured values of temperature are: $x_1 = 247.2\,°C$ and $x_2 = 125.7\,°C$. The temperature measurement uncertainties $\pm 2\sigma_T$ (half of the 95% confidence interval) are $2\sigma_{x_1} = \pm 2.5\,K$ and $2\sigma_{x_2} = \pm 1.5\,K$, respectively. Find the measurement uncertainty of the temperature difference $\Delta T = T_1 - T_2$, i.e. $\pm 2\sigma_{\Delta T}$.

Solution
Assuming that the temperature measurement random errors z_1 and z_2 are independent of each other, mean standard deviation $\sigma_{\Delta T}$ can be calculated using the error propagation rule (11.28):

$$
\sigma_{\Delta T}^2 = \sigma_{x_1}^2 + \sigma_{x_2}^2 = \left(\frac{2.5}{2}\right)^2 + \left(\frac{1.5}{2}\right)^2 = 2.125\,K^2,
$$

$$
\sigma_{\Delta T} = 1.46\,K.
$$

The measurement uncertainty of the temperature difference ΔT is: $\pm 2\sigma_{\Delta T} = 2 \cdot 1.46 = 2.92$ K. The temperature difference is then:

$$\Delta T = (x_1 - x_2) \pm 2\sigma_{\Delta T} = (247.2 - 125.7) \pm 2.92\,\text{K} = 121.5 \pm 2.92\,\text{K}.$$

If it is assumed that the maximum uncertainties of the temperature measurement are: $\Delta x_1 = \pm 2.5$ K and $\Delta x_2 = \pm 1.5$ K, the maximum uncertainty of the measurement of the temperature difference ΔT_{max} calculated according to formula (11.42) is:

$$\Delta T_{max} = \left(\left| \frac{\partial (\Delta T)}{\partial x_1} \Delta x_1 \right| + \left| \frac{\partial (\Delta T)}{\partial x_2} \Delta x_2 \right| \right) = |\Delta x_1| + |-\Delta x_2| = (2.5 + 1.5) = 4\,\text{K}.$$

It can be seen that the error propagation rule gives a less pessimistic estimation of the uncertainty of the temperature difference measurement.

Generators of pseudorandom numbers are commonly used in inverse heat conduction problems [329] to assess the impact of the temperature measurement random errors on the heat flux determination on the body surface.

11.3 Least Squares Method

The least squares method is used in overdetermined problems, when the number of sought parameters $(a_0, a_1,\ldots,a_N)$ is smaller than the number of equations n. The values of y_i, $i = 1, 2,\ldots,n$ are known from measurement, and the form of function $\eta(\mathbf{x}, \mathbf{a})$, which is to best approximate the measured values, is also known.

Measured value y_i can be expressed as a sum of function $\eta_i(\mathbf{a})$ and measuring error ε_i:

$$y_i = \eta_i(\mathbf{a}) + \varepsilon_i, \quad i = 1, 2, \ldots, n, \tag{11.56}$$

where $E(\varepsilon_i) = 0$, $\mathbf{a} = (a_0, a_1, \ldots, a_N)^{\mathrm{T}}$. Vector $\tilde{\mathbf{a}}$ minimizes the sum of squares of errors:

$$S(\mathbf{a}) = \sum_{i=1}^{n} [y_i - \eta_i(\mathbf{a})]^2. \tag{11.57}$$

If function $\eta(\mathbf{x}, \mathbf{a})$ is linear with respect to sought parameters $\mathbf{a}$, it is said that a linear least squares problem is the case. If function $\eta(\mathbf{x}, \mathbf{a})$ is nonlinear, so is the least squares problem. The downside of criterion (11.57) is the strong impact of measuring points which are substantially different from other measuring data on obtained values of function $\eta(\mathbf{x}, \mathbf{a})$. In order to eliminate this disadvantage of the ordinary least squares method defined by formula (11.57), a weighted least squares method is introduced with coefficients $w_i = 1/\sigma_i^2$:

$$S(\mathbf{a}) = \sum_{i=1}^{n} \frac{[y_i - \eta_i(\mathbf{a})]^2}{\sigma_i^2}, \tag{11.58}$$

where $E(\varepsilon_i^2) = \sigma_i^2$ is the variance characterizing the deviation of measurement results y_i from expected value $E(y_i)$. Standard deviation σ_i is often equated with the measuring error. The sum of squares presented in expression (11.57) can also be written in a matrix form:

$$S(\mathbf{a}) = [\mathbf{y} - \boldsymbol{\eta}(\mathbf{a})]^{\mathrm{T}} \, \mathbf{C}_y^{-1} [\mathbf{y} - \boldsymbol{\eta}(\mathbf{a})], \tag{11.59}$$

where:

$$\boldsymbol{\eta}(\mathbf{a}) = \left\{ \begin{matrix} \eta_1(\mathbf{a}) \\ \eta_2(\mathbf{a}) \\ \vdots \\ \eta_n(\mathbf{a}) \end{matrix} \right\}, \quad \mathbf{y} = \left\{ \begin{matrix} y_1 \\ y_2 \\ \vdots \\ y_n \end{matrix} \right\}. \tag{11.60}$$

The covariance matrix of measured quantities y_i or random errors ε_i is diagonal.

$$\mathbf{C}_y = \begin{bmatrix} \sigma_1^2 & & & 0 \\ & \sigma_2^2 & & \\ & & \ddots & \\ 0 & & & \sigma_n^2 \end{bmatrix}_{n \times n}. \tag{11.61}$$

Inverse matrix $\mathbf{C}_y^{-1}$, which is also referred to as weight matrix $\mathbf{G}_y$, has the following form:

$$\mathbf{C}_y^{-1} = \mathbf{G}_y = \begin{bmatrix} \frac{1}{\sigma_1^2} & & & 0 \\ & \frac{1}{\sigma_2^2} & & \\ & & \ddots & \\ 0 & & & \frac{1}{\sigma_n^2} \end{bmatrix}. \tag{11.62}$$

If it is assumed that all variances are equal to each other, i.e.: $\sigma_1 = \sigma_2 = \cdots = \sigma_n = \sigma$, covariance matrix $\mathbf{C}_y$ is equal to the product of unit matrix $\mathbf{I}_n$ and variance σ^2, i.e. $\mathbf{C}_y = \sigma^2 \mathbf{I}_n$, and then the weighted least squares method comes down to the ordinary least squares method as expressed by (11.57). Finding vector $\tilde{\mathbf{a}}$ for which function $S(\tilde{\mathbf{a}})$ reaches the global minimum value, it is possible to calculate the sample standard deviation (mean square error) s_y:

$$s_y = \sqrt{\frac{S(\tilde{\mathbf{a}})}{n - N - 1}}, \tag{11.63}$$

where $S(\tilde{\mathbf{a}})$ is calculated from formula (11.57), $(n - N - 1)$ is the number of freedom degrees, N—the number of sought parameters, n—the number of measurements.

In practice, the weighted least squares method is often used without knowing variance $\sigma_i^2, i = 1, \ldots, n$. Weight coefficients w of weight matrix

$$\mathbf{G_y} = \begin{bmatrix} w_1 & & & \\ & w_2 & & \\ & & \ddots & \\ & & & w_n \end{bmatrix} \tag{11.64}$$

are then not the covariance inverse, but they are selected based on other criteria. Demanding, for example, that approximation function $\eta(\mathbf{a})$ should pass through a given measuring point (x_i, y_i), a high value of the weight coefficient has to be selected in the weighted sum:

$$S(\mathbf{a}) = \sum_{i=1}^{n} w_i [y_i - \eta_i(\mathbf{a})]^2. \tag{11.65}$$

In the case of the weighted least squares method (11.58), the estimator of mean standard deviation σ is the following expression [261]:

$$\sigma \cong s_y = \sqrt{\frac{1}{n - N - 1} [\mathbf{y} - \eta(\tilde{\mathbf{a}})]^{\mathrm{T}} \mathbf{G_y} [\mathbf{y} - \eta(\tilde{\mathbf{a}})]}. \tag{11.66}$$

Mean square error s_y (11.66), which is an approximation of mean standard deviation σ_y, is used to determine the confidence interval of coefficients $\tilde{a}_0, \tilde{a}_1, \ldots, \tilde{a}_N$ and the confidence intervals of function $\tilde{\eta}$ obtained by means of the least squares method.

The weighted least squares method expressed by formula (11.65) can be easily reduced to the ordinary least squares method by writing formula (11.65) in the following form:

$$S(\mathbf{a}) = \sum_{i=1}^{n} \left[\frac{y_i - \eta_i(\mathbf{a})}{w_i^{-1/2}} \right]^2. \tag{11.67}$$

In order to transform the weighted least squares method defined by formula (11.59) into the ordinary least squares method, matrix $\mathbf{Cy}$ can be decomposed using the Cholesky decomposition, according to which $\mathbf{C_y} = \mathbf{U}^{\mathrm{T}} \mathbf{U}$, where $\mathbf{U}$ is the upper triangular matrix. Writing the system of Eq. (11.56) in the vector form:

$$\mathbf{y} = \eta(\mathbf{a}) + \varepsilon \tag{11.68}$$

and multiplying the nonlinear model (11.68) by $\mathbf{R} = (\mathbf{U}^T)^{-1}$, the following is obtained:

$$\mathbf{z} = \boldsymbol{\psi}(\mathbf{a}) + \boldsymbol{\delta}, \tag{11.69}$$

where $\mathbf{z} = \mathbf{R}\mathbf{y}$, $\boldsymbol{\psi}(\mathbf{a}) = \mathbf{R}\boldsymbol{\eta}(\mathbf{a})$ and $\boldsymbol{\delta} = \mathbf{R}\boldsymbol{\varepsilon}$.

The weighted sum of squares expressed in (11.59) can be written as the ordinary least squares method:

$$\begin{aligned} S(\mathbf{a}) &= [\mathbf{y} - \boldsymbol{\eta}(\mathbf{a})]^T \mathbf{C}_y^{-1}[\mathbf{y} - \boldsymbol{\eta}(\mathbf{a})] \\ &= [\mathbf{y} - \boldsymbol{\eta}(\mathbf{a})]^T \mathbf{R}^T \mathbf{R}[\mathbf{y} - \boldsymbol{\eta}(\mathbf{a})] \\ &= [\mathbf{z} - \boldsymbol{\psi}(\mathbf{a})]^T [\mathbf{z} - \boldsymbol{\psi}(\mathbf{a})]. \end{aligned} \tag{11.70}$$

Transformation of the weighted least squares method into the ordinary least squares method simplifies calculations, including calculations of the confidence interval. Mean square error s_y (11.66) or mean standard deviation σ_y are used to determine the confidence intervals of coefficients $\tilde{a}_0, \tilde{a}_1, \ldots, \tilde{a}_N$ and of function $\tilde{\eta}$ obtained by means of the least squares method.

If mean standard deviations σ_i of the measured quantities are known, 95% confidence intervals for coefficients $a_0, \ldots, a_N$ and function η_i, $i = 1, \ldots, n$ are found according to the error propagation rule (11.27) from the following formula:

$$\tilde{a}_j - 2\sigma_{\tilde{a}_j} \le a_j \le \tilde{a}_j + 2\sigma_{\tilde{a}_j}, \quad j = 0, \ldots, N, \tag{11.71}$$

$$\tilde{\eta}_i - 2\sigma_{\tilde{\eta}_i} \le \eta_i \le \tilde{\eta}_i + 2\sigma_{\tilde{\eta}_i}, \quad i = 1, \ldots, n, \tag{11.72}$$

where mean standard deviations $\sigma_{\tilde{a}_j}$ and $\sigma_{\tilde{\eta}_i}$ are defined by the following formulae:

$$\sigma_{\tilde{a}_j}^2 = \sum_{i=1}^{n} \left(\frac{\partial \tilde{a}_j}{\partial y_i}\right)^2 \sigma_{y_i}^2, \tag{11.73}$$

$$\sigma_{\tilde{\eta}_j}^2 = \sum_{i=1}^{n} \left(\frac{\partial \tilde{\eta}_j}{\partial y_i}\right)^2 \sigma_{y_i}^2, \tag{11.74}$$

where $\sigma_i = \sigma_{y_i}$ is the mean standard deviation of the i-th measured quantity y_i.

If standard deviations σ_i are unknown, confidence intervals are determined using the following formulae:

$$\tilde{a}_j - t_{n-N-1,\alpha} \, s_y \sqrt{c'_{jj}} \le a_j \le \tilde{a}_j + t_{n-N-1,\alpha} \, s_y \sqrt{c'_{jj}}, \quad j = 0, \ldots, N, \tag{11.75}$$

where:

$$c'_{jj} = \sum_{i=1}^{n} \left(\frac{\partial \tilde{a}_j}{\partial y_i}\right)^2 \tag{11.76}$$

are diagonal elements of matrix $\mathbf{C}'_{\tilde{\mathbf{a}}}$.

If coefficients $\tilde{a}_j$ are found by means of the weighted least squares method, formula (11.76) takes the following form:

$$c'_{jj} = \sum_{i=1}^{n} \frac{1}{w_i} \left(\frac{\partial \tilde{a}_j}{\partial y_i}\right)^2, \quad i = 1, \ldots, n, \quad j = 0, \ldots, N. \tag{11.77}$$

A similar procedure is applied to find confidence intervals $(1 - \alpha)$ for approximation function η

$$\tilde{\eta}_i - t_{n-N-1,\alpha}\, s_y \sqrt{c''_{ii}} \leq \eta_i \leq \tilde{\eta}_i + t_{n-N-1,\alpha}\, s_y \sqrt{c''_{ii}}, \quad i = 1, \ldots, n, \tag{11.78}$$

where:

$$c''_{jj} = \sum_{j=1}^{n} \left(\frac{\partial \tilde{\eta}_i}{\partial y_j}\right)^2 \tag{11.79}$$

are diagonal elements of matrix $\mathbf{C}''_{\tilde{\eta}}$. Symbol $t_{n-N-1,\alpha}$ denotes the quantile of the t-Student distribution for the $(1 - \alpha)$ 100% confidence interval (for the 95% confidence interval $\alpha = 0.05$).

If coefficients $\tilde{a}_j$ are found by means of the weighted least squares method, formula (11.79) takes the following form:

$$c''_{ii} = \sum_{j=1}^{n} \frac{1}{w_i} \left(\frac{\partial \tilde{\eta}_i}{\partial y_j}\right)^2. \tag{11.80}$$

Mean square error s_y is calculated from formula (11.63) or formula (11.66). Partial derivatives (sensitivity coefficients) $\partial a_j/\partial y_i$ and $\partial \eta_i/\partial y_i$ are calculated using difference quotients:

$$\frac{\partial \tilde{a}_j}{\partial y_i} = \frac{\tilde{a}_j(y_1, \ldots, y_i + \delta, \ldots, y_N) - \tilde{a}_j(y_1, \ldots, y_i - \delta, \ldots, y_N)}{2\delta},$$
$$j = 0, \ldots, N, \quad i = 1, \ldots, n. \tag{11.81}$$

Derivatives $\partial \tilde{\eta}_i/\partial y_j$ are calculated similarly. Symbol δ denotes a small number, e.g. $\delta = 0.1$. The quantities in the difference quotient numerator are calculated using the least squares method. If function $\eta(\mathbf{x}, \tilde{\mathbf{a}})$ has a simple form, the derivatives can be calculated analytically.

11.4   Linear Problem of the Least Squares Method

In a linear problem, also referred to as linear regression, measuring points

$$(x_1, y_1), (x_2, y_2), \ldots, (x_n, y_n) \tag{11.82}$$

are approximated using a function with a single variable x.

$$\eta = \eta(x, a_0, a_1) = a_0 + a_1 x. \tag{11.83}$$

Parameters a_0 and a_1 are found from the condition of the minimum of the sum of squares (11.57), which in this case has the following form:

$$S(a_0, a_1) = \sum_{i=1}^{n} (y_i - a_0 - a_1 x_i)^2 \rightarrow \min. \tag{11.84}$$

The necessary conditions for the existence of the minimum of sum S:

$$\frac{\partial S}{\partial a_0} = 0 \quad \text{and} \quad \frac{\partial S}{\partial a_1} = 0, \tag{11.85}$$

result in:

$$\sum_{i=1}^{n} (y_i - a_0 - a_1 x_i) = 0, \tag{11.86}$$

$$\sum_{i=1}^{n} \left(x_i y_i - a_0 x_i - a_1 x_i^2 \right) = 0. \tag{11.87}$$

Solving the system of Eqs. (11.86) and (11.87) with respect to a_0 and a_1, the following is obtained:

$$\tilde{a}_0 = \frac{\sum_{i=1}^{n} y_i - a_1 \sum_{i=1}^{n} x_i}{n} = \frac{\sum_{i=1}^{n} x_i \sum_{i=1}^{n} x_i y_i - \sum_{i=1}^{n} y_i \sum_{i=1}^{n} x_i^2}{\left(\sum_{i=1}^{n} x_i \right)^2 - n \sum_{i=1}^{n} x_i^2}, \tag{11.88}$$

$$\tilde{a}_1 = \frac{\sum_{i=1}^{n} x_i \sum_{i=1}^{n} y_i - n \sum_{i=1}^{n} x_i y_i}{\left(\sum_{i=1}^{n} x_i \right)^2 - n \sum_{i=1}^{n} x_i^2}. \tag{11.89}$$

The confidence intervals for the found parameters a_0 and a_1 can be calculated from formula (11.75), and for the determined function η—from formula (11.78).

Formula (11.78) makes it possible to calculate the confidence interval of function η only in measuring points $x_1, \ldots, x_n$. Modifying expression (11.78), function η confidence interval can be calculated easily for any x_k, $k = 1, 2, \ldots$ using the following relation:

$$\tilde{\eta}_k - t_{n-N-1,\alpha} \, s_y \sqrt{c''_{kk}} \leq \eta_k \leq \tilde{\eta} + t_{n-N-1,\alpha} \, s_y \sqrt{c''_{kk}}, \qquad (11.90)$$

where:

$$c''_{kk} = \sum_{j=1}^{n} \left(\frac{\partial \tilde{\eta}_k}{\partial y_j} \right)^2. \qquad (11.91)$$

In this case, derivatives $\partial a_0/\partial y_i$ and $\partial a_1/\partial y_i$ can be determined analytically, which gives:

$$\frac{\partial \tilde{a}_0}{\partial y_i} = \frac{x_i \sum_{i=1}^{n} x_i - \sum_{i=1}^{n} x_i^2}{\left(\sum_{i=1}^{n} x_i \right)^2 - n \sum_{i=1}^{n} x_i^2}, \qquad (11.92)$$

$$\frac{\partial \tilde{a}_1}{\partial y_i} = \frac{\sum_{i=1}^{n} x_i - n x_i}{\left(\sum_{i=1}^{n} x_i \right)^2 - n \sum_{i=1}^{n} x_i^2}, \qquad i = 1, \ldots, n. \qquad (11.93)$$

Derivative $\partial \tilde{\eta}/\partial y_j$ can be determined from the following formula:

$$\frac{\partial \tilde{\eta}}{\partial y_j} = \frac{\partial (\tilde{a}_0 + \tilde{a}_1 x)}{\partial y_j} = \frac{\partial \tilde{a}_0}{\partial y_j} + x \frac{\partial \tilde{a}_1}{\partial y_j}, \qquad j = 1, \ldots, n, \qquad (11.94)$$

where $\partial \tilde{a}_0/\partial y_i$ and $\partial \tilde{a}_1/\partial y_i$ are calculated from formulae (11.92) and (11.93), or numerically—approximating the derivatives by means of central difference quotients.

Example 11.2 **Heat flux determination using a conductometric sensor**
The least squares method application is presented on the example of indirect measurement of the heat flux.

The heat flux in a steel wall with thickness $L = 0.016$ m is found based on the temperature measurement performed in three points with the following coordinates: $x_1 = 0.004$ m, $x_2 = 0.008$ m and $x_3 = 0.012$ m. The measured temperature values are: $y_1 = 310.6$ °C, $y_2 = 246.6$ °C and $y_3 = 191.0$ °C. The sensor steel thermal conductivity is $\lambda = 18.31$ W/(m K).

Find heat flux $\dot{q}$ transferred through the wall using the least squares method. Perform the calculations for the following four cases:

(a) Apply the ordinary least squares method to determine $\dot{q}$.
(b) Apply the weighted least squares method to determine $\dot{q}$ and find uncertainty using the variance transposition method formulated by Gauss. Assume that standard deviations of temperature measured in points x_1, x_2 and x_3 are known : $\sigma_1 = 1$ K, $\sigma_2 = 2$ K, $\sigma_3 = 1.5$ K, respectively.
(c) Apply the weighted least squares method to determine $\dot{q}$ assuming the following weight coefficients:

$$w_1 = 1/\sigma_1^2 = 1/1^2 = 1\,\mathrm{K}^{-2},$$
$$w_2 = 1/\sigma_2^2 = 1/2^2 = 0.25\,\mathrm{K}^{-2},$$
$$w_3 = 1/\sigma_3^2 = 1/1.5^2 = 0.4444(4)\,\mathrm{K}^{-2}.$$

Assume that the weights have no physical interpretation, i.e. they are not the variance inverse. Standard deviations of measured temperatures are unknown.

(d) Solve the problem formulated in (c) above assuming the following weight coefficients:

$$w_1 = s_y^2/\sigma_1^2,$$
$$w_2 = s_y^2/\sigma_2^2,$$
$$w_3 = s_y^2/\sigma_3^2.$$

Standard deviations of measured temperatures are unknown.

Solution

The temperature distribution on the wall thickness can be approximated by a straight line:

$$\eta = a_0 + a_1 x,$$

and the heat flux can be calculated using the Fourier law:

$$\dot{q} = -\lambda \frac{dT}{dx} = -\lambda a_1.$$

Case (a)

The sum of squares of the measured and calculated temperature differences has the following form:

$$S = (310.6 - a_0 - 0.004a_1)^2 + (246.6 - a_0 - 0.008a_1)^2 + (191.0 - a_0 - 0.012a_1)^2.$$

The necessary conditions for the existence of the minimum of the sum of squares:

$$\frac{\partial S}{\partial a_0} = 0 \quad \text{and} \quad \frac{\partial S}{\partial a_1} = 0$$

give a system of two equations, the solution of which provides the sought coefficients

$$\tilde{a}_0 = 369.0000\;^\circ\mathrm{C},$$
$$\tilde{a}_1 = -14950.0000\;^\circ\mathrm{C/m}.$$

Identical values of coefficients $\tilde{a}_0$ and $\tilde{a}_1$ are obtained from formulae (11.88) and (11.89). Substitution of $\tilde{a}_0$ and $\tilde{a}_1$ in S results in the minimum value of the sum of squares $S_{min} = 11.76 \ \mathrm{K}^2$. The mean square deviation totals:

$$s_y = \sqrt{S_{min}/(n-N-1)} = \sqrt{\left[1.4^2 + (-2.8)^2 + 1.4^2\right]/(3-1-1)}$$

$$= \sqrt{11.76/(3-1-1)} = 3.4293 \ \mathrm{K}.$$

Partial derivatives (sensitivity coefficients) are calculated using the central difference quotient (11.81) for $\delta = 0.2$ K. The obtained results are as follows:

$$\frac{\partial \tilde{a}_0}{\partial y_1} = 1.3333, \frac{\partial \tilde{a}_0}{\partial y_2} = 0.3333, \frac{\partial \tilde{a}_0}{\partial y_3} = -0.6666,$$

$$\frac{\partial \tilde{a}_1}{\partial y_1} = -125.0000\frac{1}{\mathrm{m}}, \frac{\partial \tilde{a}_1}{\partial y_2} = 0.0000\frac{1}{\mathrm{m}}, \frac{\partial \tilde{a}_1}{\partial y_3} = 125.0000\frac{1}{\mathrm{m}}.$$

Coefficients c'_{11} and c'_{22} calculated from the following relation

$$c'_{jj} = \sum_{i=1}^{n} \left(\frac{\partial \tilde{a}_j}{\partial y_i}\right)^2$$

are

$$c'_{11} = \sum_{i=1}^{3} \left(\frac{\partial \tilde{a}_0}{\partial y_i}\right)^2 = 1.3333^2 + 0.3333^2 + (-0.6666)^2 = 2.3333,$$

$$c'_{22} = \sum_{i=1}^{3} \left(\frac{\partial \tilde{a}_1}{\partial y_i}\right)^2 = (-125)^2 + 0.0^2 + 125^2 = 31250.0000\frac{1}{\mathrm{m}^2}.$$

Considering that the number of the degrees of freedom is $(n - N - 1) = (3 - 1 - 1) = 1$, the t-Student quantile for the 95% confidence interval $t_{1;0.05}$ (cf. Table 11.2) is: $t_{1;0.05} = 12.7062$.

The confidence intervals calculated from formula (11.75) are:

$$369 - 12.7062 \cdot 3.4293 \cdot \sqrt{2.3333} \le a_0 \le 369$$

$$+ 12.7062 \cdot 3.4293 \cdot \sqrt{2.3333}, 302.4411 \ ^{\circ}\mathrm{C} \le a_0 \le 435.5589 \ ^{\circ}\mathrm{C},$$

$$- 14950 - 12.7062 \cdot 3.4293 \cdot \sqrt{31250} \le a_1 \le -14950 + 12.7062 \cdot 3.4293 \cdot \sqrt{31250},$$

$$- 22652.7244\frac{^{\circ}\mathrm{C}}{\mathrm{m}} \le a_1 \le -7247.2756\frac{^{\circ}\mathrm{C}}{\mathrm{m}}.$$

Considering that

$$a_1 = -14950 \pm \frac{-7247.2756 - (-22652.7244)}{2} = -14950 \pm 7702.7244 \frac{°C}{m}$$

it is possible to determine the heat flux value:

$$\dot{q} = -\lambda a_1 = -18.31 \cdot (-14950 \pm 7702.7244) = 273734.5 \pm 141036.9 \frac{W}{m^2}.$$

95% confidence intervals can also be determined for function η using formula
(11.78) in points x_1, x_2 and x_3 or formula (11.90) for any x. Coefficients c_{ii}'' and c_{kk}''
are calculated from the following formulae:

$$c_{ii}'' = \sum_{j=1}^{n} \left(\frac{\partial \tilde{\eta}_i}{\partial y_j}\right)^2, \quad i = 1, \dots, n,$$

$$c_{kk}'' = \sum_{j=1}^{n} \left(\frac{\partial \tilde{\eta}_k}{\partial y_j}\right)^2 \quad \text{for any } x_k.$$

Confidence intervals for η_i, $i = 1,\dots,3$ are as follows:

$$250.2016 \text{ °C} \leq \eta_1 \leq 368.1984 \text{ °C},$$
$$199.0860 \text{ °C} \leq \eta_2 \leq 299.7140 \text{ °C},$$
$$130.6016 \text{ °C} \leq \eta_3 \leq 248.5984 \text{ °C}.$$

The history of function $\tilde{\eta}(x)$, the measuring points y_i, $i = 1,\dots,3$ and the limits of
the 95% confidence interval determined from formula (11.90) are listed in
Table 11.3 and in Fig. 11.5.

Analysing the obtained results, it can be seen that the 95% confidence intervals
for coefficients a_0 and a_1 and for function $\eta(x)$ are wide due to the small number of
the measuring points: $n = 3$.

Case (b)
The values of coefficients $\tilde{a}_0$ and $\tilde{a}_1$ are obtained from formulae (11.88) and (11.89),
and they are the same as in Case (a):

$$\tilde{a}_0 = 369 \text{ °C}, \tilde{a}_1 = -14950 \text{ °C/m}.$$

The sensor temperature in points x_1, x_2 and x_3 totals:

$$\tilde{\eta}_1 = \tilde{a}_0 + \tilde{a}_1 x_1 = 369 - 14950 \cdot 0.004 = 309.2 \text{ °C},$$
$$\tilde{\eta}_2 = \tilde{a}_0 + \tilde{a}_1 x_2 = 369 - 14950 \cdot 0.008 = 249.4 \text{ °C},$$
$$\tilde{\eta}_3 = \tilde{a}_0 + \tilde{a}_1 x_3 = 369 - 14950 \cdot 0.012 = 189.6 \text{ °C}.$$

Table 11.3 Values of temperature $\tilde{\eta}$ in points x_k and the upper and the lower limit of the 95% confidence interval for determined function $\tilde{\eta}(x)$

k	x_k (m)	$\tilde{\eta}(x_k)$ (°C)	Lower limit of the 95% confidence interval (°C)	Upper limit of the 95% confidence interval (°C)
1	0.004	309.20	250.2016	368.1984
2	0.005	294.25	238.8832	349.6168
3	0.006	279.30	226.6804	331.9196
4	0.007	264.35	213.4498	315.2502
5	0.008	249.40	199.0860	299.7140
6	0.009	234.45	183.5498	285.3502
7	0.010	219.50	166.8804	272.1196
8	0.011	204.55	149.1832	259.9168
9	0.012	189.60	130.6016	248.5984

Fig. 11.5 Straight-line approximation of temperatures measured at three points inside the sensor by means of the ordinary least squares method

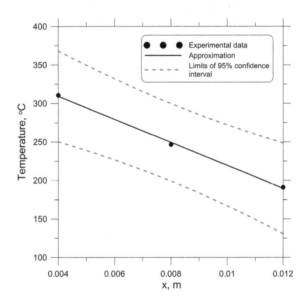

The 95% confidence intervals of coefficients $\tilde{a}_0$ and $\tilde{a}_1$ and of functions $\tilde{\eta}_1, \tilde{\eta}_2$ and $\tilde{\eta}_3$ are determined using formulae (11.71) and (11.72), respectively.

Variance $\sigma_{\tilde{a}_0}^2$ found from the following relation:

$$\sigma_{\tilde{a}_0}^2 = \sum_{j=1}^{n} \left(\frac{\partial \tilde{a}_0}{\partial y_j} \right)^2 \sigma_{y_j}^2$$

totals:

$$\sigma_{\tilde{a}_0}^2 = \left(\frac{\partial \tilde{a}_0}{\partial y_1}\right)^2 \sigma_{y_1}^2 + \left(\frac{\partial \tilde{a}_0}{\partial y_2}\right)^2 \sigma_{y_2}^2 + \left(\frac{\partial \tilde{a}_0}{\partial y_3}\right)^2 \sigma_{y_3}^2$$
$$= 1.3333^2 \cdot 1^2 + 0.3333^2 \cdot 2^2 + (-0.6666)^2 \cdot 1.5^2 = 3.2222 \; {}^{\circ}\text{C}^2,$$

which gives:

$$\sigma_{\tilde{a}_0} = 1.7950 \; {}^{\circ}\text{C}.$$

The 95% confidence interval determined from formula (11.71) is:

$$369 - 2 \cdot 1.795 \leq a_0 \leq 369 + 2 \cdot 1.795, \, 365.41 \; {}^{\circ}\text{C} \leq a_0 \leq 372.59 \; {}^{\circ}\text{C}.$$

The 95% confidence interval of coefficient a_1 are determined in a similar way from the following formula:

$$\sigma_{\tilde{a}_1}^2 = \sum_{j=1}^{n} \left(\frac{\partial \tilde{a}_1}{\partial y_j}\right)^2 \sigma_{y_j}^2,$$

which gives:

$$\sigma_{\tilde{a}_1}^2 = \left(\frac{\partial \tilde{a}_1}{\partial y_1}\right)^2 \sigma_{y_1}^2 + \left(\frac{\partial \tilde{a}_1}{\partial y_2}\right)^2 \sigma_{y_2}^2 + \left(\frac{\partial \tilde{a}_1}{\partial y_3}\right)^2 \sigma_{y_3}^2$$
$$= (-125)^2 \cdot 1^2 + 0^2 \cdot 2^2 + (125)^2 \cdot 1.5^2 = 50781.25 \; {}^{\circ}\text{C}^2/\text{m}^2,$$
$$\sigma_{\tilde{a}_1} = 225.3470 \; {}^{\circ}\text{C/m}.$$

The 95% confidence interval determined from formula (11.71) is:

$$-14950 - 2 \cdot 225.347 \leq a_1 \leq -14950 + 2 \cdot 225.347,$$
$$-15400.694 \; {}^{\circ}\text{C/m} \leq a_1 \leq -14499.306 \; {}^{\circ}\text{C/m}.$$

Partial derivatives (sensitivity coefficients) were calculated using the central difference quotient similar to (11.81) for $\delta = 0.2$ K:

$$\frac{\partial \tilde{\eta}_1}{\partial y_1} = \frac{309.36667 - 309.03333}{0.4} = 0.83335,$$

$$\frac{\partial \tilde{\eta}_1}{\partial y_2} = \frac{309.26667 - 309.13333}{0.4} = 0.33335,$$

$$\frac{\partial \tilde{\eta}_1}{\partial y_3} = \frac{309.16667 - 309.23333}{0.4} = 0.16665.$$

Variance $\sigma_{\tilde{\eta}_1}^2$ is calculated from the following formula:

$$\sigma_{\tilde{\eta}_1}^2 = \left(\frac{\partial \tilde{\eta}_1}{\partial y_1}\right)^2 \sigma_{y_1}^2 + \left(\frac{\partial \tilde{\eta}_1}{\partial y_2}\right)^2 \sigma_{y_2}^2 + \left(\frac{\partial \tilde{\eta}_1}{\partial y_3}\right)^2 \sigma_{y_3}^2$$

$$= 0.83335^2 \cdot 1^2 + 0.33335^2 \cdot 2^2 + 0.16665^2 \cdot 1.5^2 = 1.2014 \text{ K}^2,$$

which gives:

$$\sigma_{\tilde{\eta}_1} = 1.0961 \text{ °C.}$$

Standard deviation $\sigma_{\tilde{\eta}_2}$ is evaluated in a similar way:

$$\frac{\partial \tilde{\eta}_2}{\partial y_1} = \frac{249.46667 - 249.33333}{0.4} = 0.33335,$$

$$\frac{\partial \tilde{\eta}_2}{\partial y_2} = \frac{249.46667 - 249.33333}{0.4} = 0.33335,$$

$$\frac{\partial \tilde{\eta}_2}{\partial y_3} = \frac{249.46667 - 249.33333}{0.4} = 0.33335,$$

$$\sigma_{\tilde{\eta}_2}^2 = \left(\frac{\partial \tilde{\eta}_2}{\partial y_1}\right)^2 \sigma_{y_1}^2 + \left(\frac{\partial \tilde{\eta}_2}{\partial y_2}\right)^2 \sigma_{y_2}^2 + \left(\frac{\partial \tilde{\eta}_2}{\partial y_3}\right)^2 \sigma_{y_3}^2$$

$$= 0.33335^2 \cdot 1^2 + 0.33335^2 \cdot 2^2 + 0.33335^2 \cdot 1.5^2 = 0.8056 \text{ °C}^2,$$

$$\sigma_{\tilde{\eta}_2} = 0.8976 \text{ °C.}$$

Standard deviation $\sigma_{\tilde{\eta}_3}$ is calculated using a similar procedure:

$$\sigma_{\tilde{\eta}_3}^2 = \left(\frac{\partial \tilde{\eta}_3}{\partial y_1}\right)^2 \sigma_{y_1}^2 + \left(\frac{\partial \tilde{\eta}_3}{\partial y_2}\right)^2 \sigma_{y_2}^2 + \left(\frac{\partial \tilde{\eta}_3}{\partial y_3}\right)^2 \sigma_{y_3}^2$$

$$= (-0.16665)^2 \cdot 1^2 + 0.33335^2 \cdot 2^2 + 0.83335^2 \cdot 1.5^2 = 2.0348 \text{ °C}^2,$$

$$\sigma_{\tilde{\eta}_3} = 1.4265 \text{ °C.}$$

The 95% confidence intervals calculated from formula (11.72) are:

$$309.2 - 2 \cdot 1.0961 \leq \eta_1 \leq 309.2 + 2 \cdot 1.0961,$$
$$305.278 \text{ °C} \leq \eta_1 \leq 311.3922 \text{ °C},$$
$$249.4 - 2 \cdot 0.8976 \leq \eta_2 \leq 249.4 + 2 \cdot 0.8976,$$
$$247.6048 \text{ °C} \leq \eta_2 \leq 251.1952 \text{ °C},$$
$$189.6 - 2 \cdot 1.4265 \leq \eta_3 \leq 189.6 + 2 \cdot 1.4265,$$
$$186.747 \text{ °C} \leq \eta_3 \leq 192.453 \text{ °C}.$$

It can be seen that the 95% confidence intervals both for the coefficients and for the determined function are much more narrow compared to Case (a). This is the effect of the assumption that the variances in measured temperatures $y_1 y_2$ and y_3 are known.

Case (c)

The sum of squares of the measured and calculated temperature differences has the following form:

$$S = (310.6 - a_0 - 0.004a_1)^2 + 0.25(246.6 - a_0 - 0.008a_1)^2$$
$$+ 0.4444(191.0 - a_0 - 0.012a_1)^2.$$

The necessary conditions for the existence of the minimum of the sum of squares:

$$\frac{\partial S}{\partial a_0} = 0 \quad \text{and} \quad \frac{\partial S}{\partial a_1} = 0$$

give a system of two equations, the solution of which provides the sought coefficients

$$\tilde{a}_0 = 370.2364 \,^\circ C,$$
$$\tilde{a}_1 = -15018.1818 \frac{^\circ C}{m}.$$

Substitution of $\tilde{a}_0$ and $\tilde{a}_1$ in S results in the minimum value of the sum of squares $S_{min} = 3.6655 \, K^2$. The mean square deviation is:

$$s_y = \sqrt{S_{min}/(n - N - 1)} = \sqrt{3.6655(3 - 1 - 1)} = 1.9148 \, K.$$

Partial derivatives (sensitivity coefficients) are calculated using the central difference quotient (11.81) for $\delta = 0.2$ K. The obtained results are as follows:

$$\frac{\partial \tilde{a}_0}{\partial y_1} = 1.48052, \quad \frac{\partial \tilde{a}_0}{\partial y_2} = 0.03896, \quad \frac{\partial \tilde{a}_0}{\partial y_3} = -0.51948,$$
$$\frac{\partial \tilde{a}_1}{\partial y_1} = -133.11683 \frac{1}{m}, \quad \frac{\partial \tilde{a}_1}{\partial y_2} = 16.23375 \frac{1}{m}, \quad \frac{\partial \tilde{a}_1}{\partial y_3} = 116.88300 \frac{1}{m}.$$

Due to the application of the weighted least squares method, c'_{jj} is calculated using formula (11.77):

$$c'_{jj} = \sum_{i=1}^{n} \left(\frac{\partial \tilde{a}_j}{\partial y_i} \right)^2 \frac{1}{w_i}, \quad j = 1, 2.$$

The values of c'_{11} and c'_{22} calculated according to the formula are:

$$c'_{11} = 2.8052 \text{ K}^2,$$

$$c'_{22} = 49512.9772 \frac{\text{K}^2}{\text{m}^2}.$$

Considering that the number of the degrees of freedom is $(n - N - 1) = (3 - 1 - 1) = 1$, the t-Student quantile for the 95% confidence interval $t_{1;0.05}$ (cf. Table 11.2) is: $t_{1;0.05} = 12.7062$.

The confidence intervals calculated from formula (11.75) are:

$$370.2365 - 12.7062 \cdot 1.9148 \cdot \sqrt{2.8052} \le a_0 \le 370.2365$$

$$+ 12.7062 \cdot 1.19148 \cdot \sqrt{2.8052}, \, 329.49 \, {}^\circ\text{C} \le a_0 \le 410.98 \, {}^\circ\text{C},$$

$$- 15018.1781 - 12.7062 \cdot 1.9148 \cdot \sqrt{49512.9772} \le a_1 \le - 15018.1781$$

$$+ 12.7062 \cdot 1.19148 \cdot \sqrt{49512.9772}, \, -20431.2 \frac{{}^\circ\text{C}}{\text{m}} \le a_1 \le - 9605.2 \frac{{}^\circ\text{C}}{\text{m}}.$$

Considering that

$$a_1 = -15018.18 \pm \frac{-9604.4 - (-20431.9)}{2} = -15018.18 \pm 5413.75 \frac{{}^\circ\text{C}}{\text{m}}$$

it is possible to determine the heat flux value:

$$\dot{q} = -\lambda a_1 = -18.31 \cdot (-15018.18 \pm 5413.75) = 274982.9 \pm 99125.9 \frac{\text{W}}{\text{m}^2}.$$

95% confidence intervals can also be determined for function η using formula (11.78) in points x_1, x_2 and x_3 or formula (11.90) for any x. In both cases the application of the weighted least squares method should be taken into account, calculating c'_{ii} from formula (11.80) and the confidence intervals for function $\tilde{\eta}$ from the modified formula (11.90):

$$c''_{ii} = \sum_{j=1}^{n} \left(\frac{\partial \tilde{\eta}_i}{\partial y_j} \right)^2 \frac{1}{w_j}, \quad i = 1, \dots, n,$$

$$c''_{ii} = \sum_{j=1}^{n} \left(\frac{\partial \tilde{\eta}}{\partial y_j} \right)^2 \frac{1}{w_j} \quad \text{for any } x.$$

The 95% confidence interval is determined for $\tilde{\eta}_3$. Derivatives $\partial \tilde{\eta}_3 / \partial y_i, i = 1, \dots, 3$ are calculated by means of central difference quotients for $\delta = 0.2$ K:

$$\frac{\partial \tilde{\eta}_3}{\partial y_1} = \frac{189.99481 - 190.04156}{0.4} = -0.11688,$$

$$\frac{\partial \tilde{\eta}_3}{\partial y_2} = \frac{190.06493 - 189.97143}{0.4} = 0.23375,$$

$$\frac{\partial \tilde{\eta}_3}{\partial y_3} = \frac{190.19481 - 189.84156}{0.4} = 0.88313.$$

First, c_{33}'' is found using the following formula:

$$
\begin{aligned}
c_{33}'' &= \left(\frac{\partial \tilde{\eta}_3}{\partial y_1}\right)^2 \frac{1}{w_1} + \left(\frac{\partial \tilde{\eta}_3}{\partial y_2}\right)^2 \frac{1}{w_2} + \left(\frac{\partial \tilde{\eta}_3}{\partial y_3}\right)^2 \frac{1}{w_3} \\
&= (-0.11688)^2 \cdot \frac{1}{1} + 0.23375^2 \cdot \frac{1}{0.25} + 0.88313^2 \cdot \frac{1}{0.44444} \\
&= 1.98701,
\end{aligned}
$$

Then, the 95% confidence interval is calculated for η_3:

$$190.01818 - 12.7062 \cdot 1.9148 \cdot \sqrt{1.98701} \leq \eta_3 \leq 190.01818$$
$$+ 12.7062 \cdot 1.9148 \cdot \sqrt{1.98701}, 155.72\ °C \leq \eta_3 \leq 224.31\ °C.$$

The 95% confidence intervals for η_i, $i = 1, 2$ are found in a similar way, which gives:

$$286.48\ °C \leq \eta_1 \leq 333.85\ °C, 230.10\ °C \leq \eta_2 \leq 270.08\ °C.$$

The function $\tilde{\eta}(x)$, the measuring points y_i, $i = 1,\ldots,3$ and the limits of the 95% confidence interval determined from formula (11.90) are depicted in Fig. 11.6.

Analysing the obtained results, it can be seen that the 95% confidence intervals for coefficients a_0 and a_1 and for function $\eta(x)$ are wide due to the small number of the measuring points: $n = 3$.

Case (d)

This example presents an analysis of the impact of weight coefficients w_i on obtained results [300]. The weight coefficients are determined using the following relationship:

$$w_i = \frac{s_y^2}{\sigma_i^2}, \quad i = 1, 2, 3.$$

Assuming $s_y^2 = 3.6655\ K^2$ obtained in Case (c) and adopting variances the same as in Case (a): $\sigma_1^2 = 1\ K^2, \sigma_2^2 = 2^2 K^2 = 4\ K^2, \sigma_3^2 = 1.5^2 = 2.25\ K^2$, the weight coefficients total:

Fig. 11.6 Straight-line approximation of temperatures measured in three points inside the sensor by means of the least squares method

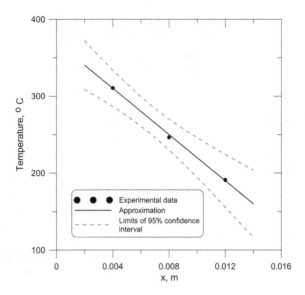

$$w_1 = s_y^2/\sigma_1^2 = 3.6655/1^2 = 3.6655,$$
$$w_2 = s_y^2/\sigma_2^2 = 3.6655/2^2 = 0.9164,$$
$$w_3 = s_y^2/\sigma_3^2 = 3.6655/1.5^2 = 1.6291.$$

Using the weighted least squares method, the following is obtained:

$$\tilde{a}_0 = 370.24 \,^\circ\mathrm{C},$$
$$\tilde{a}_1 = -15018.18 \frac{^\circ\mathrm{C}}{\mathrm{m}}.$$

Also the 95% confidence intervals for coefficients a_0 and a_1

$$329.49 \,^\circ\mathrm{C} < a_0 < 410.98 \,^\circ\mathrm{C}, \; -20431.2 \frac{^\circ\mathrm{C}}{\mathrm{m}} < a_1 < -9605.2 \frac{^\circ\mathrm{C}}{\mathrm{m}},$$

and for temperatures η_1, η_2 and η_3

$$286.48 \,^\circ\mathrm{C} < \eta_1 < 333.85 \,^\circ\mathrm{C},$$
$$230.10 \,^\circ\mathrm{C} < \eta_2 < 270.08 \,^\circ\mathrm{C},$$
$$155.73 \,^\circ\mathrm{C} < \eta_3 < 224.31 \,^\circ\mathrm{C}$$

are the same as in Case (c).

The convergence of the results obtained for Case (c) and Case (d) comes as no surprise, considering the equality of the quotients of weight coefficients w_2/w_1 and w_3/w_1 in the two cases.

The least squares method is discussed in [1, 25, 29, 35, 38, 70, 73, 74, 82, 87, 95, 96, 174, 215, 239, 243, 261, 288, 292, 293, 338, 356, 375]. There are also many commercial computer programs available on the market that enable statistical development of measurement results, e.g. Table Curve, Origin, Statgraphics or Statistica.

11.5   Indirect Measurements

In indirect measurements physical quantities are determined based on quantities which are measured directly. The measurement of the heat flux on the sensor surface based on the values of temperature measured in several points located inside the sensor can serve as an example. The temperatures measured inside the sensor are considered as direct measurement results. Another example is indirect measurement of mean heat transfer coefficients on two sides of the wall based on directly measured physical quantities, which in this case are the mass flow rates of the two mediums and the mediums temperatures measured at the exchanger inlet and outlet.

11.5.1   Indirect Measurements. Linear Problem of the Least Squares Method—Multiple Regression

In indirect measurements, unknown parameters a_j, $j = 0,1,2,...,N$ are found based on measurements of quantity η_j. It is assumed that quantity $\eta = \eta(\mathbf{x}, \mathbf{a})$ is linear

$$\eta = a_0 + a_1 x_1 + a_2 x_2 + \cdots + a_N x_N, \tag{11.95}$$

where $x_1, x_2, \ldots, x_N$ denote independent variables

$$
\begin{aligned}
y_1 &= \eta_1 + \varepsilon_1 = a_0 + a_1 x_{1,1} + a_2 x_{2,1} + \cdots + a_N x_{N,1} + \varepsilon_1, \\
y_2 &= \eta_2 + \varepsilon_2 = a_0 + a_1 x_{1,2} + a_2 x_{2,2} + \cdots + a_N x_{N,2} + \varepsilon_2, \\
&\quad\vdots \qquad\qquad\qquad \vdots \qquad\qquad \vdots \\
y_j &= \eta_j + \varepsilon_j = a_0 + a_1 x_{1,j} + a_2 x_{2,j} + \cdots + a_N x_{N,j} + \varepsilon_j, \\
&\quad\vdots \qquad\qquad\qquad \vdots \qquad\qquad \vdots \\
y_n &= \eta_n + \varepsilon_n = a_0 + a_1 x_{1,n} + a_2 x_{2,n} + \cdots + a_N x_{N,n} + \varepsilon_n.
\end{aligned}
\tag{11.96}
$$

It is also taken into account that the measurement result of quantity y_j is the sum of the true quantity η_j and random measuring error ε_j:

$$y_j = \eta_j + \varepsilon_j, \tag{11.97}$$

where errors ε_j are random numbers with a normal distribution and the mean equal to zero, i.e.:

$$E\left[\varepsilon_j\right] = 0, \tag{11.98}$$

$$E\left[\varepsilon_j^2\right] = \sigma_j^2. \tag{11.99}$$

Defining vectors $\boldsymbol{\eta}$, $\boldsymbol{\varepsilon}$, $\mathbf{y}$ and $\mathbf{a}$

$$\boldsymbol{\eta} = \left\{\begin{array}{c} \eta_1 \\ \eta_2 \\ \vdots \\ \eta_n \end{array}\right\}, \boldsymbol{\varepsilon} = \left\{\begin{array}{c} \varepsilon_1 \\ \varepsilon_2 \\ \vdots \\ \varepsilon_n \end{array}\right\}, \mathbf{y} = \left\{\begin{array}{c} y_1 \\ y_2 \\ \vdots \\ y_n \end{array}\right\}, \mathbf{a} = \left\{\begin{array}{c} a_1 \\ a_2 \\ \vdots \\ a_n \end{array}\right\} \tag{11.100}$$

and the coefficient matrices $\mathbf{X}$ and covariance matrix $\mathbf{C_y}$

$$\mathbf{X} = \begin{bmatrix} x_{1,1} & x_{2,1} & \cdots & x_{N,1} \\ x_{1,2} & x_{2,2} & \cdots & x_{N,2} \\ \vdots & \vdots & \ddots & \vdots \\ x_{1,n} & x_{2,n} & \cdots & x_{N,n} \end{bmatrix}, \mathbf{C_y} = \mathbf{C_\varepsilon} = \begin{bmatrix} \sigma_1^2 & & & 0 \\ & \sigma_2^2 & & \\ & & \ddots & \\ 0 & & & \sigma_n^2 \end{bmatrix}, \tag{11.101}$$

the system of Eq. (11.97) can be written as:

$$\mathbf{y} = \boldsymbol{\eta} + \boldsymbol{\varepsilon}, \tag{11.102}$$

where:

$$\boldsymbol{\eta} = \mathbf{X}\mathbf{a}. \tag{11.103}$$

Coefficients $\tilde{a}_0, \tilde{a}_1, \ldots \tilde{a}_N$ are determined from the necessary condition for the existence of the minimum of function S defined by formula (11.59). Substituting (11.102) in (11.59), the necessary condition for the existence of the function minimum gives:

$$\frac{\partial S}{\partial \tilde{\mathbf{a}}} = -\mathbf{X}^{\mathrm{T}}\mathbf{C_y}^{-1}(\mathbf{y} - \mathbf{X}\mathbf{a}) - (\mathbf{y} - \mathbf{X}\mathbf{a})^{\mathrm{T}}\mathbf{C_y}^{-1}\mathbf{X} = 0. \tag{11.104}$$

Considering that both S and $\partial S/\partial \tilde{\mathbf{a}}$ are scalar quantities for which the relation $e = e^{\mathrm{T}}$ is true, and using the matrix transposition rules:

$$(\mathbf{A_1}\mathbf{A_2})^{\mathrm{T}} = \mathbf{A_2^T}\mathbf{A_1^T}, \tag{11.105}$$

$$(\mathbf{A_1}\mathbf{A_2}\ldots\mathbf{A_n})^{\mathrm{T}} = \mathbf{A_n^T}\ldots\mathbf{A_2^T}\mathbf{A_1^T} \tag{11.106}$$

the second term in derivative (11.104) can be written as:

$$
\begin{aligned}
(\mathbf{y} - \mathbf{X\tilde{a}})^{\mathrm{T}} \mathbf{C}_y^{-1} \mathbf{X} &= \left[(\mathbf{y} - \mathbf{X\tilde{a}})^{\mathrm{T}} \mathbf{C}_y^{-1} \mathbf{X} \right]^{\mathrm{T}} \\
&= \mathbf{X}^{\mathrm{T}} \left(\mathbf{C}_y^{-1} \right)^{\mathrm{T}} \left[(\mathbf{y} - \mathbf{X\tilde{a}})^{\mathrm{T}} \right]^{\mathrm{T}} = \mathbf{X}^{\mathrm{T}} \mathbf{C}_y^{-1} (\mathbf{y} - \mathbf{X\tilde{a}}).
\end{aligned}
\tag{11.107}
$$

It is taken into account in the penultimate constituent of (11.107) that equality $\left(\mathbf{C}_y^{-1} \right)^{\mathrm{T}} = \mathbf{C}_y^{-1}$ holds for diagonal matrix $\mathbf{C}_y^{-1}$. Substitution of (11.107) in (11.104) gives normal equations:

$$
2\mathbf{X}^{\mathrm{T}} \mathbf{C}_y^{-1} (\mathbf{y} - \mathbf{X\tilde{a}}) = 0,
\tag{11.108}
$$

which give the vector of sought parameters

$$
\tilde{\mathbf{a}} = \left(\mathbf{X}^{\mathrm{T}} \mathbf{C}_y^{-1} \mathbf{X} \right)^{-1} \mathbf{X}^{\mathrm{T}} \mathbf{C}_y^{-1} \mathbf{y}.
\tag{11.109}
$$

Substituting (11.109) in (11.103), the following is obtained:

$$
\tilde{\boldsymbol{\eta}} = \mathbf{X} \left[\left(\mathbf{X}^{\mathrm{T}} \mathbf{C}_y^{-1} \mathbf{X} \right)^{-1} \mathbf{X}^{\mathrm{T}} \mathbf{C}_y^{-1} \mathbf{y} \right].
\tag{11.110}
$$

Vector $\tilde{\boldsymbol{\eta}}$ is the vector of corrected measurements.
Vector $\tilde{\boldsymbol{\varepsilon}}$

$$
\tilde{\boldsymbol{\varepsilon}} = \mathbf{y} - \tilde{\boldsymbol{\eta}}
\tag{11.111}
$$

is the estimator of measuring errors (corrections).

In the case of the ordinary least squares method, when all variances $\sigma_i^2 = \sigma^2$, $i = 1,\ldots, n$ are constant, covariance matrix (11.101) has the following form:

$$
\mathbf{C}_y = \begin{bmatrix} \sigma^2 & & & 0 \\ & \sigma^2 & & \\ & & \ddots & \\ 0 & & & \sigma^2 \end{bmatrix} = \sigma^2 \mathbf{I}.
\tag{11.112}
$$

Considering (11.112) in (11.109), the following is obtained:

$$
\tilde{\mathbf{a}} = \left(\mathbf{X}^{\mathrm{T}} \mathbf{X} \right)^{-1} \mathbf{X}^{\mathrm{T}} \mathbf{y}.
\tag{11.113}
$$

Formula (11.110) takes the following form:

$$\tilde{\boldsymbol{\eta}} = \mathbf{X}\left[\left(\mathbf{X}^T\mathbf{X}\right)^{-1}\mathbf{X}^T\mathbf{y}\right]. \tag{11.114}$$

Variances in regression coefficients $\tilde{\mathbf{a}}$ can be calculated determining covariance matrix $\mathbf{C}_{\tilde{\mathbf{a}}}$.

$$\mathbf{C}_{\tilde{\mathbf{a}}} = E\left[(\tilde{\mathbf{a}} - E(\tilde{\mathbf{a}}))(\tilde{\mathbf{a}} - E(\tilde{\mathbf{a}}))^T\right] = E\left[(\tilde{\mathbf{a}} - \mathbf{a})(\tilde{\mathbf{a}} - \mathbf{a})^T\right]. \tag{11.115}$$

Considering that the vector of coefficients $\tilde{\mathbf{a}}$ is defined by relation (11.109):

$$\tilde{\mathbf{a}} = \left(\mathbf{X}^T\mathbf{C}_{\mathbf{y}}^{-1}\mathbf{X}\right)^{-1}\mathbf{X}^T\mathbf{C}_{\mathbf{y}}^{-1}\mathbf{y}$$

and noting that

$$E(\tilde{\mathbf{a}}) = \mathbf{a} = \left(\mathbf{X}^T\mathbf{C}_{\mathbf{y}}^{-1}\mathbf{X}\right)^{-1}\mathbf{X}^T\mathbf{C}_{\mathbf{y}}^{-1}\boldsymbol{\eta} \tag{11.116}$$

and introducing the following relation:

$$\mathbf{T} = \left(\mathbf{X}^T\mathbf{C}_{\mathbf{y}}^{-1}\mathbf{X}\right)^{-1}\mathbf{X}^T\mathbf{C}_{\mathbf{y}}^{-1} \tag{11.117}$$

formula (11.115) can be written as:

$$\begin{aligned}\mathbf{C}_{\tilde{\mathbf{a}}} &= E\left[(\mathbf{T}\mathbf{y} - \mathbf{T}\boldsymbol{\eta})\,(\mathbf{T}\mathbf{y} - \mathbf{T}\boldsymbol{\eta})^T\right] = E\left[\mathbf{T}(\mathbf{y} - \boldsymbol{\eta})\,(\mathbf{y} - \boldsymbol{\eta})^T\mathbf{T}^T\right] \\ &= E\left(\mathbf{T}\,\boldsymbol{\varepsilon}\,\boldsymbol{\varepsilon}^T\mathbf{T}^T\right) = \mathbf{T}\,\mathbf{C}_{\boldsymbol{\varepsilon}}\mathbf{T}^T = \mathbf{T}\,\mathbf{C}_{\mathbf{y}}\mathbf{T}^T.\end{aligned} \tag{11.118}$$

Considering (11.117) in (11.118), the following is obtained:

$$\begin{aligned}\mathbf{C}_{\tilde{\mathbf{a}}} &= \left(\mathbf{X}^T\mathbf{C}_{\mathbf{y}}^{-1}\mathbf{X}\right)^{-1}\mathbf{X}^T\mathbf{C}_{\mathbf{y}}^{-1}\mathbf{C}_{\mathbf{y}}\left[\left(\mathbf{X}^T\mathbf{C}_{\mathbf{y}}^{-1}\mathbf{X}\right)^{-1}\mathbf{X}^T\mathbf{C}_{\mathbf{y}}^{-1}\right]^T \\ &= \left(\mathbf{X}^T\mathbf{C}_{\mathbf{y}}^{-1}\mathbf{X}\right)^{-1}\left[\mathbf{X}^T\mathbf{C}_{\mathbf{y}}^{-1}\mathbf{C}_{\mathbf{y}}\left(\mathbf{C}_{\mathbf{y}}^{-1}\right)^T\mathbf{X}\right]\left[\left(\mathbf{X}^T\mathbf{C}_{\mathbf{y}}\mathbf{X}\right)^{-1}\right]^T \\ &= \left(\mathbf{X}^T\mathbf{C}_{\mathbf{y}}^{-1}\mathbf{X}\right)^{-1}\left(\mathbf{X}^T\mathbf{C}_{\mathbf{y}}^{-1}\mathbf{X}\right)^T\left[\left(\mathbf{X}^T\mathbf{C}_{\mathbf{y}}\mathbf{X}\right)^{-1}\right]^T,\end{aligned} \tag{11.119}$$

from which it follows that

$$\mathbf{C}_{\tilde{\mathbf{a}}} = \left(\mathbf{X}^T\mathbf{C}_{\mathbf{y}}^{-1}\mathbf{X}\right)^{-1}. \tag{11.120}$$

Transforming expression (11.119), it is taken into account that $\mathbf{C}_y^{-1} = (\mathbf{C}_y^{-1})^T$ because matrix $\mathbf{C}_y^{-1}$ is diagonal. If the variances in matrix $\mathbf{C}_y$ are identical, covariance matrix $\mathbf{C}_y^{-1}$ is defined by formula (11.112) and matrix (11.120) is defined as:

$$\mathbf{C}_{\tilde{\mathbf{a}}} = \sigma^2 (\mathbf{X}^T \mathbf{X})^{-1}. \tag{11.121}$$

Diagonal element c_{ii} of matrix (11.120) or (11.121) is equal to the variance in coefficient a_i, $i = 1,...,N$. The values of c_{ii} are used to determine confidence intervals of regression coefficients a_i.

In order to find covariance matrix $\mathbf{C}_{\tilde{\boldsymbol{\eta}}}$, expression (11.110) can be written in the following form:

$$\tilde{\boldsymbol{\eta}} = \mathbf{X}\tilde{\mathbf{a}}, \tag{11.122}$$

where vector $\tilde{\mathbf{a}}$ is defined by formula (11.109).

Covariance matrix $\mathbf{C}_{\tilde{\boldsymbol{\eta}}}$ is found from the following formula:

$$\mathbf{C}_{\tilde{\boldsymbol{\eta}}} = E\left[(\tilde{\boldsymbol{\eta}} - \boldsymbol{\eta}) (\tilde{\boldsymbol{\eta}} - \boldsymbol{\eta})^T \right], \tag{11.123}$$

which, after transformations, gives:

$$\begin{aligned} \mathbf{C}_{\tilde{\boldsymbol{\eta}}} &= E\left[(\mathbf{X}\tilde{\mathbf{a}} - \mathbf{X}\mathbf{a}) (\mathbf{X}\tilde{\mathbf{a}} - \mathbf{X}\mathbf{a})^T \right] = E\left[\mathbf{X}(\tilde{\mathbf{a}} - \mathbf{a}) (\tilde{\mathbf{a}} - \mathbf{a})^T \mathbf{X}^T \right] \\ &= \mathbf{X}E[\mathbf{X}(\tilde{\mathbf{a}} - \mathbf{a}) (\tilde{\mathbf{a}} - \mathbf{a})]\mathbf{X}^T. \end{aligned} \tag{11.124}$$

Considering formulae (11.115) and (11.120) in expression (11.124), covariance matrix $\mathbf{C}_{\tilde{\boldsymbol{\eta}}}$ is defined as:

$$\mathbf{C}_{\tilde{\boldsymbol{\eta}}} = \mathbf{X}\mathbf{C}_{\tilde{\mathbf{a}}}\mathbf{X}^T. \tag{11.125}$$

Considering formula (11.120) in expression (11.125), the following is obtained:

$$\mathbf{C}_{\tilde{\boldsymbol{\eta}}} = \mathbf{X}\left(\mathbf{X}^T \mathbf{C}_y^{-1} \mathbf{X} \right)^{-1} \mathbf{X}^T. \tag{11.126}$$

If variances $\sigma_i^2 = \sigma^2$ are constant, $i = 1,...,n$, formula (11.126) takes the following form:

$$\mathbf{C}_{\tilde{\boldsymbol{\eta}}} = \sigma^2 \mathbf{X}(\mathbf{X}^T \mathbf{X})^{-1} \mathbf{X}^T. \tag{11.127}$$

The square roots of diagonal elements are "residual errors" $\Delta\eta_i$ [35]. If the number of measuring points n is much bigger than the number of sought parameters $(N + 1)$, residual errors are much smaller compared to measuring errors σ_i, i.e.: $\Delta\eta_i < \sigma_i$, $i = 1,...,n$.

Therefore, it may be assumed that owing to the least squares method application, the original measurement results y_i, $i = 1,...,n$ are corrected,

The determined function $\tilde{\eta}(\tilde{\mathbf{a}})$ is close to the true relation $\eta(\mathbf{a})$ if the number of sought parameters $(N + 1)$ is much smaller than the number of measurements n and if measuring errors ε_i are random.

Knowing matrices $\mathbf{C}_{\tilde{\mathbf{a}}}$ and $\mathbf{C}_{\tilde{\eta}}$ defined by formulae (11.120) and (11.126), respectively, the 68 and 95% confidence intervals can be defined for a_i and η_i:

$$\tilde{a}_i - \sqrt{c_{ii}^a} \leq a_i \leq \tilde{a}_i + \sqrt{c_{ii}^a} \quad \text{dla } P = 68\%,$$
$$\tilde{a}_i - 2\sqrt{c_{ii}^a} \leq \eta_i \leq \tilde{a}_i + 2\sqrt{c_{ii}^a} \quad \text{dla } P = 95\%,$$

(11.128)

where c_{ii} are diagonal elements of matrix $\mathbf{C}_{\tilde{\mathbf{a}}}$ defined by formula (11.120). The respective confidence intervals for η_i are as follows:

$$\tilde{\eta}_i - \sqrt{c_{ii}^\eta} \leq \eta_i \leq \tilde{\eta}_i + \sqrt{c_{ii}^\eta} \quad \text{dla } P = 68\%,$$
$$\tilde{\eta}_i - 2\sqrt{c_{ii}^\eta} \leq \eta_i \leq \tilde{\eta}_i + 2\sqrt{c_{ii}^\eta} \quad \text{dla } P = 95\%,$$

(11.129)

where c_{ii}^η are diagonal elements of matrix $\mathbf{C}_{\tilde{\eta}}$ (11.126). If variances $\sigma_i = \sigma_1 = ...\sigma_n = \sigma$ are constant, formulae (11.128) and (11.129) are reduced.

For η_i, formulae (11.129) take the following form:

$$\tilde{\eta}_i - \sigma\sqrt{c_{ii}^{\prime\prime}} < \eta_i < \tilde{\eta}_i + \sigma\sqrt{c_{ii}^{\prime\prime}} \quad \text{dla } P = 68\%$$

(11.130)

and

$$\tilde{\eta}_i - 2\sigma\sqrt{c_{ii}^{\prime\prime}} < \eta_i < \tilde{\eta}_i + 2\sigma\sqrt{c_{ii}^{\prime\prime}} \quad \text{dla } P = 95\%,$$

(11.131)

where $c_{ii}^{\prime\prime}$ are diagonal elements of matrix

$$\mathbf{C}_{\eta}^{\prime\prime} = \frac{\mathbf{C}_{\tilde{\eta}}}{\sigma^2} = \mathbf{X}(\mathbf{X}^T\mathbf{X})^{-1}\mathbf{X}^T.$$

The above-described procedure for determining confidence intervals may be used if variances σ_i^2 are known. In the case of the ordinary least squares method (11.57) or the weighted method (11.65) with weights w_i which are not the inverse of variance σ_i^2, the confidence interval determination method for $\tilde{a}_i$, $i = 0,...,N$ and η_i, $i = 1,...,n$ will be discussed separately.

In practical calculations it is often the case that due to a one-off measurement of y_i, $i = 1,...,n$, variances σ_i^2 are unknown. Consequently, covariance matrix $\mathbf{C_y}$ is unknown, too. A constant value of variance $\sigma_i^2 = \sigma^2$, $i = 1,...,n$ is then usually assumed for all measured values y_i. Considering that the weighted minimum sum of squares totals:

$$S(\tilde{\mathbf{a}}) = \sum_{i=1}^{n} w_i [y_i - \eta_i(\tilde{\mathbf{a}})]^2, \qquad (11.132)$$

where $w_i \geq 0$ are any selected weight coefficients, variance σ^2 is approximated using mean square error s_y:

$$\sigma \cong s_y = \sqrt{\frac{\sum_{i=1}^{n} w_i [y_i - \eta_i(\tilde{a})]^2}{n - N - 1}}. \qquad (11.133)$$

For the ordinary least squares method, in formulae (11.132) and (11.133) it is assumed that $w_i = 1$, $i = 1,...,n$. If random errors $\varepsilon_i = (y_i - \eta_i)$ have a normal distribution $N(0, \sigma^2)$, i.e. if the mean of error ε_i is zero and the variance is equal to σ^2, random variables have a t-Student distribution with $(n - N - 1)$ degrees of freedom. In order to calculate the confidence interval of coefficients $\tilde{a}_i$ determined by means of the least squares method as defined in (11.65), the following can be written:

$$t = \frac{\tilde{a}_i - a_i}{s_y \sqrt{c'_{ii}}} \cong t_{n-N-1,\alpha}, \qquad (11.134)$$

where t—t-Student random variable, a_i—expected true value of a_i, $\tilde{a}_i$—a_i value found by means of the least squares method.

In the denominator, c'_{ii} is the i-th diagonal element of covariance matrix $\mathbf{C}_{\tilde{a}}$ defined by formula (11.120), where variance σ is approximated using formula (11.133). The true values of coefficients a_i are included with probability $P = (1 - \alpha)100\%$ in the interval:

$$\tilde{a}_i - t_{n-N-1,\alpha}s_y\sqrt{c'_{ii}} \leq a_i \leq \tilde{a}_i + t_{n-N-1,\alpha}s_y\sqrt{c'_{ii}}, \quad i = 1,\ldots,n, \qquad (11.135)$$

where c'_{ii} is the i-th diagonal element of the matrix defined by formula (11.120), where covariance matrix $\mathbf{C}_y^{-1}$ is replaced by weight matrix $\mathbf{G}_y$:

$$\mathbf{C}_{\tilde{a}} = \left(\mathbf{X}^T\mathbf{G}_y\mathbf{X}\right)^{-1}. \qquad (11.136)$$

The t-Student distribution quantiles are given in Table 11.2.

A similar procedure can be used to calculate the confidence interval for η_i, according to the following formula:

$$\tilde{\eta}_i - t_{n-N-1,\alpha}s_y\sqrt{c''_{ii}} \leq \eta_i \leq \tilde{\eta}_i + t_{n-N-1,\alpha}s_y\sqrt{c''_{ii}}, \quad i = 1,\ldots,n, \qquad (11.137)$$

where c''_{ii} are diagonal elements of matrix

$$\mathbf{C}_{\tilde{\eta}} = \mathbf{X}(\mathbf{X}^{\mathrm{T}}\mathbf{G_y}\mathbf{X})^{-1}\mathbf{X}^{\mathrm{T}}. \tag{11.138}$$

The matrix is very similar to the one defined by expression (11.126). The difference is that matrix $\mathbf{C}_{\mathbf{y}}^{-1}$ in (11.126) is replaced by weight matrix $\mathbf{G_y}$ in (11.138). The method of the confidence interval determination not only for η_i, $i = 1,...,n$ but also for other values of η is presented below. In order to establish the confidence area of the linear function $\tilde{\eta}$ where coefficients $\tilde{a}$ are found using the least squares method:

$$\tilde{\eta} = \tilde{a}_0 + \tilde{a}_1 x_1 + \tilde{a}_2 x_2 + \cdots + \tilde{a}_N x_N = \mathbf{x}^{\mathrm{T}}\tilde{a}, \tag{11.139}$$

where $\mathbf{x}^{\mathrm{T}} = (1, x_1, x_2, \ldots, x_N)$, the confidence interval for function $\tilde{\eta}$ was calculated at one of the points with the following coordinates:

$$\left(\mathbf{x}^0\right)^{\mathrm{T}} = \begin{bmatrix} 1 \\ x_1^0 \\ x_2^0 \\ \vdots \\ x_N^0 \end{bmatrix} = \left(1, x_1^0, x_2^0, \ldots, x_N^0\right)^{\mathrm{T}}.$$

The value of function $\tilde{\eta}$ in point $\left(\mathbf{x}^0\right)^{\mathrm{T}}$, marked as $\tilde{\eta}^0$, is:

$$\tilde{\eta}^0 = (\tilde{\mathbf{a}})^{\mathrm{T}}\mathbf{x}^0 = \left(\mathbf{x}^0\right)^{\mathrm{T}}\tilde{\mathbf{a}}, \tag{11.140}$$

where:

$$\tilde{\mathbf{a}} = (\tilde{a}_0, \tilde{a}_1, \tilde{a}_2, \ldots, \tilde{a}_N)^{\mathrm{T}}.$$

It is assumed that regression coefficients $\tilde{\mathbf{a}} = (\tilde{a}_0, \tilde{a}_1, \tilde{a}_2, \ldots, \tilde{a}_N)^{\mathrm{T}}$ are determined by means of the least squares method and are known. The expected values of regression coefficients $\mathbf{a} = (a_0, a_1, a_2, a_3, \ldots, a_N)^{\mathrm{T}}$ are also known. It is now possible to calculate the approximation function true value from the following relation:

$$\eta = \mathbf{x}^{\mathrm{T}}\mathbf{a}.$$

Variance var[η] is:

$$\begin{aligned} \mathrm{var}\left[\tilde{\eta}^0\right] &= E\left\{\left[\tilde{\eta}^0 - E\left(\tilde{\eta}^0\right)\right]\left[\tilde{\eta}^0 - E\left(\tilde{\eta}^0\right)\right]^{\mathrm{T}}\right\} \\ &= E\left\{\left[\left(\mathbf{x}^0\right)^{\mathrm{T}}\tilde{\mathbf{a}} - \left(\mathbf{x}^0\right)^{\mathrm{T}}\mathbf{a}\right]\left[\left(\mathbf{x}^0\right)^{\mathrm{T}}\tilde{\mathbf{a}} - \left(\mathbf{x}^0\right)^{\mathrm{T}}\mathbf{a}\right]^{\mathrm{T}}\right\} \\ &= \left(\mathbf{x}^0\right)^{\mathrm{T}} E\left\{(\tilde{\mathbf{a}} - \mathbf{a})(\tilde{\mathbf{a}} - \mathbf{a})^{\mathrm{T}}\right\}\mathbf{x}^0. \end{aligned} \tag{11.141}$$

Considering covariance matrix $\mathbf{C_{\tilde{a}}}$ defined by formula (11.120) and noting that $\mathbf{C_y^{-1}} = \mathbf{G_y}$, expression (11.141) gives:

$$\text{var}[\tilde{\eta}_0] = (\mathbf{x}^0)^{\mathrm{T}} \mathbf{C_{\tilde{a}}} \mathbf{x}^0. \tag{11.142}$$

Considering (11.120) in (11.142) and taking account of the fact that $\sigma \cong s_y$, the following is obtained from (11.142):

$$\text{var}[\tilde{\eta}_0] = s_y^2 (\mathbf{x}^0)^{\mathrm{T}} (\mathbf{X}^{\mathrm{T}} \mathbf{G_y} \mathbf{X})^{-1} \mathbf{x}^0. \tag{11.143}$$

Knowing variance var$[\eta]$, it is possible to determine the confidence interval for the found simple regression $\tilde{\eta}$ in point $\mathbf{x}^0$:

$$\tilde{\eta}_0 - t_{n-N-1,\alpha} s_y \sqrt{(\mathbf{x}^0)^{\mathrm{T}} (\mathbf{X}^{\mathrm{T}} \mathbf{G_y} \mathbf{X})^{-1} \mathbf{x}^0} \le \eta_0 \le \tilde{\eta}_0$$
$$+ t_{n-N-1,\alpha} s_y \sqrt{(\mathbf{x}^0)^{\mathrm{T}} (\mathbf{X}^{\mathrm{T}} \mathbf{G_y} \mathbf{X})^{-1} \mathbf{x}^0}, \tag{11.144}$$

where s_y is defined by formula (11.133). It should be remembered that η_0 denotes the true value of function η in point $\mathbf{x}^0$.

Next, the prediction (forecast) interval was determined for quantity y measured in point $\mathbf{x} = \mathbf{x}^0$, in which the measured quantity y_0 is included with probability P. According to formula (11.102), the following is obtained:

$$y_0 = \eta(\mathbf{a}, \mathbf{x}^0) + \varepsilon_0, \tag{11.145}$$

where $\varepsilon_0 \approx N(0, \sigma^2)$ is independent of the vector of random errors ε. For a large number of measuring points n it may be assumed that the value of function $\tilde{\eta}_0 = \eta(\tilde{\mathbf{a}}, \mathbf{x}^0)$ determined by means of the least squares method represents the true function $\eta(\tilde{\mathbf{a}}, \mathbf{x}^0)$. It may therefore be written that

$$E[y_0 - \tilde{\eta}_0] = E[y_0 - \eta(\mathbf{x}^0, \tilde{\mathbf{a}})] = E[\eta(\mathbf{x}^0, \mathbf{a}) + \varepsilon_0 - \eta(\mathbf{x}^0, \tilde{\mathbf{a}})] \cong 0. \tag{11.146}$$

Similarly:

$$\text{var}[y_0 - \tilde{\eta}_0] = \text{var}[\eta(\mathbf{x}^0, \mathbf{a}) + \varepsilon_0 - \eta(\mathbf{x}^0, \tilde{\mathbf{a}})]$$
$$= \text{var}[\varepsilon_0] + \text{var}[\eta(\mathbf{x}^0, \mathbf{a}) - \eta(\mathbf{x}^0, \tilde{\mathbf{a}})] \tag{11.147}$$

which, taking account of (11.140), results in:

$$\text{var}[y_0 - \tilde{\eta}_0] = \text{var}[\varepsilon_0] + \text{var}\left[(\tilde{\mathbf{x}}^0)^{\mathrm{T}} (\mathbf{a} - \tilde{\mathbf{a}})\right] \cong \sigma_0^2 + (\tilde{\mathbf{x}}^0)^{\mathrm{T}} \mathbf{C_{\tilde{a}}} \tilde{\mathbf{x}}^0. \tag{11.148}$$

If $\sigma^2 = \sigma_0^2$ is constant, formula (11.148) is reduced to the following form (cf. Appendix E):

$$\text{var}[y_0 - \tilde{\eta}_0] = \sigma^2 \left[1 + \left(\mathbf{x}^0\right)^T \left(\mathbf{X}^T \mathbf{G_y} \mathbf{X}\right)^{-1} \mathbf{x}^0\right]. \tag{11.149}$$

For a great number of measurements n, the following occurs in approximation $\left(\sigma \cong s_y\right)$:

$$\frac{y_0 - \tilde{\eta}_0}{s_y \sqrt{1 + \left(\mathbf{x}^0\right)^T \left(\mathbf{X}^T \mathbf{G_y} \mathbf{X}\right)^{-1} \mathbf{x}^0}} \cong t_{n-N-1,\alpha}. \tag{11.150}$$

Assuming the confidence level $P = (1 - \alpha) \, 100\%$, the y_0 prediction interval is defined as:

$$\tilde{\eta}_0 - t_{n-N-1,\alpha} s_y \sqrt{1 + \left(\mathbf{x}^0\right)^T \left(\mathbf{X}^T \mathbf{G_y} \mathbf{X}\right)^{-1} \mathbf{x}^0} \le y_0 \le \tilde{\eta}_0$$
$$+ t_{n-N-1,\alpha} s_y \sqrt{1 + \left(\mathbf{x}^0\right)^T \left(\mathbf{X}^T \mathbf{G_y} \mathbf{X}\right)^{-1} \mathbf{x}^0}, \tag{11.151}$$

where s_y is defined by formula (11.66) or (11.133).

The t-Student distribution quantiles for $t_{n-N-1,\alpha}$ are listed in Table 11.2. Using formula (11.151), it is possible to determine the prediction area in which measuring quantities y_i, $i = 1,2,\ldots,n$ are included with probability P. Comparing (11.151) with (11.144), it can be seen that the y_0 prediction interval is wider than the η_0 confidence interval.

As mentioned before, mean square error s_y is usually found from formula (11.66), which for the ordinary least squares method, when $w_i = 1, i = 1,\ldots,n$, takes the following form:

$$s_y = \sqrt{\frac{\sum_{i=1}^n \left[y_i - \eta_i(\tilde{\mathbf{a}})\right]^2}{n - N - 1}}. \tag{11.152}$$

A different way of deriving the prediction interval formula (11.151) for the measured value y_0 is presented in Appendix E.

Example 11.3 The heat flux value is determined using the sensor presented in Fig. 11.4, where the metal temperature is measured at three and not at two points. The sensor thickness is $L = 0.016$ m. The sensor material thermal conductivity totals 18.31 W/(m K). The thermocouples measuring the sensor temperature are fixed at the following three points: $x_1 = d_1 = 0.004$ m; $x_2 = d_2 = 0.008$ m; $x_3 = d_3 = 0.012$ m. In order to perform testing calculations, "measuring data" was generated using an analytical exact solution for the following data: $\dot{q} = 274800$ W/m², the temperature of the medium flowing along the sensor rear surface $T_{cz} = 15$ °C, the heat transfer coefficient on the sensor rear surface $\alpha_w = 2400$ W/(m² K). Considering that the temperature distribution on the sensor thickness is defined by the following function:

$$T(x) = \frac{-\dot{q}}{\lambda}x + \dot{q}\left(\frac{L}{\lambda} + \frac{1}{\alpha}\right) + T_{cz},$$

after substitution of relevant data, the result is:

$$T(x) = -15008.19224x + 369.63108.$$

First, the wall temperature in points x_1, x_2 and x_3 is found by means of the obtained formula (cf. Table 11.4).

Then "measuring errors": $\varepsilon_1 = 1.0017$ °C, $\varepsilon_2 = -2.96554$ °C and $\varepsilon_3 = 1.4672$ °C are added to so-obtained "exact measuring data": $\eta(x_1) = T(x_1) = 309.5983$ °C, $\eta(x_2) = T(x_2) = 249.5655$ °C and $\eta(x_3) = T(x_3) = 189.5328$ °C, which gives:

$$x_1 = d_1 = 0.004 \text{ m}; y_1 = \eta(x_1) + \varepsilon_1 = 310.6 \text{ °C},$$
$$x_2 = d_2 = 0.008 \text{ m}; y_2 = \eta(x_2) + \varepsilon_2 = 246.6 \text{ °C},$$
$$x_3 = d_3 = 0.012 \text{ m}; y_3 = \eta(x_3) + \varepsilon_3 = 191.0 \text{ °C}.$$

Like in Case (c) in Example 11.2, it is assumed that weights $w_1 = 1/\sigma_1^2 = 1$ K^2, $w_2 = 1/\sigma_2^2 = 0.25$ K^2, $w_3 = 1/\sigma_3^2 = 0.444(4)$ K^2, where $\sigma_1^2 = 1$ K^2, $\sigma_2^2 = 4$ K^2 and $\sigma_3^2 = 2.25$ K^2 are not the variance inverse, i.e. the variances in temperatures y_1, y_2 and y_3 measured in points x_1, x_2 and x_3 are unknown. Because the smallest weight is adopted for the second measuring point, the reliability of the temperature measurement at this point is assumed as the lowest.

Table 11.4 Temperature distribution on the sensor thickness

No.	$x = d$ (m)	$\eta = T$ (°C)
1.	0.000	369.6311
2.	0.001	354.6229
3.	0.002	339.6147
4.	0.003	324.6065
5.	0.004	309.5983
6.	0.005	294.5901
7.	0.006	279.5819
8.	0.007	264.5737
9.	0.008	249.5655
10.	0.009	234.5574
11.	0.010	219.5492
12.	0.011	204.5410
13.	0.012	189.5328
14.	0.013	174.5246
15.	0.014	159.5164
16.	0.015	144.5082
17.	0.016	129.5000

The temperature distribution was approximated by means of the following linear function:

$$\eta = a_0 + a_1 x.$$

Matrix $\mathbf{C_y}$ has the following form:

$$\mathbf{C_y} = \begin{bmatrix} 1 & 0 & 0 \\ 0 & 4 & 0 \\ 0 & 0 & 2.25 \end{bmatrix}.$$

Considering that

$$\mathbf{X} = \begin{bmatrix} 1 & 0.004 \\ 1 & 0.008 \\ 1 & 0.012 \end{bmatrix}, \quad \mathbf{y} = \begin{Bmatrix} 310.6 \\ 246.6 \\ 191.0 \end{Bmatrix},$$

the following is obtained:

$$\tilde{\mathbf{a}} = \begin{Bmatrix} 370.2365 \\ -15018.1781 \end{Bmatrix}.$$

The vector of corrected measurements is:

$$\tilde{\boldsymbol{\eta}} = \mathbf{X}\tilde{\mathbf{a}} = \begin{bmatrix} 1 & 0.004 \\ 1 & 0.008 \\ 1 & 0.012 \end{bmatrix} \begin{Bmatrix} 370.2365 \\ -15018.1781 \end{Bmatrix} = \begin{Bmatrix} 310.163 \\ 250.091 \\ 190.018 \end{Bmatrix}.$$

Knowing $\tilde{\boldsymbol{\eta}}$ and $\mathbf{y}$, it is possible to estimate measuring errors (corrections).

$$\tilde{\boldsymbol{\varepsilon}} = \mathbf{y} - \tilde{\boldsymbol{\eta}} = \begin{Bmatrix} 310.6 \\ 246.6 \\ 191.0 \end{Bmatrix} - \begin{Bmatrix} 310.163 \\ 250.091 \\ 190.018 \end{Bmatrix} = \begin{Bmatrix} 0.437 \\ -3.491 \\ 0.982 \end{Bmatrix}.$$

Considering that

$$\tilde{a}_1 = -\frac{\tilde{q}}{\lambda},$$

the heat flux value is:

$$\tilde{q} = -\lambda \tilde{a}_1 = -18.31 \cdot (-15018.1781),$$
$$\tilde{q} = 274982.84 \ \text{W/m}^2.$$

Covariance matrix $\mathbf{C_{\tilde{a}}}$ of determined coefficients $\tilde{a}$ has the following form:

$$\mathbf{C_{\tilde{a}}} = \left(\mathbf{X^T G_y X}\right)^{-1}$$

$$= \left\{ \begin{bmatrix} 1 & 1 & 1 \\ 0.004 & 0.008 & 0.012 \end{bmatrix} \begin{bmatrix} 1 & 0 & 0 \\ 0 & 0.25 & 0 \\ 0 & 0 & 0.44(4) \end{bmatrix} \begin{bmatrix} 1 & 0.004 \\ 1 & 0.008 \\ 1 & 0.012 \end{bmatrix} \right\}^{-1}$$

$$= \begin{bmatrix} 2.8052 & -331.1688 \\ -331.1688 & 49512.9772 \end{bmatrix}.$$

Square roots of the diagonal elements of matrix $\mathbf{C_{\tilde{a}}}$ total:

$$\sqrt{c'_{11}} = \sqrt{2.8052} = 1.6749 \text{ K},$$

$$\sqrt{c'_{22}} = \sqrt{49412.9772} = 222.5151 \ \frac{K}{m}.$$

Confidence intervals for $P = 95\%$ can be calculated using formula (11.135), which gives:

$$370.2365 - 12.7062 \cdot 1.9148 \cdot \sqrt{2.8052} \le a_0 \le 370.2365$$

$$+ 12.7062 \cdot 1.19148 \cdot \sqrt{2.8052}, 329.49 \ ^{\circ}\text{C} \le a_0 \le 410.98 \ ^{\circ}\text{C},$$

$$- 15018.1781 - 12.7062 \cdot 1.9148 \cdot \sqrt{49512.9772} \le a_1 \le -15018.1781$$

$$+ 12.7062 \cdot 1.19148 \cdot \sqrt{49512.9772}, -20431.2 \frac{^{\circ}\text{C}}{m} \le a_1 \le -9605.2 \frac{^{\circ}\text{C}}{m}.$$

Comparing the obtained results, it can be seen that they are the same as in Case (c) in Example 11.2. The mean square deviation calculated from formula (11.133) is:

$$s_y = \sqrt{\frac{1 \cdot (0.437)^2 + 0.25 \cdot (-3.491)^2 + 0.4444 \cdot (0.982)^2}{3 - 1 - 1}} = 1.9148 \text{ K}.$$

Matrix $\mathbf{C_{\tilde{\eta}}}$ defined by formula (11.126) has the following form:

$$\mathbf{C_{\tilde{\eta}}} = \begin{bmatrix} 0.94805 & 0.41558 & -0.11688 \\ 0.41558 & 0.67532 & 0.93506 \\ -0.11688 & 0.93506 & 1.987011 \end{bmatrix}.$$

Confidence intervals for $P = 95\%$ are calculated using formula (11.144) and they are as follows:

$$310.163 - 12.7062 \cdot 1.9145 \cdot \sqrt{0.94805} < \eta_1 < 310.163$$
$$+ 12.7062 \cdot 1.9145 \cdot \sqrt{0.94805},$$
$$286.48 \,^\circ C < \eta_1 < 333.85 \,^\circ C,$$
$$250.091 - 12.7062 \cdot 1.9145 \cdot \sqrt{0.67532} < \eta_2 < 250.091$$
$$+ 12.7062 \cdot 1.9145 \cdot \sqrt{0.67532},$$
$$230.10 \,^\circ C < \eta_2 < 270.08 \,^\circ C,$$
$$190.018 - 12.7062 \cdot 1.9145 \cdot \sqrt{1.987011} < \eta_3 < 190.018$$
$$+ 12.7062 \cdot 1.9145 \cdot \sqrt{1.987011},$$
$$230.10 \,^\circ C < \eta_3 < 270.08 \,^\circ C.$$

Formula (11.144) can also be used to calculate the confidence intervals for function $\eta(\tilde{\mathbf{a}})$. For point $x_2 = d_2 = 0.008$ m, i.e. for the second measuring point, the following is obtained:

$$\left(\mathbf{x}^0 \right)^T = (1.0 \ 0.008) \,,$$

$$\left(\mathbf{x}^0 \right)^T \left(\mathbf{X}^T \mathbf{G_y} \mathbf{X} \right)^{-1} \mathbf{x}^0 = (1.0 \ 0.008) \begin{bmatrix} 2.8052 & -331.1688 \\ -331.1688 & 19512.9772 \end{bmatrix} \cdot \left\{ \begin{matrix} 1 \\ 0.008 \end{matrix} \right\} = 0.67533.$$

The (95%) confidence interval for $\tilde{\eta}_2$ calculated according to formula (11.144) is:

$$250.091 - 12.7062 \cdot 1.9145 \cdot \sqrt{0.67532} < \eta_2 < 250.091$$
$$+ 12.7062 \cdot 1.9145 \cdot \sqrt{0.67532}, 230.1 \,^\circ C < \eta_2 < 270.08 \,^\circ C.$$

The confidence intervals for η_1 and η_3 for coordinate $x_1 = d_1 = 0.004$ m and $x_3 = d_3 = 0.012$ m, and for any coordinate x_i, could also be calculated identically.

The y_2 prediction interval for $x_2 = d_2 = 0.008$ m found by means of formula (11.151) is:

$$250.091 - 12.7062 \cdot 1.9145 \cdot \sqrt{1 + 0.67533} < y_2 < 250.091$$
$$+ 12.7062 \cdot 1.9145 \cdot \sqrt{1 + 0.67533}, 218.6 \,^\circ C < y_2 < 281.6 \,^\circ C.$$

The same example was calculated using the Table Curve program [293].

Figure 11.7 presents approximation results with marked confidence intervals. Table 11.5 presents the calculation results obtained by means of the Table Curve program, including the 95% confidence interval. The (95%) confidence intervals of parameters $\tilde{a}_0$ and $\tilde{a}_1$ determined by means of the Table Curve program are as follows:

Fig. 11.7 Results of
straight-line approximation of
experimental data: $\tilde{\eta} = \tilde{a}_0 + \tilde{a}_1 x$, $\tilde{a}_0 = 370.236\ °C$;
$\tilde{a}_1 = -15018.182\ °C/m$ with
marked 95% confidence and
prediction intervals

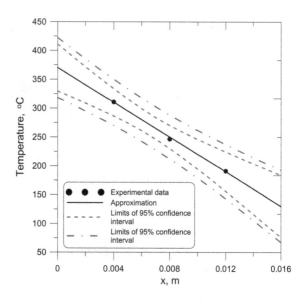

Table 11.5 Results of experimental data approximation by means of the Table Curve program

No.	$x_i = d_i$ (m)	y_i (°C)	Weight coefficients w_i	$\tilde{\eta}_i$ (°C)	$y_i - \tilde{\eta}_i$ (K)
1.	0.0040000	310.60000	1	310.16364	0.4363636
2.	0.0080000	246.60000	0.25	250.09091	−3.490909
3.	0.0120000	191.00000	0.444(4)	190.01818	0.9818182
No.	$100(y_i - \eta_i)/y_i$ %	Confidence interval for η_i (°C)			
1.	0.1404905	286.12743	334.19984		
2.	−1.415616	229.80445	270.37736		
3.	0.5140409	155.22051	224.81585		

$$328.89\ °C \le \tilde{a}_0 \le 411.58\ °C,\ -20511.18\ °C \le \tilde{a}_1 \le\ < -9525.18\ °C/m.$$

Sum of squares $S = \sum_{i=1}^{n} w_i(y_i - \tilde{\eta}_i)^2$ is $S = 3.6655\ K^2$.

It can be seen that the results produced by the Table Curve program are close to
those obtained in the example. The slight differences are the effect of rounding errors.

Variances $\sigma_{\tilde{a}_i}^2$, $i = 0, 1, \ldots, N$ and standard deviations $\sigma_{\tilde{\eta}_i}$, $i = 1, \ldots, n$ as well as
the respective confidence intervals, are calculated by means of the matrix calculus.
However, this is not the only method of calculating these quantities because they
can be determined directly from the variance transposition (error propagation) rule,
which is demonstrated in Example 11.2, where in Case (c) the obtained results are
almost the same as in this example.

Nevertheless, it should be emphasized that at a great number of measuring data, the matrix calculus application enables a fast solution of the least squares problem, especially considering the fact that mathematical operations on matrices can be performed easily using commercial software packages such as Mathcad, Matlab, Mathematica and many others.

11.5.2 Indirect Measurements. Nonlinear Problem of the Least Squares Method

The **y** measuring data in the nonlinear least squares method was approximated using function $\eta(\mathbf{x}, \mathbf{a})$:

$$\mathbf{y} = \eta(\mathbf{x}, \mathbf{a}) + \varepsilon, \tag{11.153}$$

where:

$$\mathbf{y} = \begin{bmatrix} y_1 \\ y_2 \\ \vdots \\ y_{n-1} \\ y_n \end{bmatrix}, \eta = \begin{bmatrix} \eta_1 \\ \eta_2 \\ \vdots \\ \eta_{n-1} \\ \eta_n \end{bmatrix}, \varepsilon = \begin{bmatrix} \varepsilon_1 \\ \varepsilon_2 \\ \vdots \\ \varepsilon_{n-1} \\ \varepsilon_n \end{bmatrix}. \tag{11.154}$$

The sought parameters vector $\mathbf{a} = (a_0, a_1, \ldots, a_N)^{\mathrm{T}}$ was determined from the condition that the sum of squares

$$S(\mathbf{a}) = \sum_{i=1}^{n} w_i[y_i - \eta_i(\mathbf{a})]^2 \tag{11.155}$$

should reach the minimum. Expression (11.155) can be written in the following matrix form:

$$S(\mathbf{a}) = [\mathbf{y} - \eta(\mathbf{a})]^{\mathrm{T}} \mathbf{C}_{\mathbf{y}}^{-1} [\mathbf{y} - \eta(\mathbf{a})], \tag{11.156}$$

where weight matrix $\mathbf{G}_{\mathbf{y}} = \mathbf{C}_{\mathbf{y}}^{-1}$ is defined by formula (11.64). Due to the fact that function $\eta(\mathbf{a})$ is nonlinear, the least squares problem can be solved iteratively, in this case—by means of the Gauss-Newton method. Assuming that solution $\mathbf{a}^{(s)}$ is known, solution $\mathbf{a}^{(s+1)}$ can be estimated in the next iteration step $(s + 1)$. According to the Gauss-Newton method, function $\eta_i(\mathbf{a})$ was expanded into a Taylor series around $\mathbf{a}^{(s)}$. Limiting the considerations to linear terms only, the following is obtained:

$$\eta_i(\mathbf{a}) \cong \eta_i(\mathbf{a}^{(s)}) + \sum_{j=0}^{N} \left. \frac{\partial \eta_i}{\partial a_j} \right|_{\mathbf{a}^{(s)}} \left(a_j - a_j^{(s)} \right), \quad i = 0, \ldots, n, \quad j = 1, \ldots, N. \tag{11.157}$$

Expression (11.157) can be written in the following matrix form:

$$\boldsymbol{\eta}(\mathbf{a}) \cong \boldsymbol{\eta}\left(\mathbf{a}^{(s)}\right) + \mathbf{J}^{(s)}\left(\mathbf{a} - \mathbf{a}^{(s)}\right),$$

(11.158)

where $\mathbf{J}^{(s)}$ is a Jacobian matrix defined as:

$$\mathbf{J}^{(s)} = \left[\frac{\partial \eta_i(\mathbf{a})}{\partial a_j}\right]\Bigg|_{\tilde{\mathbf{a}}^{(s)}}, \quad i = 0, \ldots, n, \quad j = 1, \ldots, N$$

(11.159)

or

$$\mathbf{J}^{(s)} = \begin{bmatrix} \frac{\partial \eta_1}{\partial a_0} & \frac{\partial \eta_1}{\partial a_1} & \cdots & \frac{\partial \eta_1}{\partial a_N} \\ \frac{\partial \eta_2}{\partial a_0} & \frac{\partial \eta_2}{\partial a_1} & \cdots & \frac{\partial \eta_2}{\partial a_N} \\ \vdots & \vdots & & \vdots \\ \frac{\partial \eta_n}{\partial a_0} & \frac{\partial \eta_n}{\partial a_1} & \cdots & \frac{\partial \eta_n}{\partial a_N} \end{bmatrix}_{\mathbf{a}^{(s)}}.$$

(11.160)

Substituting (11.158) in (11.156), the necessary condition for the existence of the minimum of sum $S(\mathbf{a})$ gives a new parameters vector $(\mathbf{a})$, which constitutes the next approximation $\mathbf{a}^{(s+1)}$. According to the necessary condition for the existence of the minimum of function (11.156):

$$\frac{\partial S(\mathbf{a})}{\partial \mathbf{a}} = 0,$$

(11.161)

where:

$$\frac{\partial S}{\partial \mathbf{a}} = \left(\frac{\partial S(\mathbf{a})}{\partial a_0}, \frac{\partial S(\mathbf{a})}{\partial a_1}, \ldots, \frac{\partial S(\mathbf{a})}{\partial a_N}\right)^{\mathrm{T}}.$$

Considering that

$$S(\mathbf{a}) = \left[\mathbf{y} - \boldsymbol{\eta}\left(\mathbf{a}^{(s)}\right) - \mathbf{J}^{(s)}\left(\mathbf{a} - \mathbf{a}^{(s)}\right)\right]^{\mathrm{T}} \mathbf{C}_{\mathbf{y}}^{-1}\left[\mathbf{y} - \boldsymbol{\eta}\left(\mathbf{a}^{(s)}\right) - \mathbf{J}^{(s)}\left(\mathbf{a} - \mathbf{a}^{(s)}\right)\right]$$

(11.162)

the following is obtained:

$$\mathbf{a} = \mathbf{a}^{(s)} + \left[\left(\mathbf{J}^{(s)}\right)^{\mathrm{T}} \mathbf{C}_{\mathbf{y}}^{-1} \mathbf{J}^{(s)}\right]^{-1} \left(\mathbf{J}^{(s)}\right)^{\mathrm{T}} \mathbf{C}_{\mathbf{y}}^{-1}\left[\mathbf{y} - \boldsymbol{\eta}\left(\mathbf{a}^{(s)}\right)\right] = \mathbf{a}^{(s)} + \boldsymbol{\delta}^{(s)}.$$

(11.163)

Considering that the determined vector $\mathbf{a}$ defined by formula (11.163) is the $(s + 1)$ approximation of vector $\tilde{\mathbf{a}}$, which satisfies condition $S(\tilde{\mathbf{a}}) = \min$, it is assumed that $\mathbf{a}^{(s+1)} = \mathbf{a}$. Expression (11.163) gives the following iteration scheme:

$$\mathbf{a}^{(s+1)} = \mathbf{a} + \boldsymbol{\delta}^{(s)}, \tag{11.164}$$

where:

$$\boldsymbol{\delta}^{(s)} = \left[\left(\mathbf{J}^{(s)} \right)^{\mathrm{T}} \mathbf{C}_{\mathbf{y}}^{-1} \mathbf{J}^{(s)} \right]^{-1} \left(\mathbf{J}^{(s)} \right)^{\mathrm{T}} \mathbf{C}_{\mathbf{y}}^{-1} \left[\mathbf{y} - \boldsymbol{\eta} \left(\mathbf{a}^{(s)} \right) \right]. \tag{11.165}$$

The iteration process is continued as long as the following inequality holds:

$$\left| \frac{a_j^{(s+1)} - a_j^{(s)}}{a_j^{(s)}} \right| > \varepsilon_t, \quad j = 0, 1, \ldots, N, \tag{11.166}$$

where ε_t is the iteration process set accuracy.

The number of iterations depends on correct selection of initial approximation $\mathbf{a}^{(0)}$. If $\mathbf{a}^{(0)}$ differs from the solution substantially, the iteration process expressed by (11.164) may be divergent. A comparison between (11.165) and (11.109) indicates that the two formulae are very similar. The difference is that matrix $\mathbf{X}$ in formula (11.109) is replaced in expression (11.165) by Jacobian matrix $\mathbf{J}$. This enables fast calculations of covariance matrices $\mathbf{C}_{\tilde{\mathbf{a}}}$ and $\mathbf{C}_{\tilde{\eta}}$, the confidence interval for the determined function $\eta(\tilde{\mathbf{a}})$ and the prediction interval for the measured value y_i. After the iteration process stops when inequality (11.166) no longer holds and the final solution $\tilde{\mathbf{a}}$ is found such that sum $S(\tilde{\mathbf{a}})$ reaches its minimum, an assumption is made that

$$\tilde{\mathbf{a}} = \mathbf{a}^{(s+1)}. \tag{11.167}$$

Covariance matrix $\mathbf{C}_{\tilde{\mathbf{a}}}$ can be calculated by means of a formula which is similar to (11.136):

$$\mathbf{C}_{\tilde{\mathbf{a}}}^{(s)} = \left[\left(\mathbf{J}^{(s)} \right)^{\mathrm{T}} \mathbf{C}_{\mathbf{y}}^{-1} \mathbf{J}^{(s)} \right]^{-1}. \tag{11.168}$$

Like in (11.138), covariance matrix $\mathbf{C}_{\tilde{\eta}}$ needed to find the confidence interval for η_i, $i = 1,\ldots,n$ is determined:

$$\mathbf{C}_{\tilde{\eta}}^{(s)} = \mathbf{J}^{(s)} \left[\left(\mathbf{J}^{(s)} \right)^{\mathrm{T}} \mathbf{C}_{\mathbf{y}}^{-1} \mathbf{J}^{(s)} \right]^{-1} \left(\mathbf{J}^{(s)} \right)^{\mathrm{T}}. \tag{11.169}$$

In order to determine the confidence and the prediction intervals, the measured values y_i can be written as follows:

$$y_i = \eta(\mathbf{x_i}, \mathbf{a}) + \varepsilon_i, \quad i = 1, 2, \ldots, n, \tag{11.170}$$

where $\mathbf{x_i}$ is the vector of coordinates of measuring points, e.g. time or space coordinates. The confidence and the prediction intervals can be determined for any measuring point $\mathbf{x} = \mathbf{x_0}$. Introducing vector:

$$\mathbf{f^o} = \left(\frac{\partial \eta(\mathbf{x_0}, \tilde{\mathbf{a}})}{\partial \tilde{a}_0}, \frac{\partial \eta(\mathbf{x_0}, \tilde{\mathbf{a}})}{\partial \tilde{a}_1}, \ldots, \frac{\partial \eta(\mathbf{x_0}, \tilde{\mathbf{a}})}{\partial \tilde{a}_N} \right)^{\mathrm{T}}, \qquad (11.171)$$

it is possible to calculate the confidence intervals for function $\eta(\mathbf{x}, \tilde{\mathbf{a}})$ determined by means of the least squares method in point $\mathbf{x} = \mathbf{x_0}$:

$$\tilde{\eta}_0 - t_{n-N-1,\alpha} s_y \sqrt{\left[(\mathbf{f^o})^{\mathrm{T}} \left(\mathbf{J}^{\mathrm{T}} \mathbf{C_y^{-1}} \mathbf{J} \right)^{-1} \mathbf{f^o} \right]^{(s)}} \leq \eta_0 \leq \tilde{\eta}_0$$

$$+ t_{n-N-1,\alpha} s_y \sqrt{\left[(\mathbf{f^o})^{\mathrm{T}} \left(\mathbf{J}^{\mathrm{T}} \mathbf{C_y^{-1}} \mathbf{J} \right)^{-1} \mathbf{f^o} \right]^{(s)}}, \qquad (11.172)$$

where $\tilde{\eta}_0 = \eta(\mathbf{x_0}, \tilde{\mathbf{a}})$.

A similar procedure can be used to find the y prediction interval for the set confidence level $P = (1 - \alpha)\ 100\%$:

$$\tilde{\eta}_0 - t_{n-N-1,\alpha} s_y \sqrt{1 + \left[(\mathbf{f^o})^{\mathrm{T}} \left(\mathbf{J}^{\mathrm{T}} \mathbf{C_y^{-1}} \mathbf{J} \right)^{-1} \mathbf{f^o} \right]^{(s)}} \leq y_0 \leq \tilde{\eta}_0$$

$$+ t_{n-N-1,\alpha} s_y \sqrt{1 + \left[(\mathbf{f^o})^{\mathrm{T}} \left(\mathbf{J}^{\mathrm{T}} \mathbf{C_y^{-1}} \mathbf{J} \right)^{-1} \mathbf{f^o} \right]^{(s)}}. \qquad (11.173)$$

Weight matrix $\mathbf{G_y} = \mathbf{C_y^{-1}}$ is a diagonal matrix where the diagonal elements are weights for individual measuring points.

Coordinate $\mathbf{x_0}$ may differ from the coordinate of the i-th measuring point $\mathbf{x_i}$, $i = 1,2,\ldots,n$.

Example 11.4 The heat flux value is measured using the sensor presented in Fig. 11.4. The sensor dimensions and the location of the measuring points are the same as in Example 11.2. The "exact measuring data" was calculated using an analytical exact solution for the following data: $\dot{q} = 274800$ W/m^2, $T_{cz} = 15\ ^\circ$C and $\alpha_w = 2400$ W/(m^2 K). The temperature distribution on the thickness of the sensor wall was calculated taking account of the thermal conductivity dependence on temperature:

$$\lambda(T) = a + bT, \text{W}/(\text{m K}),$$

where temperature T is expressed in $^\circ$C,

$$a = 14.64\ \text{W}/(\text{m K}), b = 0.0144\ \text{W}/(\text{m K}^2).$$

The values of coefficients a and b are typical of austenitic steels.

Assume that the measuring errors and the variances at temperatures measured at points $x_1 = d_1$, $x_2 = d_2$ and $x_3 = d_3$ are as follows:

$$x_1 = d_1 = 0.004 \text{ m}, \ \varepsilon_1 = 1.0116 \text{ K}, \ \sigma_1^2 = 1 \text{ K}^2,$$
$$x_2 = d_2 = 0.008 \text{ m}, \ \varepsilon_2 = -2.9761 \text{ K}, \ \sigma_2^2 = 4 \text{ K}^2,$$
$$x_3 = d_3 = 0.012 \text{ m}, \ \varepsilon_3 = 1.4683 \text{ K}, \ \sigma_3^2 = 2.25 \text{ K}^2.$$

First, in order to generate exact measuring data, the temperature distribution in the sensor was determined. According to the Fourier law:

$$\dot{q} = -\lambda(T)\frac{dT}{dx}$$

which, after the expression describing $\lambda(T)$ is substituted, gives:

$$\dot{q} = -(a+bT)\frac{dT}{dx}.$$

Separation of variables and integration result in:

$$\dot{q}x = -\left(aT + \frac{1}{2}bT^2\right) + C.$$

The solution of the obtained quadratic equation is expressed by the following function:

$$T(x) = -\frac{a}{b} + \sqrt{\left(\frac{a}{b}\right)^2 - \frac{2(\dot{q}x - C)}{b}}.$$

Considering that the temperature of the sensor rear surface washed by water is

$$T|_{d=L} = \frac{\dot{q}}{\alpha_w} + T_{cz} = \frac{274800}{2400} + 15 = 129.5 \ ^\circ\text{C},$$

it is possible to find constant C:

$$C = \dot{q}L + aT|_{d=L} + \frac{1}{2}b(T|_{x=L})^2.$$

Substituting relevant data, the following is obtained:

$$C = 274800 \cdot 0.016 + 14.65 \cdot 129.5 + 1/2 \cdot 0.0144 \cdot 129.5^2 = 6414.72 \text{ W/m}.$$

The temperature distribution on the sensor thickness is presented in Fig. 11.5. Comparing Tables 11.4 and 11.6, it can be seen that the temperatures of the front and the rear surface of the sensor are close to each other. In the linear case (cf. Table 11.4), the temperature distribution on the sensor thickness is calculated (Example 11.2) assuming mean thermal conductivity λ:

$$\lambda = \lambda(T_m) = 18.31 \ \text{W/(m K)},$$

where mean temperature T_m is calculated as the arithmetic mean of temperatures in $x_1 = d_1 = 0.004$ m and $x_2 = d_2 = 0.008$ m (cf. Table 11.6):

$$T_m = 0.5(314.29 + 194.23) = 254.26 \ ^\circ\text{C}.$$

The following temperatures are adopted as the "noisy" measuring data:

$$x_1 = d_1 = 0.004 \ \text{m}, \ y_1 = \eta_1 + \varepsilon_1 = 314.2884 + 1.0116 = 315.3 \ ^\circ\text{C},$$
$$x_2 = d_2 = 0.008 \ \text{m}, \ y_2 = \eta_2 + \varepsilon_2 = 255.6761 - 2.9761 = 252.7 \ ^\circ\text{C},$$
$$x_3 = d_3 = 0.012 \ \text{m}, \ y_3 = \eta_3 + \varepsilon_3 = 194.2317 + 1.4683 = 195.7 \ ^\circ\text{C}.$$

Using these "measuring data", temperature distribution $T(x)$ was used as the approximation function:

$$\eta(x, \mathbf{a}) = -\frac{a}{b} + \sqrt{\left(\frac{a}{b}\right)^2 - \frac{2(a_0 x - a_1)}{b}},$$

where a_0 and a_1 are the sought parameters, heat flux in W/m^2 and constant C in W/m, respectively.

The Jacobian matrix elements can be determined using the following derivatives:

Table 11.6 Temperature distribution on the sensor thickness determined taking account of temperature-dependent thermal conductivity

No.	$x = d$ (m)	T (°C)
1.	0.000	370.4274
2.	0.001	356.6077
3.	0.002	342.6476
4.	0.003	328.5427
5.	0.004	314.2884
6.	0.005	299.8799
7.	0.006	285.3119
8.	0.007	270.5793
9.	0.008	255.6761
10.	0.009	240.5964
11.	0.010	225.3337
12.	0.011	209.8812
13.	0.012	194.2317
14.	0.013	178.3773
15.	0.014	162.3099
16.	0.015	146.0206
17.	0.016	129.5000

$$\frac{\partial \eta_i}{\partial a_0} = \frac{-\frac{x_i}{b}}{\sqrt{\left(\frac{a}{b}\right)^2 - \frac{2(a_0 x_i - a_1)}{b}}}, \quad i = 1, 2, 3,$$

$$\frac{\partial \eta_i}{\partial a_1} = \frac{\frac{1}{b}}{\sqrt{\left(\frac{a}{b}\right)^2 - \frac{2(a_0 x_i - a_1)}{b}}}, \quad i = 1, 2, 3.$$

The following is adopted as the first approximation of $\mathbf{a}^{(0)} = \left(a_0^{(0)}, a_1^{(0)} \right)^{\mathrm{T}}$:

$$a_0^{(0)} = \frac{\lambda(y_1) + \lambda(y_3)}{2} \cdot \frac{y_3 - y_1}{x_3 - x_1} = \frac{19.19 + 17.47}{2} \cdot \frac{315.3 - 195.7}{0.012 - 0.004} = 274021.54 \text{ W/m}^2,$$

$$a_1^{(0)} = C^{(0)} = a_0^{(0)} + a y_3 + \frac{b}{2} y_3^2 = 274021.54 \cdot 0.012 + 14.65 \cdot 195.7$$

$$+ \frac{0.0144}{2} \cdot 195.7^2 = 6431 \text{ W/m}.$$

The first approximation of the Jacobian matrix has the following form:

$$\mathbf{J}^{(0)} = \begin{bmatrix} -208.4385 & 52\,109.634 \\ -435.9815 & 54\,497.6825 \\ -686.9679 & 57\,247.3251 \end{bmatrix} \cdot 10^{-6}.$$

Vector $\boldsymbol{\eta}^{(0)} = \boldsymbol{\eta}\left(\mathbf{a}^{(0)} \right)$ is:

$$\boldsymbol{\eta}^{(0)} = \left\{ \begin{array}{c} \eta\left(d_1, \mathbf{a}^{(0)}\right) \\ \eta\left(d_2, \mathbf{a}^{(0)}\right) \\ \eta\left(d_3, \mathbf{a}^{(0)}\right) \end{array} \right\} = \left\{ \begin{array}{c} 315.2992\,°C \\ 256.9031\,°C \\ 195.6990\,°C \end{array} \right\}.$$

Considering that matrix $\mathbf{C}_y^{-1}$ has the following form:

$$\mathbf{C}_y^{-1} = \begin{bmatrix} 1 & 0 & 0 \\ 0 & \frac{1}{4} & 0 \\ 0 & 0 & \frac{1}{2.25} \end{bmatrix}$$

covariance matrix $\mathbf{C}_a^{(0)}$ can be determined, calculating first:

$$\left(\mathbf{J}^{(0)}\right)^{\mathrm{T}}\mathbf{C}_{\mathbf{y}}^{(0)}\mathbf{J}^{(0)} = \begin{bmatrix} -208.4385 & -435.9815 & -686.9679 \\ 52109.634 & 54497.6825 & 57247.3251 \end{bmatrix} \times 10^{-6}$$

$$\times \begin{bmatrix} 1 & 0 & 0 \\ 0 & \frac{1}{4} & 0 \\ 0 & 0 & \frac{1}{2.25} \end{bmatrix} \times \begin{bmatrix} -208.4385 & 52109.634 \\ -435.9815 & 54497.6825 \\ -686.9679 & 57247.3251 \end{bmatrix} \times 10^{-6}$$

$$= \begin{bmatrix} 300710.98 & -34280350.59 \\ -34280350.59 & 4914471629 \end{bmatrix} \times 10^{-12}.$$

The covariance matrix is:

$$\mathbf{C_a} = \left[\left(\mathbf{J}^{(0)}\right)^{\mathrm{T}}\mathbf{C}_{\mathbf{y}}^{(0)}\mathbf{J}^{(0)}\right]^{-1} = \begin{bmatrix} 16235820.56 & 113251.16 \\ 113251.16 & 993.45 \end{bmatrix}.$$

The diagonal elements are variances $\sigma_{\tilde{a}_0}^2$ and $\sigma_{\tilde{a}_1}^2$:

$$\left(\sigma_{\tilde{a}_0}^{(0)}\right)^2 = 16235820.56 \ \mathrm{W}^2/\mathrm{m}^4, \ \sigma_{\tilde{a}_0}^{(0)} = 4029.37 \ \mathrm{W/m}^2,$$

$$\left(\sigma_{\tilde{a}_1}^{(0)}\right)^2 = 993.45 \ \mathrm{W/m}^2, \ \sigma_{\tilde{a}_1}^{(0)} = 31.52 \ \mathrm{W/m}^2.$$

Coefficients $a_0^{(1)}$ and $a_1^{(1)}$ total:

$$a_0^{(1)} = 274974.36 \ \mathrm{W/m}^2, \ a_1^{(1)} = 6426 \ \mathrm{W/m}.$$

The first approximation of vector **a** has the following form:

$$\mathbf{a}^{(1)} = \left\{ \begin{array}{c} 274974.36 \\ 6426.00 \end{array} \right\}.$$

The solution is very close to the input data and to solutions $\tilde{a}_0 = 274974.36 \ \mathrm{W/m}^2$ and $\tilde{a}_1 = 6426 \ \mathrm{W/m}$ obtained by means of the Levenberg-Marquardt method using the Table Curve program. Covariance matrix $\mathbf{C}_{\eta}^{(0)}$ calculated using formula (11.169) is:

$$\mathbf{C}_\eta^{(0)} = \begin{bmatrix} -208.4385 & 52109.634 \\ -435.9815 & 54497.6825 \\ -686.9679 & 57247.3251 \end{bmatrix} \times 10^{-6} \cdot \begin{bmatrix} 16235820.56 & 113251.16 \\ 113251.16 & 993.45 \end{bmatrix}$$

$$\times \begin{bmatrix} -208.4385 & -435.9815 & -686.9679 \\ 52109.634 & 54497.6825 & 57247.3251 \end{bmatrix} \times 10^{-6}$$

$$= \begin{bmatrix} 0.942834 & 0.437292 & -0.117081 \\ 0.437292 & 0.654960 & 0.895605 \\ -0.117081 & 0.895605 & 2.010209 \end{bmatrix},$$

which gives:

$$\sigma_{\eta_1}^{(0)} = \sqrt{0.942834} = 0.971 \text{ K},$$

$$\sigma_{\eta_2}^{(0)} = \sqrt{0.65496014} = 0.8093 \text{ K},$$

$$\sigma_{\eta_3}^{(0)} = \sqrt{2.01021} = 1.4178 \text{ K}.$$

The values of function $\eta^{(1)}$ in points: $x_1 = d_1$, $x_2 = d_2$ and $x_3 = d_3$ were calculated using formula (11.158):

$$\eta^{(1)} = \eta^{(0)} + \mathbf{J}^{(s)} \left(\mathbf{a}^{(1)} - \mathbf{a}^{(0)} \right)$$

$$= \begin{Bmatrix} 315.2992 \\ 256.9031 \\ 195.6990 \end{Bmatrix} + \begin{bmatrix} -208.4385 & 52109.634 \\ -435.9815 & 54497.6825 \\ -686.9679 & 57247.3251 \end{bmatrix} \times 10^{-6} \cdot$$

$$\cdot \begin{Bmatrix} 274974.36 - 274021.54 \\ 6426 - 6431 \end{Bmatrix} = \begin{Bmatrix} 314.840 \text{ °C} \\ 256.21 \text{ °C} \\ 194.759 \text{ °C} \end{Bmatrix}.$$

Vector $\varepsilon^{(1)}$ is:

$$\varepsilon^{(1)} = \mathbf{y}^{(1)} - \varepsilon^{(1)} = \begin{Bmatrix} 315.3 - 314.840 \\ 252.7 - 256.215 \\ 195.7 - 194.759 \end{Bmatrix} = \begin{Bmatrix} 0.460 \text{ K} \\ -3.515 \text{ K} \\ 0.941 \text{ K} \end{Bmatrix}.$$

The confidence interval of coefficients $a_0^{(1)}$ and $a_1^{(1)}$ for P = 95% determined from formula (11.71) is:

$$274974.36 - 2 \cdot 4029.37 \le a_0^{(1)} \le 274974.36 + 2 \cdot 4029.37,$$

$$266915.6 \text{ W/m}^2 \le a_0^{(1)} \le 283033.1 \text{ W/m}^2,$$

$$6426 - 2 \cdot 31.52 \le a_1^{(1)} \le 6426 + 2 \cdot 31.52,$$

$$6362.96 \text{ W/m} < a_1^{(1)} < 6489.04 \text{ W/m}.$$

The approximation function η confidence interval for P = 95%, calculated using formula (11.72) is:

$$314.84 - 2 \cdot 0.971 \le \eta_1^{(1)} \le 314.84 + 2 \cdot 0.971,$$

$$312.9 \text{ °C} \le \eta_1^{(1)} \le 316.8 \text{ °C},$$

$$256.215 - 2 \cdot 0.8093 \le \eta_2^{(1)} \le 256.215 + 2 \cdot 0.8093,$$

$$254.6 \text{ °C} \le \eta_2^{(1)} \le 257.8 \text{ °C},$$

$$194.759 - 2 \cdot 1.4178 \le \eta_3^{(1)} \le 194.759 + 2 \cdot 1.4178,$$

$$191.9 \text{ °C} \le \eta_3^{(1)} \le 197.6 \text{ °C}.$$

Variances $\sigma_{\tilde{a}_0}^{(2)}$ and $\sigma_{\tilde{a}_1}^{(2)}$ are calculated determining covariance matrix $\mathbf{C}_{\tilde{a}}$. But they can also be found using directly the variance transposition rule (11.28), which in this case takes the following form:

$$\sigma_{\tilde{a}_0}^2 = \sum_{i=1}^{n} \left(\frac{\partial \tilde{a}_0}{\partial y_i} \right)^2 \sigma_{y_i}^2$$

and

$$\sigma_{\tilde{a}_1}^2 = \sum_{i=1}^{n} \left(\frac{\partial \tilde{a}_1}{\partial y_i} \right)^2 \sigma_{y_i}^2, \quad n = 3.$$

Using the Levenberg-Marquardt method to determine $\tilde{a}_0$ and $\tilde{a}_1$, and approximating the derivatives using central difference quotients, $\sigma_{\tilde{a}_0}^{(2)}$ and $\sigma_{\tilde{a}_1}^{(2)}$ are found. After relevant partial derivatives (sensitivity coefficients) are determined:

$$\frac{\partial \tilde{a}_0}{\partial y_i} = 2514.989 \ \mathrm{W}/\left(\mathrm{m}^2 \ \mathrm{K}\right),$$

$$\frac{\partial \tilde{a}_0}{\partial y_2} = 927.862 \ \mathrm{W}/\left(\mathrm{m}^2 \ \mathrm{K}\right),$$

$$\frac{\partial \tilde{a}_0}{\partial y_{3i}} = -2272.939 \ \mathrm{W}/\left(\mathrm{m}^2 \ \mathrm{K}\right),$$

$$\frac{\partial \tilde{a}_1}{\partial y_1} = 28.143 \ \mathrm{W}/(\mathrm{m} \ \mathrm{K}),$$

$$\frac{\partial \tilde{a}_1}{\partial y_2} = 1.194 \ \mathrm{W}/(\mathrm{m} \ \mathrm{K}),$$

$$\frac{\partial \tilde{a}_1}{\partial y_{13}} = -9.287 \ \mathrm{W}/(\mathrm{m} \ \mathrm{K})$$

it is possible to calculate $\sigma_{\tilde{a}_0}^2$ and $\sigma_{\tilde{a}_1}^2$:

$$\sigma_{\tilde{a}_0}^2 = 16200618.2 \ \mathrm{W}^2/\mathrm{m}^4, \ \sigma_{\tilde{a}_0} = 4025.0 \ \mathrm{W}/\mathrm{m}^2,$$
$$\sigma_{\tilde{a}_1}^2 = 993.452 \ \mathrm{W}^2/\mathrm{m}^2, \ \sigma_{\tilde{a}_1} = 31.52 \ \mathrm{W}/\mathrm{m}.$$

Comparing the determined values of $\sigma_{\tilde{a}_1}$ and $\sigma_{\tilde{a}_2}$ with the values obtained in the first approximation using the Gauss-Newton method, it can be seen that the Gauss-Newton method accuracy is very good.

11.5.3 Dependent Measurements. Least Squares Method with Equality Constraints

In indirect measurements, the determined function $\eta(\mathbf{a})$ and the sought parameters $\mathbf{a}$ were not subject to any additional constraints. This is the effect of the known form of approximation function $\eta(\mathbf{a})$, which exactly satisfies the steady-state heat conduction equation.

If the temperature field is determined by means of numerical methods, e.g. using the control (finite) volume method or the finite element method, there is no single analytical $\eta(\mathbf{a})$ solution which can be used to approximate experimental data in the entire area. Solving the boundary problem by means of numerical methods, temperature values are obtained in selected points of the area under analysis.

The temperatures at the points (nodes) are found from a system of equations resulting from the numerical method application. At the same time, the system constitutes the constraint equations. Linear constraints with the form as presented below [35] was analysed first:

$$b_{10} + b_{11}\eta_1 + b_{12}\eta_2 + \cdots + b_{1n}\eta_n = 0,$$
$$b_{20} + b_{21}\eta_1 + b_{22}\eta_2 + \cdots + b_{2n}\eta_n = 0,$$
$$\vdots \qquad\qquad\qquad \vdots \tag{11.174}$$
$$b_{q0} + b_{q1}\eta_1 + b_{q2}\eta_2 + \cdots + b_{qn}\eta_n = 0,$$

which can also be written in a matrix form:

$$\mathbf{B}\boldsymbol{\eta} + \mathbf{b_0} = 0. \tag{11.175}$$

Quantities η_i (temperatures at nodes) are determined using the least squares method so that the sum of squares

$$S = \sum_{i=1}^{n} \frac{(y_i - \eta_i)^2}{\sigma_{y_i}^2} \tag{11.176}$$

should reach the minimum. Writing sum S in the following matrix form:

$$S = \boldsymbol{\varepsilon}^{\mathrm{T}} \mathbf{C_y}^{-1} \boldsymbol{\varepsilon} \tag{11.177}$$

the Lagrange function is created:

$$L = \boldsymbol{\varepsilon}^{\mathrm{T}} \mathbf{C_y}^{-1} \boldsymbol{\varepsilon} + 2\boldsymbol{\mu}^{\mathrm{T}}(\mathbf{B}\boldsymbol{\eta} + \mathbf{b_0}). \tag{11.178}$$

Noting that $\boldsymbol{\eta} = \mathbf{y} - \boldsymbol{\varepsilon}$, the Lagrange function (11.178) can be expressed as:

$$L = \boldsymbol{\varepsilon}^{\mathrm{T}} \mathbf{C_y}^{-1} \boldsymbol{\varepsilon} + 2\boldsymbol{\mu}^{\mathrm{T}}(\mathbf{c} - \mathbf{B}\boldsymbol{\varepsilon}), \tag{11.179}$$

where:

$$\mathbf{c} = \mathbf{B}\mathbf{y} + \mathbf{b_0}, \quad \boldsymbol{\mu} = \left(\mu_1, \mu_2, \ldots, \mu_q\right)^{\mathrm{T}}. \tag{11.180}$$

The necessary condition for the existence of the minimum of the Lagrange function:

$$\frac{\mathrm{d}L}{\mathrm{d}\boldsymbol{\varepsilon}} = 0 \tag{11.181}$$

gives:

$$\boldsymbol{\varepsilon}^{T} \mathbf{C_y}^{-1} - \boldsymbol{\mu}^{\mathrm{T}}\mathbf{B} = 0. \tag{11.182}$$

Noting that Eq. (11.182) can be written as:

$$\mathbf{C_y}^{-1}\boldsymbol{\varepsilon} = \mathbf{B}^{\mathrm{T}}\boldsymbol{\mu} \tag{11.183}$$

the following is obtained:

$$\boldsymbol{\varepsilon} = \mathbf{C}_\mathbf{y}\mathbf{B}^\mathrm{T}\boldsymbol{\mu}. \tag{11.184}$$

Substitution of Eq. (11.184) in constraint Eq. (11.175) results in the following relation:

$$\mathbf{B}(\mathbf{y} - \boldsymbol{\varepsilon}) + \mathbf{b_0} = 0, \mathbf{B}\left(\mathbf{y} - \mathbf{C}_\mathbf{y}\mathbf{B}^\mathrm{T}\boldsymbol{\mu}\right) + \mathbf{b_0} = 0 \tag{11.185}$$

which is used to determine μ:

$$\tilde{\boldsymbol{\mu}} = \left(\mathbf{BC}_\mathbf{y}\mathbf{B}^\mathrm{T}\right)^\mathrm{T}\mathbf{c} \tag{11.186}$$

Considering (11.186) in (11.184), $\tilde{\boldsymbol{\varepsilon}} = \mathbf{y} - \tilde{\boldsymbol{\eta}}$ vectors are obtained:

$$\tilde{\boldsymbol{\varepsilon}} = \mathbf{C}_\mathbf{y}\mathbf{B}^\mathrm{T}\left(\mathbf{BC}_\mathbf{y}\mathbf{B}^\mathrm{T}\right)^{-1}\mathbf{c}. \tag{11.187}$$

The $\eta_1, \eta_2, \ldots, \eta_n$ values can be calculated from the following formula:

$$\tilde{\boldsymbol{\eta}} = \mathbf{y} - \tilde{\boldsymbol{\varepsilon}}, \tag{11.188}$$

where $\tilde{\boldsymbol{\varepsilon}}$ is defined by formula (11.187).

Covariance matrices $\mathbf{C}_{\tilde{\mu}}$ and $\mathbf{C}_{\tilde{\eta}}$ determined according to the variance transposition rule have the following form:

$$\mathbf{C}_{\tilde{\mu}} = \left(\mathbf{BC}_\mathbf{y}^{-1}\mathbf{B}^\mathrm{T}\right)^{-1} = \mathbf{G_B}, \tag{11.189}$$

$$\mathbf{C}_{\tilde{\eta}} = \mathbf{C}_\mathbf{y} - \mathbf{C}_\mathbf{y}\mathbf{B}^\mathrm{T}\mathbf{G_B}\mathbf{BC}_\mathbf{y}^{-1}. \tag{11.190}$$

The Lagrange multiplier method is used in inverse heat conduction problems, where the temperature distribution is found using numerical methods. The algebraic equations resulting from the finite volume method or from the finite element method, which enable determination of the temperature distributions in nodes, are at the same time the equations of constraints.

Example 11.5 Computing and measuring data are the same as in Example 11.2. In contrast to Example 11.3, the temperature distribution on the sensor thickness was determined by means of the finite volume method. The division into finite volumes is presented in Fig. 11.8. The temperature distribution on the wall thickness is determined in nodal points only and, according to the least squares method, the temperatures in nodes 2, 3 and 4 should be selected so that the sum

$$S = \sum_{i=1}^{3} \frac{(y_i - \eta_i)^2}{\sigma_{y_i}^2}$$

Fig. 11.8 Heat flux sensor division into control (finite) volumes; filled circle—nodes of finite volumes; temperature measuring points are located in nodes: 1, 2 and 3

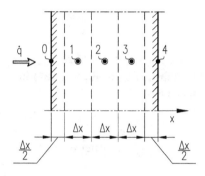

should reach the minimum. The symbols y_i and η_i denote the measured and calculated temperatures, respectively.

According to the finite volume method, temperature in point 2 should satisfy the following energy balance equation:

$$\frac{\lambda}{\Delta x}(\eta_1 - \eta_2) + \frac{\lambda}{\Delta x}(\eta_3 - \eta_2) = 0,$$

which gives the equation of constraints:

$$\eta_1 - 2\eta_2 + \eta_3 = 0.$$

Considering that

$$\mathbf{B} = (1, -2, 1), \mathbf{b_0} = 0,$$

$$\mathbf{c} = \mathbf{By} + \mathbf{b_0} = (1, -2, 1) \begin{Bmatrix} 310.6 \\ 246.6 \\ 191.0 \end{Bmatrix} = 8.4\text{ K}$$

and then calculating:

$$\mathbf{C_y}\mathbf{B}^\mathsf{T} = \begin{bmatrix} 1 & 0 & 0 \\ 0 & 4 & 0 \\ 0 & 0 & 2.25 \end{bmatrix} \begin{Bmatrix} 1 \\ -2 \\ 1 \end{Bmatrix} = \begin{Bmatrix} 1 \\ -8 \\ 2.25 \end{Bmatrix},$$

$$\mathbf{G_B} = \left(\mathbf{BC_y}\mathbf{B}^\mathsf{T}\right)^{-1} = \left((1, -2, 1) \begin{Bmatrix} 1 \\ -8 \\ 2.25 \end{Bmatrix} \right)^{-1} = \frac{1}{19.25},$$

the error vector can be determined:

$$\tilde{\boldsymbol{\varepsilon}} = \mathbf{C_y B^T \left(B C_y B^T \right)^{-1} c} = \left\{ \begin{array}{c} 1 \\ -8 \\ 2.25 \end{array} \right\} \cdot \frac{1}{19.25} \cdot 8.4,$$

$$\tilde{\boldsymbol{\varepsilon}} = \left\{ \begin{array}{c} 0.4364 \\ -3.4909 \\ 0.9818 \end{array} \right\}.$$

The vector of calculated temperatures is:

$$\tilde{\boldsymbol{\eta}} = \mathbf{y} - \tilde{\boldsymbol{\varepsilon}} = \left\{ \begin{array}{c} 310.6 \\ 246.6 \\ 191.0 \end{array} \right\} - \left\{ \begin{array}{c} 0.4364 \\ -3.4909 \\ 0.9818 \end{array} \right\} = \left\{ \begin{array}{c} 310.1636 \\ 250.0909 \\ 190.0182 \end{array} \right\}.$$

Temperatures $\tilde{\eta}_1$, $\tilde{\eta}_2$ and $\tilde{\eta}_3$ are expressed in degrees Celsius. Next, calculations are performed of covariance matrices $\mathbf{C}_{\tilde{\mu}}$ and $\mathbf{C}_{\tilde{\eta}}$:

$$\mathbf{C}_{\tilde{\mu}} = \mathbf{G_B} = \frac{1}{19.25} = 0.05195 \text{ K},$$

$$\mathbf{C}_{\tilde{\eta}} = \mathbf{C_y} - \mathbf{C_y B^T G_B B C_y^{-1}}$$

$$= \begin{bmatrix} 1 & 0 & 0 \\ 0 & 4 & 0 \\ 0 & 0 & 2.25 \end{bmatrix} - \begin{bmatrix} 1 & 0 & 0 \\ 0 & 4 & 0 \\ 0 & 0 & 2.25 \end{bmatrix} \cdot \left\{ \begin{array}{c} 1 \\ -2 \\ 1 \end{array} \right\} \cdot \frac{1}{19.25} \cdot (1, -2, 1) \cdot$$

$$\cdot \begin{bmatrix} 1 & 0 & 0 \\ 0 & a4 & 0 \\ 0 & 0 & 2.25 \end{bmatrix} = \begin{bmatrix} 0.94805 & 0.41558 & -0.11688 \\ 0.41558 & 0.67532 & 0.93506 \\ -0.11688 & 0.93506 & 1.98701 \end{bmatrix}$$

It can be seen that the diagonal elements of matrix $\mathbf{C}_{\tilde{\eta}}$ obtained in this example and in Example 11.2 are the same. Identical results are obtained using the variance transposition rule directly:

$$\sigma_{\tilde{\eta}_i} = \sqrt{\sum_{j=1}^{3} \left(\frac{\partial \tilde{\eta}_i}{\partial y_j} \right)^2 \sigma_{y_j}^2}, \quad i = 2, 3, 4,$$

where sensitivity coefficients $\frac{\partial \tilde{\eta}_i}{\partial y_j}$ are calculated using central difference quotients. Confidence intervals for $\tilde{\eta}_1$, $\tilde{\eta}_2$ and $\tilde{\eta}_3$ are defined by formulae (11.72). In order to determine temperature at nodes 0 and 4, heat balance equations were written for nodes 1 and 3. As a result, the following equations were obtained:

$$\tilde{\eta}_0 - 2\tilde{\eta}_1 + \tilde{\eta}_2 = 0,$$
$$\tilde{\eta}_2 - 2\tilde{\eta}_3 + \tilde{\eta}_4 = 0.$$

The first equation results in:

$$\tilde{\eta}_0 = 2\tilde{\eta}_1 - \tilde{\eta}_2 = 2 \cdot 310.1636 - 250.0909 = 370.2363 \text{ °C}.$$

The second equation gives the temperature in node 4.

$$\tilde{\eta}_4 = 2\tilde{\eta}_3 - \tilde{\eta}_2 = 2 \cdot 3190.0182 - 250.0909 = 129.9455 \text{ °C}.$$

Temperatures $\tilde{\eta}_0$ and $\tilde{\eta}_4$ can be determined using the following formula: $\eta = a_0 + a_1 x$

$$\tilde{\eta} = a_0 + a_1 x = 370.2365 - 15018.1781\, x.$$

Assuming $x_1 = 0$ m, the result is $\tilde{\eta}_0 = 370.2365$ °C.
Temperature $\tilde{\eta}_4$ calculated for $x_4 = 0.016$ m is equal to $\tilde{\eta}_4 = 129.9456$ °C. This means that the results demonstrate very good agreement. The heat flux on the sensor front surface can be found from the heat balance equation for node 0:

$$\dot{\tilde{q}} + \lambda \frac{\tilde{\eta}_1 - \tilde{\eta}_0}{\Delta x} = 0,$$

from which it follows that

$$\dot{\tilde{q}} = -\lambda \frac{\tilde{\eta}_1 - \tilde{\eta}_0}{\Delta x} = -18.31 \frac{310.1636 - 370.2363}{0.04} = 274982.78 \text{ W/m}^2.$$

Taking account of rounding errors, the results are identical to those obtained in Example 11.2.

The calculations also indicate that in the Lagrange multiplier method the temperature measuring points should lie on a closed curve or surface (cf. Fig. 11.9).

Fig. 11.9 Location of the temperature measuring points on the surface of and inside the area marked with lines

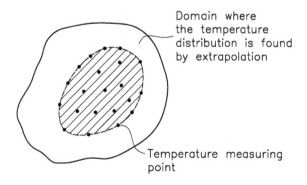

Domain where the temperature distribution is found by extrapolation

Temperature measuring point

Using the Lagrange multiplier method, the temperature distribution is first determined inside and on the surface of the area marked with lines. Then, marching in space from the marked area surface to the surface of the body, the temperature distribution is found in the other area.

Like in Example 11.5, energy balance equations are written for points located on the boundary of the area marked with lines, from which temperatures are then determined in the nodes adjacent to the marked area. Continuing like with the newly determined temperatures, the temperature distribution is found in the entire remaining area of the body.

11.5.4   Dependent Measurements. Nonlinear Problem

In this case, equality constraints

$$f_k(\boldsymbol{\eta}) = 0, \quad k = 1, 2, \ldots, q \tag{11.191}$$

are nonlinear. In order to find temperatures $\boldsymbol{\eta} = (\eta_1, \eta_2, \ldots, \eta_n)^{\mathrm{T}}$, Eq. (11.191) were linearized using a linear Taylor series. Expanding Eq. (11.191) into a Taylor series and limiting the considerations to linear terms, the following is obtained:

$$f_k(\boldsymbol{\eta}) = f_k\left(\boldsymbol{\eta}^{(s)}\right) + \left(\frac{\partial f_k}{\partial \eta_1}\right)_{\boldsymbol{\eta}^{(s)}}\left(\eta_1 - \eta_1^{(s)}\right) + \cdots + \left(\frac{\partial f_k}{\partial \eta_n}\right)_{\boldsymbol{\eta}^{(s)}}\left(\eta_n - \eta_n^{(s)}\right),$$

$$\tag{11.192}$$

where $\boldsymbol{\eta}^{(s)}$ is the vector of the s-th approximations of temperatures η_i. Considering that sought temperatures η are the difference between measured quantities $\mathbf{y}$ and measuring errors $\boldsymbol{\varepsilon}$, i.e.:

$$\boldsymbol{\eta} = \mathbf{y} - \boldsymbol{\varepsilon}, \tag{11.193}$$

equality constraints (11.191) can be written as:

$$f_k(\mathbf{y} - \boldsymbol{\varepsilon}) = 0 \tag{11.194}$$

or in a slightly different form:

$$f_k(y_1 - \varepsilon_1, y_2 - \varepsilon_2, \ldots, y_3 - \varepsilon_3) = 0. \tag{11.195}$$

Estimators of measuring errors ε_i are usually very small, and functions f_k can be expanded into a Taylor series around point $\mathbf{y} = (y_1, y_2, \ldots, y_n)^{\mathrm{T}}$ using linear terms only:

$$f_k(\boldsymbol{\eta}) = f_k(\mathbf{y}) - \left(\frac{\partial f_k}{\partial \eta_1}\right)_{\mathbf{y}} \varepsilon_1 - \left(\frac{\partial f_k}{\partial \eta_2}\right)_{\mathbf{y}} \varepsilon_2 -, \cdots, - \left(\frac{\partial f_k}{\partial \eta_n}\right)_{\mathbf{y}} \varepsilon_n = 0,$$

$$k = 1, \ldots, q. \tag{11.196}$$

Introducing the following quantities:

$$b_{kl} = \left(\frac{\partial f_k}{\partial \eta_l}\right)_{\mathbf{y}}, \mathbf{B} = \mathbf{J} = \begin{bmatrix} b_{11} & b_{12} & \cdots & b_{1n} \\ b_{21} & b_{22} & \cdots & b_{2n} \\ \vdots & & & \\ b_{q1} & b_{q2} & \cdots & b_{qn} \end{bmatrix}, \tag{11.197}$$

where: $c_k = f_k(\mathbf{y}), \mathbf{c} = (c_1, c_2, \ldots, c_q)^{\mathrm{T}}, \boldsymbol{\varepsilon} = (\varepsilon_1, \varepsilon_2, \ldots, \varepsilon_q)^{\mathrm{T}}$, the system of Eq. (11.196) can be written as:

$$\mathbf{B}\boldsymbol{\varepsilon} - \mathbf{c} = 0. \tag{11.198}$$

The least squares problem comes down to finding the vector of errors (corrections) $\boldsymbol{\varepsilon}$ for which sum

$$S = \boldsymbol{\varepsilon}^{\mathrm{T}} \mathbf{C}_{\mathbf{y}}^{-1} \boldsymbol{\varepsilon} \tag{11.199}$$

reaches the minimum and, at the same time, constraint Eq. (11.198) are satisfied.

The necessary condition for the existence of the minimum of the Lagrange function:

$$\frac{\mathrm{d}L}{\mathrm{d}\varepsilon} = 0, \tag{11.200}$$

where:

$$L = \boldsymbol{\varepsilon}^{\mathrm{T}} \mathbf{C}_{\mathbf{y}}^{-1} \boldsymbol{\varepsilon} + 2\boldsymbol{\mu}^{\mathrm{T}} (\mathbf{c} - \mathbf{B}\boldsymbol{\varepsilon}),$$

$$\boldsymbol{\mu}^{T} = (\mu_1, \mu_2, \ldots, \mu_q), \tag{11.201}$$

gives:

$$\tilde{\boldsymbol{\varepsilon}} = \mathbf{C}_{\mathbf{y}} \mathbf{B}^{\mathrm{T}} \left(\mathbf{B}\mathbf{C}_{\mathbf{y}}\mathbf{B}^{\mathrm{T}}\right)^{-1} \mathbf{c}. \tag{11.202}$$

The vector $\boldsymbol{\eta}$ can be calculated from formula (11.193). Usually, the calculations can be stopped once the first approximation defined by formula (11.202) is obtained because the accuracy of the solution is very good. Covariance matrices $\mathbf{C}_{\tilde{\mu}}$ and $\mathbf{C}_{\tilde{\eta}}$ are defined by formulae (11.189) and (11.190). Matrix $\mathbf{B}$ and vector $\mathbf{c}$ are taken into account. If measuring errors are bigger, the first approximation $\boldsymbol{\eta}^{(1)} = \mathbf{y}$ is far from the real solution $\tilde{\boldsymbol{\eta}}$. Assuming that solution $\boldsymbol{\eta}^{(s)}$ is known in the s-th iteration, and assuming further that solution $\boldsymbol{\eta}^{(s+1)}$ in the next iteration satisfies equation $f_k(\boldsymbol{\eta}^{(s+1)}) = 0$, the following is obtained from expression (11.196):

$$f_k\left(\boldsymbol{\eta}^{(s+1)}\right) = f_k\left(\boldsymbol{\eta}^{(s)}\right) + \left(\frac{\partial f_k}{\partial \eta_1}\right)_{\boldsymbol{\eta}^{(s)}}\left(\varepsilon_1^{(s)} - \varepsilon_1^{(s+1)}\right) + \cdots$$

$$+ \left(\frac{\partial f_k}{\partial \eta_n}\right)_{\boldsymbol{\eta}^{(s)}}\left(\varepsilon_n^{(s)} - \varepsilon_n^{(s+1)}\right) = 0, \tag{11.203}$$

where $\varepsilon_i^{(s)} = y_i - \eta_i^{(s)}$, $i = 1,\dots,n$.

Equation (11.203) can be transformed into:

$$\left(\frac{\partial f_k}{\partial \eta_1}\right)_{\boldsymbol{\eta}^{(s)}} \varepsilon_1^{(s+1)} + \cdots + \left(\frac{\partial f_k}{\partial \eta_n}\right)_{\boldsymbol{\eta}^{(s)}} \varepsilon_n^{(s+1)}$$

$$= f_k\left(\boldsymbol{\eta}^{(s)}\right) + \left(\frac{\partial f_k}{\partial \eta_1}\right)_{\boldsymbol{\eta}^{(s)}} \varepsilon_1^{(s)} + \cdots + \left(\frac{\partial f_k}{\partial \eta_n}\right)_{\boldsymbol{\eta}^{(s)}} \varepsilon_n^{(s)}. \tag{11.204}$$

Introducing the following quantities:

$$c_k^{(s)} = f_k(\boldsymbol{\eta}(s)) + \sum_{i=1}^{n} \left(\frac{\partial f_k}{\partial \eta_i}\right)_{\boldsymbol{\eta}^{(s)}} \varepsilon_i^{(s)}, \tag{11.205}$$

$$b_{kl} = \left(\frac{\partial f_k}{\partial \eta_l}\right)_{\boldsymbol{\eta}^{(s)}}, \tag{11.206}$$

$$\mathbf{B} = \mathbf{J} = \begin{bmatrix} b_{11} & b_{12} & \cdots & b_{1n} \\ b_{21} & b_{22} & \cdots & b_{2n} \\ \vdots & \vdots & & \vdots \\ b_{q1} & b_{q2} & \cdots & b_{qn} \end{bmatrix}. \tag{11.207}$$

the system of Eq. (11.205) can be written as:

$$\mathbf{B}\varepsilon^{(s+1)} = \mathbf{c}^{(s)}. \tag{11.208}$$

The Lagrange function has the following form:

$$L = \left(\varepsilon^{\mathrm{T}}\right)^{(s+1)} \mathbf{C}_y^{-1} \varepsilon^{(s+1)} + 2\left(\boldsymbol{\mu}^{\mathrm{T}}\right)^{(s+1)}\left(\mathbf{c}^{(s)} - \mathbf{B}\varepsilon^{(s+1)}\right). \tag{11.209}$$

The necessary condition for the existence of the Lagrange function minimum gives:

$$\frac{dL}{d\varepsilon^{(s+1)}} = 2\left(\varepsilon^T\right)^{(s+1)}\mathbf{C_y}^{-1} - 2\left(\boldsymbol{\mu}^T\right)^{(s+1)}\mathbf{B} \tag{11.210}$$

from which it follows that

$$\left(\varepsilon^T\right)^{(s+1)} = \mathbf{C_y}^{-1} - \left(\boldsymbol{\mu}^T\right)^{(s+1)}\mathbf{B}, \tag{11.211}$$

$$\tilde{\varepsilon}^{(s+1)} = \mathbf{C_y}^{-1}\mathbf{B}^T\tilde{\boldsymbol{\mu}}^{(s+1)}. \tag{11.212}$$

Substituting (11.212) in (11.208) and performing appropriate transformations, the formula defining the vector of the Lagrange multipliers is obtained:

$$\tilde{\boldsymbol{\mu}}^{(s+1)} = \left(\mathbf{BC_yB}^T\right)^{-1}\mathbf{c}^{(s)}, \tag{11.213}$$

which, after it is substituted in (11.212), gives:

$$\tilde{\varepsilon}^{(s+1)} = \mathbf{C_yB}^T\left(\mathbf{BC_y}^{-1}\mathbf{B}^T\right)^{-1}\mathbf{c}^{(s)}. \tag{11.214}$$

The sought distribution of temperature $\tilde{\boldsymbol{\eta}}^{(s+1)}$ is determined using the following formula:

$$\tilde{\boldsymbol{\eta}}^{(s+1)} = \mathbf{y} - \tilde{\varepsilon}^{(s+1)} = \mathbf{y} - \mathbf{C_yB}^T\left(\mathbf{BC_y}^{-1}\mathbf{B}^T\right)^{-1}\mathbf{c}^{(s)}, \tag{11.215}$$

which, introducing the following quantity:

$$\mathbf{G_B} = \left(\mathbf{BC_yB}^T\right)^{-1} \tag{11.216}$$

can be written as:

$$\tilde{\boldsymbol{\eta}}^{(s+1)} = \mathbf{y} - \mathbf{C_yB}^T\mathbf{G_B}\mathbf{c}^{(s)}, \quad s = 0, 1, 2, \ldots \tag{11.217}$$

The following is assumed in the first iteration $s = 0$:

$$\varepsilon^{(0)} = \left\{\begin{matrix} 0 \\ 0 \\ \vdots \\ 0 \end{matrix}\right\}, \quad \text{tj. } \varepsilon_1^{(0)} = 0, \varepsilon_2^{(0)} = 0, \ldots, \varepsilon_n^{(0)} = 0. \tag{11.218}$$

Using the error propagation rule upon completion of the iteration process, when the following condition is satisfied:

$$\left| \frac{\varepsilon_i^{(s+1)} - \varepsilon_i^{(s)}}{\varepsilon_i^{(s)}} \right| < e, \quad i = 1, \ldots, n, \tag{11.219}$$

where e is the set tolerance of the computations, the following covariance matrices are calculated:

$$\left(\mathbf{C}_{\hat{\mu}} \right)^{(s+1)} = \left(\mathbf{B} \mathbf{C}_{\mathbf{y}} \mathbf{B}^{\mathrm{T}} \right)^{-1}, \tag{11.220}$$

$$\left(\mathbf{C}_{\hat{\eta}} \right)^{(s+1)} = \mathbf{C}_{\mathbf{y}} - \mathbf{C}_{\mathbf{y}} \mathbf{B}^{\mathrm{T}} \mathbf{G}_{\mathbf{B}} \mathbf{B} \mathbf{C}_{\mathbf{y}}. \tag{11.221}$$

Example 11.6 Computing and measuring data are the same as in Example 11.5. The temperature distribution on the wall thickness was determined by means of the finite volume method. The sensor (the flat wall) division into finite volumes is presented in Fig. 11.8. The location of measuring points is the same as in Example 11.5. The temperature distribution on the wall thickness is determined so that the following sum of squares:

$$S = \sum_{i-1}^{3} \frac{(y_i - \tilde{\eta}_i)^2}{\sigma_i^2}$$

should reach the minimum. Furthermore, the condition resulting from the need to satisfy the energy balance equation for node 2, has to be met:

$$f_1(\mathbf{\eta}) \equiv \frac{\lambda(\eta_1) + \lambda(\eta_2)}{2\lambda_0} (\eta_1 - \eta_2) + \frac{\lambda(\eta_3) + \lambda(\eta_2)}{2\lambda_0} (\eta_3 - \eta_2) = 0.$$

Changes in thermal conductivity $\lambda(\eta)$ are defined by the formula presented in Example 11.4, which gives: $\lambda_0 = \lambda(\eta = 0) = 14.65$ W/(m K). Assuming the first approximation as:

$$\mathbf{\varepsilon}^{(0)} = \left\{ \begin{array}{c} 0 \\ 0 \\ 0 \end{array} \right\}, \mathbf{\eta}^{(0)} = \mathbf{y} - \mathbf{\varepsilon}^{(0)} = \mathbf{y} = \left\{ \begin{array}{c} 315.3 \\ 252.7 \\ 195.7 \end{array} \right\}$$

the following is obtained:

$$c_1^{(0)} = f_1\left(\boldsymbol{\eta}^{(0)}\right) = \frac{19.1903 + 18.2888}{2 \cdot 14.65}(315.3 - 252.7) + \frac{17.4681 + 18.2888}{2 \cdot 14.65} \cdot$$
$$\cdot (195.7 - 252.7) = 80.07484 - 69.5612 = 10.5137 \text{ K},$$

$$\left.\frac{\partial f_1}{\partial \eta_1}\right|_{\boldsymbol{\eta}^{(0)}} = \frac{b}{2\lambda_0}\left(\eta_1^{(0)} - \eta_2^{(0)}\right) + \frac{\lambda\left(\eta_1^{(0)}\right) + \lambda\left(\eta_2^{(0)}\right)}{2\lambda_0}$$
$$= \frac{0.0144}{2 \cdot 14.65} \cdot (315.3 - 252.7) + \frac{19.1903 + 18.2888}{2 \cdot 14.65} = 1.3099,$$

$$\left.\frac{\partial f_1}{\partial \eta_2}\right|_{\boldsymbol{\eta}^{(0)}} = \frac{b}{2\lambda_0}\left(\eta_1^{(0)} - \eta_2^{(0)} + \eta_3^{(0)}\right) + \frac{\lambda\left(\eta_1^{(0)}\right) + \lambda\left(\eta_2^{(0)}\right) + \lambda\left(\eta_3^{(0)}\right)}{2\lambda_0}$$
$$= \frac{0.0144}{2 \cdot 14.65} \cdot (315.3 - 252.7 + 195.7) - \frac{19.1903 + 2 \cdot 18.2888 + 17.4681}{2 \cdot 14.65} = -2.4968,$$

$$\left.\frac{\partial f_1}{\partial \eta_3}\right|_{\boldsymbol{\eta}^{(0)}} = \frac{b}{2\lambda_0}\left(\eta_3^{(0)} - \eta_2^{(0)}\right) + \frac{\lambda\left(\eta_3^{(0)}\right) + \lambda\left(\eta_2^{(0)}\right)}{2\lambda_0}$$
$$= \frac{0.0144}{2 \cdot 14.65} \cdot (195.7 - 252.7) + \frac{17.4681 + 18.2888}{2 \cdot 14.65} = 1.1924,$$
$$\mathbf{B}^{(0)} = (1.3099 \quad -2.4968 \quad 1.1924),$$

$$\mathbf{G}_{\mathbf{B}}^{(0)} = \left(\mathbf{B}^{(0)}\mathbf{C}_{\mathbf{y}}\left(\mathbf{B}^{\mathbf{T}}\right)^{(0)}\right)^{-1} = \left\{(1.3099 \quad -2.4968 \quad 1.1924) \times \begin{bmatrix} 1 & 0 & 0 \\ 0 & 4 & 0 \\ 0 & 0 & 2.25 \end{bmatrix}\right.$$

$$\left.\times \left\{\begin{matrix} 1.3099 \\ -2.4968 \\ 1.1924 \end{matrix}\right\}\right\}^{-1} = \frac{1}{29.8502} = 0.0335,$$

$$\left(\mathbf{C}_{\mathbf{y}}\mathbf{B}^{\mathbf{T}}\right)^{(0)} = \begin{bmatrix} 1 & 0 & 0 \\ 0 & 4 & 0 \\ 0 & 0 & 2.25 \end{bmatrix} \times \left\{\begin{matrix} 1.3099 \\ -2.4968 \\ 1.1924 \end{matrix}\right\} = \left\{\begin{matrix} 1.3099 \\ -9.9871 \\ 2.6828 \end{matrix}\right\},$$

$$\boldsymbol{\varepsilon}^{(1)} = \left(\mathbf{C}_{\mathbf{y}}\mathbf{B}^{\mathbf{T}}\right)^{(0)}\mathbf{G}_{\mathbf{B}}^{(0)}c^{(0)} = \left\{\begin{matrix} 1.3099 \\ -9.9871 \\ 2.6828 \end{matrix}\right\} \times \frac{1}{29.8502} \times 10.537 = \left\{\begin{matrix} 0.4614 \\ -3.5176 \\ 0.9449 \end{matrix}\right\},$$

$$\boldsymbol{\eta}^{(1)} = \mathbf{y} - \tilde{\boldsymbol{\varepsilon}}^{(1)} = \left\{\begin{matrix} 315.3 \\ 252.7 \\ 195.7 \end{matrix}\right\} - \left\{\begin{matrix} 0.4614 \\ -3.5176 \\ 0.9449 \end{matrix}\right\} = \left\{\begin{matrix} 314.8386 \\ 256.2176 \\ 194.7551 \end{matrix}\right\}.$$

The second approximation $\mathbf{\eta}^{(2)}$ was determined similarly. Calculating

$$f_1\left(\mathbf{\eta}^{(1)}\right) = 1.28066 \cdot 58.62103 + 1.22164 \cdot (-61.46252) = -0.011626,$$

$$\left.\frac{\partial f_1}{\partial \eta_1}\right|_{\mathbf{\eta}^{(1)}} = 1.30947,$$

$$\left.\frac{\partial f_1}{\partial \eta_2}\right|_{\mathbf{\eta}^{(1)}} = -2.50369,$$

$$\left.\frac{\partial f_1}{\partial \eta_3}\right|_{\mathbf{\eta}^{(1)}} = 1.19143,$$

$$c_1^{(1)} = -0.011626 + 1.30947 \cdot 0.4614 - 2.50369 \cdot (-3.5176) + 1.19143 \cdot 0.9449$$
$$= 10.525307 \text{ K},$$

$$\mathbf{B}^{(1)} = (\, 1.3095 \quad -2.5037 \quad 1.1914 \,),$$

$$\mathbf{G}_\mathbf{B}^{(1)} = \left(\mathbf{B}^{(1)}\mathbf{C}_\mathbf{y}(\mathbf{B}^\mathbf{T})^{(1)}\right)^{-1} = \left\{ (\, 1.3095 \quad -2.5037 \quad 1.1914\,) \times \begin{bmatrix} 1 & 0 & 0 \\ 0 & 4 & 0 \\ 0 & 0 & 2.25 \end{bmatrix} \right.$$

$$\left. \times \left\{ \begin{matrix} 1.3095 \\ -2.5037 \\ 1.1914 \end{matrix} \right\} \right\}^{-1} = \frac{1}{29.98247},$$

$$(\mathbf{C}_\mathbf{y}\mathbf{B}^\mathbf{T})^{(1)} = \left\{ \begin{matrix} 1.30947 \\ -10.01476 \\ 2.68072 \end{matrix} \right\},$$

it is possible to determine the vector of errors (corrections):

$$\mathbf{\epsilon}^{(2)} = \left(\mathbf{C}_\mathbf{y}\mathbf{B}^\mathbf{T}\mathbf{G}_\mathbf{B}^{-1}\mathbf{c}\right)^{(1)},$$

$$\mathbf{\epsilon}^{(2)} = \left\{ \begin{matrix} 0.4597 \\ -3.5157 \\ 0.9411 \end{matrix} \right\}$$

and the distribution of temperature $\mathbf{\eta}^{(2)}$:

$$\mathbf{\eta}^{(2)} = \mathbf{y} - \mathbf{\epsilon}^{(2)} = \left\{ \begin{matrix} 315.3 \\ 252.7 \\ 195.7 \end{matrix} \right\} - \left\{ \begin{matrix} 0.4597 \\ -3.5157 \\ 0.9411 \end{matrix} \right\} = \left\{ \begin{matrix} 314.8403 \\ 256.2157 \\ 194.7589 \end{matrix} \right\}.$$

It can be seen that the temperature distribution found both in the first and in the second iteration ($\mathbf{\eta}^{(1)}$ and $\mathbf{\eta}^{(2)}$, respectively) is almost identical to the distribution determined by means of the Gauss-Newton method in Example 11.4. The covariance matrix has the following form:

$$\mathbf{C}_{\tilde{\eta}}^{(2)} = \mathbf{C}_y - \mathbf{C}_y \left(\mathbf{B}^\mathsf{T} \mathbf{G}_\mathbf{B} \mathbf{B} \right)^{(1)} \mathbf{C}_y = \begin{bmatrix} 1 & 0 & 0 \\ 0 & 4 & 0 \\ 0 & 0 & 2.25 \end{bmatrix} - \left\{ \begin{matrix} 1.3099 \\ -2.4968 \\ 1.1924 \end{matrix} \right\} \times \frac{1}{29.9825}$$

$$\cdot \left(1.3099 \quad -2.4968 \quad 1.1924 \right) \times \begin{bmatrix} 1 & 0 & 0 \\ 0 & 4 & 0 \\ 0 & 0 & 2.25 \end{bmatrix} = \begin{bmatrix} 0.9428 & 0.4374 & -0.1171 \\ 0.4374 & 0.6549 & 0.8954 \\ -0.1171 & 0.8954 & 2.0103 \end{bmatrix}.$$

The calculated covariance matrix is almost the same as in Example 11.4. Adopting the same procedure as in Examples 11.3 and 11.4, confidence intervals can be determined for temperatures $\tilde{\eta}$ obtained from approximations.

Analysing the results obtained in Examples 11.3–11.5, it may be concluded that the presented methods are very effective in solving linear and nonlinear least squares problems. They also make it possible to determine confidence intervals for the found coefficients $\tilde{\mathbf{a}}$ and function $\tilde{\eta}(\tilde{\mathbf{a}})$ if the variances in measured temperatures $y_i, i = 1, \ldots, n$ are known or unknown. Performing the computations, special attention should be focused on the process of determination of inverse matrices because the matrix inversion operation is often inaccurate.

11.6   Final Comments

In practical calculations it is best to use the error propagation rule [cf. formula (11.28)], which can be applied to assess both the uncertainty of quantity z_i determined indirectly based on directly measured quantities x_j and the uncertainty of coefficients $\tilde{a}_j, j = 1, \ldots, N$ found by means of the least squares method. Partial derivatives $\partial z_i / \partial x_j$, also referred to as sensitivity coefficients, can be calculated analytically for a simple form of function $z_i(\mathbf{x})$ measured indirectly or a simple form of approximation function $\eta(\mathbf{x}, \mathbf{a})$ used in the least squares method. If functions $z_i(\mathbf{x})$ or $\eta(\mathbf{x}, \mathbf{a})$ are composite, sensitivity coefficients can be calculated using difference quotients, e.g. the central difference quotient defined by formula (11.35). If function $\eta(\mathbf{x}, \mathbf{a})$ is nonlinear with respect to sought parameters $\mathbf{a}$, the nonlinear least squares problem is solved iteratively, e.g. using the Gauss-Newton or the Levenberg-Marquardt method [25, 29, 261].

As the methods produce approximate results, sensitivity coefficients $\partial \eta(\mathbf{x_i}, \mathbf{a}) / \partial y_j$ and $\partial a_i / \partial y_j$, where y_j denotes temperature measured directly, should be determined with great accuracy. Sensitivity coefficients are usually calculated by means of a central difference quotient with a form similar to formula (11.35). A change in the measured quantity y_j by a slight value δ may result in a slight change in a_i, comparable to the error of this parameter determination by means of the iterative optimization method used to solve the least squares problem.

Solving inverse problems by means of the least squares method involves a great risk that sensitivity coefficients will be found with inadequate accuracy. The problems are often ill-posed, and a slight change in the measured quantity y_j, typically—the temperature inside a solid body or the fluid temperature at the exchanger outlet, may make the values of parameters found at a disturbed piece of data $y_j \pm \delta$ differ substantially from those obtained if y_j is not disturbed. Another method of finding sensitivity coefficients in inverse linear initial-boundary value problems is to differentiate the partial equation, the boundary conditions and the initial condition with respect to a given parameter, and then solve the obtained new initial-boundary value problem [118, 136, 223, 239]. In this way, it is possible to determine distributions of sensitivity coefficients in time and space in the entire area under analysis.

The confidence intervals determined from formulae (11.75) or (11.135) for coefficients a_i, or from formulae (11.78) or (11.144) for function η_i, are too wide, especially at a small number of the degrees of freedom $(n - N - 1)$, because the t-Student distribution quantile for a small number of freedom degrees $(n - N - 1)$ is big.

Also the mean square error s_y defined by formula (11.63) and being an approximation of standard deviation σ is in many cases too high.

The advantage of adopting s_y to estimate σ is that this move makes it possible to take account of the inaccuracy caused by inappropriate selection of approximation function $\eta(\mathbf{x}, \mathbf{a})$. If the uncertainty of the indirectly measured (determined) quantity η is found directly from the error propagation rule (11.28), which is also the basis for formula (D1) presented in Appendix D and recommended by the ASME, the error caused by wrong selection of function $\eta(\mathbf{x}, \mathbf{a})$ is not taken into consideration. This is due to the fact that if the mean square error, or any other measure of the measurement uncertainty, is calculated from the error propagation rule, function η is assumed to be accurate.

The disadvantage of formulae (11.78) or (11.144) is that during the assessment of the confidence interval (the measurement uncertainty) for function $\eta(\mathbf{x}, \mathbf{a})$, they take account only of the measurement uncertainty of the physical quantity the function describes. If heat flux $\dot{q}$ is found based on temperatures measured inside the sensor, s_y includes only the uncertainties related to the sensor temperature measurement. No account is taken of the uncertainty of thermal conductivity λ, the sensor thickness L or the temperature sensors location $d_i, i = 1, \ldots, n$. The above-mentioned measurement uncertainties have an impact on the value of the minimum sum of squares S, and consequently—on the value of s_y.

However, it is impossible to make an explicit evaluation of the effect of the measurement uncertainty of these quantities on the uncertainty of determination of function $\eta(\mathbf{x}, \mathbf{a})$.

In practical calculations, it is best to use the error propagation rule expressed by formula (11.28) or the ASME formula (D1) from Appendix D to assess the uncertainty of the quantities or coefficients determined indirectly. The formulae can be applied both in the case when the explicit strict dependence of indirectly determined quantity z on directly measured quantities $x_i, i = 1, \ldots, N$ is known and in the situation where function $z = \eta$ is determined using the least squares method, and the form of function $\eta = \eta(\mathbf{a})$ is selected by the person performing the calculations.

Owing to the application of the Gaussian error propagation rule as defined in (11.28), it is easy to take account of the quantities that affect the uncertainty of the quantity determined indirectly. More information on determination of the uncertainty of the results of indirect measurements can be found in [63, 91, 126, 127, 168].

The error propagation rule can also be used to assess the uncertainty of the characteristics of devices made of elements or subassemblies with known uncertainties of the parameters they reach [126].

It should be added that normal equations written in a matrix form are the basis for the science referred to as econometrics [33, 54, 73, 74, 96, 113, 215, 375]. In Polish literature, econometrics is generally understood as a science dealing with special statistical methods applied to statistical (empirical) data which, however, are not obtained from the design of experiments.

The values of coefficients $\bar{a}$ found from normal equations are burdened with rounding errors, especially if the number of determined parameters (N) is high. For this reason, a number of effective methods [25, 29, 95, 174, 239, 242, 261, 288, 293] have been developed to solve overdetermined systems of linear equations in terms of the least squares method. The most important of them are the modified Gram-Schmidt orthogonalization method [29, 94, 174, 239, 288] or the matrix singular value decomposition (SVD) [94, 174, 239, 261].

It has to be emphasized that the matrix calculus application to find unknown coefficients and determine the uncertainty of their determination is very useful, if not indispensable, when the number of measuring data is very high. In this case, using the error propagation rule as defined by formula (11.28) is troublesome due to the need to calculate a great number of sensitivity coefficients. In the matrix calculus, there are many library programs available to perform various mathematical operations on matrices, which makes the computation process much easier.

Chapter 12
Measurements of Basic Parameters in Experimental Testing of Heat Exchangers

The primary aim of the heat exchanger testing is to measure the exchanger pressure drops Δp_h and Δp_c on the side of the hot and the cold medium, respectively. The heat flow rate $\dot{Q}$ transferred in the exchanger from the hot to the cold fluid, and overall heat transfer k are also determined. The above-mentioned physical quantities are found based on measurements of the mass flow rate and the inlet and outlet temperatures of the two fluids.

12.1 Determination of the Heat Flow Rate Exchanged Between Fluids and the Overall Heat Transfer Coefficient

The quantities to be found in the plate fin-and-tube heat exchanger thermal and flow testing are the pressure drop on the side of the two mediums and the heat flow rate transferred from the hot fluid $(\dot{Q}_h)$ to the cold fluid $(\dot{Q}_c)$. If the heat exchanger outer surface thermal insulation is perfect, the two heat flow rates are equal to each other, i.e. $\dot{Q}_h = \dot{Q}_c$. In reality, heat flow rates $\dot{Q}_h$ and $\dot{Q}_c$ determined based on measurements are different, which results from the inaccuracy of the measurement of the medium mass flow rate $\dot{m}$ and the fluid mass-averaged temperature $\bar{T}$. Heat flow rates $\dot{Q}_h$ and $\dot{Q}_c$ can be found using the following formulae:

$$\dot{Q}_h = \dot{m}_h\left(i_{h,wlot} - i_{h,wylot}\right) = \dot{m}_h\bar{c}_{ph}\left(\bar{T}_{h,wlot} - \bar{T}_{h,wylot}\right), \tag{12.1}$$

$$\dot{Q}_c = \dot{m}_c\left(i_{c,wylot} - i_{c,wlot}\right) = \dot{m}_c\bar{c}_{pw}\left(\bar{T}_{c,wylot} - \bar{T}_{c,wlot}\right), \tag{12.2}$$

where: $\dot{m}_h$ and $\dot{m}_c$—hot and cold medium mass flow rate, respectively, kg/s; $i_{h,wlot}$ and $i_{h,wylot}$—hot medium enthalpy at the exchanger inlet and outlet, respectively,

© Springer International Publishing AG, part of Springer Nature 2019
D. Taler, *Numerical Modelling and Experimental Testing of Heat Exchangers*,
Studies in Systems, Decision and Control 161,
https://doi.org/10.1007/978-3-319-91128-1_12

J/kg; $i_{c,wlot}$ and $i_{c,wylot}$—cold medium enthalpy at the exchanger inlet and outlet, respectively, J/kg; $\bar{T}_{h,wlot}$ and $\bar{T}_{h,wylot}$—hot medium mass-averaged temperature at the exchanger inlet and outlet, respectively, °C; $\bar{T}_{c,wlot}$ and $\bar{T}_{c,wylot}$—cold medium mass-averaged temperature at the exchanger inlet and outlet, respectively, °C; $\bar{c}_{ph} = c_{ph}\Big|_{\bar{T}_{h,wylot}}^{\bar{T}_{h,wlot}}$—hot medium mean specific heat at constant pressure in the temperature range from $\bar{T}_{h,wylot}$ to $\bar{T}_{h,wlot}$, J/(kg K); $\bar{c}_{pc} = c_{pc}\Big|_{\bar{T}_{c,wlot}}^{\bar{T}_{c,wylot}}$—cold medium mean specific heat at constant pressure in the temperature range from $\bar{T}_{c,wlot}$ to $\bar{T}_{c,wylot}$, J/(kg K).

The mean heat flow rate from the hot to the cold medium $\dot{Q}_m$ can be found using the following formula:

$$\dot{Q}_m = \frac{\dot{Q}_c + \dot{Q}_h}{2}. \tag{12.3}$$

Overall heat transfer coefficient k_A, related to the heat transfer surface area A, is defined as:

$$k_A = \frac{\dot{Q}_m}{F A \Delta T_m}, \tag{12.4}$$

where: $F \leq 1$—correction factor taking account of the fact that the exchanger under analysis is not a countercurrent one; for a countercurrent heat exchanger, it should be assumed that $F = 1$.

Logarithmic mean temperature difference ΔT_m is calculated assuming that the exchanger is a countercurrent one, regardless of the exchanger actual flow system. The formula for ΔT_m calculation is as follows:

$$\Delta T_m = \frac{\left(\bar{T}_{h,wlot} - \bar{T}_{c,wylot}\right) - \left(\bar{T}_{h,wylot} - \bar{T}_{c,wlot}\right)}{\ln\frac{\bar{T}_{h,wlot} - \bar{T}_{c,wylot}}{\bar{T}_{h,wylot} - \bar{T}_{c,wlot}}}. \tag{12.5}$$

It should be noted that formula (12.5) may be used for exchangers where no big changes occur in the physical properties of the mediums.

This mainly concerns cases where changes in the specific heat of the mediums with temperature are not too dramatic. The heat transfer coefficients on the hot and the cold medium side should not change substantially, either. For example, if superheated steam is cooled in one part of the exchanger and saturated steam is condensed in the other, each of the two parts should be considered as a separate heat exchanger due to the big differences between the heat transfer coefficient values. For each part, the mean temperature difference is determined using formula (12.5). Correction factor F in formula (12.4) can be found for different heat exchanger types from the charts presented in some books on the heat transfer, e.g. [122, 123, 169, 347].

12.2 Measurement of the Fluid Mean Velocity in the Channel

The fluid mass flow rate $\dot{m}$ can be calculated using the following formula:

$$\dot{m} = \dot{V} \rho(\bar{T}), \qquad (12.6)$$

where $\dot{V}$ is the fluid volume flow rate (m^3/s) and ρ is the fluid density (kg/m^3) in the fluid mass-averaged temperature $\bar{T}$ determined in the measuring cross-section.

It should be emphasized that the fluid density should be determined for the same cross-section of the channel where the fluid volume flow rate is measured. In steady states, mass flow rate $\dot{m}$ is constant, i.e.

$$\dot{m}_{wlot} = \dot{m}_{wylot} = \dot{m} = \text{const.} \qquad (12.7)$$

The fluid mass flow rate measurement creates no problems—the quantity can be measured using various instruments, e.g., metering orifices or flow meters [93, 175, 300, 332, 343]. The measurement of the gas mass flow rate is difficult and laborious due to the large surface area of the channel cross-section and considerable differences in the velocity field in the measuring cross-section, especially in the case of short ducts, where the flow is not fully developed hydraulically.

12.2.1 Measurement of the Fluid Volume Flow Rate Using the Velocity Distribution Integration

The fluid volume flow rate can be found provided the distribution of the fluid velocity in the channel cross-section is known. This measurement method is particularly useful for ducts with large cross-sections, where it is difficult to keep straight sections of pipelines up- and downstream the measuring instrument, as required by relevant standards. In such cases, the velocity distribution in the channel cross-section is complex and can only be determined experimentally. Flue gas stacks and ducts in large power boilers are an example.

This section presents a method where mean velocity $\bar{w}_z$ is determined by measuring the maximum velocity in the duct axis and a method where the mean velocity value is calculated based on the velocity measurements performed at selected points. The measuring points different arrangement in circular and rectangular cross-sections, the calculation principles and the instruments used to perform the measurements will be presented. The fluid velocity measurement using averaging probes enabling direct measurement of the mean velocity value will also be demonstrated.

The fluid volume or mass flow rates can be found by integrating the velocity distribution along the surface. The volume flow rate in the cross-section of a circular or rectangular duct can be determined if the function describing the velocity distribution is known. The function can be given in an analytical or a discrete form.

The mass flow rate through the channel cross-section with surface area $A_{c.s.}$ is defined as:

$$\dot{m} = - \int\limits_{A_{c.s.}} \rho \mathbf{w} \cdot \mathbf{n} \, dA. \tag{12.8}$$

The minus sign before the integral means that if the fluid mass flow rate enters the control area, quantity $\dot{m}$ is positive. In such a case, the angle between vectors $\mathbf{w}$ and $\mathbf{n}$ is obtuse, and the cosine of the angle between the vectors is negative.

If the fluid density is constant, i.e., the fluid is incompressible, or the changes in the fluid density in the measuring cross-section are slight, formula (12.8) can be written in the following form:

$$\dot{V} = - \int\limits_{A_{c.s.}} \mathbf{w} \cdot \mathbf{n} \, dA = \bar{w}_n A_{c.s.}, \tag{12.9}$$

where $\bar{w}_n$ is the mean value of the normal component on surface area $A_{c.s.}$

Mean normal velocity $\bar{w}_n$ is defined as:

$$\bar{w}_n = - \frac{1}{A_{c.s.}} \int\limits_{} \mathbf{w} \cdot \mathbf{n} \, dA = \frac{1}{A_{c.s.}} \int\limits_{} w_n \, dA. \tag{12.10}$$

Expressions (12.9) and (12.10) enable correct determination of the fluid volume flow rate and mean velocity even if eddies occur in the measuring cross-section, i.e., if the velocity vector senses are different. Still, for the sake of the measurement accuracy, it is better to choose measuring cross-sections where eddies do not arise, and the velocity vector sense is the same in the entire measuring cross-section. If the measurement location is selected according to binding standards, no eddies arise in the measuring cross-section, and there are no backflows of the fluid.

The fluid volume flow rate in a tube with a circular cross-section with the inner surface radius r_w, assuming that the fluid flow is one-dimensional, is calculated from the following formula:

$$\dot{V} = \int\limits_0^{r_w} \int\limits_0^{2\pi} w_z(r, \phi) r \, d\phi \, dr. \tag{12.11}$$

The fluid volume flow rate can be calculated from formula (12.11) if velocity profile $w_z(r, \phi)$ is known.

In long, straight sections of the pipeline, the velocity profile in the cross-section is fully developed, and it can be described by an analytical function.

In short pipelines with valves or knees, the velocity profile in the cross-section is not fully developed, and it cannot be described using an analytical function. In such a case, the fluid velocity measurement has to be performed in points located in the pipeline cross-section. The number of measuring points may vary from several to a few dozen, depending on the measurement required accuracy and the actual velocity profile.

For laminar flows, the velocity distribution in a circular tube cross-section is defined by the following formula:

$$w_z = w_{max}\left(1 - \frac{r^2}{r_w^2}\right). \tag{12.12}$$

Mean velocity is defined as:

$$\bar{w}_z = \frac{1}{\pi r_w^2} \int_0^{r_w} 2\pi r\, w_{max}\left(1 - \frac{r^2}{r_w^2}\right) dr = \frac{w_{max}}{2}. \tag{12.13}$$

Based on the measurement of velocity w_{max} in the duct axis, the volume and the mass flow rates $\left(\dot{V} \text{ and } \dot{m}\right)$ can be determined:

$$\dot{V} = \pi r_w^2\, \bar{w}_z = \frac{1}{2}\pi r_w^2\, w_{max}, \tag{12.14}$$

$$\dot{m} = \rho \dot{V} = \rho \pi r_w^2\, \bar{w}_z = \frac{1}{2}\pi r_w^2\, \rho\, w_{max}. \tag{12.15}$$

If the fluid flow is turbulent, the velocity distribution in the tube cross-section can be approximated using the following function [66, 361]:

$$w_z = w_{max}\left(1 - \frac{r}{r_w}\right)^{1/n}, \tag{12.16}$$

where w_{max} is the maximum velocity in the duct axis for $r = 0$. For smooth tubes, exponent n in formula (12.16) varies from $n = 6$ for $\text{Re} = 4 \times 10^3$ to $n = 10$ for $\text{Re} = 2 \times 10^6$. For ducts with rough walls, exponent n varies from $n = 4$ to $n = 5$. Formula (12.16) is not valid close to the channel surface. The fluid flow mean velocity in the tube $\bar{w}_z$ is determined by the following formula:

$$\bar{w}_z = \frac{1}{\pi r_w^2} \int_0^{r_w} 2\pi w_z r\, dr = \frac{2}{r_w^2} \int_0^{r_w} w_z r\, dr. \tag{12.17}$$

Substituting (12.16) in (12.17) and performing integration, the following is obtained:

$$\bar{w}_z = \frac{2w_{\max}}{r_w^2}\left[\frac{1}{\frac{1}{r_w^2}\left(\frac{1}{n}+2\right)}\left(1-\frac{r}{r_w}\right)^{\frac{1}{n}} - \frac{1}{\frac{1}{r_w^2}\left(\frac{1}{n}+1\right)}\left(1-\frac{r}{r_w}\right)^{\frac{1}{n}+1}\right]_0^{r_w}$$

$$= -2\left(\frac{n}{2n+1}-\frac{n}{n+1}\right)w_{\max} = \frac{2n^2}{(2n+1)(n+1)}w_{\max}. \tag{12.18}$$

Assuming that $n = 7$, formula (12.18) gives:

$$\bar{w}_z = \frac{49}{60}w_{\max} = 0.8167w_{\max}.$$

Exponent n for smooth tubes can also be calculated from the following formula [123]:

$$n = 0.2944 + 1.71367\ln\text{Re} - 0.1890(\ln\text{Re})^2 + 0.007937(\ln\text{Re})^3. \tag{12.19}$$

The mean velocity value in smooth tubes is also calculated using a different formula:

$$\bar{w}_z = \frac{w_{\max}}{1+3.45\sqrt{\frac{\xi}{8}}}, \tag{12.20}$$

where friction factor ξ can be found using the formulae given in Chap. 4. Table 12.1 presents the $\bar{w}_z/w_{\max}$ ratio values for a duct with a circular cross-section for different values of the Reynolds number.

The method of determining mean velocity $\bar{w}_z$ based on the measurement of maximum velocity $w_{\max}$ is not easy to apply in practice because the maximum velocity measuring point should be located at a distance of $(40 \div 50)\,d_w$ from a straight duct inlet. Real ducts have bends, valves, fans, narrowings or widenings, all of which disturb the fully developed velocity profile. The velocity measurement results obtained for air ducts with circular cross-sections indicate that the velocity profile may be highly non-uniform, especially in the vicinity of bends, and almost flat—downstream fans.

The measurement of the maximum velocity in a rectangular duct makes it also possible to find the mean velocity (in a rectangular ($2a \times 2b$) channel). After the fully developed velocity profile is approximated using the following function [163]:

Table 12.1 Values of the mean-to-maximum velocity ratio ($\bar{w}_z/w_{max}$) in a circular duct depending on the Reynolds number

Re	2.3×10^3	$(4.5-5.8) \times 10^4$	2×10^5	6.4×10^5	2×10^6
$\frac{\bar{w}_z}{w_{max}}$	0.791	0.817	0.837	0.853	0.866

$$\frac{w_z}{w_{max}} = \left[\left(1 - \frac{x}{a}\right)\left(1 - \frac{y}{b}\right)\right]^{1/n} \tag{12.21}$$

mean velocity can be calculated using the following formula:

$$\bar{w}_z = \frac{1}{ab} \int_0^a \int_0^b w_z \, dx \, dy. \tag{12.22}$$

The fluid volume flow rate can, therefore, be calculated as follows:

$$\dot{V} = 4ab\bar{w}_z. \tag{12.23}$$

The volume flow rate can be determined based on the velocity measurement in the duct cross-section center only in the case of a fully developed laminar or hydraulic flow at a considerable distance from the duct inlet or other obstacles disturbing the flow.

Finding the fluid flow mean velocity by measuring velocity at selected points of the cross-section is much more accurate.

The fluid volume flow rate $\dot{V}$ based on velocity measurements in selected points located in the duct cross-section is usually determined by calculating integral (12.9) using the rectangle method. If a duct with a circular cross-section is divided into n areas in the radial direction (one circle in the center of the duct and $n - 1$ annuli) and m sectors in the circumferential direction, using the rectangle method, integral (12.1) can be approximated as follows:

$$\dot{V} = \sum_{j=1}^m \sum_{i=1}^n (w_z)_{i,j} \frac{2\pi}{m} r_{pi} \, \Delta r_i, \tag{12.24}$$

where $(w_z)_{i,j}$ is the fluid velocity in point (i, j) located in the quadrangle center of gravity, and r_{pi} is the radius of point (i, j) where local velocity is measured (cf. Fig. 12.1).

If velocity changes in the radial direction only, the fluid volume flow rate $\dot{V}$ is defined as:

$$\dot{V} = 2\pi \int_0^{r_w} w_{zi} r_{pi} \, dr \simeq \sum_{i=1}^n w_{zi} \left(2\pi r_{pi} \, \Delta r_i\right). \tag{12.25}$$

Fig. 12.1 Illustration of the
volume flow rate calculation
using formula (12.24),
$\Delta\phi = 2\pi/m$, m—number of
circle sectors in the
cross-section division

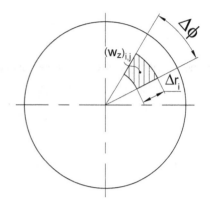

The fluid mean velocity in the measuring cross-section is:

$$\bar{w}_z = \frac{\dot{V}}{\pi r_w^2} = \frac{\sum\limits_{i=1}^{n} w_{zi}\left(2\,\pi\,r_{pi}\,\Delta r_i\right)}{\sum\limits_{i=1}^{n} \left(2\,\pi\,r_{pi}\,\Delta r_i\right)}. \tag{12.26}$$

If the tube cross-section is divided into annuli with equal surface areas, i.e., if the following equality holds:

$$2\,\pi\,r_{pi}\Delta r_i = 2\,\pi\,r_{pi+1}\Delta r_{i+1} = \text{const}, \quad i = 1,\ldots,\, n-1, \tag{12.27}$$

Formula (12.26) is reduced to the following form:

$$\bar{w}_z = \frac{\dot{V}}{\pi r_w^2} = \frac{\sum\limits_{i=1}^{n} w_{zi}}{n}. \tag{12.28}$$

Point r_{pi}, where local velocity w_{zi} is measured, is located in the middle of the radius width, i.e., the coordinate of the velocity measuring point calculated from the following expression:

$$r_{pi} = (r_{i-1} + r_i)/2, \quad i = 1,\ldots,\, n. \tag{12.29}$$

The location of point r_{pi} can be selected so that the surface areas of annuli with widths $(r_{pi}-r_{i-1})$ and (r_i-r_{pi}) should be identical. The duct cross-section is usually divided into n annuli with equal surface areas (cf. Fig. 12.2).

If the i-th annulus outer radius is marked as r_i and the inner radius as r_{i-1}, the radii of individual annuli can be calculated from the condition of equality of the surface areas of all the annuli that the circular duct entire cross-section is divided into.

Fig. 12.2 Division of a circular duct cross-section into annuli with identical surface areas

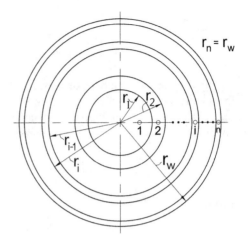

For the first measuring point, the circle radius r_1 is found from the following equation:

$$\pi r_1^2 = \frac{\pi r_w^2}{n},$$

which gives:

$$r_1 = r_w\sqrt{\frac{1}{n}}. \tag{12.30}$$

Radius r_2 of annulus 2 is determined using the following equation:

$$\pi\left(r_2^2 - r_1^2\right) = \frac{\pi\, r_w^2}{n}. \tag{12.31}$$

Considering (12.30) in (12.31), the following is obtained:

$$r_2 = r_w\sqrt{\frac{2}{n}}. \tag{12.32}$$

The i-th annulus outer radius r_i can be found in the same way:

$$r_i = r_w\sqrt{\frac{i}{n}}. \tag{12.33}$$

Dividing the duct cross-section into six equal areas $(n = 6)$, where the center area is a circle with radius r_1, the radii of individual annuli are as follows:

$$r_1 = 0.4082\, r_w \quad r_4 = 0.8165\, r_w$$
$$r_2 = 0.5774\, r_w \quad r_5 = 0.9129\, r_w$$
$$r_3 = 0.7071\, r_w \quad r_6 = r_w.$$

There are several methods of selecting the velocity measuring point inside the i-th annulus. One of the more popular ways is to measure velocity in the point lying on radius r_{pi} which divides the i-th annulus into two equal surface areas. The condition of equality of the surface areas:

$$\pi\left(r_i^2 - r_{pi}^2\right) = \pi\left(r_{pi}^2 - r_{i-1}^2\right) \tag{12.34}$$

gives:

$$r_{pi}^2 = \frac{r_i^2 + r_{i-1}^2}{2}. \tag{12.35}$$

Substituting (12.33) in (12.34), the following is obtained after transformations:

$$r_{pi} = r_w\sqrt{\frac{2i - 1}{2n}} = \frac{d}{2}\sqrt{\frac{2i - 1}{2n}}, \quad i = 1, \ldots, n, \tag{12.36}$$

where r_{pi} denotes the coordinate of the i-th velocity measuring point and $d = 2r_w$ is the inner diameter of the circular duct.

In practice, it is more convenient to measure the velocity measuring point distance from the duct inner surface:

$$x_i = r_w\left(1 - \sqrt{\frac{2i - 1}{2n}}\right), \quad x_{n+i} = r_w\left(1 + \sqrt{\frac{2i - 1}{2n}}\right), i = 1, \ldots, n, \tag{12.37}$$

where i is the measuring point number and n—the number of all the elements with equal surface areas that the tube cross-section is divided into. The number of areas n should not be smaller than $n = 3$ for tubes with a diameter of (150–300) mm; for tubes with a diameter of 300–900 mm, it should be from $n = 4$ to $n = 10$ [200].

The fluid flow mean velocity in the channel cross-section is equal to the arithmetic mean of velocities measured in all measuring points.

The location of velocity measuring points x_i, $i = 1, \ldots, 8$ for $n = 4$ is presented in Fig. 12.3.

Dividing the cross-section into four areas ($n = 4$), coordinates x_i, $i = 1, \ldots, 8$ are as follows (cf. Fig. 12.3):

Fig. 12.3 Division of the circular duct cross-section into annuli with identical surface areas (the measuring points are marked with circles), $n = 4$

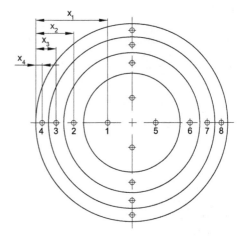

$$x_1 = \frac{d}{2}\left(1 - \sqrt{\frac{1}{8}}\right) = 0.32322d, \quad x_2 = \frac{d}{2}\left(1 - \sqrt{\frac{3}{8}}\right) = 0.19381d,$$

$$x_3 = \frac{d}{2}\left(1 - \sqrt{\frac{5}{8}}\right) = 0.10472d, \quad x_4 = \frac{d}{2}\left(1 - \sqrt{\frac{7}{8}}\right) = 0.03229d,$$

$$x_5 = \frac{d}{2}\left(1 + \sqrt{\frac{1}{8}}\right) = 0.67678d, \quad x_6 = \frac{d}{2}\left(1 + \sqrt{\frac{3}{8}}\right) = 0.80619d,$$
(12.38)

$$x_7 = \frac{d}{2}\left(1 + \sqrt{\frac{5}{8}}\right) = 0.89528d, \quad x_8 = \frac{d}{2}\left(1 + \sqrt{\frac{7}{8}}\right) = 0.96771d.$$

where $d = 2r_w$ is the duct inner diameter.

Velocity is measured in all the points lying on two lines perpendicular to each other which pass through the circular duct axis.

Figure 12.3 presents the location of velocity measuring points for $n = 4$.

If the velocity profile in the tube cross-section is not fully developed and the velocity field asymmetry occurs with respect to the horizontal or vertical plane passing through the duct axis, it is better to locate the points measuring the flow local velocity in three different planes (cf. Figs. 12.4 and 12.5).

Figure 12.5 presents the location of measuring points as recommended by the ANSI/ASHRAE 84-1991 standard [200]. The coordinates of the velocity measuring points x_i for $n = 4$ are shown in Fig. 12.5. They are a bit different from those defined by formulae (12.38). Velocity is measured in measuring points lying on three lines passing through the duct axis. The fluid flow mean velocity is equal to the arithmetic mean of velocities measured in all the measuring points.

If the duct cross-section is divided into n equal surface areas with outer radii defined by formulae (12.33), it is also possible to locate the velocity measuring points at an equal distance from the outer and the inner radius of a given annulus.

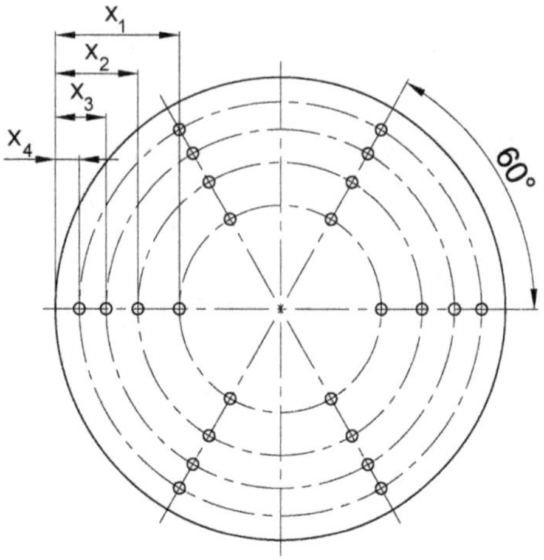

Fig. 12.4 Division of the circular duct cross-section into annuli with identical surface areas (the measuring points are marked with circles), n = 4

Fig. 12.5 Location of the velocity measuring points in a circular duct cross-section as recommended by the ANSI/ASHRAE 84-1991 standard [200]

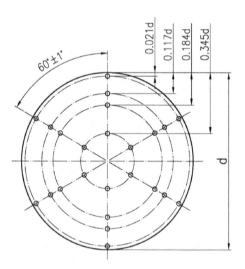

For annulus i with an inner radius r_{i-1} and outer radius r_i, the velocity measuring point coordinate is defined by the following formula:

$$r_{pi} = \frac{r_{i-1} + r_i}{2} = r_{i-1} + \frac{\Delta r_i}{2}, \tag{12.39}$$

Substituting (12.33) in (12.39), the following is obtained after transformations:

$$r_{pi} = \frac{d}{4}\sqrt{\frac{i}{n}}\left(1 + \sqrt{\frac{i-1}{i}}\right), \quad i = 1, \dots, n. \tag{12.40}$$

Coordinates x_i of the velocity measuring points counted from the duct inner surface (cf. Fig. 12.3) and calculated using formula (12.40) for $n = 4$ are as follows:

$$x_1 = 0.375d, \quad x_2 = 0.1982d, \quad x_3 = 0.1067d, \quad x_4 = 0.0335d,$$
$$x_5 = 0.625d, \quad x_6 = 0.8018d, \quad x_7 = 0.8933d, \quad x_8 = 0.9665d. \quad (12.41)$$

It should be added that dividing the duct cross-section into $n \geq 4$ annuli, high accuracy is achieved of the fluid volume flow rate measurement for all three ways of locating the velocity measuring points: defined by formulae (12.38) or (12.41), or as presented in Fig. 12.5.

If the mean velocity is measured in a rectangular channel, the channel cross-section is divided into rectangles with an identical surface area. Velocity is measured at the center of each rectangle. For channels with the cross-section surface area of 0.35 m², the cross-section should be divided into at least 16 equal elements. Measuring the fluid local velocities in the center of each rectangle, the arithmetic mean of all measured values is calculated, which is equal to the fluid mean velocity

Fig. 12.6 Measurement of the fluid flow mean velocity in a rectangular channel cross-section using sensors located at the centres of rectangles; **a** division of the rectangular channel cross-section into *m n* rectangles with equal surface areas, **b** mean velocity measurement using thermo-anemometer probes placed at the centers of rectangles with equal surface areas [84]

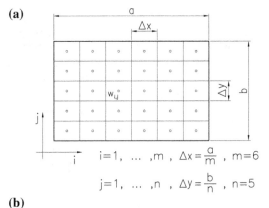

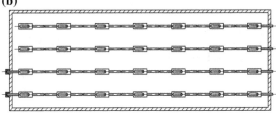

in the measuring cross-section. Figure 12.6a presents an example division of a rectangular channel into rectangles with equal surface areas. The mean velocity of air is calculated from the following formula:

$$\bar{w}_z = \frac{1}{mn} \sum_{i=1}^{m} \sum_{j=1}^{n} (w_z)_{i,j}.$$ (12.42)

The mean velocity measurement can be performed using some sensors located in the centers of the rectangles with an identical surface area (cf. Fig. 12.6b) [84].

It should be emphasized that the mean velocity determination based on the measurement of the local velocity values performed in many points of the cross-section is difficult as the procedure is somewhat laborious. If all local velocities are not measured at the same time instant, more significant measuring errors should be anticipated because in real conditions it is very difficult to make sure that the spatial profile of velocity does not change over time.

12.2.2 Averaging Probes

Mean velocity can be measured directly using averaging probes. Figure 12.7 shows an Annubar^{TM} averaging probe [146]. Total pressure is measured in six orifices arranged uniformly on the probe face. On each of the two lateral surfaces on the two sides, there are six orifices measuring static pressure. Averaging total and mean pressure in the channel cross-section based on the measured pressure difference Δp, it is possible to find the fluid flow mean velocity:

$$\bar{w}_z = \sqrt{\frac{2\Delta p}{\rho}}.$$ (12.43)

Fig. 12.7 Measurement of the fluid flow mean velocity using the Annubar^{TM} probe [146]

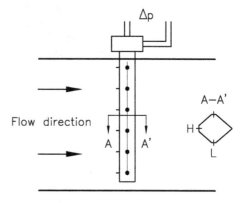

A different structure of the averaging probe is shown in Fig. 12.8. Total pressure is measured in six points located on the front part of the tube on the inflow side. Static pressure is measured in the orifice in the rear part of the tube. The probe presented in Fig. 12.8 [343] was used to measure the mean velocity of water with the pressure of 11 MPa and temperature of 320 °C in the power boiler downcomer tube in one of the power plants in Poland.

Averaging Prandtl tubes are often used to measure the mean velocity of air in the boiler installation ducts.

The dynamic pressure value measured using averaging tubes usually does not exceed 50 mm H_2O (about 500 Pa).

Averaging Pitot tubes can also be used to measure the fluid mean velocity in channels with a rectangular cross-section.

The pressure difference $\Delta p = p_c - p_s$ is measured using two averaging probes measuring mean dynamic pressure p_d and four orifices in the channel wall measuring static pressure p_s. The averaging probes are located on the rectangular channel diagonals.

An example will be presented for determining the mean velocity of air in a channel with a circular cross-section.

The air volume flow rate $\dot{V}$ will be found in a duct with a circular cross-section with inner diameter $d_w = 100$ mm. The air pressure and temperature are as follows: $p_s = 101,353$ Pa, $T_s = 20$ °C. Dynamic pressure $p_d = p_c - p_s$ is measured in five different points with coordinates as listed in Table 12.2.

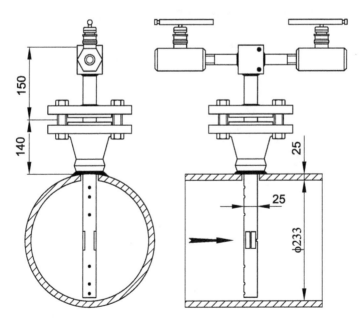

Fig. 12.8 Diagram of the TORBAR flow meter [339]

Table 12.2 Dynamic pressure measurement results

I	1	2	3	4	5	6	7
r_i, mm	0	12.5	25	37.5	43.75	46.875	50
H, mm H_2O	74	69	61	48	41	33	0

Air velocities in individual points will be calculated first:

$$w_i = \sqrt{\frac{2 p_{d,i}}{\rho}}.$$

Considering that air density totals:

$$\rho = \frac{p_s}{RT_s} = \frac{101353}{287.1 \cdot 293.15} = 1.204 \text{ kg/m}^3$$

it is possible to calculate the values of w_i. The results are listed in Table 12.3. It is taken into account that

$$p_{d,i} = \rho_w g H_i,$$

where: $\rho_w = 998.23$ kg/m^3 is the density of water in the temperature of 20 °C and H_i is the height of the water column in a U-tube manometer. The air volume flow rate is calculated from the following formula:

$$\dot{V} = \int_0^{r_w} 2\pi r w \, dr \cong \sum_{i=1}^{6} w_i \, \Delta A_i,$$

where (cf. Fig. 12.9):

$$\Delta A_i = \pi \left(r_{m,i} - r_{m,i-1} \right), \quad i = 1, \ldots, 6,$$

$$r_{m,0} = 0,$$

$$r_{m,i} = \frac{r_i + r_{i+1}}{2}, \quad i = 1, \ldots, 6.$$

The fluid mean velocity in the tube cross-section totals:

$$\bar{w} = \frac{\dot{V}}{\pi r_w^2} = \frac{0.21206}{7.85398 \cdot 10^{-3}} = 27 \text{ m/s.}$$

Table 12.3 Air mean velocity calculation results

i	1	2	3	4	5	6	7
w_i, m/s	34.7	33.5	31.5	27.9	25.8	26.2	0
$r_{m,i}$, m	0.00625	0.01875	0.03125	0.040625	0.0453125	0.0484375	
ΔA_i, m²	1.2272×10^{-4}	9.8175×10^{-4}	1.9635×10^{-3}	2.1169×10^{-3}	1.2655×10^{-3}	9.2039×10^{-4}	
$w_i \cdot \Delta A_i$, ms	4.2584×10^{-3}	3.2889×10^{-2}	6.1850×10^{-2}	5.9062×10^{-2}	3.265×10^{-2}	2.1353×10^{-2}	
Δr_i, m	0.0125	0.0125	0.0125	0.00625	0.003125	0.003125	
$r_{m,i} \cdot \Delta r_i$, m²	7.8125×10^{-5}	2.3438×10^{-4}	3.9063×10^{-4}	2.5391×10^{-4}	1.4160×10^{-4}	1.5137×10^{-4}	
$w_{m,i}$, m/s	34.1	32.5	29.7	26.85	24.5	11.6	

Fig. 12.9 Illustration of approximate calculation of the volume flow rate

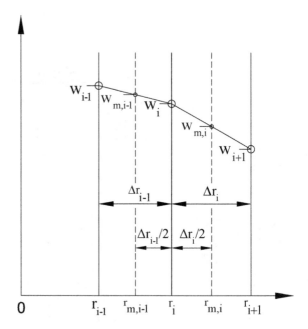

The quantity can also be calculated using the following formula:

$$\bar{w} = \frac{\dot{V}}{\pi r_w^2} = \frac{\int_0^{r_w} 2\pi r\, w\, dr}{\pi r_w^2} \simeq \frac{2\sum_{i=1}^{6} r_{m,i}\, \Delta r_i\, w_{m,i}}{r_w^2} = \frac{0.06785}{0.05^2} = 27.14 \text{ m/s.}$$

In the second formula above, $w_{m,i}$ is the flow velocity in point $r_{m,i}$, $i = 1,...,6$ determined from the following equation:

$$w_{m,i} = \frac{w_i + w_{i+1}}{2}, \quad i = 1,...,6.$$

$\Delta r_i = r_{i+1} - r_i$, $i = 1,...,6$ denotes the distance between neighbouring points of the velocity measurement (cf. Fig. 12.9).

It can be seen that both methods of calculating mean velocity $\bar{w}$ produce similar results.

12.3 Measurement of the Mass-Averaged Temperature of a Fluid Flowing Through a Channel

Heat flow rate $\dot{Q}$ cannot be determined without knowing the fluid mass-averaged temperature $\bar{T}$. The necessity of finding this temperature results from the differences between velocity and temperature values in the channel cross-section. Considering that heat flow rate $\dot{Q}$ is defined as:

$$\dot{Q} = \dot{m}\, i = \dot{m} c_p \big|_0^{\bar{T}}\, \bar{T} = - \int\limits_{A_{c.s.}} c_p\, T\, \rho\, \mathbf{w} \cdot \mathbf{n}\, dA \qquad (12.44)$$

mass-averaged temperature $\bar{T}$ is determined using the following equality:

$$- \int\limits_{A_{c.s.}} c_p\, T\, \rho\, \mathbf{w} \cdot \mathbf{n}\, dA = -\bar{T} \int\limits_{A_{c.s.}} c_p\, \rho\, \mathbf{w} \cdot \mathbf{n}\, dA, \qquad (12.45)$$

which gives:

$$\bar{T} = \frac{\int\limits_{A_{c.s.}} c_p\, T\, \rho\, \mathbf{w} \cdot \mathbf{n}\, dA}{\int\limits_{A_{c.s.}} c_p\, \rho\, \mathbf{w} \cdot \mathbf{n}\, dA} = \frac{\int\limits_{A_{c.s.}} \rho\, c_p\, T\, w_n\, dA}{\int\limits_{A_{c.s.}} \rho\, c_p\, w_n\, dA}. \qquad (12.46)$$

In the case of a duct with a circular cross-section, where the flow is circularly symmetric, and no eddies occur, considering that:

$$dA = 2\,\pi\, r\, dr, \quad w_n = w_z \qquad (12.47)$$

formula (12.46) can be written as

$$\bar{T} = \frac{\int\limits_0^{r_w} \rho\, c_p\, T\, w_z\, r\, dr}{\int\limits_0^{r_w} \rho\, c_p\, w_z\, r\, dr}. \qquad (12.48)$$

The changes in density ρ and specific heat c_p in the channel cross-section are usually slight, and therefore the quantities may be considered as constant. In such a situation, formula (12.46) is reduced to the following form:

$$\bar{T} = \frac{\int\limits_{A_{c.s.}} T\, w_n\, dA}{\int\limits_{A_{c.s.}} w_n\, dA}. \qquad (12.49)$$

One of the ways of measuring the fluid mass-averaged temperature in the heat exchanger testing is to mix the fluid stream using a fan (in the case of air or another gas) or a pump (in the case of a liquid) so that the fluid temperature in the entire cross-section should be the same. Despite the differences in the fluid velocity in the cross-section, the expression is reduced to the following form:

$$\bar{T} = T. \tag{12.50}$$

In this case, it is enough to measure the fluid temperature T at one point only, e.g., in the channel cross-section center.

If the fluid velocity w_n and temperature T in the measuring cross-section are not uniform, the cross-section surface area is divided into n elements with surface areas ΔA_i, $i = 1,\ldots,n$, and temperature T_i and velocity $w_{n,i}$ are measured in the centers of gravity of individual elements. The mean temperature $\bar{T}$ is calculated using formula (12.50), approximating integrals using sums:

$$\bar{T} = \frac{\sum\limits_{i=1}^{n} w_{n,i}\, T_i \Delta A_i}{\sum\limits_{i=1}^{n} w_{n,i}\Delta A_i}. \tag{12.51}$$

If all the elements have an identical surface area, i.e., $\Delta A_i = $ const, formula (12.51) is reduced to the following form:

$$\bar{T} = \frac{\sum\limits_{i=1}^{n} w_{n,i}\, T_i}{\sum\limits_{i=1}^{n} w_{n,i}} = \frac{\sum\limits_{i=1}^{n} w_{n,i}\, T_i}{n\, \bar{w}_n}, \tag{12.52}$$

where $\bar{w}_n$ is the fluid flow mean velocity in the channel.

The coordinates of the temperature and velocity measuring points may be the same as in the mean velocity measurement. The formulae for the calculation of the radii on which the fluid velocities are measured in circular ducts are presented in subsection 12.2.1. Many instruments for the measurement of the fluid local velocity are equipped with temperature sensors, which makes it possible to measure $w_{n,i}$ and T_i at the same time.

Chapter 13
Determination of the Local and the Mean Heat Transfer Coefficient on the Inner Surface of a Single Tube and Finding Experimental Correlations for the Nusselt Number Calculation

The local heat transfer coefficient cannot be determined experimentally if the temperature and heat flux values on the channel inner surface and the fluid mass-averaged temperature are unknown. The heat transfer coefficient can be determined directly only in single tubes with an easily accessible outer surface. This method is particularly suitable for determination of changes in the heat transfer coefficient on the length of a tube where the heat transfer is intensified by flow turbulizers in the form of inserts made of twisted tapes for example. Direct measurement of the heat transfer coefficient can also be used to find the heat transfer coefficient value on the inner surface of a tube if the medium flowing inside is a fluid-particles body mixture (suspension) or a mixture of two-phase suspensions being a mixture of ice and liquid particles.

This type of measurement is difficult to apply to determine the local or the mean heat transfer coefficient in heat exchangers due to the difficulty of calculation or measurement of the heat flux on the tube inner surface. The tube with a fluid flowing inside is heated on the outer surface by a resistance heater. The heater can be in the form of an electrically insulated heater wire or heating mats. The measurement will not be correct unless heat flux $\dot{q}_w$ on the tube inner surface, both along the tube perimeter and length, is constant. Still, it has to be remembered that the constant heat flux on the tube outer surface ($\dot{q}_z$) does not guarantee that the heat flux on the inner surface ($\dot{q}_w$) will be constant too. This is the case for the fluid flow through horizontal or vertical channels with a rectangular cross-section or through horizontal channels with a circular cross-section, where changes occur in the heat transfer coefficient on the channel perimeter. Similarly, in two-phase flows through horizontal channels, a liquid flows in the channel lower part and steam flows through the upper part, which involves substantial changes in the heat transfer coefficient values on the channel inner surface. In such situations, it is difficult to determine the local heat flux on the inner surface because even if the heat flux value is constant at the channel outer surface. There is intense circumferential heat flow in the channel wall, and the local heat flux at the inner surface cannot be found from

© Springer International Publishing AG, part of Springer Nature 2019
D. Taler, *Numerical Modelling and Experimental Testing of Heat Exchangers*,
Studies in Systems, Decision and Control 161,
https://doi.org/10.1007/978-3-319-91128-1_13

the following simple relation which holds for one-dimensional heat conduction in the channel wall:

$$\dot{q}_w = \dot{q}_z \frac{U_z}{U_w},\qquad(13.1)$$

where U_z and U_w denote the channel outer and inner perimeter, respectively.

If heat flux $\dot{q}_z$ and the heat transfer coefficient α on the channel inner surface are constant, relation (13.1) is correct. For a channel with a circular cross-section, where $U_z = 2\pi\, r_z$ and $U_w = 2\pi\, r_w$, formula (13.1) takes the following form:

$$\dot{q}_w = \dot{q}_z \frac{r_z}{r_w}.\qquad(13.2)$$

The fluid mass-averaged temperature $\bar{T}_f\big|_x$ in the cross-section with coordinate x, measured from the channel inlet $x = 0$, can be calculated from the energy balance equation:

$$\dot{m}\bar{c}\big|_0^x \bar{T}_f\big|_{x=0} + 2\,\pi\,r_w\,x\,\dot{q}_w = \dot{m}\bar{c}\big|_0^x \bar{T}_f\big|_x,\qquad(13.3)$$

which, after simple transformations, gives:

$$\bar{T}_f\big|_x = \bar{T}_f\big|_{x=0} + \frac{2\,\pi\,r_w\,\dot{q}_w}{\dot{m}\,\bar{c}\big|_0^x}\,x,\qquad(13.4)$$

where $\dot{m}$ is the fluid mass flow rate and $\bar{c}\big|_0^x$ is the fluid mean specific heat in the temperature range from $\bar{T}_f\big|_{x=0}$ to $\bar{T}_f\big|_x$.

The heat flux on the tube inner surface is found from the following simple relation:

$$\dot{q}_w = \frac{\dot{Q}}{2\,\pi\,r_w\,L_c},\qquad(13.5)$$

where L_c is the length of the heated section of the tube.

Heat flow rate $\dot{Q}$ can be found by measuring the electric power consumed by the resistance heater or using the following simple expression:

$$\dot{Q} = \dot{m}\,\bar{c}\big|_0^{L_c} \left(\bar{T}_f\big|_{x=L_c} - \bar{T}_f\big|_{x=0}\right),\qquad(13.6)$$

where $\bar{c}\big|_0^{L_c}$ is the fluid mean specific heat at constant pressure in the temperature range from $\bar{T}_f\big|_{x=0}$ to $\bar{T}_f\big|_{x=L_c}$.

The mean heat transfer coefficient α_x on the tube perimeter in the cross-section with coordinate x is found from the following formula:

$$\alpha_x = \frac{\dot{q}_w}{\left[\bar{T}_w(x)|_{r=r_w} - \bar{T}_f(x)\right]}, \tag{13.7}$$

where $\bar{T}_w(x)|_{r=r_w}$ is the mean temperature on the circumference of the tube inner surface in the cross-section with coordinate x.

If the temperature of the flowing fluid changes only slightly, e.g., by 1 K, it may be assumed that temperature $\bar{T}_f(x)$ is constant and equal to the mean temperature value, i.e. $\bar{T}_f = \left(\bar{T}_f|_{x=0} + \bar{T}_f|_{x=L_c}\right)/2$. The mean heat transfer coefficient $\bar{\alpha}$ on the tube length L_c is found from the following equation:

$$\dot{Q} = \int_0^{L_c} \alpha_x \left[\bar{T}_w(x)|_{r=r_w} - \bar{T}_f(x)\right] 2\pi\, r_w dx = \bar{\alpha} \int_0^{L_c} \left[\bar{T}_w(x)|_{r=r_w} - \bar{T}_f(x)\right] 2\pi\, r_w dx,$$

$$\tag{13.8}$$

which gives:

$$\bar{\alpha} = \frac{\dot{Q}}{\int_0^{L_c} \left[\bar{T}_w(x)|_{r=r_w} - \bar{T}_f(x)\right] 2\pi\, r_w\, dx}. \tag{13.9}$$

Formula (13.9) can be simplified calculating the wall mean temperature $\bar{T}_{wm}(r = r_w)$ on the entire inner surface of the tube and the fluid mean temperature $\bar{T}_{fm}$ on the tube length.

$$\bar{\alpha} = \frac{\dot{Q}}{2\pi\, r_w L_c \left(\bar{T}_{wm}|_{r=r_w} - \bar{T}_{fm}\right)}. \tag{13.10}$$

The tube inner surface mean temperature is defined as:

$$\bar{T}_{wm}|_{r=r_w} = \frac{1}{2\pi r_w L_c} \int_0^{L_c} 2\pi r_w \bar{T}_w(x)|_{r=r_w} dx \cong \frac{\frac{1}{2}\bar{T}_{w,1} + \frac{1}{2}\bar{T}_{w,N+1} + \sum_{i=2}^{N} \bar{T}_{w,i}}{N},$$

$$\tag{13.11}$$

where $\bar{T}_{w,i}$, $i = 1,\ldots,N+1$ are means of the tube inner surface temperature measured on the tube perimeter at points with the following coordinates:

$$x_i = (i-1)\Delta x, \quad i = 1,\ldots,N+1, \quad \Delta x = L_c/N,$$

where N is the number of intervals with a length Δx that the tube heated section with length L_c is divided into.

The fluid mean temperature on the tube heated length L_c can be approximated using the arithmetic mean of the fluid inlet and outlet temperatures because, as indicated by formula (13.4), the changes in temperature are approximately linear:

$$\bar{T}_{fm} = \frac{1}{2\pi r_w L_c} \int_0^{L_c} 2\pi r_w \bar{T}_f(x) dx = \frac{1}{L_c} \int_0^{L_c} \bar{T}_f(x) dx \cong \left(\bar{T}_f|_{x=0} + \bar{T}_f|_{x=L_c}\right)/2. \quad (13.12)$$

The measurements are usually performed using copper tubes with the wall thickness of 1–1.5 mm. The tube temperature is measured at $N+1$ equidistant points using jacket thermocouples arranged on the radius r_m. The thermocouples can be placed in grooves milled on the tube perimeter and soft-soldered to the tube surface to eliminate measuring errors caused by contact resistance or those arising due to heat conduction through the thermocouple from the measuring point to the environment. The tube inner surface temperature $\bar{T}_w(x)|_{r=r_w}$ can be calculated using the following formula:

$$\bar{T}_w(x)|_{r=r_w} = \bar{T}_w(x)|_{r=r_m} - \frac{\dot{Q}}{2\pi L_c \lambda_w} \ln \frac{r_m}{r_w}. \quad (13.13)$$

The temperature drop on thickness $(r_m - r_w)$ is usually slight, and it may be assumed that the tube inner surface temperature is equal to the measured temperature of the wall. The fluid temperature at the control section inlet and outlet $\bar{T}_f|_{x=0}$ and $\bar{T}_f|_{x=L_c}$, respectively, can be measured using resistance thermometers, as they are more precise compared to thermocouples. The protruding parts of the resistance thermometers should be insulated thermally to minimize measuring errors resulting from heat conduction from the fluid to the environment through the thermometer casing.

This type of measurement is difficult to apply to determine the local or the mean heat transfer coefficient in heat exchangers due to the difficulty of calculation or measurement of the heat flux on the tube inner surface.

Having found the local and the mean heat transfer coefficient for different Reynolds and Prandtl numbers, it is possible to determine the Nusselt number formula as a function of the Reynolds and the Prandtl number. The fluid thermophysical properties are usually established in the mean temperature calculated as the arithmetic mean of the fluid inlet and outlet temperatures. The form of formula $\text{Nu} = f(\text{Re}, \text{Pr}, x/d_w)$ or $\overline{\text{Nu}} = f(\text{Re}, \text{Pr}, L_c/d_w)$ can be selected based on the analytical or numerical formulae obtained in Chap. 3 for the laminar flow or in Chap. 4—for the turbulent flow. The unknown coefficients in the adopted form of the formula for the calculation of Nu and $\overline{\text{Nu}}$ are found using the least squares method.

Useful tools for determination of dimensionless numbers which are typical of a given phenomenon are the dimensionless notation of the mass, momentum, and energy conservation equations and the dimensional analysis.

13.1 Determination of Dimensionless Numbers from Boundary Conditions and Differential Equations

It is possible to find dimensionless numbers that can be used later on in the formulae approximating experimental data using a dimensionless notation of the boundary conditions and differential equations describing the phenomenon under analysis,

This will be demonstrated on the example of the third-type boundary condition written for the inner surface of a tube with radius, where the fluid flowing in the tube has a temperature T. The heat transfer coefficient α is defined as:

$$\alpha = \frac{\dot{q}_w}{T_w|_{r=r_w} - T_{fm}},\qquad(13.14)$$

where: $\dot{q}_w$—heat flux at the tube inner surface, the temperature of the tube inner surface, T_{fm}—the mass-averaged temperature of the fluid in the cross-section under consideration.

Considering that heat flux $\dot{q}_w$ on the tube inner surface found based on the temperature distribution inside the tube $(T_w(r))$ is defined as:

$$\dot{q}_w = \lambda_w \frac{\partial T_w}{\partial r}\Big|_{r=r_w},\qquad(13.15)$$

the coefficient can be written as:

$$\alpha = \lambda_w \frac{\frac{\partial T_w}{\partial r}\big|_{r=r_w}}{T_w|_{r=r_w} - T_{fm}}.\qquad(13.16)$$

Introducing dimensionless radius $R = r/r_w$ and considering that

$$\lambda_w \frac{\partial T_w}{\partial r}\Big|_{r=r_w} = \lambda_w \frac{\partial T_w}{\partial R}\frac{\partial R}{\partial r}\Big|_{r=r_w} = \frac{\lambda_w}{r_w}\frac{\partial T_w}{\partial R}\Big|_{R=1}\qquad(13.17)$$

the following is obtained from formula (13.16):

$$\frac{\partial T_w}{\partial R}\Big|_{R=1} = \frac{\alpha\, r_w}{\lambda_w}\left(T_w|_{R=1} - T_{fm}\right).\qquad(13.18)$$

Introducing dimensionless Biot number (Bi):

$$\mathrm{Bi} = \frac{\alpha\, r_w}{\lambda_w}.\qquad(13.19)$$

boundary condition (13.18) can be written as:

$$\frac{\partial T_w}{\partial R}\Big|_{R=1} = \text{Bi}\left(T_w\big|_{R=1} - T_{fm}\right). \tag{13.20}$$

The Nusselt number (Nu) formula can be derived similarly, determining the heat flux on the liquid side based on the temperature distribution in the flowing fluid $(T(r))$:

$$\dot{q}_w = \lambda \frac{\partial T}{\partial r}\Big|_{r=r_w} = \frac{\lambda}{r_w}\frac{\partial T}{\partial R}\Big|_{R=1}. \tag{13.21}$$

Substituting (13.21) in (13.14), the following is obtained after transformations:

$$\frac{\partial T}{\partial R}\Big|_{R=1} = \frac{\text{Nu}}{2}\left(T\big|_{R=1} - T_{fm}\right), \tag{13.22}$$

where:

$$\text{Nu} = \frac{\alpha\, d_w}{\lambda}, \quad d_w = 2\, r_w. \tag{13.23}$$

The Biot number Bi appears in solutions of heat conduction problems formulated for solid bodies, and the Nusselt number Nu—in solutions describing the convective heat transfer.

Dimensionless numbers will now be determined by introducing dimensionless variables into the Navier-Stokes differential equation.

The Navier-Stokes Eq. (2.80) for velocity component w_z can be written as:

$$\rho\left(\frac{\partial w_z}{\partial t} + w_x\frac{\partial w_z}{\partial x} + w_y\frac{\partial w_z}{\partial y} + w_z\frac{\partial w_z}{\partial z}\right)$$
$$= -\rho g - \frac{\partial p}{\partial z} + \mu\left(\frac{\partial^2 w_z}{\partial x^2} + \frac{\partial^2 w_z}{\partial y^2} + \frac{\partial^2 w_z}{\partial z^2}\right). \tag{13.24}$$

Introducing reference length L, reference velocity w_∞ and dimensionless variables:

$$x^* = \frac{x}{L}, \quad y^* = \frac{y}{L}, \quad w_x^* = \frac{w_x}{w_\infty}, \quad w_y^* = \frac{w_y}{w_\infty}, \quad w_z^* = \frac{w_z}{w_\infty}, \quad t^* = \frac{t\, w_\infty}{L} \tag{13.25}$$

the terms on the left side of Eq. (13.24) can be written in the following form:

$$\frac{\partial w_z}{\partial t} = \frac{\partial w_z}{\partial t^*}\frac{\partial t^*}{\partial t} = \frac{w_\infty}{L}\frac{\partial w_z}{\partial t^*} = \frac{w_\infty}{L}\frac{\partial w_z^*}{\partial t^*}\frac{\partial w_z}{\partial w_z^*} = \frac{w_\infty^2}{L}\frac{\partial w_z^*}{\partial t^*}, \tag{13.26}$$

$$\frac{\partial w_z}{\partial x} = \frac{\partial w_z}{\partial x^*}\frac{\partial x^*}{\partial x} = \frac{1}{L}\frac{\partial w_z}{\partial x^*} = \frac{1}{L}\frac{\partial w_z^*}{\partial x^*}\frac{\partial w_z}{\partial w_z^*} = \frac{w_\infty}{L}\frac{\partial w_z^*}{\partial x^*},\tag{13.27}$$

$$\frac{\partial w_z}{\partial y} = \frac{\partial w_z}{\partial y^*}\frac{\partial y^*}{\partial y} = \frac{1}{L}\frac{\partial w_z}{\partial y^*} = \frac{1}{L}\frac{\partial w_z^*}{\partial y^*}\frac{\partial w_z}{\partial w_z^*} = \frac{w_\infty}{L}\frac{\partial w_z^*}{\partial y^*},\tag{13.28}$$

$$\frac{\partial w_z}{\partial z} = \frac{\partial w_z}{\partial z^*}\frac{\partial z^*}{\partial z} = \frac{1}{L}\frac{\partial w_z}{\partial z^*} = \frac{1}{L}\frac{\partial w_z^*}{\partial z^*}\frac{\partial w_z}{\partial w_z^*} = \frac{w_\infty}{L}\frac{\partial w_z^*}{\partial z^*}.\tag{13.29}$$

The Navier-Stokes equation can be expressed using a similar procedure to transform the terms on the right side of Eq. (13.24) as:

$$\frac{\rho w_\infty^2}{L}\left(\frac{\partial w_z^*}{\partial t^*} + w_x^*\frac{\partial w_z^*}{\partial x^*} + w_y^*\frac{\partial w_z^*}{\partial y^*} + w_z^*\frac{\partial w_z^*}{\partial z^*}\right)$$
$$= -\rho g - \frac{1}{L}\frac{\partial p}{\partial z^*} + \frac{\mu w_\infty}{L^2}\left(\frac{\partial^2 w_z^*}{\partial x^{*2}} + \frac{\partial^2 w_z^*}{\partial y^{*2}} + \frac{\partial^2 w_z^*}{\partial z^{*2}}\right).\tag{13.30}$$

Dividing both sides of Eq. (13.30) by $(\rho w_\infty^2)/L$ and introducing dimensionless pressure

$$p^* = \frac{p}{\rho w_\infty^2},\tag{13.31}$$

Equation (13.30) is now as follows:

$$\frac{\partial w_z^*}{\partial t^*} + w_x^*\frac{\partial w_z^*}{\partial x^*} + w_y^*\frac{\partial w_z^*}{\partial y^*} + w_z^*\frac{\partial w_z^*}{\partial z^*}$$
$$= -\frac{gL}{w_\infty^2} - \frac{\partial p^*}{\partial z^*} + \frac{\nu}{w_\infty L}\left(\frac{\partial^2 w_z^*}{\partial x^{*2}} + \frac{\partial^2 w_z^*}{\partial y^{*2}} + \frac{\partial^2 w_z^*}{\partial z^{*2}}\right).\tag{13.32}$$

The dimensionless parameters in Eq. (13.33) are named after the famous scientists:

$$\text{the Reynolds number—Re} = \frac{w_\infty L}{\nu},\tag{13.33}$$

$$\text{the Froude number—Fr} = \frac{w_\infty^2}{gL}.\tag{13.34}$$

Dimensionless pressure p^* corresponds to the Euler number, which is usually defined as:

$$\text{Eu} = \frac{\Delta p}{\rho w_\infty^2},\tag{13.35}$$

where Δp is the pressure drop (the difference in pressure).

If dynamic similarity occurs, dimensionless solutions are identical and so are local variables at points corresponding to each other, i.e., for points with identical coordinates x/L, where vector $\mathbf{x}$ (the radius vector of the point) is defined as: $\mathbf{x} = (x, y, z)$. Local pressure p^* or local velocity is thus a function of the Reynolds number Re, the Froude number Fr and coordinate x/L. By expressing differential equations in a dimensionless form, dimensionless numbers are obtained that can be used to determine an experimental relation, e.g., to find the fluid flow pressure drop on a cylinder or sphere. It should be noted that dimensionless numbers can be determined even if the solution of the equation under analysis is unknown.

13.2 Dimensional Analysis

In complex flow problems, differential equations describing a given phenomenon may be unknown. In such cases, dynamic similarity conditions can be determined using a dimensional analysis of the variables needed to describe the phenomenon [36, 94, 160, 167, 202, 353, 357, 358].

13.2.1 Matrix of Dimensions

The dimensional analysis involves making the so-called matrix of dimensions. In flow problems, the primary (fundamental) dimensions are the units of mass, length and time—M, L and t, respectively. One more unit is essential in thermal problems: temperature—T. The dimensions of any other physical quantity can be expressed using the fundamental dimensions (cf. Table 13.1). The first step in the dimensional analysis is to determine the matrix of dimensions and its rank, which will be shown on the example of finding the formula describing the pressure drop Δp in a tube

Table 13.1 Physical variables and their dimensions

Physical quantity	Symbol	Dimensions
Tube diameter	d_w	L
Length	l	L
Surface roughness	e	L
Velocity	w	L/t
Density	ρ	M/L^3
Pressure	p	$M/(Lt^2)$
Dynamic viscosity	μ	$M/(Lt)$
Specific heat at constant pressure	c_p	$L^2/(t^2 T)$
Thermal conductivity	λ	$M\,L/(t^3 T)$
Heat transfer coefficient	α	$M/(t^3 T)$

with length l and inner diameter d_w. The tube inner surface roughness is e. A fluid with density ρ and dynamic viscosity μ flows through a tube with mean velocity w.

The pressure drop can be expressed using a function including the other physical quantities. The relation can be written in the following form:

$$f(\Delta p, d_w, l, e, w, \rho, \nu) = 0. \tag{13.36}$$

The matrix of dimensions is presented in Table 13.2. The columns present exponents of powers to which the fundamental dimensions should be raised to obtain the dimension (unit) of a given physical quantity, e.g., $m^1 s^{-1}$ = m/s is the dimension (unit) of velocity w.

The matrix rank r is defined as the dimension of the greatest square submatrix whose determinant is different from zero. In the case of the matrix of dimensions presented in Table 13.2, the determinant of the following submatrix is different from zero:

$$\begin{vmatrix} 0 & 1 & 1 \\ 1 & -3 & -1 \\ -1 & 0 & -1 \end{vmatrix} = -1.$$

It can be seen that in this case, the rank of the matrix is $r = 3$.

13.2.2 Buckingham Theorem

If in experimental testing the experimentally determined pressure drop Δp or the heat transfer coefficient α depends on $(n-1)$ physical quantities, the number of columns in the matrix of dimensions is n. In the case of the matrix shown in Table 13.2, $n = 7$. If in the general case there are n variables: $q_1, q_2, \ldots, q_n$, the Buckingham theorem states that the following functional relation exists:

$$f(q_1, q_2, \ldots, q_n) = 0, \tag{13.37}$$

where n variables q_i can be grouped in $(n-r)$ dimensionless variables, where r is the rank of the matrix of dimensions. The relation (13.37) can be written taking π_i to denote a dimensionless parameter, referred to as a dimensionless number in the heat transfer literature, in the following form:

$$f(\pi_1, \pi_2, \ldots, \pi_{n-r}) = 0. \tag{13.38}$$

Table 13.2 Matrix of dimensions for the pressure drop in a rough tube

	Δp	d_w	l	e	U	ρ	μ
M	1	0	0	0	0	1	1
L	−1	1	1	1	1	−3	−1
t	−2	0	0	0	−1	0	−1

The set of parameters π is non-unique. Some of them can be combined to create new dimensionless numbers. The number of independent dimensionless parameters is $(n - r)$, and the parameters make up a complete system.

13.3 Examples of the Dimensional Analysis Application

Two examples of the dimensional analysis application will be presented. In the first, the analysis will cover the fluid pressure drop Δp for a flow with mean velocity w inside a tube with inner diameter d_w and length l. The tube inner surface roughness is e.

In the second example, dimensionless numbers will be determined which are needed to describe the turbulent heat transfer in a tube with an inner diameter d_w and a hydraulically smooth inner surface. The considerations concern a fully developed turbulent flow at a great distance from the tube inlet.

13.3.1 Pressure Drop in the Fluid Flow Through a Rough Tube

It follows from Table 13.2 that four of the seven variables presented therein make up the core group. The core group variables must have different dimensions, and each of them must include fundamental dimensions M, L and t.

Velocity w, the tube inner diameter d_w and the fluid density ρ are selected as core physical quantities. Each of the four dimensionless parameters π will be a product of the three basic core quantities, supplemented with one of the other parameters. The first dimensionless parameter π_1 is the following product:

$$\pi_1 = w^a d_w^b \rho^c \Delta p. \tag{13.39}$$

The condition for non-dimensionality of π_1 gives:

$$M^0 L^0 t^0 = \left(L t^{-1}\right)^a (L)^b \left(M L^{-3}\right)^c \left(M L^{-1} t^{-2}\right) = M^{c+1} L^{a+b-3c-1} t^{-a-2}. \tag{13.40}$$

The M, L and t exponents should be equal to zero. As a result, the following system of equations is obtained:

$$
\begin{aligned}
c + 1 &= 0, \\
a + b - 3c - 1 &= 0, \\
-a - 2 &= 0.
\end{aligned} \tag{13.41}
$$

Solving the system of Eq. (13.42), the results are as follows:

$$a = -2, \quad b = 0, \quad c = -1. \tag{13.42}$$

Substitution of exponents (13.43) in (13.40) gives:

$$\pi_1 = w^{-2} d_w^0 \rho^{-1} \Delta p = \frac{\Delta p}{\rho w^2} \equiv \text{Eu.} \tag{13.43}$$

In literature, the $\Delta p/(\rho w^2)$ ratio is referred to as the Euler number (Eu). Using a similar procedure, the following dimensionless parameters are obtained:

$$\pi_2 = w^a d_w^b \rho^c l = \frac{l}{d_w}, \tag{13.44}$$

$$\pi_3 = w^a d_w^b \rho^c e = \frac{e}{d_w}, \tag{13.45}$$

$$\pi_4 = w^a d_w^b \rho^c \mu = \frac{\mu}{\rho w d_w} \equiv \frac{1}{\text{Re}}. \tag{13.46}$$

The dimensionless pressure drop can be expressed using the following formula:

$$\frac{\Delta p}{\rho w^2} = f_1 \left(\frac{l}{d_w}, \frac{e}{d_w}, \frac{\mu}{\rho w d_w} \right), \tag{13.47}$$

which can also be written as:

$$\text{Eu} = f_1 \left(\frac{l}{d_w}, \frac{e}{d_w}, \text{Re} \right). \tag{13.48}$$

In engineering practice, the friction-related pressure drop in the tube is calculated using the Darcy-Weisbach equation:

$$\Delta p = \xi \left(\text{Re}, \frac{e}{d_w} \right) \frac{l}{d_w} \frac{\rho w^2}{2}, \tag{13.49}$$

where ξ is the friction factor. Many empirical relations for the calculation of friction factor ξ are presented in Chap. 4. In British and American literature in particular, the Fanning friction factor f is also used. The factor expresses shear stress τ_w on the tube inner surface.

$$\tau_w = f \frac{\rho w^2}{2}. \tag{13.50}$$

Considering the following relation between the pressure drop on length l and shear stress:

$$\pi r_w^2 \Delta p = 2\pi r_w l \tau_w,$$ (13.51)

the following is obtained:

$$\Delta p = \frac{2}{r_w} l \tau_w = 4\frac{l}{d_w}\tau_w.$$ (13.52)

Substitution of (13.51) in (13.53) gives:

$$\Delta p = 4f\frac{l}{d_w}\frac{\rho w^2}{2}.$$ (13.53)

Comparing formulae (13.50) and (13.54), it can be seen that:

$$\xi = 4f,$$ (13.54)

i.e., the Darcy-Weisbach friction factor is four times higher than the Fanning friction factor.

13.3.2   Convective Heat Transfer in the Fluid Flow Through a Tube

The matrix of dimensions for the fluid forced flow through a tube is presented in Table 13.3.

In this case, the number of variables is $n = 7$ and the rank of the matrix is $r = 4$. It is, therefore, possible to determine three dimensionless numbers $(n - r = 7 - 4 = 3)$.

The following four physical quantities are selected as the core quantities: the tube inner diameter d_w, density ρ, dynamic viscosity μ and thermal conductivity λ. The following will be added one by one to the product of the core quantities raised to appropriate powers: the fluid flow mean velocity w, specific heat at constant pressure c_p and the heat transfer coefficient α.

Table 13.3 Matrix of dimensions for the fluid forced flow through a tube

	w	d_w	ρ	μ	c_p	λ	α
M	0	0	1	1	0	1	1
L	1	1	-3	-1	2	1	0
t	-1	0	0	-1	-2	-3	-3
T	0	0	0	0	-1	-1	-1

The first dimensionless number π_1 has the following form:

$$\pi_1 = d_w^a \, \rho^b \mu^c \, \lambda^d w. \tag{13.55}$$

Substituting the units of all the physical quantities in the dimensionless product (13.56), the following is obtained:

$$M^0 \, L^0 \, t^0 \, T^0 = L^a \left(M L^{-3}\right)^b \left(M L^{-1} t^{-1}\right)^c \left(M L t^{-3} T^{-1}\right)^d \left(L t^{-1}\right). \tag{13.56}$$

After the dimensions are grouped on the right side of Eq. (13.56), the equation can be transformed into the following form:

$$M^0 \, L^0 \, t^0 \, T^0 = M^{b+c+d} \, L^{a-3b-c+d+1} \, t^{-c-3d-1} \, T^{-d}. \tag{13.57}$$

Because the M, L, t and T exponents should be equal to zero, the following system of equations is obtained:

$$\begin{aligned} b + c + d &= 0, \\ a - 3b - c + d + 1 &= 0, \\ -c - 3d - 1 &= 0, \\ -d &= 0. \end{aligned} \tag{13.58}$$

Solving the system of Eq. (13.58) gives:

$$a = 1, \quad b = 1, \quad c = -1, \quad d = 0. \tag{13.59}$$

Considering solution (13.59) in (13.55) yields the following dimensionless number:

$$\pi_1 = \frac{d_w \rho \, w}{\mu} \equiv \mathrm{Re}. \tag{13.60}$$

The next dimensionless number is given by the product of the four core quantities raised to appropriate powers and supplemented with c_p (specific heat at constant pressure):

$$\pi_2 = d_w^e \, \rho^f \mu^g \lambda^h c_p. \tag{13.61}$$

Substituting the units of the physical quantities in expression (13.61) and considering that π_2 is a dimensionless number, the result is:

$$M^0 \, L^0 \, t^0 \, T^0 = L^e \left(M L^{-3}\right)^f \left(M L^{-1} t^{-1}\right)^g \left(M L t^{-3} T^{-1}\right)^h \left(L^{-2} t^{-2} T^{-1}\right). \tag{13.62}$$

Expression (13.62) can be transformed into:

$$M^0 L^0 t^0 T^0 = M^{f+g+h} L^{e-3f-g+h+2} t^{-g-3h-2} T^{-h-1}. \tag{13.63}$$

Considering that the M, L, t and T exponents should be equal to zero, the following system of equations is obtained:

$$\begin{aligned} f+g+h &= 0, \\ e-3f-g+h+2 &= 0, \\ -g-3h-2 &= 0, \\ -h-1 &= 0. \end{aligned} \tag{13.64}$$

The system of Eq. (13.64) yields the following solution:

$$e = 0, \quad f = 0, \quad g = 1, \quad h = -1. \tag{13.65}$$

Considering solution (13.65) in (13.61) gives a dimensionless number π_2:

$$\pi_2 = \frac{\mu c_p}{\lambda} \equiv \text{Pr}. \tag{13.66}$$

The Prandtl number can also be written in a slightly different form:

$$\text{Pr} \equiv \frac{\mu c_p}{\lambda} = \frac{\nu}{a}, \tag{13.67}$$

where the thermal diffusivity a is defined as $a = \lambda / (\rho c_p)$.

The third dimensional product π_3 is made by the four core variables supplemented with the fifth variable α:

$$\pi_3 = d_w^i \rho^j \mu^k \lambda^l \alpha. \tag{13.68}$$

Substituting the units of the physical quantities in expression (13.68) yields:

$$M^0 L^0 t^0 T^0 = L^i \left(M L^{-3}\right)^j \left(M L^{-1} t^{-1}\right)^k \left(M L t^{-3} T^{-1}\right)^l \left(M t^{-3} T^{-1}\right). \tag{13.69}$$

After simple transformations of the right side of expression (13.69), the following is obtained:

$$M^0 L^0 t^0 T^0 = M^{j+k+l+1} L^{i-3j-k+l} t^{-k-3l-3} T^{-l-1}. \tag{13.70}$$

The condition for zeroing power exponents in (13.70) gives the following equations:

$$\begin{aligned} j+k+l+1 &= 0, \\ i-3j-k+l &= 0, \\ -k-3l-3 &= 0, \\ -l-1 &= 0. \end{aligned} \qquad (13.71)$$

The solution of the system of equations is:

$$i = 1, \quad j = 0, \quad k = 0, \quad l = -1. \qquad (13.72)$$

Considering solution (13.72), the dimensionless number described by (13.68) has the following form:

$$\pi_3 = \frac{\alpha\, d_w}{\lambda} \equiv \mathrm{Nu}. \qquad (13.73)$$

The dimensionless product π_3 is the Nusselt number (Nu).
Experimental data are most often approximated using the following function:

$$\mathrm{Nu} = f(\mathrm{Re}, \mathrm{Pr}). \qquad (13.74)$$

The Dittus-Boelter formula, which is common in the literature [160, 202, 357, 358], is an example relation of the form of (13.74).

Another form of expression (13.61) are the formulae proposed by Prandtl, Petukhov, and Gnielinski

$$\mathrm{Nu} = f[\xi(\mathrm{Re}), \ \mathrm{Re}, \ \mathrm{Pr}], \qquad (13.75)$$

where ξ is the friction factor for tubes with a smooth inner surface, i.e., for tubes with the inner surface roughness e equal to zero.

For the laminar fluid flow, the friction factor is defined as $\xi = 64/\mathrm{Re}$. In the case of the turbulent fluid flow, the friction factor formulae $\xi(\mathrm{Re})$ can be found in Chap. 4. It should be added that the set of numbers π_1, π_2 and π_3 is non-unique. The Stanton number St can be introduced, defined as:

$$\mathrm{St} \equiv \frac{\mathrm{Nu}}{\mathrm{Re}\,\mathrm{Pr}} = \frac{\pi_3}{\pi_1 \pi_2}. \qquad (13.76)$$

Experimental data can then be approximated using the following formula:

$$\mathrm{St} = f(\mathrm{Re}, \mathrm{Pr}). \qquad (13.77)$$

Another example of a newly introduced dimensionless quantity is the Colburn parameter j:

$$j = \frac{\mathrm{Nu}}{\mathrm{Re}\,\mathrm{Pr}^{1/3}}. \qquad (13.78)$$

Using the analogy between the heat and the momentum transfer proposed by Chilton and Colburn [160, 202, 357, 358] experimental data for the fluid forced flow through a tube can be expressed using the following formula:

$$j = \frac{\xi(\mathrm{Re})}{8}.$$
(13.79)

The modified version of formula (13.79):

$$\frac{\mathrm{Nu}}{\mathrm{Re}\,\mathrm{Pr}^{1/3}} = f(\mathrm{Re})$$
(13.80)

is widely used in practice, e.g., to determine the air-side heat transfer coefficient in cross-flow heat exchangers.

Chapter 14
Determination of Mean Heat Transfer Coefficients Using the Wilson Method

Formulae for the calculation of mean heat transfer coefficients on the surface of ducts and in heat exchangers are commonly used in practice [107, 122, 123, 125, 127, 160, 241, 267, 268, 297, 306, 314].

The method of the heat transfer coefficient measurement described in Chap. 14 is challenging to apply to determine the local or the mean heat transfer coefficient in heat exchangers due to the difficulty of calculation or measurement of the heat flux on the tube inner surface. One popular way of establishing the mean heat transfer coefficient values in heat exchangers is the Wilsons method [365]. Easy to use, the method finds broad application in testing state-of-the-art heat exchangers.

The Wilson method is used to determine correlations for the mean heat transfer coefficients (mean Nusselt numbers) in exchangers on one or both sides of the wall separating the fluids.

The use of the method is limited to power correlations defining the Nusselt numbers. It cannot be used if the Nusselt numbers are approximated as a function of the Reynolds and the Prandtl numbers by means of formulae of a different form, such as the Prandtl [237, 309], the Petukhov [226] or the Gnielinski [107] correlations for example. During measurements, the heat transfer coefficient on one side of the wall is varied through changes in the fluid flow velocity, but it should remain constant on the other side. The heat transfer coefficient constancy conditions are satisfied well in the case of a gaseous medium because the gas viscosity, as well as other physical properties, are not strongly temperature-dependent. The constant heat transfer coefficient is difficult to keep on the liquid side, since the physical properties of liquids usually depend strongly on the temperature. Wilson proposed his method in 1915 [365]. Since then, a great number of the method modifications have been put forward [270, 282]. Herein, the Wilson method application is illustrated by the case of determining the water- and the air-side Nusselt numbers in a plate-finned tube heat exchanger which is a car radiator with continuous (plate) fins, with an in-line tube arrangement, made of $(n_u + n_l) = 38$ tubes with an oval cross-section; $n_u = 20$ tubes are placed in the upper pass, 10 tubes per row. In the lower pass there are $n_l = 18$ tubes, 9 tubes per row. The tube pitches are as follows: transverse to the

© Springer International Publishing AG, part of Springer Nature 2019 485
D. Taler, *Numerical Modelling and Experimental Testing of Heat Exchangers*,
Studies in Systems, Decision and Control 161,
https://doi.org/10.1007/978-3-319-91128-1_14

air flow $p_1 = 18.5$ mm, longitudinal to the flow $p_2 = 17$ mm. The oval tube outer diameters are: $d_{min} = 6.35$ mm, $d_{max} = 11.82$ mm. The wall thickness is $\delta_r = 0.4$ mm. The plate fins (lamellae) are strips of thin aluminium sheets with the width of 34 mm, length—359 mm and thickness $\delta_{\dot{z}} = 0.08$ mm. There are 520 lamellae spaced with pitch $s = 1$ mm. Both the tubes and the lamellae are made of aluminium. The exchanger width is 520 mm. Thermal conductivity of the material that the tubes and the fins are made of (aluminium) is $\lambda_r = 207$ W/(m K). The fluid flows along a U-shaped path. The correlations for the water- and the air-side Nusselt numbers were determined based on thermal measurements of the exchanger upper pass. Because the water flow velocity in the lower pass is different, the Wilson method cannot be used for the entire exchanger. The outer surface area of the bare (unfinned) tubes totals: $A_{zrg} = 0.30436$ m^2. The surface area of the fluid flow cross-section in a tube is $A_{rw} = 9.4434 \times 10^{-4}$ m^2. The equivalent hydraulic diameter on the water and on the air side is $d_{h,w} = 7.06$ mm and $d_{h,a} = 1.41$ mm, respectively. The ratio between the surface area of the fins $A_{\dot{z}}$ and the surface area of the bare tubes A_{zrg} totals $A_{\dot{z}}/A_{zrg} = 17.551$, and the ratio between the surface area in between the tubes $A_{m\dot{z}}$ and the surface area of bare tubes A_{zrg} is $A_{m\dot{z}}/A_{zrg} = 0.92$. The efficiency of a plate fin fixed on an oval tube and characterized by the following dimensions: width—$p_2 = 17$ mm, height—$p_1 = 18.5$ mm, thickness—$\delta_{\dot{z}} = 0.08$ mm and thermal conductivity $\lambda_{\dot{z}} = 207$ W/(m K) is determined by means of the finite element method. The calculation results are approximated using the following relation:

$$\eta_{\dot{z}} = 0.99620135 - 0.0020913668 \cdot \alpha_a + 50.4768993 \cdot 10^{-6} \cdot \alpha_a^2 \\ - 10.4791026 \cdot 10^{-7} \cdot \alpha_a^{2,5} + 0.0037986964 \cdot \exp(-\alpha_a), \tag{14.1}$$

where α_a is expressed in W/(m^2 K).

Based on experimental testing, heat flow rate $\dot{Q}$ transferred from hot water flowing inside the tubes to cooler air flowing outside is determined first.

$$\dot{Q} = \frac{\dot{Q}_a + \dot{Q}_w}{2}, \tag{14.2}$$

where the heat flow rate from water $\dot{Q}_w$ and the heat flow rate absorbed by air $\dot{Q}_a$ are defined by the following formulae:

$$\dot{Q}_w = \dot{m}_w \bar{c}_w \left(T_w' - T_{wm} \right), \tag{14.3}$$

$$\dot{Q}_a = \dot{m}_u \bar{c}_{pa} \left(T_{um}''' - T_{am}' \right), \tag{14.4}$$

where $\dot{m}_w$ and $\dot{m}_u$ are the water mass flow rates in the upper pass, respectively, $\bar{c}_w$—is the water mean specific heat in the temperature range from T_w'' to T_w' (cf. Fig. 10.8), T_w' and T_{wm}—water temperature at the exchanger inlet and the first pass outlet, respectively, $\bar{c}_{pa}$—air mean specific heat at constant pressure in the temperature range from T'_{am} to T''_{am}, T'_{am} and T'''_{um}—mean temperatures of air at the exchanger inlet and the first pass outlet, respectively.

The markings adopted in formulae (14.2)–(14.4) are shown in Fig. 10.8.

Then the overall heat transfer coefficient k_g is found from the following formula

$$k_g = \frac{\dot{Q}}{F A_{zrg} \Delta T_m},$$ (14.5)

where the logarithmic mean temperature difference between the two mediums ΔT_m in the upper pass is determined in the same way as for a countercurrent heat exchanger, using the following formula:

$$\Delta T_m = \frac{\left(T''_{wm} - T'_{am}\right) - \left(T'_w - T'''_{um}\right)}{\ln\left(\frac{T''_{wm} - T'_{am}}{T'_w - T'''_{um}}\right)}.$$ (14.6)

Correction factor F [122, 127, 160, 169, 271] for a two-row cross-flow heat exchanger, where the first and the second row are supplied in parallel with water, is found as a function of parameters P and R, which are defined as:

$$P = \frac{T'''_{um} - T'_{am}}{T'_w - T'_{am}}, \quad R = \frac{\dot{m}_u \bar{c}_{pa}}{\dot{m}_w \bar{c}_w}.$$ (14.7)

where the overall heat transfer coefficient k_g related to the bare tube outer surface area A_{zrg} is defined as:

$$\frac{1}{k_g} = \frac{1}{\alpha_{zr}} + \frac{A_m}{A_{wrg}} \frac{\delta_r}{\lambda_r} + \frac{A_{zrg}}{A_{wrg}} \frac{1}{\alpha_w},$$ (14.8)

where: $A_m = (A_{zrg} + A_{wrg})/2$—mean surface area of the exchanger tubes, A_{wrg}—inner surface area of the exchanger tubes, A_{zrg}—outer surface area of the exchanger bare tubes, δ_r—tube wall thickness, λ_r—tube material thermal conductivity, α_w—heat transfer coefficient on the tube inner surface.

For the exchanger under analysis, the surface area ratios in expression (14.8) are as follows: $A_m/A_{wrg} = 1.055$ and $A_{zrg}/A_{wrg} = 1.11$. The weighted heat transfer coefficient related to the outer surface area A_{zrg} is determined by the following formula:

$$\alpha_{zr} = \alpha_a \left(\frac{A_{m\dot{z}}}{A_{zrg}} + \eta_{\dot{z}} \frac{A_{\dot{z}}}{A_{zrg}}\right),$$ (14.9)

where: α_a—heat transfer coefficient on the air side, $\eta_{\dot{z}}$—fin efficiency.

The heat transfer coefficient on the tube inner surface will be determined using the following correlation:

$$\mathrm{Nu}_w = C \, \mathrm{Re}_w^{0.8083} \, \mathrm{Pr}_w^{0.333} \left[1 + \left(\frac{d_{h,w}}{L_{ch}} \right)^{2/3} \right], \qquad (14.10)$$

where: $\mathrm{Re}_w = \frac{w_w \, d_{h,w}}{v_w}$ —the Reynolds number, $\mathrm{Pr}_w = \frac{c_{pw} \, \mu_w}{\lambda_w}$ —the Prandtl number, w_w—water flow velocity in tubes, v_w—water kinematic viscosity, c_{pw}—water specific heat, μ_w—dynamic viscosity, λ_w—water thermal conductivity, L_{ch}—the length of the exchanger tubes.

All the thermophysical properties of water are calculated in mean temperature $\bar{T}_w = (T'_w + T''_w)/2$ (cf. Fig. 10.8). The unknown coefficient C will be found using the Wilson method. Substituting formula (14.10) in (14.18), gives the following relationship after transformations:

$$\frac{1}{k_g \frac{A_{zrg}}{A_{wrg}}} = \frac{1}{C \, \mathrm{Re}_w^{0.8083} \, \mathrm{Pr}_w^{0.333} \left[1 + \left(\frac{d_{h,w}}{L_{ch}} \right)^{2/3} \right] \frac{\lambda_w}{d_{h,w}}} + \frac{1}{\alpha_{zr} \frac{A_{zrg}}{A_{wrg}}} + \frac{A_{wrg}}{A_m} \frac{\delta_r}{\lambda_r}. \qquad (14.11)$$

Introducing the following relations:

$$y = \frac{1}{k_g \frac{A_{zrg}}{A_{wrg}}}, \qquad (14.12)$$

$$b = \frac{1}{C}, \qquad (14.13)$$

$$x = \frac{1}{\mathrm{Re}_w^{0.8083} \, \mathrm{Pr}_w^{0.333} \left[1 + \left(\frac{d_{h,w}}{L_{ch}} \right)^{2/3} \right] \frac{\lambda_w}{d_{h,w}}}, \qquad (14.14)$$

$$a = \frac{1}{\alpha_{zr} \frac{A_{zrg}}{A_{wrg}}} + \frac{A_{wrg}}{A_m} \frac{\delta_r}{\lambda_r}, \qquad (14.15)$$

Equation (14.11) can be written as follows:

$$y = bx + a. \qquad (14.16)$$

Constants a and b are determined based on the exchanger thermal measurements carried out for at least two different water velocity values at a constant velocity of the air flow. The value of constant a depends on the air velocity at which the test is performed. If measurements are made at different air velocity values, it is also possible to find correlations for the air-side heat transfer coefficient. It should be added that the value of constant b should be the same regardless of changes in the velocity of air.

The exchanger thermal measurements will be performed for m different velocities of water: $w_{w,i}$, $i = 1,..., m$ and for a constant air velocity upstream the exchanger—$w_{0,j}$. Constants b_j and a_j for the constant value $w_{0,j}$ will be found using the least squares method. If the measurements are made for $n \geq 2$ different air velocity values, it is also possible to find the air-side Nusselt number correlation. For each constant value $w_{0,j}$, constants b_j and a_j are found from the condition of the minimum of the sum of squares:

$$S_w = \sum_{i=1}^{m} (y_i - b_j x_i - a_j)^2 = \min, \quad j = 1,...,n. \tag{14.17}$$

If a_j is known, it is possible to determine the air-side heat transfer coefficient $\alpha_{a,j}$ for the set constant air velocity $w_{0,j}$. The Reynolds number $Re_{a,j}$, and the Prandtl number $Pr_{a,j}$ are also known. They are defined as follows:

$$Re_a = \frac{w_{max} d_{h,a}}{v_a}, \quad Pr_a = \frac{c_{pa} \mu_a}{\lambda_a}. \tag{14.18}$$

Symbol w_{max} denotes the air flow maximum velocity in the exchanger that occurs in the narrowest cross-section. It is defined by the following formula:

$$w_{max} = \frac{s p_1}{(s - \delta_{\dot{z}})(p_1 - d_{min})} \frac{\bar{T}_{am} + 273}{T'_{am} + 273} w_0, \tag{14.19}$$

where $\bar{T}_{am}$ is the air mean temperature on the exchanger thickness (cf. Fig. 10.8):

$$\bar{T}_{am} = (T'_{am} + T''_{am})/2. \tag{14.20}$$

Symbol $s = 1$ mm is the fin pitch, and d_{min} denotes the oval tube minimum outer diameter.

Heat transfer coefficient $\alpha_{a,j}$ is found solving the following nonlinear algebraic equation:

$$a_j - \left\{ \frac{1}{\frac{A_{zrg}}{A_{wrg}} \alpha_{a,j} \left[\frac{A_{m \dot{z}}}{A_{zrg}} + \eta_{\dot{z}}(\alpha_{a,j}) \frac{A_{\dot{z}}}{A_{zrg}} \right]} + \frac{A_{wrg}}{A_m} \frac{\delta_r}{\lambda_r} \right\} = 0, \quad j = 1,...,n. \tag{14.21}$$

Constant a_j is determined by means of the least squares method for constant air velocity $w_{0,j}$. Equation (14.21) is solved using the interval search method with the following step: $\Delta \alpha_a = 0.02$ W/(m² K). The values of the Colburn parameter are calculated next:

$$j_{a,j}^e = Nu_{a,j}/\left(Re_{a,j} Pr_{a,j}^{1/3} \right). \tag{14.22}$$

The $\left(\mathrm{Re}_{a,j}, j^e_{a,j}\right)$, $j = 1,\ldots,n$ measuring points determined experimentally are approximated by the following function:

$$j_a = C_1 \mathrm{Re}_a^{C_2}. \tag{14.23}$$

Constants C_1 and C_2 are found by means of the least squares method from the following condition:

$$S_a = \sum_{j=1}^{n} \left(j^e_{a,j} - C_1 \mathrm{Re}_{a,j}^{C_2}\right)^2 = \min. \tag{14.24}$$

After constants C_1 and C_2 are determined, they are substituted in (14.23), which gives:

$$\mathrm{Nu}_a = C_1 \mathrm{Re}_a^{(1+C_2)} \mathrm{Pr}_a^{0.333}. \tag{14.25}$$

The procedure described above is applied in $n = 4$ measuring series. The "measuring data" are generated artificially using the exchanger analytical model presented in [296, 297]. The water temperature at the exchanger inlet is $T'_w = 75\ ^\circ$C; the air temperature is $T'_{am} = 15\ ^\circ$C. The "tests" were carried out for the following four values of the air flow velocity upstream the radiator: $w_{0,1} = 0.75$ m/s, $w_{0,2} = 1.25$ m/s, $w_{0,3} = 1.75$ m/s and $w_{0,4} = 2.25$ m/s. For each air velocity value, measurements were performed at $m = 7$ different water volume flow rates: $\dot{V}_{w,1} = 700$ l/h, $\dot{V}_{w,2} = 1000$ l/h, $\dot{V}_{w,3} = 1300$ l/h, $\dot{V}_{w,4} = 1600$ l/h, $\dot{V}_{w,5} = 1900$ l/h, $\dot{V}_{w,6} = 2200$ l/h, $\dot{V}_{w,7} = 2500$ l/h. The following water velocity values at the exchanger inlet correspond to the water volume flow rates given above: $w_{w,1} = 0.2059$ m/s, $w_{w,2} = 0.2942$ m/s, $w_{w,3} = 0.3824$ m/s, $w_{w,4} = 0.4706$ m/s, $w_{w,5} = 0.5589$ m/s, $w_{w,6} = 0.6471$ m/s, $w_{w,7} = 0.7354$ m/s. The measured mean temperatures of water and air at the exchanger outlet are listed in Table 14.1. The values of x and y for air velocity values $w_0 = 0.75$, 1.25, 1.75 and 2.25 m/s are presented in Tables 14.2, 14.3, 14.4 and 14.5, respectively. The results of the $y(x)$ approximation by means of function (14.16) are shown in Figs. 14.1, 14.2, 14.3 and 14.4.

The values of the constants in function (14.16) for air velocity $w_0 = 0.75$ m/s are as follows: $a = 0.0012256$; $b = 69.02955$. The determination coefficient is $r^2 = 0.9937$. The following 95% confidence intervals are also determined:

$$0.001187 \le a \le 0.001265, \quad 62.7145 \le b \le 75.3446.$$

Constant C in correlation (14.10) is $C = 1/b = 0.014487$.

The values of the constants in function (14.16) for air velocity $w_0 = 1.25$ m/s are as follows: $a = 0.00079777$; $b = 68.7131$. The determination coefficient is $r^2 = 0.9993$. Also the following 95% confidence intervals are determined:

Table 14.1 Measured mean temperatures of water and air at the exchanger outlet

$\dot{V}_w$ (l/h)	$j = 1$, $w_{0,1} = 0.75$ m/s		$j = 2$, $w_{0,2} = 1.25$ m/s		$j = 3$, $w_{0,3} = 1.75$ m/s		$j = 4$, $w_{0,4} = 2.25$ m/s	
	$T_{1,w}''$ (°C)	$T_{1,am}''$ (°C)	$T_{2,w}''$ (°C)	$T_{2,am}''$ (°C)	$T_{3,w}''$ (°C)	$T_{3,am}''$ (°C)	$T_{4,w}''$ (°C)	$T_{4,am}''$ (°C)
700	65.48	59.37	61.44	52.89	58.52	47.88	56.33	43.97
1000	67.93	62.04	64.60	56.53	62.02	52.01	59.98	48.30
1300	69.38	63.61	66.56	58.80	64.29	54.70	62.43	51.23
1600	70.34	64.65	67.90	60.34	65.88	56.60	64.19	53.35
1900	71.02	65.38	68.88	61.47	67.06	58.00	65.52	54.95
2200	71.52	65.93	69.61	62.32	67.97	59.09	66.56	56.21
2500	71.91	66.36	70.19	62.99	68.70	59.95	67.39	57.22

Table 14.2 Values of individual quantities determined based on measurements—air mean velocity in the channel upstream the exchanger $w_0 = 0.75$ m/s

$j = 1$, $w_0 = 0.75$ m/s

$w_{0,i}$ (m/s)	$\dot{Q}$ (W)	Re_w	Pr_w	ΔT_m (K)	F	k_g, [W/(m² K)]	x [(m² K)/W]	y [(m² K)/W]
0.75	4122.9	3522	2.56	29.96	0.9654	468.4	1.00616E-05	0.001928
0.76	4378.4	5109	2.52	28.49	0.9704	520.4	7.4812E-06	0.001730
0.74	4400.5	6709	2.49	27.32	0.9744	543.1	6.0196E-06	0.001657
0.76	4583.2	8300	2.47	26.77	0.9772	575.6	5.0753E-06	0.001564
0.74	4531.9	9905	2.46	26.10	0.9798	582.2	4.4058E-06	0.001546
0.75	4627.9	11,502	2.45	25.77	0.9816	601.2	3.9075E-06	0.001497
0.76	4716.4	13,099	2.45	25.51	0.9831	618.0	3.5198E-06	0.001457

Table 14.3 Values of individual quantities determined based on measurements—air mean velocity in the channel upstream the exchanger $w_0 = 1.25$ m/s

$j = 2$, $w_0 = 1.25$ m/s

$w_{0,i}$ (m/s)	$\dot{Q}$ (W)	Re_w	Pr_w	ΔT_m (K)	F	k_g [W/(m² K)]	x [(m² K)/W]	y [(m² K)/W]
1.25	5957.6	3432	2.63	33.80	0.9650	600.2	1.0202E-05	0.001500
1.26	6486.7	5002	2.58	32.10	0.9683	685.7	7.5656E-06	0.001313
1.25	6750.4	6587	2.54	30.83	0.9714	740.6	6.0780E-06	0.001215
1.24	6909.4	8179	2.51	29.89	0.9740	779.8	5.1150E-06	0.001154
1.24	7054.2	9772	2.50	29.21	0.9762	812.7	4.4373E-06	0.001108
1.25	7211.9	11,365	2.48	28.73	0.9780	843.3	3.9323E-06	0.001067
1.26	7348.5	12,959	2.48	28.34	0.9795	869.6	3.5398E-06	0.001035

Table 14.4 Values of individual quantities determined based on measurements—air mean velocity in the channel upstream the exchanger $w_0 = 1.75$ m/s

$w_{0,i}$ (m/s)	$\dot{Q}$ (W)	Re_w	Pr_w	ΔT_m (K)	F	k_g [W/(m² K)]	x [(m² K)/W]	y [(m² K)/W]
1.75	7323.1	3366	2.69	36.40	0.9668	683.8	1.0307E-05	0.001316
1.76	8153.2	4919	2.62	34.70	0.9688	797.0	7.6326E-06	0.001129
1.75	8636.5	6492	2.58	33.38	0.9709	875.4	6.1251E-06	0.001028
1.74	8956.1	8074	2.55	32.38	0.9730	934.0	5.1501E-06	0.000964
1.74	9216.8	9661	2.53	31.63	0.9749	982.1	4.4644E-06	0.000917
1.75	9459.0	11,248	2.51	31.06	0.9765	1024.6	3.9539E-06	0.000879
1.76	9662.9	12,838	2.50	30.60	0.9779	1060.9	3.5574E-06	0.000848

$j = 3$, $w_0 = 1.75$ m/s

Table 14.5 Values of individual quantities determined based on measurements—air mean velocity in the channel upstream the exchanger $w_0 = 2.25$ m/s

$w_{0,i}$ (m/s)	$\dot{Q}$ (W)	Re_w	Pr_w	ΔT_m (K)	F	k_g [W/(m² K)]	x [(m² K)/W]	y [(m² K)/W]
2.25	8370.3	3318	2.74	38.27	0.9690	741.6	1.0387E-05	0.001214
2.26	9493.4	4854	2.66	36.65	0.9700	877.3	7.6866E-06	0.001026
2.25	10200.6	6414	2.62	35.37	0.9715	975.5	6.1643E-06	0.000923
2.24	10691.1	7987	2.58	34.35	0.9731	1050.8	5.1799E-06	0.000857
2.24	11080.8	9565	2.56	33.57	0.9746	1112.7	4.4878E-06	0.000809
2.25	11421.0	11,147	2.54	32.96	0.9760	1166.4	3.9728E-06	0.000772
2.26	11704.5	12,732	2.52	32.46	0.9773	1212.4	3.5730E-06	0.000742

$j = 4$, $w_0 = 2.25$ m/s

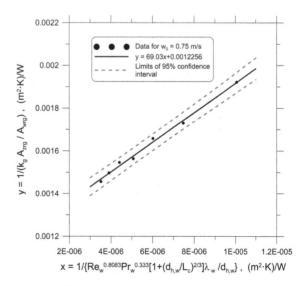

Fig. 14.1 Approximation of experimental results presented in Table 14.1 for $w_0 = 0.75$ m/s

Fig. 14.2 Approximation of experimental results presented in Table 14.3 for $w_0 = 1.25$ m/s

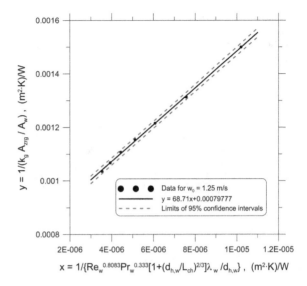

Fig. 14.3 Approximation of experimental results presented in Table 14.4 for $w_0 = 1.75$ m/s

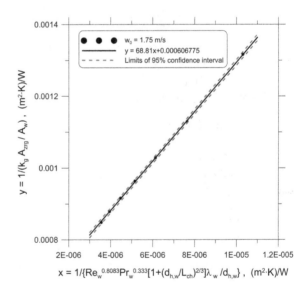

$$0.0007848 \leq a \leq 0.0008107, \quad 66.6364 \leq b \leq 70.7898.$$

Constant C in correlation (14.10) is $C = 1/b = 0.01455$.

The values of the constants in function (14.16) for air velocity $w_0 = 1.75$ m/s are as follows: $a = 0.00060678$; $b = 68.8074$. The determination coefficient is $r^2 = 0.9998$. The following 95% confidence intervals are also determined:

Fig. 14.4 Approximation of experimental results presented in Table 14.5 for $w_0 = 2.25$ m/s

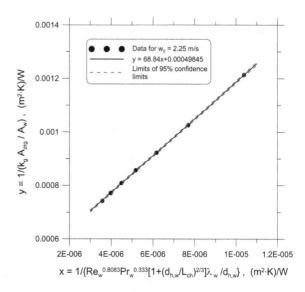

$x = 1/\{Re_w^{0.8083} Pr_w^{0.333}[1 + (d_{h,w}/L_{ch})^{2/3}]\lambda_w / d_{h,w}\}$, $(m^2 \cdot K)/W$

$$0.0006000 \le a \le 0.0006135, \quad 67.7355 \le b \le 69.8792.$$

Constant C in correlation (14.10) is $C = 1/b = 0.01453$.

The values of the constants in function (14.16) for air velocity $w_0 = 2.25$ m/s are as follows: $a = 0.0004985$; $b = 68.8350$. The determination coefficient is $r^2 = 0.9999$. The following 95% confidence intervals are also determined:

$$0.0004943 \le a \le 0.0005026, \quad 68.1783 \le b \le 69.4917.$$

Constant C in correlation (14.10) is $C = 1/b = 0.01453$.

Based on the determined values of a_j, $j = 1,...,4$, the values of the heat transfer coefficient α_a are found for different air flow velocities w_0. The calculation results are listed in Table 14.6 constants C_1 and C_2 determined by means of the least squares method are as follows: $C_1 = 0.0301$, $C_2 = -0.06207$ (determination coefficient $r^2 = 0.9987$). The 95% confidence intervals are (Fig. 14.5):

$$0.02897 \le C_1 \le 0.03117, \quad -0.0689 \le C_2 \le -0.05523.$$

Table 14.6 Determined values of the Colburn parameter j_a as a function of the Reynolds number Re_a on the air side for $\dot{V}_w = 1600$ l/h

j	$w_{0,j}$ (m/s)	Re_a	j_a
1	0.75	112.12	0.022452
2	1.25	187.84	0.021694
3	1.75	264.19	0.021271
4	2.25	341.03	0.020954

Fig. 14.5 Approximation of the air-side experimental data using function (14.23): $j_a = C_1 \, \mathrm{Re}_a^{C_2}$

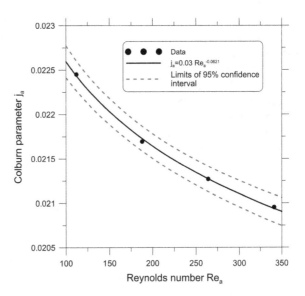

In this case, correlation (14.25) takes the following form:

$$\mathrm{Nu}_a = 0.0301 \, \mathrm{Re}_a^{0.9379} \, \mathrm{Pr}_a^{0.333}. \tag{14.26}$$

In the water-side Nusselt number correlation (14.10), coefficient $C = 1/b$ varies from $b = 0.014487$ to $b = 0.01453$ for the air inlet velocity $w_0 = 0.75$ m/s and $w_0 = 2.25$ m/s, respectively. The obtained results are very well approximated by the empirical correlations determined based on 73 measured data sets, where power exponents are found using the Levenberg-Marquardt method:

$$\mathrm{Nu}_w = 0.01454 \, \mathrm{Re}_w^{0.8083} \, \mathrm{Pr}_w^{0.333} \left[1 + \left(\frac{d_{h,w}}{L_{ch}} \right)^{2/3} \right], \tag{14.27}$$

$$\mathrm{Nu}_a = 0.02975 \, \mathrm{Re}_a^{0.9398} \, \mathrm{Pr}_a^{0.333}. \tag{14.28}$$

Correlations (14.27) and (14.28) were used in the exchanger analytical mathematical model [297, 298] to generate the "measuring data" presented in Tables 14.1, 14.2, 14.3, 14.4 and 14.5. The conducted theoretical analyses and the performed calculations indicate that the Wilson method enables determination of power correlations defining the water- and the air-side Nusselt numbers. The Wilson method accuracy would be lower if both working mediums were liquids. It can also be noticed that the accuracy of the Nusselt number correlation found on the water side is higher compared to the air side. This is the effect of the air-side heat transfer coefficient constancy at the air constant velocity, despite changes in the water volume flow rate. Variations in the water volume flow rate involve changes in the temperatures of the tube wall and air, but these have no significant impact on the air-side heat transfer coefficient because of the Prandtl number Pr_a and the λ_a/ν_a ratio for air change very little due to changes in the air temperature. The accuracy of the correlation determined for the air side is lower because, despite the water constant volume flow rate, the water temperature varies with changes in the air velocity, which in turn involves changes in the heat transfer coefficient on the water side. In such a situation, the Wilson method conditions are not fully satisfied because the method assumes that during the experiment the heat transfer coefficient on one side of the wall is constant, while on the wall other side the coefficient varies due to changes in the flow velocity of the medium.

Chapter 15
Determination of Correlations for the Heat Transfer Coefficient on the Air Side Assuming a Known Heat Transfer Coefficient on the Tube Inner Surface

A car radiator experimental testing results will be used to find the correlation for the air-side Nusselt number Nu_a assuming that the water-side Nusselt number correlation is known. The structure of the radiator intended for cooling a 1600 cm^3 spark-ignition combustion engine is discussed in [321]. It is a two-pass two-row plate fin-and-tube heat exchanger. The radiator flow system is presented in Fig. 10.8. The heat flow rates from the water and the air side—$\dot{Q}_{w,i}$ and $\dot{Q}_{a,i}$, respectively—are calculated for m measuring series using the following formulae:

$$\dot{Q}_{w,i} = \dot{V}_{w,i} \cdot \rho_{w,i} \left[\left(T'_{w,i} \right)^m \right] \cdot c_w \frac{\left| \left(T'_{w,i} \right)^m \right|}{\left(T''_{w,i} \right)^m} \cdot \left[\left(T'_{w,i} \right)^m - \left(T''_{w,i} \right)^m \right], \quad i = 1, \ldots, m,$$

(15.1)

$$\dot{Q}_{a,i} = \dot{V}_{a,i} \cdot \rho_{a,i} \left[\left(T'_{am,i} \right)^m \right] \cdot c_{pa,i} \frac{\left| \left(T''_{am,i} \right)^m \right|}{\left(T'_{am,i} \right)^m} \cdot \left[\left(T''_{am,i} \right)^m - \left(T'_{am,i} \right)^m \right], \quad i = 1, \ldots, m,$$

(15.2)

where:

$$\dot{V}_{a,i} = H_{ch} L_{ch} w_{0,i}.$$

(15.3)

The following symbols are used in formulae (15.1)–(15.3): $\dot{V}_{w,i}$, $\dot{V}_{a,i}$—water and air volume flow rates measured at the exchanger inlet, respectively, $\rho_{w,i}$, $\rho_{a,i}$—water and air density at the exchanger inlet, respectively, $c_{pa,i}$, $c_{pw,i}$—specific heat of air and water at constant pressure, respectively, $\left(T'_{am,i} \right)^m$, $\left(T'_{w,i} \right)^m$—mean temperature of air and water measured at the exchanger inlet, respectively, $\left(T''_{am,i} \right)^m$, $\left(T''_{w,i} \right)^m$—mean temperature of air and water measured at the exchanger outlet, respectively, H_{ch}, L_{ch}—the radiator active height and width, respectively, $w_{0,i}$—air velocity upstream the radiator.

© Springer International Publishing AG, part of Springer Nature 2019
D. Taler, *Numerical Modelling and Experimental Testing of Heat Exchangers*,
Studies in Systems, Decision and Control 161,
https://doi.org/10.1007/978-3-319-91128-1_15

Relative difference ε_i between the heat flow rate absorbed by water $\dot{Q}_{w,i}$ and the mean heat flow rate $\dot{Q}_{m,i}$ is calculated from the following formula:

$$\varepsilon_i = \frac{\dot{Q}_{w,i} - \dot{Q}_{m,i}}{\dot{Q}_{m,i}} \cdot 100, \tag{15.4}$$

where:

$$\dot{Q}_{m,i} = \frac{\dot{Q}_{w,i} + \dot{Q}_{a,i}}{2}. \tag{15.5}$$

The measurement results are listed in Table 15.1, whereas the heat flow rates from the water and the air side, $\dot{Q}_{w,i}$ and $\dot{Q}_{a,i}$ respectively, as well as the values of relative difference ε_i,—in Table 15.2.

Using the measuring data presented in Table 15.1, the air-side correlation coefficients x_1 and x_2 are determined:

$$\mathrm{Nu}_a = x_1 \mathrm{Re}_a^{x_2} \mathrm{Pr}_a^{1/3}, \tag{15.6}$$

Table 15.1 Measuring data

i	$w_{0,i}$, m/s	$\dot{V}_{w,i}$, l/h	$\left(T'_{w,i}\right)^m$, °C	$\left(T''_{w,i}\right)^m$, °C	$\left(T'_{am,i}\right)^m$, °C	$\left(T''_{am,i}\right)^m$, °C
1	1.00	872.40	71.08	61.83	15.23	54.98
2	1.00	949.20	70.76	62.07	14.89	55.31
3	1.00	1025.40	70.51	62.35	14.74	55.64
4	1.00	1103.40	70.30	62.65	14.59	56.03
5	1.00	1182.60	70.18	62.91	14.65	56.39
6	1.00	1258.80	69.99	63.18	14.87	56.75
7	1.00	1335.00	69.79	63.33	14.87	56.90
8	1.00	1408.80	69.68	63.51	14.71	57.15
9	1.00	1488.60	69.48	63.67	14.86	57.33
10	1.00	1564.80	69.25	63.73	14.81	57.45
11	1.00	1642.20	69.01	63.77	14.78	57.53
12	1.00	1714.80	68.82	63.83	14.77	57.53
13	1.00	1797.00	68.60	63.85	14.97	57.66
14	1.00	1892.40	68.35	63.83	14.98	57.65
15	1.00	1963.80	67.57	63.26	14.65	57.14
16	1.00	2041.20	66.96	62.80	14.24	56.72
17	1.00	2116.20	66.86	62.77	14.17	56.68
18	1.00	2190.60	66.73	62.83	14.27	56.75
19	1.27	865.80	66.33	56.74	14.11	49.56

(continued)

Table 15.1 (continued)

i	$w_{0,i}$, m/s	$\dot{V}_{w,i}$, l/h	$\left(T'_{w,i}\right)^m$, °C	$\left(T''_{w,i}\right)^m$, °C	$\left(T'_{am,i}\right)^m$, °C	$\left(T''_{am,i}\right)^m$, °C
20	1.27	942.60	66.16	56.96	13.91	49.69
21	1.27	1020.00	66.00	57.40	14.21	50.28
22	1.27	1099.20	65.82	57.66	13.91	50.60
23	1.27	1176.00	65.76	58.01	13.76	51.03
24	1.27	1252.20	65.68	58.27	13.63	51.42
25	1.27	1329.00	65.51	58.43	13.94	51.76
26	1.27	1404.00	65.46	58.71	13.83	52.02
27	1.27	1478.40	65.36	58.95	14.02	52.34
28	1.27	1557.60	65.25	59.12	13.88	52.52
29	1.27	1631.40	65.14	59.25	13.78	52.68
30	1.27	1708.80	65.05	59.35	13.58	52.83
31	1.27	1789.20	65.02	59.55	13.48	53.06
32	1.27	1882.20	65.02	59.80	13.49	53.23
33	1.27	2040.00	64.70	59.80	13.40	53.50
34	1.27	2118.00	64.70	59.80	13.40	53.41
35	1.27	2188.80	64.73	60.14	13.42	53.61
36	1.77	863.40	63.93	52.22	13.17	42.85
37	1.77	1015.80	63.65	53.18	13.21	44.23
38	1.77	1173.60	63.57	54.15	13.18	45.43
39	1.77	1249.20	63.53	54.60	13.09	45.92
40	1.77	1327.80	63.40	54.86	13.14	46.34
41	1.77	1476.60	63.36	55.44	13.00	47.11
42	1.77	1630.80	63.34	56.05	13.03	47.87
43	1.77	1789.80	63.25	56.52	13.14	48.37
44	1.77	1959.00	63.14	56.91	13.03	48.86
45	1.77	2112.60	62.91	57.10	13.00	49.12
46	1.77	2186.40	62.89	57.26	13.00	49.32
47	2.20	865.20	62.28	49.58	13.12	38.51
48	2.20	1017.00	62.24	50.64	12.91	39.83
49	2.20	1171.80	62.09	51.53	12.80	41.03
50	2.20	1251.00	61.96	51.93	12.73	41.62
51	2.20	1326.60	61.89	52.28	12.74	42.05
52	2.20	1476.60	61.65	52.85	12.73	42.82
53	2.20	1630.80	61.58	53.41	12.76	43.50
54	2.20	1788.00	61.39	53.82	12.73	44.06
55	2.20	1954.20	61.24	54.19	12.69	44.52
56	2.20	2109.60	61.18	54.56	12.69	44.94
57	2.20	2186.40	61.00	54.56	12.70	45.06

Table 15.2 Comparison of water- and air-side heat flow rates: $\dot{Q}_{w,i}$ and $\dot{Q}_{a,i}$, respectively, and relative difference ε_i between the water-side heat flow rate and the mean heat flow rate $\dot{Q}_{m,i}$

i	$\dot{Q}_{w,i}$, W	$\dot{Q}_{a,i}$, W	$\dot{Q}_{m,i} = \left(\dot{Q}_{w,i} + \dot{Q}_{a,i}\right)/2$, W	$\varepsilon_i = \frac{\dot{Q}_{w,i} - \dot{Q}_{m,i}}{\dot{Q}_{m,i}} \cdot 100$, %
1	9186.2	9031.1	9108.7	0.9
2	9390.6	9194.0	9292.3	1.1
3	9526.4	9307.8	9417.1	1.2
4	9610.8	9436.8	9523.8	0.9
5	9789.2	9502.6	9645.9	1.5
6	9761.0	9526.6	9643.8	1.2
7	9820.4	9560.9	9690.6	1.3
8	9898.3	9660.4	9779.3	1.2
9	9849.3	9661.9	9755.6	1.0
10	9837.4	9702.8	9770.1	0.7
11	9801.1	9727.8	9764.4	0.4
12	9746.7	9730.4	9738.5	0.1
13	9723.4	9708.1	9715.8	0.1
14	9744.6	9703.4	9724.0	0.2
15	9645.5	9673.9	9659.7	-0.1
16	9679.1	9684.6	9681.9	0.0
17	9866.3	9694.3	9780.3	0.9
18	9739.1	9683.2	9711.2	0.3
19	9470.8	10266.2	9868.5	−4.0
20	9891.9	10369.5	10130.7	−2.4
21	10006.3	10443.4	10224.9	−2.1
22	10232.0	10634.0	10433.0	−1.9
23	10396.8	10808.5	10602.6	−1.9
24	10584.9	10963.0	10773.9	−1.8
25	10734.2	10959.5	10846.9	−1.0
26	10811.4	11071.9	10941.7	−1.2
27	10811.1	11102.3	10956.7	−1.3
28	10892.9	11200.1	11046.5	−1.4
29	10962.7	11279.3	11121.0	−1.4
30	11112.6	11388.9	11250.8	−1.2
31	11165.9	11490.1	11328.0	−1.4
32	11209.3	11536.0	11372.6	−1.4
33	11405.5	11645.0	11525.2	−1.0
34	11841.6	11617.7	11729.6	1.0
35	11462.9	11668.4	11565.6	−0.9
36	11545.4	12015.9	11780.6	−2.0
37	12145.1	12557.8	12351.5	−1.7
38	12624.0	13058.8	12841.4	−1.7

(continued)

Table 15.2 (continued)

i	$\dot{Q}_{w,i}$, W	$\dot{Q}_{a,i}$, W	$\dot{Q}_{m,i} = (\dot{Q}_{w,i} + \dot{Q}_{a,i})/2$, W	$\varepsilon_i = \frac{\dot{Q}_{w,i} - \dot{Q}_{m,i}}{\dot{Q}_{m,i}} \cdot 100$, %
39	12738.0	13297.4	13017.7	−2.1
40	12948.5	13446.3	13197.4	−1.9
41	13353.8	13822.1	13588.0	−1.7
42	13574.7	14117.2	13846.0	−2.0
43	13753.7	14269.2	14011.5	−1.8
44	13935.7	14517.4	14226.5	−2.0
45	14016.1	14637.8	14326.9	−2.2
46	14056.2	14717.9	14387.1	−2.3
47	12556.6	12780.4	12668.5	−0.9
48	13480.4	13558.7	13519.6	−0.3
49	14139.5	14226.0	14182.8	−0.3
50	14337.8	14563.4	14450.6	−0.8
51	14567.5	14771.2	14669.4	−0.7
52	14848.6	15165.5	15007.0	−1.1
53	15224.9	15494.8	15359.9	−0.9
54	15467.2	15791.3	15629.3	−1.0
55	15744.1	16046.0	15895.1	−0.9
56	15959.4	16259.4	16109.4	−0.9
57	16091.6	16315.5	16203.5	−0.7

where the Nusselt, the Reynolds and the Prandtl numbers are defined as follows:

$$\mathrm{Nu}_a = \frac{\alpha_a d_{h,a}}{\lambda_a}, \quad \mathrm{Re}_a = \frac{w_{max} d_{h,a}}{\nu_a}, \quad \mathrm{Pr}_a = \frac{c_{pa}\mu_a}{\lambda_a}. \tag{15.7}$$

It is also assumed that the water-side heat transfer coefficient α_w is known and calculated using own formulae or formulae available from the literature. The air-side correlation unknown coefficients x_1 and x_2 are found using the least squares method searching for the minimum of the function:

$$S = \sum_{i=1}^{m} \left[\left(T''_{w,i} \right)^m - \left(T''_{w,i} \right)^c \right]^2 = \min. \tag{15.8}$$

In the general case, n unknown parameters x_1, x_2, …, x_n are determined based on $m > n$ measuring series. The unknown parameters are found using the Levenberg-Marquardt method [261].

The water temperature at the exchanger outlet $\left(T''_{w,i} \right)^c$ is calculated from the analytical model presented in [321]:

$$\left(T''_{w,i}\right)^c = \left(T''_{w,i}\right)^c \left[\left(\dot{V}_{w,i}, T'_{w,i}\right), \left(w_{0,i}, T'_{am,i}\right), x_1, x_2, \ldots, x_n\right]. \qquad (15.9)$$

$\left(T''_{w,i}\right)^m$ is the water temperature measured at the exchanger outlet. The 95% confidence intervals for coefficients $x_1, x_2, \ldots, x_n$ are calculated using the formulae presented in Chapter Eleven.

15.1   Determination of the Heat Transfer Coefficient on the Water Side

Heat transfer coefficients on the tube inner surface are found using formulae available from the literature or those derived in Chap. 3 or Chap. 4. One of the most popular correlations used for this purpose is the Dittus-Boelter formula [79, 160], determined based on experimental testing of car radiators.

$$Nu = 0.023 Re^{0.8} Pr^n, \quad 0.7 \le Pr \le 100, \quad 6000 \le Re, \quad L/d \ge 60, \qquad (15.10)$$

where:

$$n = \begin{cases} 0.4 & \text{for heated tube} \\ 0.3 & \text{for cooled tube}. \end{cases} \qquad (15.11)$$

It follows from the experiments carried out later by other researchers [8, 160, 286] and from the analyses conducted in Chap. 4 that the power form of the formula is too simple to hold in a wide range of changes in the Reynolds or the Prandtl numbers. In the transitional range of $2300 \le Re \le 10,000$, the values of the Nusselt number (Nu) are overestimated, whereas for high Reynolds numbers—they are underestimated. In Chap. 4, it is shown that exponent n in formula (15.10) should depend on the Prandtl number. For small values of the Prandtl number $(Pr < 1)$, the value of exponent n is ~ 0.7; for high and very high Prandtl numbers $(Pr > 3)$, exponent n is about 0.4.

Colburn [59] proposed a similar correlation three years later, after Dittus-Boelter's proposal, based on the analogy between the heat and the mass transfer:

$$j = \frac{\xi}{8}, \qquad (15.12)$$

where the Colburn parameter is defined as:

$$j = \frac{Nu}{Re\,Pr^{1/3}}. \qquad (15.13)$$

Considering that for circular tubes with a smooth inner surface friction factor ξ can be approximated using a simple power function:

$$\xi = \frac{0.184}{Re^{0.2}}, \quad 2 \cdot 10^4 \leq Re \leq 10^6. \tag{15.14}$$

the Colburn relation (15.12) gives:

$$Nu = 0.023 Re^{0.8} Pr^{1/3}, \ 0.7 \leq Pr \leq 160, \\ 2 \cdot 10^4 \leq Re \leq 10^6, \ L/d \geq 60. \tag{15.15}$$

Correlation (15.15) was modified by Sieder and Tate [21, 272, 357, 358] to take account of temperature-dependent changes in the fluid viscosity when the difference between the channel surface temperature T_w and the fluid mass-averaged temperature T_{fm} is more significant. The Sieder-Tate equation has the following form:

$$Nu = 0.027 Re^{0.8} Pr^{1/3} \left(\frac{\mu}{\mu_w}\right)^{0.14}, \\ 0.7 \leq Pr \leq 16700, \ 6000 \leq Re, \ L/d \geq 60. \tag{15.16}$$

Except for μ_w, all the fluid physical properties are calculated in the mean temperature, which is the arithmetic mean of temperature values at the tube inlet and outlet. Dynamic viscosity μ_w is determined in the wall temperature T_w.

Other power correlations can also be found in literature. In the correlation proposed by Li and Xuan [179, 373]:

$$Nu = 0.0059 Re^{0.9238} Pr^{0.4} \tag{15.17}$$

the power exponent at the Reynolds number is much higher than 0.8.

If the $Nu = f(Re, Pr)$ data presented in [227] are approximated using a power function using the least squares method, the obtained power formula is similar to expression (15.17):

$$Nu = 0.00685 Re^{0.904} Pr^{0.427}, \ 10^4 \leq Re \leq 5 \cdot 10^6, \\ 0.5 \leq Pr \leq 200. \tag{15.18}$$

Another modified version of the power formula is proposed in [88] to increase the accuracy of the Nusselt number calculations in a wide range of changes in the Prandtl number:

$$Nu = 0.0225 Re^{0.795} Pr^{0.495} \exp\left[-0.0225 (\ln Pr)^2\right], \\ 0.3 \leq Pr \leq 300, \quad 4 \cdot 10^4 \leq Re \leq 10^6. \tag{15.19}$$

Sleicher and Rouse [279] put forward modification of the Dittus-Boelter formula, making the exponents at the Reynolds and the Prandtl numbers dependent on the Prandtl number:

$$Nu = 5 + 0.015 Re^m Pr^n, \quad 0.1 \leq Pr \leq 10^4, \quad 10^4 \leq Re \leq 10^6, \tag{15.20}$$

where: $m = 0.88 - \frac{0.24}{4 + Pr}$, $n = \frac{1}{3} + \frac{1}{2}\exp(-0.6Pr)$.

The accuracy of the Nusselt number formulae can be improved by solving the energy conservation equation at a known universal distribution of velocity in the boundary layer. Based on his two-layer model of the boundary layer, Prandtl [16, 236] proposed the following formula:

$$Nu = \frac{\frac{\xi}{8} Re\, Pr}{1 + C\sqrt{\frac{\xi}{8}}(Pr - 1)}, \quad Pr \geq 0.5. \tag{15.21}$$

Constant C in formula (15.21) is the dimensionless coordinate y^+ of the end of the laminar sublayer, where the velocity distribution is linear. If the Prandtl two-layer model of the boundary layer is applied, $C = 11.6$, whereas, in the case of the three-layer model proposed by von Kármán, constant C is smaller: $C = 5$ [357]. Based on experimental data, Prandtl [236] proposed that $C = 8.7$ should be assumed in formula (15.21). A detailed method for formula (15.21) derivation is presented in Chap. 5. Petukhov and Kirillov [228] modified the Prandtl formula:

$$Nu = \frac{\frac{\xi}{8} Re\, Pr}{1.07 + 12.7\sqrt{\frac{\xi}{8}}(Pr^{2/3} - 1)}, \quad 10^4 \leq Re \leq 5 \cdot 10^6, \tag{15.22}$$

$$0.5 \leq Pr \leq 2000,$$

where the friction factor is defined as:

$$\xi = (1.82 \log Re - 1.64)^{-2}. \tag{15.23}$$

The method for formula (15.22) derivation and the advantages and disadvantages thereof are discussed in Chap. 6. It should be emphasized that friction factor ξ can be calculated from other formulae developed for tubes with smooth inner surfaces, e.g., from those presented in Chap. 6.

Formula (15.22) has been modified many times to extend the range of its application to cover smaller Reynolds numbers (Re > 4000) and improve its accuracy for gases when the Prandtl number is small, e.g., for air when Pr = 0.7. In the case of gases and small Reynolds numbers, formula (15.22) gives overestimated values of the Nusselt number.

The following two modifications of formula (15.22) are proposed in [229]:

$$Nu = \frac{\frac{\xi}{8} Re\, Pr}{1 + 3.4\xi + \left(11.7 + \frac{1.8}{Pr^{1/3}}\right)\sqrt{\frac{\xi}{8}}\left(Pr^{2/3} - 1\right)},$$

(15.24)

$$4 \cdot 10^3 \le Re \le 5 \cdot 10^6, 0.5 \le Pr \le 2000$$

and

$$Nu = \frac{\frac{\xi}{8} Re\, Pr}{1.07 + \frac{900}{Re} + \frac{0.63}{1 + 10\,Pr} + 12.7\sqrt{\frac{\xi}{8}}\left(Pr^{2/3} - 1\right)},$$

(15.25)

$$4 \cdot 10^3 \le Re \le 5 \cdot 10^6, 0.5 \le Pr \le 2000.$$

In [226, 227], formula (15.22) is reduced to the form expressed by formula (6.190).

Gnielinski [104] put forward modification of the Petukhov-Kirillov formula (15.22), bringing it to the following notation:

$$Nu = \frac{\frac{\xi}{8}(Re - 1000)\, Pr}{1 + 12.7\sqrt{\frac{\xi}{8}}\left(Pr^{2/3} - 1\right)}, 4 \cdot 10^3 \le Re \le 5 \cdot 10^6,$$

(15.26)

$$0.5 \le Pr \le 2000.$$

Formula (15.22) has gained considerable popularity in worldwide literature. It is now the most common tool for the calculation of the heat transfer coefficient on the inner surface of circular tubes and other channels. Nevertheless, it should be remembered that it is a modification of the Petukhov-Kirillov formula (15.22), which holds for $Re \ge 10000$. Relation (15.26) gives overestimated values of the Nusselt number for smaller Reynolds numbers included in the range of $4000 \le Re \le 10000$. Heat transfer coefficients can be calculated using correlations (6.174)–(6.175) and (6.229), which are proposed in Chap. 6. They are all characterized by excellent accuracy. As proved by the comparisons with experimental data, formulae (6.174) and (6.175) ensure very good agreement between results and experimental data in a wide range of changes in the Reynolds and the Prandtl numbers.

In the case of the transitional and turbulent flow that occurs for $Re > 2300$, the Nusselt number can be calculated using own formula (6.229), or formulae (6.197) and (6.220), recently proposed by Gnielinski [108]. For short ducts with a small L/d_w ratio, the Gnielinski formulae (6.197) and (6.220) do not give satisfactory results, which is demonstrated in Chap. 6.

15.2 Determination of Experimental Correlations on the Air Side for a Car Radiator

The water-side heat transfer coefficients are calculated using five different formulae listed in Table 15.3.

The formulae take account of the correction for the entrance length. The oval tube hydraulic diameter and the radiator tubes active length are $d_{h,w} = 7.06$ mm and $L_{ch} = 520$ mm, respectively. Numbers with the plus/minus sign at the coefficients denote half the 95% confidence interval. Symbol S_{min} denotes the value of the minimum sum of squares, and s_t is the root-mean-square error. The confidence intervals are calculated according to the formulae presented in Chap. 11. The changes in the Nusselt numbers calculated using the formulae listed in Table 15.3 are shown in Fig. 15.1a and b for $Pr_w = 3$ and $Pr_w = 5$, respectively.

Table 15.3 Formulae for the air-side Nusselt numbers determined based on experimental testing of a car radiator

No.	Formulae	Determined coefficients
1.	$Nu_a = x_1 \, Re_a^{x_2} \, Pr_a^{1/3}$ $Nu_w = 0.023 \, Re_w^{0.8} \, Pr_w^{0.3} \left[1 + \left(\dfrac{d_{h,w}}{L_{ch}}\right)^{2/3}\right]$ (15.27)	$S_{min} = 1.0549$ K^2 $s_t = 0.1385$ K $x_1 = 0.1115 \pm 0.0031$ $x_2 = 0.6495 \pm 0.0057$
2.	$Nu_a = x_1 \, Re_a^{x_2} \, Pr_a^{1/3}$ $Nu_w = 0.00685 \, Re_w^{0.904} \, Pr_w^{0.427} \left[1 + \left(\dfrac{d_{h,w}}{L_{ch}}\right)^{2/3}\right]$ (15.28)	$S_{min} = 0.5570$ K^2 $s_t = 0.1016$ K $x_1 = 0.0915 \pm 0.0219$ $x_2 = 0.7012 \pm 0.0057$
3.	$Nu_a = x_1 \, Re_a^{x_2} \, Pr_a^{1/3}$ $Nu_w = \dfrac{\frac{\xi}{8}(Re_w - 1000)\,Pr_w}{1 + 12.7\sqrt{\frac{\xi}{8}}\left(Pr_w^{2/3} - 1\right)} \left[1 + \left(\dfrac{d_{h,w}}{L_{ch}}\right)^{2/3}\right]$ (15.29)	$S_{min} = 0.6678$ K^2 $s_t = 0.1102$ K $x_1 = 0.1117 \pm 0.0024$ $x_2 = 0.6469 \pm 0.0045$
4.	$Nu_a = x_1 \, Re_a^{x_2} \, Pr_a^{1/3}$ $Nu_w = \dfrac{\frac{\xi}{8}(Re_w - 1000)\,Pr_w}{k_1 + 12.7\sqrt{\frac{\xi}{8}}\left(Pr_w^{2/3} - 1\right)} \left[1 + \left(\dfrac{d_{h,w}}{L_{ch}}\right)^{2/3}\right]$ (15.30) $k_1 = 1.07 + \dfrac{900}{Re_w} + \dfrac{0.63}{(1 + 10\,Pr_w)}$	$S_{min} = 1.2799$ K^2 $s_t = 0.1540$ K $x_1 = 0.1309 \pm 0.0418$ $x_2 = 0.6107 \pm 0.0559$
5.	$Nu_a = x_1 \, Re_a^{x_2} \, Pr_a^{1/3}$ $Nu_w = \dfrac{\frac{\xi}{8}\,Re_w\,Pr_w}{1 + 8.7\sqrt{\frac{\xi}{8}}\left(Pr_w - 1\right)} \left[1 + \left(\dfrac{d_{h,w}}{L_{ch}}\right)^{2/3}\right]$ (15.31)	$S_{min} = 1.4034$ K^2 $s_t = 0.1569$ K $x_1 = 0.1212 \pm 0.0398$ $x_2 = 0.6258 \pm 0.0595$

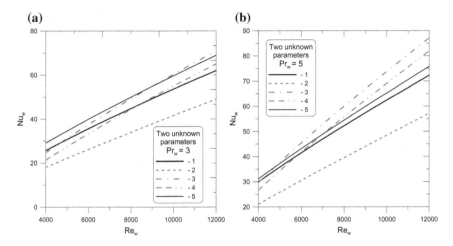

Fig. 15.1 Water-side Nusselt numbers calculated by means of the formulae listed in Table 15.3; numbers in the legend correspond to the formula number in the Table 15.3

Analysing the results presented in Fig. 15.1a, b, considerable differences can be observed between the values of the water-side Nusselt number Nu_w calculated from different formulae.

Figure 15.2 presents the air-side Nusselt numbers calculated using the formulae listed in Table 15.3.

Analysing the results presented in Fig. 15.2, it can be noticed that the determined values of the air-side Nusselt numbers differ from each other. Especially

Fig. 15.2 Nusselt numbers calculated using the formulae listed in Table 15.3; numbers in the legend correspond to the formula number in the Table 15.3

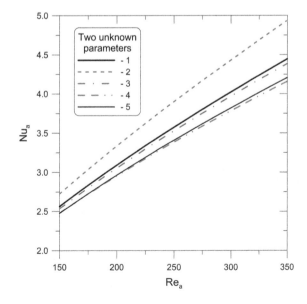

correlation 2, described by formula (15.28), differs substantially from the Nu_a values calculated from the other correlations. This confirms the high uncertainty of the power correlations used to calculate the heat transfer coefficient on the water side.

The results indicate clearly that to obtain correct Nusselt number correlations on the water and the air side (Nu_w and Nu_a, respectively), the formulae should be determined in parallel. Using literature correlations for the water-side heat transfer coefficient calculations may lead to a considerable uncertainty of the correlations determined on the air side. The water-side Nusselt number correlations offered in literature are derived for straight tubes and take no account of the heat exchanger real structure.

Chapter 16
Parallel Determination of Correlations for Heat Transfer Coefficients on the Air and Water Sides

This chapter is devoted to the parallel determination of unknown coefficients in the Nusselt number correlations on the air and the water side (Nu_a and Nu_w, respectively) based on the testing of the car radiator of a 1600 cm^3 spark-ignition combustion engine. The first step is to select the forms of the Nu_a and Nu_w approximation functions with $n \leq m$ unknown coefficients x_i, $i = 1, ..., n$. Symbol m is the number of measuring series; in this case $m = 57$. Coefficients $x_1, x_2, ..., x_n$ are found using the least squares method where the minimized sum S is defined by the following formula [309]:

$$ S = \sum_{i=1}^{m} \frac{\left[\left(T''_{w,i} \right)^m - \left(T''_{w,i} \right)^c \right]^2}{\sigma_{w,i}^2} + \sum_{i=1}^{m} \frac{\left[\left(T''_{am,i} \right)^m - \left(T''_{am,i} \right)^c \right]^2}{\sigma_{a,i}^2} = \min, \qquad (16.1) $$

where temperatures $\left(T''_{w,i} \right)^m$ and $\left(T''_{am,i} \right)^m$, $i = 1, ..., m$ are, respectively, the temperatures of water and air measured downstream the exchanger (cf. Fig. 10.8). The unknown coefficients $x_1, x_2, ..., x_n$ are determined based on the following physical quantities:

- water volume flow rate upstream the exchanger: $\dot{V}_w$,
- air velocity upstream the exchanger: w_0,
- mean temperature of water at the exchanger inlet and outlet: T'_w and T''_w, respectively,
- mean temperature of the air at the exchanger inlet and outlet: T'_{am} and T''_{am}, respectively.

The mean temperatures of water $\left(T''_{w,i} \right)^c$ and air $\left(T''_{am,i} \right)^c$ at the exchanger outlet are calculated using an analytical mathematical model [296, 297, 313]. Consequently, the temperatures are a function of measured physical quantities and sought parameters:

© Springer International Publishing AG, part of Springer Nature 2019
D. Taler, *Numerical Modelling and Experimental Testing of Heat Exchangers*,
Studies in Systems, Decision and Control 161,
https://doi.org/10.1007/978-3-319-91128-1_16

$$\left(T''_{w,i}\right)^c = \left(T''_{w,i}\right)^c \left[\left(\dot{V}_{w,i}, T'_{w,i}\right), \left(w_{0,i}, T'_{am,i}\right), x_1, x_2, \ldots, x_n\right],$$

$$\left(T''_{am,i}\right)^c = \left(T''_{am,i}\right)^c \left[\left(\dot{V}_{w,i}, T'_{w,i}\right), \left(w_{0,i}, T''_{am,i}\right), x_1, x_2, \ldots, x_n\right].$$

The sum of squares of the temperature differences (16.1) can also be written in a more concise form:

$$S(\mathbf{x}) = \left\{(\mathbf{T}'')^m - [\mathbf{T}''(\mathbf{x})]^c\right\}^{\mathrm{T}} \mathbf{W}\left\{(\mathbf{T}'')^m - [\mathbf{T}''(\mathbf{x})]^c\right\}, \tag{16.2}$$

where:

$$(\mathbf{T}'')^m = \left[\left(T''_{w,1}\right)^m, \left(T''_{w,2}\right)^m, \ldots, \left(T''_{w,m}\right)^m, \left(T''_{am,1}\right)^m, \left(T''_{am,2}\right)^m, \ldots, \left(T''_{am,m}\right)^m\right]^{\mathrm{T}},$$
$$\tag{16.3}$$

$$(\mathbf{T}'')^c = \left[\left(T''_{w,1}\right)^c, \left(T''_{w,2}\right)^c, \ldots, \left(T''_{w,m}\right)^c, \left(T''_{am,1}\right)^c, \left(T''_{am,2}\right)^c, \ldots, \left(T''_{am,m}\right)^c\right]^{\mathrm{T}},$$
$$\tag{16.4}$$

$$\mathbf{W} = \begin{bmatrix} w_{w,1} & \cdots & 0 & 0 & \cdots & 0 \\ 0 & \cdots & & & & \\ & & w_{w,m} & & & \vdots \\ \vdots & & & w_{a,1} & & \\ & & & & \cdots & 0 \\ 0 & & \cdots & & 0 & w_{a,m} \end{bmatrix}_{2m \times 2m}, \tag{16.5}$$

where weight coefficients $w_{w,i}$ and $w_{a,i}$ are equal to the inverses of variances of measured temperatures of water and air at the exchanger outlet, i.e., $w_{w,i} = 1/\sigma^2_{w,i}$, $w_{a,i} = 1/\sigma^2_{a,i}$, $i = 1, \ldots, m$.

The sum of squares (16.1) can now be written in the following form:

$$S^* = \sum_{i=1}^m w^*_{w,i}\left[\left(T''_{w,i}\right)^m - \left(T''_{w,i}\right)^c\right]^2 + \sum_{i=1}^m w^*_{a,i}\left[\left(T''_{am,i}\right)^m - \left(T''_{am,i}\right)^c\right]^2 = \min,$$
$$\tag{16.6}$$

where $w^*_{w,i}$ and $w^*_{a,i}$ are dimensionless weight coefficients defined as:

$$w^*_{w,i} = \frac{w_{w,i}}{w_m}, \quad w^*_{a,i} = \frac{w_{a,i}}{w_m}. \tag{16.7}$$

The mean weight coefficient w_m is calculated from the following relation:

$$w_m = \frac{1}{2m} \sum_{i=1}^{m} \left(w_{w,i} + w_{a,i} \right) = \frac{1}{2m} \sum_{i=1}^{m} \left(\frac{1}{\sigma_{w,i}^2} + \frac{1}{\sigma_{a,i}^2} \right). \tag{16.8}$$

The fact that dimensionless weight coefficients are used in the sum of squares facilitates determination of the uncertainty of coefficients x_1, x_2, ..., x_n for different weight coefficients $w_{w,i} = 1/\sigma_{w,i}^2$, $w_{a,i} = 1/\sigma_{a,i}^2$, $i = 1$, ..., m, but with the same variance ratio $\sigma_{a,i}^2/\sigma_{w,i}^2$.

The uncertainty of determined coefficients will be estimated using the Gaussian rule of summing variances [25, 261, 297, 310]. Confidence intervals will be calculated for coefficients determined both in the water- and in the air-side correlation. The real values of coefficients $\tilde{x}_1$, ..., $\tilde{x}_n$ determined using the least squares method are included with probability $P = (1 - \alpha) \cdot 100\%$ in the following intervals:

$$x_i - t_{2m-n}^{\alpha/2} \, s_t^* \, \sqrt{c_{ii}^*} \le \tilde{x}_i \le x_i + t_{2m-n}^{\alpha/2} \, s_t^* \, \sqrt{c_{ii}^*}, \tag{16.9}$$

where: x_i—parameter found using the least squares method, $t_{2m-n}^{\alpha/2}$—t-Student distribution quantile for the confidence level of $100(1 - \alpha)\%$ and $(2m - n)$ degrees of freedom.

Determination of the confidence interval for parameters found using the least squares method is discussed in detail in Chap. 11.

The square of the root-mean-square error $(s_t^*)^2$, which is an approximation of the variance of data, is calculated from the following formula:

$$(s_t^*)^2 = \frac{S_{\min}^*}{2m - n - 1}$$
$$= \frac{\frac{1}{2m-n-1} \left\{ \sum_{i=1}^{m} \frac{\left[\left(T_{w,i}'' \right)^m - \left(T_{w,i}'' \right)^c \right]^2}{\sigma_{w,i}^2} + \sum_{i=1}^{m} \frac{\left[\left(T_{am,i}'' \right)^m - \left(T_{am,i}'' \right)^c \right]^2}{\sigma_{a,i}^2} \right\}_{\min}}{\frac{1}{2m} \sum_{i=1}^{m} \left(\frac{1}{\sigma_{w,i}^2} + \frac{1}{\sigma_{a,i}^2} \right)}, \tag{16.10}$$

where: m—the number of measuring series, n—the number of determined parameters.

The minimum of the sum of squares of the differences between measured and calculated temperatures is found using the Levenberg-Marquardt method [261]. The following variance-covariance matrix:

$$\mathbf{D}_x^{(s)} = s_t \mathbf{C}_x^{(s)} = s_t \left[\left(\mathbf{J}^{(s)} \right)^{\mathrm{T}} \mathbf{W}^* \mathbf{J}^{(s)} \right]^{-1}, \tag{16.11}$$

is determined for the last iterative step, where the solution is found. Matrix $\mathbf{C}_x^{(s)}$ is calculated from the following formula:

$$\mathbf{C}_x^{(s)} = \left[\left(\mathbf{J}^{(s)} \right)^T \mathbf{W}^* \mathbf{J}^{(s)} \right]^{-1}. \tag{16.12}$$

The matrix of dimensionless weight coefficients is defined as:

$$\mathbf{W}^* = \begin{bmatrix} w_{w,1}^* & \cdots & 0 & 0 & \cdots & 0 \\ 0 & \cdots & & & & \\ & & w_{w,m}^* & & & \vdots \\ \vdots & & & w_{a,1}^* & & \\ & & & & \cdots & 0 \\ 0 & & \cdots & & 0 & w_{a,m}^* \end{bmatrix}_{2m \times 2m} . \tag{16.13}$$

Superscript (s) in formulae (16.11)–(16.12) denotes the number of the last iteration, and $\mathbf{J}$ is the Jacobian matrix. The symbol in expression (16.9) is the diagonal element in the matrix $\mathbf{C}_x^{(s)}$. The quantiles of the t-Student distribution $t_{m-n}^{\alpha/2}$ for the 95% confidence level ($\alpha = 0.05$) are $t_{53}^{0.025} = 2$ and $t_{110}^{0.025} = 2$. Quantile $t_{53}^{0.025} = m - n = 57 - 4 = 53$ appears in formula (16.9) if the sum of squares (16.6) takes account of the outlet temperatures of water only, and quantile $t_{110}^{0.025} = 2m - n = 114 - 4 = 110$—if both water and air outlet temperatures are taken into consideration in formula (16.6). After the minimum of the sum of squares S_{min}^* is found using the Levenberg-Marquardt method, s_t^* is determined, and then the limits of the 95% confidence interval are found from formula (16.9). Three, four or five parameters are determined in the Nusselt number correlations on the air and the water side using the measured data presented in Table 15.1.

In most cases, only the water temperatures measured at the exchanger outlet are taken into account, as these data are burdened with a much smaller measuring error compared to the results of the measurement of the mass-averaged temperature of air downstream the exchanger. The air temperatures are taken into consideration only if four parameters are to be found: two in the air-side Nusselt number correlation and two in the water-side formula.

16.1 Three Unknown Parameters

It is assumed that two unknown parameters x_1 and x_2 are used in the formula for the Nusselt number on the air side Nu_a, and the third coefficient x_3 appears in the correlation for the water-side Nusselt number Nu_w. The form of the air-side formula results from the Chilton-Colburn analogy.

Water physical properties and the Prandtl number values on the water side change substantially with temperature, and the power form of the formula defining the water-side Nusselt number Nu_w is not always able to reflect changes in Nu_w as a function of Re_w and Pr_w. The air- and the water-side correlations established based on the measurement of the water temperature at the exchanger outlet are listed in Table 16.1.

A power form of the correlation for the water-side Nusselt number Nu_w is adopted in formulae (16.14)–(16.16) and in formula (16.19). Formula (16.15) is similar in form to expression (15.18), which in turn is found based on the Petukhov-Kirillov data. It should be noted that right agreement characterizes the two formulae. However, in formula (15.18) the coefficient before the Reynolds number is 0.00685, whereas in formula (16.15) it is equal to 0.00655. Expression (16.19) is similar to the Dittus-Boelter formula, but the Reynolds number raised to the power of 0.8 ($Re_w^{0.8}$) is replaced by $(Re_w - x_3)^{0.8}$. Coefficient x_3 determined using the least squares method is $x_3 = 1000.5$. This value is very close to the number 1000 assumed in the Gnielinski formula, which can be explained by the analyzed spectrum of the Reynolds numbers included in the range of $4000 \le Re \le 12{,}000$. Gnielinski replaced Re_w with $(Re_w - 1000)$ so that the Petukhov-Kirillov formula should hold for the Reynolds numbers higher than 4000.

Expression (16.17) has the structure of the formula derived by Prandtl for the boundary layer divided into two sublayers. Similarly, expressions (16.18) and (16.21) are modifications of the Prandtl formula. In formula (16.18), the Prandtl number Pr_w in the numerator is replaced by $Pr_w^{4/3}$, and in formula (16.21)— $(Re_w - x_3)$ is used instead of the Reynolds number Re_w. Also, in this case, the value of coefficient $x_3 = 1095.9$ is close to the number 1000 in the Gnielinski formula. However, it has to emphasized that the analysis of the data presented in Table (16.1) in and Figs. 16.1, 16.2 and 16.3 does not give satisfactory results for formulae (16.17)–(16.18) and (16.21). This is the effect of the term $(Pr_w - 1)$, which appears in the denominator of all the three correlations mentioned above. By replacing $(Pr_w - 1)$ in the Prandtl formula with $(Pr_w^{2/3} - 1)$, Petukhov substantially improved the Petukhov-Kirillov Nusselt number correlation agreement with experimental data. Correlation (16.20) is similar in form to the Gnielinski formula. However, coefficient x_3 determined using the least squares method is 17.47, whereas in the correlation proposed by Gnielinski it equals 12.7.

Analysing the comparisons presented in Figs. 16.1, 16.2 and 16.3, considerable differences can be noticed between the values of both the air- and the water-side Nusselt numbers Nu_a and Nu_w. Comparing the results for $Pr_w = 3$ (cf. Fig. 16.2) with those for $Pr_w = 5$, it can be seen that for the latter ($Pr_w = 5$) the differences between the Nu_w values obtained using different correlations from Table 16.1 are more significant. On the one hand, the differences in the Nusselt numbers on both the air- and the water-side result from the different forms of the functions approximating the water-side Nusselt number. The differences are also because in the water-side Nusselt number correlations there is only one parameter x_3, which is

Table 16.1 Air- and water-side Nusselt number formulae—two parameters (x_1 and x_2) are determined in the correlation on the air side and one (x_3)—on the water side

No.	Formula	Weights	Determined parameters
1.	$Nu_a = x_1\, Re_a^{x_2}\, Pr_a^{1/3}$ $Nu_w = x_3 Re_w^{0.8} Pr_w^{0.3}\left[1+\left(\dfrac{d_{h,w}}{L_{ch}}\right)^{2/3}\right]$ (16.14)	$w_{w,i}=1$ $w_{a,i}=0$ $i=1,\ldots,m$	$S_{min}=0.5642\ \mathrm{K}^2$ $s_t=0.1032\ \mathrm{K}$ $x_1=0.0635\pm0.0047$ $x_2=0.7890\pm0.0212$ $x_3=0.0161\pm0.0969$
2.	$Nu_a = x_1\, Re_a^{x_2}\, Pr_a^{1/3}$ $Nu_w = x_3 Re_w^{0.9} Pr_w^{0.43}\left[1+\left(\dfrac{d_{h,w}}{L_{ch}}\right)^{2/3}\right]$ (16.15)	$w_{w,i}=1$ $w_{a,i}=0$ $i=1,\ldots,m$	$S_{min}=0.5325\ \mathrm{K}^2$ $s_t=0.1002\ \mathrm{K}$ $x_1=0.0814\pm0.0010$ $x_2=0.7307\pm0.0139$ $x_3=0.00655\pm0.0452$
3.	$Nu_a = x_1\, Re_a^{x_2}\, Pr_a^{1/3}$ $Nu_w = 0.023\, Re_w^{x_3} Pr_w^{0.3}\left[1+\left(\dfrac{d_{h,w}}{L_{ch}}\right)^{2/3}\right]$ (16.16)	$w_{w,i}=1$ $w_{a,i}=0$ $i=1,\ldots,m$	$S_{min}=0.5824\ \mathrm{K}^2$ $s_t=0.1048\ \mathrm{K}$ $x_1=0.0574\pm0.0023$ $x_2=0.8150\pm0.0350$ $x_3=0.7530\pm0.0124$
4.	$Nu_a = x_1\, Re_a^{x_2}\, Pr_a^{1/3}$ $Nu_w = \dfrac{\frac{\xi}{8} Re_w\, Pr_w}{1+x_3\sqrt{\frac{\xi}{8}}\,(Pr_w-1)}\left[1+\left(\dfrac{d_{h,w}}{L_{ch}}\right)^{2/3}\right]$ (16.17)	$w_{w,i}=1$ $w_{a,i}=0$ $i=1,\ldots,m$	$S_{min}=0.5791\ \mathrm{K}^2$ $s_t=0.1045\ \mathrm{K}$ $x_1=0.0549\pm0.0198$ $x_2=0.8122\pm0.0349$ $x_3=17.6914\pm0.2083$
5.	$Nu_a = x_1\, Re_a^{x_2}\, Pr_a^{1/3}$ $Nu_w = \dfrac{\frac{\xi}{8} Re_w\, Pr_w^{4/3}}{1+x_3\sqrt{\frac{\xi}{8}}\,(Pr_w-1)}\left[1+\left(\dfrac{d_{h,w}}{L_{ch}}\right)^{2/3}\right]$ (16.18)	$w_{w,\,i}=1$ $w_{a,i}=0$ $i=1,\ldots,m$	$S_{min}=0.5638\ \mathrm{K}^2$ $s_t=0.1031\ \mathrm{K}$ $x_1=0.0618\pm0.0026$ $x_2=0.7902\pm0.0081$ $x_3=28.6270\pm0.2059$
6.	$Nu_a = x_1\, Re_a^{x_2}\, Pr_a^{1/3}$ $Nu_w = 0.023(Re_w-x_3)^{0.8} Pr_w^{0.3}\left[1+\left(\dfrac{d_{h,w}}{L_{ch}}\right)^{2/3}\right]$ (16.19)	$w_{w,i}=1$ $w_{a,i}=0$ $i=1,\ldots,m$	$S_{min}=0.5118\ \mathrm{K}^2$ $s_t=0.0983\ \mathrm{K}$ $x_1=0.0850\pm0.0022$ $x_2=0.7139\pm0.0051$ $x_3=1000.5\pm0.1967$
7.	$Nu_a = x_1\, Re_a^{x_2}\, Pr_a^{1/3}$ $Nu_w = \dfrac{\frac{\xi}{8}(Re_w-1000)\, Pr_w}{1+x_3\sqrt{\frac{\xi}{8}}\,(Pr_w^{2/3}-1)}\left[1+\left(\dfrac{d_{h,w}}{L_{ch}}\right)^{2/3}\right]$ (16.20)	$w_{w,i}=1$ $w_{a,i}=0$ $i=1,\ldots,m$	$S_{min}=0.5109\ \mathrm{K}^2$ $s_t=0.0973\ \mathrm{K}$ $x_1=0.0873\pm0.0036$ $x_2=0.7060\pm0.0078$ $x_3=17.47\pm0.1940$
8.	$Nu_a = x_1\, Re_a^{x_2}\, Pr_a^{1/3}$ $Nu_w = \dfrac{\frac{\xi}{8}(Re_w-x_3)\, Pr_w}{1+8.7\sqrt{\frac{\xi}{8}}\,(Pr_w-1)}\left[1+\left(\dfrac{d_{h,w}}{L_{ch}}\right)^{2/3}\right]$ (16.21)	$w_{w,i}=1$ $w_{a,i}=0$ $i=1,\ldots,m$	$S_{min}=0.5150\ \mathrm{K}^2$ $s_t=0.0986\ \mathrm{K}$ $x_1=0.0864\pm0.0021$ $x_2=0.7059\pm0.0049$ $x_3=1095.9\pm0.1972$

Table 16.2 Air- and water-side Nusselt number formulae—two parameters (x_1 and x_2) are determined in the correlation on the air side and two (x_3 and x_4)—on the water side

No.	Formula	Weights	Determined parameters
1.	$$\mathrm{Nu}_a = x_1\,\mathrm{Re}_a^{x_2}\,\mathrm{Pr}_a^{1/3}$$ $$\mathrm{Nu}_w = x_3\mathrm{Re}_w^{x_4}\mathrm{Pr}_w^{0.3}\left[1+\left(\frac{d_{h,w}}{L_{ch}}\right)^{2/3}\right]\quad(16.22)$$	$w_{w,i}=1$ $w_{a,i}=0$ $i=1,\ldots,m$	$S_{\min}=0.5237\ \mathrm{K}^2$ $s_t=0.1004\ \mathrm{K}$ $x_1=0.0824\pm0.0042$ $x_2=0.7221\pm0.0146$ $x_3=0.0052\pm0.0180$ $x_4=0.9450\pm0.0699$
2.	$$\mathrm{Nu}_a = x_1\,\mathrm{Re}_a^{x_2}\,\mathrm{Pr}_a^{1/3}$$ $$\mathrm{Nu}_w = \frac{\frac{\xi}{8}\left(\mathrm{Re}_w-x_3\right)\mathrm{Pr}_w}{1+x_4\,\sqrt{\frac{\xi}{8}}\left(\mathrm{Pr}_w^{2/3}-1\right)}\left[1+\left(\frac{d_{h,w}}{L_{ch}}\right)^{2/3}\right]\quad(16.23)$$	$w_{w,i}=1$ $w_{a,i}=0$ $i=1,\ldots,m$	$S_{\min}=0.5085\ \mathrm{K}^2$ $s_t=0.0989\ \mathrm{K}$ $x_1=0.0899\pm0.0028$ $x_2=0.6990\pm0.0061$ $x_3=1078.9\pm0.1993$ $x_4=16.38\pm0.2016$
3.	$$\mathrm{Nu}_a = x_1\,\mathrm{Re}_a^{x_2}\,\mathrm{Pr}_a^{1/3}$$ $$\mathrm{Nu}_w = \frac{\frac{\xi}{8}\left(\mathrm{Re}_w-x_3\right)\mathrm{Pr}_w}{1+x_4\,\sqrt{\frac{\xi}{8}}\left(\mathrm{Pr}_w-1\right)}\left[1+\left(\frac{d_{h,w}}{L_{ch}}\right)^{2/3}\right]\quad(16.24)$$	$w_{w,i}=1$ $w_{a,i}=0$ $i=1,\ldots,m$	$S_{\min}=0.5098\ \mathrm{K}^2$ $s_t=0.0990\ \mathrm{K}$ $x_1=0.0988\pm0.0082$ $x_2=0.6741\pm0.0153$ $x_3=1422.2\pm0.1967$ $x_4=6.22\pm0.2008$
4.	$$\mathrm{Nu}_a = x_1\,\mathrm{Re}_a^{x_2}\,\mathrm{Pr}_a^{1/3}$$ $$\mathrm{Nu}_w = \frac{\frac{\xi}{8}\left(\mathrm{Re}_w-x_3\right)\mathrm{Pr}_w}{x_4+8.7\,\sqrt{\frac{\xi}{8}}\left(\mathrm{Pr}_w-1\right)}\left[1+\left(\frac{d_{h,w}}{L_{ch}}\right)^{2/3}\right]\quad(16.25)$$	$w_{w,i}=1$ $w_{a,i}=0$ $i=1,\ldots,m$	$S_{\min}=0.5149\ \mathrm{K}^2$ $s_t=0.0995\ \mathrm{K}$ $x_1=0.0869\pm0.0146$ $x_2=0.7041\pm0.0328$ $x_3=1112.0\pm0.1432$ $x_4=0.9756\pm0.1998$
5.	$$\mathrm{Nu}_a = x_1\,\mathrm{Re}_a^{x_2}\,\mathrm{Pr}_a^{1/3}$$ $$\mathrm{Nu}_w = \frac{\frac{\xi}{8}\left(\mathrm{Re}_w-x_3\right)\mathrm{Pr}_w^{4/3}}{1+x_4\,\sqrt{\frac{\xi}{8}}\left(\mathrm{Pr}_w-1\right)}\left[1+\left(\frac{d_{h,w}}{L_{ch}}\right)^{2/3}\right]\quad(16.26)$$	$w_{w,i}=1$ $w_{a,i}=0$ $i=1,\ldots,m$	$S_{\min}=0.5030\ \mathrm{K}^2$ $s_t=0.0983\ \mathrm{K}$ $x_1=0.1022\pm0.0066$ $x_2=0.6675\pm0.0119$ $x_3=1389.9\pm0.1971$ $x_4=12.28\pm0.1994$
6.	$$\mathrm{Nu}_a = x_1\,\mathrm{Re}_a^{x_2}\,\mathrm{Pr}_a^{1/3}$$ $$\mathrm{Nu}_w = \frac{\frac{\xi}{8}\left(\mathrm{Re}_w-x_3\right)\mathrm{Pr}_w}{1+x_4\,\sqrt{\frac{\xi}{8}}\left(\mathrm{Pr}_w^{2/3}-1\right)}\left[1+\left(\frac{d_{h,w}}{L_{ch}}\right)^{2/3}\right]\quad(16.27)$$	$w_{w,i}=11.1111$ $(\sigma_{w,i}=0.3)$ $w_{a,i}=0.1111$ $(\sigma_{a,i}=3)$ $i=1,\ldots,m$	$S_{\min}^*=1.5870\,\mathrm{K}^2$ $s_t^*=0.1207\ \mathrm{K}$ $x_1=0.0849\pm0.0049$ $x_2=0.7122\pm0.0107$ $x_3=1144.2\pm0.2401$ $x_4=16.21\pm0.2418$
7.	$$\mathrm{Nu}_a = x_1\,\mathrm{Re}_a^{x_2}\,\mathrm{Pr}_a^{1/3}$$ $$\mathrm{Nu}_w = \frac{\frac{\xi}{8}\left(\mathrm{Re}_w-x_3\right)\mathrm{Pr}_w}{1+x_4\,\sqrt{\frac{\xi}{8}}\left(\mathrm{Pr}_w^{2/3}-1\right)}\left[1+\left(\frac{d_{h,w}}{L_{ch}}\right)^{2/3}\right]\quad(16.28)$$	$w_{w,i}=4.0$ $(\sigma_{w,i}=0.5)$ $w_{a,i}=0.04$ $(\sigma_{a,i}=5)$ $i=1,\ldots,m$	$S_{\min}^*=1.5870\,\mathrm{K}^2$ $s_t^*=0.1207\ \mathrm{K}$ $x_1=0.0849\pm0.0047$ $x_2=0.7123\pm0.0104$ $x_3=1145.1\pm0.2398$ $x_4=16.21\pm0.2420$

Fig. 16.1 Comparison of the air-side Nusselt number (Nu_a) correlations; the numbers of the curves correspond to the numbers of the correlations in Table 16.1

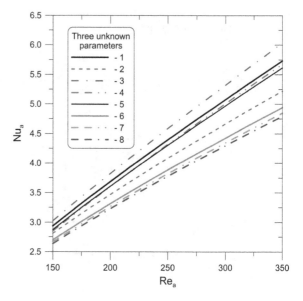

Fig. 16.2 Comparison of the water-side Nusselt number (Nu_w) correlations for the Prandtl number $Pr_w = 3$; the numbers of the curves correspond to the numbers of the correlations in Table 16.1

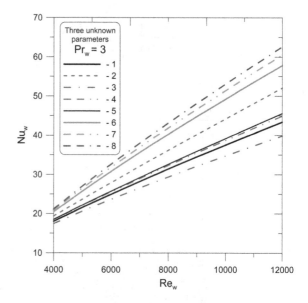

insufficient to ensure proper matching of the water-side correlation with experimental data.

It should also be noted that if the water-side Nusselt number formulae give higher values of Nu_w, the Nusselt numbers on the air side are lower. Similarly, if the Nu_w values are lower, then Nu_a values are higher. Formulae (16.14) and (16.16)–(16.18) give underestimated values of Nu_w (cf. Figs. 16.2 and 16.3), which in turn

Fig. 16.3 Comparison of the water-side Nusselt number (Nu_w) correlations for the Prandtl number $\mathrm{Pr}_w = 5$; the numbers of the curves correspond to the numbers of the correlations in Table 16.1

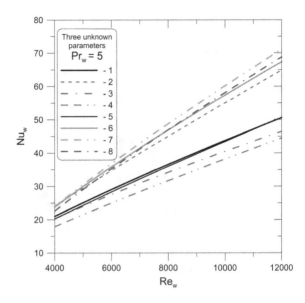

leads to overestimated values of Nu_a. For each measuring series, it is possible to determine an overall heat transfer coefficient being a function of the heat transfer coefficient both on the air and on the water side. Consequently, if the Nu_a value is overestimated, Nu_w is underestimated and vice versa. It should be noted that correlation (16.15), which defines the Nusselt number on the water side, gives results lying in between the groups of curves 1-3-4-5 and 6-7-8.

16.2 Four Unknown Parameters

It is assumed that two unknown parameters x_1 and x_2 are used in the formula for the Nusselt number on the air side Nu_a, and the other two coefficients x_3 and x_4 appear in the correlation for the water-side Nusselt number Nu_w. The air- and the water-side Nusselt number correlations defined by formulae (16.22)–(16.26) are established based on the temperature of water measured at the radiator outlet. Coefficients $x_1, \ldots, x_4$ in formulae (16.27)–(16.28) are found based on the temperatures of water and air measured at the exchanger outlet. Weight coefficients for water $w_{w,i}$ are a hundred times bigger than weight coefficients for air $w_{a,i}$ due to the higher uncertainty of the air temperature measured downstream the radiator. In both cases, the $\sigma_{w,i}/\sigma_{a,i}$ ratio is the same, and therefore the determined coefficients $x_1, \ldots, x_4$ in formulae (16.27) and (16.28) are almost identical. The slight differences between the values of individual coefficients are due to numerical computations inaccuracy.

The changes in the air-side Nusselt number Nu_a depending on the Reynolds number Re_a are shown in Fig. 16.4. Figures 16.5 and 16.6 illustrate changes in the

Fig. 16.4 Comparison of the air-side Nusselt number Nu_a correlations; the numbers of the curves correspond to the numbers of the correlations in Table 16.2

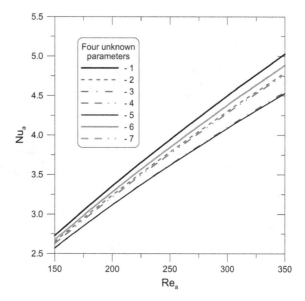

Fig. 16.5 Comparison of the water-side Nusselt number (Nu_w) correlations for $Pr_w = 3$; the numbers of the curves correspond to the numbers of the correlations in Table 16.2

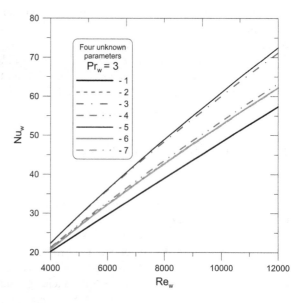

water-side Nusselt number Nu_w depending on the Reynolds number Re_w for the Prandtl numbers $Pr_w = 3$ and $Pr_w = 5$, respectively.

The results presented in Figs. 16.4 and 16.5 indicate that if the water-side Nusselt number is approximated using the power-type formula (16.22) (curve 1 in Figs. 16.5 and 16.6), the obtained values of the Nusselt number on the water side Nu_w are underestimated, whereas the air-side values of the Nusselt number Nu_a are

Fig. 16.6 Comparison of the
water-side Nusselt number
(Nu_w) correlations for
$Pr_w = 5$; the numbers of the
curves correspond to the
numbers of the correlations in
Table 16.2

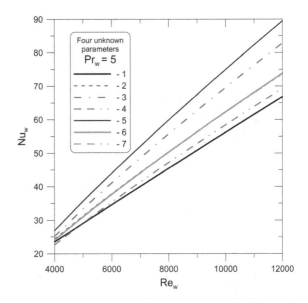

overestimated. If the water-side Nusselt number is approximated using formula
(16.26) (curve 5), which is a modification of the Prandtl formula where the Prandtl
number Pr_w in the numerator is replaced by $Pr_w^{4/3}$, the opposite is the case—the Nu_a
values are underestimated, and the values of Nu_w are overestimated. If the original
form of the Prandtl formula defined by expression (16.24) is kept, the obtained
water-side Nusselt number Nu_w correlation is also very sensitive to changes in the
Prandtl number on the water side Pr_w (cf. curve 3 in Figs. 16.4, 16.5 and 16.6).
Good agreement between the water- and the air-side Nusselt numbers Nu_w and Nu_a
is obtained if the Nusselt number on the water side is approximated using the
modified Prandtl formula (16.25) (curve 4) or formulae (16.23, 16.27 and 16.28),
which are modifications of the Petukhov-Kirillov-Gnielinski formulae (curves 2, 6
and 7).

16.3   Five Unknown Parameters

Table 16.3 presents correlations for the air- and the water-side Nusselt numbers,
Nu_a and Nu_w respectively, where it is assumed that the experimental data
approximation formulae include five unknown coefficients $x_1, ..., x_5$. Two of them
—x_1 and x_2—are used in the air-side Nusselt number correlation, whereas the other
three: $x_3, ..., x_5$ are used in the correlation for the water-side Nusselt number (Nu_w).
    The changes in the air-side Nusselt number Nu_a depending on the Reynolds
number Re_a and in the water-side Nusselt number Nu_w depending on the Reynolds

Table 16.3 Air- and water-side Nusselt number formulae—two parameters (x_1 and x_2) are determined in the correlation on the air side and three (x_3, x_4 and x_5)—on the water side

No.	Formula	Weights	Determined parameters
1.	$Nu_a = x_1 \, Re_a^{x_2} \, Pr_a^{1/3}$ $Nu_w = x_3 Re_w^{x_4} Pr_w^{x_5}\left[1 + \left(\dfrac{d_{h,w}}{L_{ch}}\right)^{2/3}\right]$ (16.29)	$w_{w,i} = 1$ $w_{a,i} = 0$ $i = 1,\ldots,m$	$S_{min} = 0.5075$ K² $s_t = 0.0998$ K $x_1 = 0.0916 \pm 0.0021$ $x_2 = 0.6942 \pm 0.0062$ $x_3 = 0.00264 \pm 0.0114$ $x_4 = 1.0407 \pm 0.0773$ $x_5 = 0.2873 \pm 0.2112$
2.	$Nu_a = x_1 \, Re_a^{x_2} \, Pr_a^{1/3}$ $Nu_w = \dfrac{\frac{\xi}{8}(Re_w - x_3)\, Pr_w}{x_4 + x_5\sqrt{\frac{\xi}{8}}\left(Pr_w^{2/3} - 1\right)}\left[1 + \left(\dfrac{d_{h,w}}{L_{ch}}\right)^{2/3}\right]$ (16.30)	$w_{w,i} = 1$ $w_{a,i} = 0$ $i = 1,\ldots,m$	$S_{min} = 0.5042$ K² $s_t = 0.0994$ K $x_1 = 0.0947 \pm 0.0053$ $x_2 = 0.6889 \pm 0.0134$ $x_3 = 1179.4 \pm 0.1449$ $x_4 = 1.1785 \pm 0.2417$ $x_5 = 13.0574 \pm 0.2292$
3.	$Nu_a = x_1 \, Re_a^{x_2} \, Pr_a^{1/3}$ $Nu_w = \dfrac{\frac{\xi}{8}(Re_w - x_3)\, Pr_w}{x_4 + x_5\sqrt{\frac{\xi}{8}}\left(Pr_w - 1\right)}\left[1 + \left(\dfrac{d_{h,w}}{L_{ch}}\right)^{2/3}\right]$ (16.31)	$w_{w,i} = 1$ $w_{a,i} = 0$ $i = 1,\ldots,m$	$S_{min} = 0.5126$ K² $s_t = 0.1002$ K $x_1 = 0.0920 \pm 0.0094$ $x_2 = 0.7025 \pm 0.0214$ $x_3 = 986.2427 \pm 0.1516$ $x_4 = 1.7054 \pm 0.2084$ $x_5 = 5.3820 \pm 0.2015$
4.	$Nu_a = x_1 \, Re_a^{x_2} \, Pr_a^{1/3}$ $Nu_w = \dfrac{\frac{\xi}{8}(Re_w - x_5)\, Pr_w^{4/3}}{x_3 + x_4\sqrt{\frac{\xi}{8}}\left(Pr_w - 1\right)}\left[1 + \left(\dfrac{d_{h,w}}{L_{ch}}\right)^{2/3}\right]$ (16.32)	$w_{w,i} = 1$ $w_{a,i} = 0$ $i = 1,\ldots,m$	$S_{min} = 0.4989$ K² $s_t = 0.0989$ K $x_1 = 0.1046 \pm 0.0096$ $x_2 = 0.6678 \pm 0.0183$ $x_3 = 1.4823 \pm 0.1594$ $x_4 = 10.3555 \pm 0.1988$

(continued)

Table 16.3 (continued)

No.	Formula	Weights	Determined parameters
5.	$\mathrm{Nu}_a = x_1\,\mathrm{Re}_a^{x_2}\,\mathrm{Pr}_a^{1/3}$ $\mathrm{Nu}_w = \dfrac{\frac{\xi}{8}\left(\mathrm{Re}_w - x_5\right)\mathrm{Pr}_w^{x_3}}{1 + x_4\sqrt{\frac{\xi}{8}}\left(\mathrm{Pr}_w - 1\right)}\left[1 + \left(\dfrac{d_{h,w}}{L_{ch}}\right)^{2/3}\right]$ (16.33)	$w_{w,i} = 1$ $w_{a,i} = 0$ $i = 1,\ldots,m$	$x_5 = 1302.7510 \pm 0.1986$ $S_{\min} = 0.5016\ \mathrm{K}^2$ $s_t = 0.0992\ \mathrm{K}$ $x_1 = 0.1000 \pm 0.0070$ $x_2 = 0.6729 \pm 0.01761$ $x_3 = 1.3815 \pm 0.0768$ $x_4 = 14.1688 \pm 0.2085$ $x_5 = 1309.4260 \pm 0.2012$

number Re_w are shown, respectively, in Fig. 16.7 and in Fig. 16.8 (for $\text{Pr}_w = 3$) and Fig. 16.9 (for $\text{Pr}_w = 5$).

Analysing the results presented in Figs. 16.7, 16.8 and 16.9, it can be seen that all the correlations under consideration demonstrate better agreement compared with correlations when three or four unknown coefficients are assumed. Only correlation (16.31), which has the Prandtl formula structure, gives results that differ

Fig. 16.7 Comparison of the air-side Nusselt number (Nu_a) correlations; the numbers of the curves correspond to the numbers of the correlations in Table 16.3

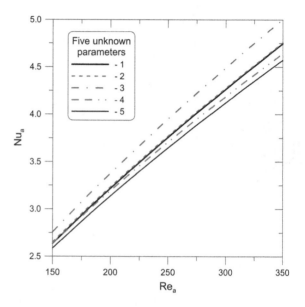

Fig. 16.8 Comparison of the water-side Nusselt number (Nu_w) correlations for $\text{Pr}_w = 3$; the numbers of the curves correspond to the numbers of the correlations in Table 16.3

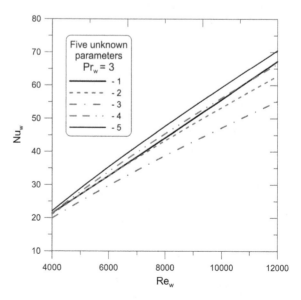

Fig. 16.9 Comparison of the water-side Nusselt number (Nu_w) correlations for $Pr_w = 5$; the numbers of the curves correspond to the numbers of the correlations in Table 16.3

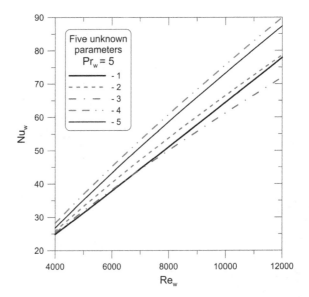

from those obtained using the other correlations (curve 3). The Prandtl formula modification where the Prandtl number Pr_w in the numerator of formula (16.31) is replaced by $Pr_w^{4/3}$ in formula (16.32) or by $Pr_w^{1.3815}$ in formula (16.33) improves the agreement of the obtained correlations with the other formulae presented in Table 16.3. However, it can be noticed that the water-side Nusselt number Nu_w is highly sensitive to changes in the Prandtl number Pr_w (cf. Figs. 16.8 and 16.9).

The calculations performed at different numbers of unknown coefficients prove that the adequate form of the water-side Nusselt number correlation is of great importance for the correct approximation of experimental data. The Nusselt number correlations used to calculate the heat transfer coefficient in straight tubes cannot usually be applied to find the heat transfer coefficient in heat exchangers, which are usually characterized by a complex flow system. The correlations for the Nusselt numbers on the water and the air side should be determined in parallel. Adopting the formulae derived for straight tubes to calculate the mean heat transfer coefficient on the water side in a heat exchanger may result in an uncertain correlation for the Nusselt number on the air side.

Chapter 17
Determination of Correlations for the Heat Transfer Coefficient on the Air Side Using CFD Simulations

Correlations describing the air-side heat transfer coefficient in cross-flow finned-tube exchangers can be found using the CFD (Computational Fluid Dynamics) tools [302, 306, 314, 315, 318, 319, 328]. An inverse problem is solved in [185] where the mean heat transfer coefficient on the flue-gas side and the steam temperature at the superheater inlet are determined based on the steam temperature measurement performed in selected inner points located along the flue gas path. In every iteration step, the temperature field in the superheater tubes and in passing steam is found using the ANSYS-CFX program [11]. Due to the rather complex structure of the analysed heat exchangers and insufficient computing power of computers, it is impossible to model the device in its entirety. Therefore, a selected repetitive fragment is usually used to model thermal and flow phenomena on the air side only, whereas the heat transfer inside the tube is not modelled. The heat transfer coefficient on the tube inner surface is calculated using the Nusselt number correlations established experimentally, e.g. the Dittus-Boelter or the Gnielinski formulae [34, 79, 104–108, 122–123, 128, 142, 160, 190, 257]. As it is, CFD modelling is still an imperfect tool. Despite numerous efforts, researchers find it difficult to model the convective heat transfer because the fluid flow is either transitional (laminar to turbulent) or turbulent. The turbulence models incorporated in commercial CFD packages lead to different results of the heat transfer coefficient calculations on the surface of a solid body. Nonetheless, a significant advantage of CFD modelling is the opportunity to simulate thermal and flow phenomena for fluids flowing past/through solid bodies with complex shapes. The fluid and the solid body surface temperature distributions are often difficult to determine due to problems posed by the measurement itself. This chapter presents the determination of correlations for the air-side Nusselt number using the ANSYS-CFX program.

© Springer International Publishing AG, part of Springer Nature 2019   525
D. Taler, *Numerical Modelling and Experimental Testing of Heat Exchangers*,
Studies in Systems, Decision and Control 161,
https://doi.org/10.1007/978-3-319-91128-1_17

17.1   Mass, Momentum and Energy Conservation Equations and the Turbulence Model

Modelling the turbulent flow of air passing transversely to the axis of tubes is based on the mass, momentum and energy conservation equations supplemented with equations used to model the flow turbulence. The temperature distribution in tubes and plate fins is found from the energy conservation equation taking account of the tube air- and water-side boundary conditions.

Time-averaged, three-dimensional equations of the mass, momentum and energy conservation are solved using a hybrid method combining the finite volume and the finite element methods (FVM-FEM). The heat transfer in the air flow is modelled using the mass, momentum and energy conservation equations written below [11]:

$$\frac{\partial \rho}{\partial t} + \nabla \cdot (\rho \, \mathbf{U}) = 0, \tag{17.1}$$

$$\frac{\partial (\rho \, \mathbf{U})}{\partial t} + \nabla \cdot (\rho \, \mathbf{U} \otimes \mathbf{U}) = -\nabla p + \nabla \cdot \tau + \mathbf{S}_M, \tag{17.2}$$

$$\frac{\partial (\rho \, i)}{\partial t} - \frac{\partial p}{\partial t} + \nabla \cdot (\rho \, \mathbf{U} \, i) = \nabla \cdot (\lambda \nabla T) + \mathbf{U} \cdot \nabla p + \tau : \nabla \mathbf{U} + S_E, \tag{17.3}$$

where stress tensor τ is the function of the strain rate:

$$\tau = \mu \left(\nabla \mathbf{U} + (\nabla \mathbf{U})^{\mathrm{T}} - \frac{2}{3} \delta \nabla \cdot \mathbf{U} \right). \tag{17.4}$$

The following marking is adopted in formulae (17.1)–(17.4): ρ—density, $\mathbf{U}$—velocity vector, t—time, p—pressure, i—specific enthalpy, $\mathbf{S}_M$—momentum source unit power, T—temperature, λ—thermal conductivity, S_E—continuous energy source unit power, μ—dynamic viscosity, δ—unit matrix.

The temperature distribution in tube walls and plate fins satisfies the heat conduction equation:

$$c_w \, \rho_w \, \frac{\partial T_w}{\partial t} = \nabla \cdot [\lambda_w(T_w) \, \nabla T_w], \tag{17.5}$$

where: c_w—wall material specific heat, ρ_w—wall material density, λ_w—wall material thermal conductivity, T_w—wall temperature.

The turbulent flow is modelled using the two-equation SST (Shear Stress Transport) turbulence model developed based on the Wilcox k–ω and the k–ε models [11, 170, 198]. The standard k–ω model is a two-equation turbulence model that offers a compromise between computational accuracy and efficiency. It is a combination of the k–ω and the k–ε models.

The Wilcox k–ω model [362–364] is described by the following two equations:

$$\frac{\partial(\rho k)}{\partial t} + \frac{\partial}{\partial x_j}(\rho U_j k) = \frac{\partial}{\partial x_j}\left[\left(\mu + \frac{\mu_t}{\sigma_{k1}}\right)\frac{\partial k}{\partial x_j}\right] + P_k - \beta' \rho k \omega, \qquad (17.6)$$

$$\frac{\partial(\rho \omega)}{\partial t} + \frac{\partial}{\partial x_j}(\rho U_j \omega) = \frac{\partial}{\partial x_j}\left[\left(\mu + \frac{\mu_t}{\sigma_{\omega1}}\right)\frac{\partial \omega}{\partial x_j}\right] + \alpha_1 \frac{\omega}{k} P_k - \beta_1' \rho \omega^2, \qquad (17.7)$$

where: $\omega = \frac{1}{\beta'}\frac{\varepsilon}{k}$, β'—constant.

In order to improve the standard k–ω turbulence model, a transformed k–ε model is also used:

$$\frac{\partial(\rho k)}{\partial t} + \frac{\partial}{\partial x_j}(\rho U_j k) = \frac{\partial}{\partial x_j}\left[\left(\mu + \frac{\mu_t}{\sigma_{k2}}\right)\frac{\partial k}{\partial x_j}\right] + P_k - \beta' \rho k \omega, \qquad (17.8)$$

$$\begin{aligned}\frac{\partial(\rho \omega)}{\partial t} + \frac{\partial}{\partial x_j}(\rho U_j \omega) = {} & \frac{\partial}{\partial x_j}\left[\left(\mu + \frac{\mu_t}{\sigma_{\omega2}}\right)\frac{\partial \omega}{\partial x_j}\right] \\ & + 2\rho\,\frac{1}{\sigma_{\omega2}\,\omega}\frac{\partial k}{\partial x_j}\frac{\partial \omega}{\partial x_j} + \alpha_2 \frac{\omega}{k} P_k - \beta_2 \rho \omega^2.\end{aligned} \qquad (17.9)$$

The modified k–ω model was obtained by multiplying the Wilcox model Eqs. (17.6–17.7) by weight function F_1 and the transformed k–ε Eq. (17.8–17.9) by weight function $(1 - F_1)$ and adding appropriate equations for k and ω:

$$\frac{\partial(\rho k)}{\partial t} + \frac{\partial}{\partial x_j}(\rho U_j k) = \frac{\partial}{\partial x_j}\left[\left(\mu + \frac{\mu_t}{\sigma_{k3}}\right)\frac{\partial k}{\partial x_j}\right] + P_k - \beta' \rho k \omega + P_{kb}, \qquad (17.10)$$

$$\begin{aligned}\frac{\partial(\rho \omega)}{\partial t} + \frac{\partial}{\partial x_j}(\rho U_j \omega) = {} & \frac{\partial}{\partial x_j}\left[\left(\mu + \frac{\mu_t}{\sigma_{\omega3}}\right)\frac{\partial \omega}{\partial x_j}\right] + (1 - F_1)2\rho\,\frac{1}{\sigma_{\omega2}\,\omega}\frac{\partial k}{\partial x_j}\frac{\partial \omega}{\partial x_j} \\ & + \alpha_3 \frac{\omega}{k} P_k - \beta_3 \rho \omega^2 + P_{\omega b}.\end{aligned}$$

$$(17.11)$$

The improved k–ω model has the advantages of the Wilcox k–ω and the k–ε models, but it is unable to predict the onset or the size of the flow separation from smooth surfaces. This weakness results from ignoring the transport of turbulent shear stresses, which leads to overestimation of turbulent viscosity. In order to eliminate this downside, the Shear Stress Transport (SST) turbulence model was developed. In it, turbulent viscosity v_t is limited by the following equation:

$$v_t = \frac{\alpha_1 k}{\max(\alpha_1 \omega, S F_2)}. \qquad (17.12)$$

The following symbols are used in formulae (17.1)–(17.11): F_1, F_2—weight functions, k—turbulence kinetic energy, P_k—rate of changes in turbulence kinetic energy, P_ω—turbulence frequency source term, S—strain rate invariant measure, U_j—velocity vector j-th component, x_j—j-th component, α, β, β'—constants in the Wilcox k–ω turbulence model, ε—turbulence dissipation, μ_t—dynamic turbulent viscosity, σ_k—turbulence model constant in the kinetic energy equation, σ_ω—constant in the k–ω turbulence model, Ω—turbulence frequency, ∇—gradient operator (nabla), $\otimes$—dyadic product.

17.2 Heat Transfer on the Tube Inner Surface

In CFD modelling of the air-side heat transfer, the temperature of the water flowing inside the tube is assumed as constant because in real heat exchangers the water temperature drop on the tube length is slight. The assumption is entirely justified considering that the tube surface temperature has very little impact on physical properties of air. The Nusselt number for water is found from the following formula [313]:

$$
\mathrm{Nu}_w = \frac{\frac{\xi}{8}\,(\mathrm{Re}_w - 1404.486)\,\mathrm{Pr}_w}{1 + 11.9166\,\sqrt{\frac{\xi}{8}}\,(\mathrm{Pr}_w^{2/3} - 1)}\left[1 + \left(\frac{d_{h,w}}{L_{ch}}\right)^{2/3}\right],
$$
$$
4000 \le \mathrm{Re}_w \le 12000,
$$
(17.13)

where friction factor ξ is defined as:

$$
\xi = (1.82\,\log \mathrm{Re}_w - 1.64)^{-2}.
$$
(17.14)

The Nusselt, the Reynolds and the Prandtl numbers Nu_w, Pr_w and Re_w are defined as follows: $\mathrm{Nu}_w = \alpha_w\,d_{h,w}/\lambda_w$, $\mathrm{Re}_w = w_w\,d_{h,w}/\nu_w$ and $\mathrm{Pr}_w = \mu_w\,c_w/\lambda_w$.

Hydraulic diameter $d_{h,w}$ is described by the following equation:

$$
d_{h,w} = \frac{4A_{wr}}{U_w},
$$
(17.15)

where U_w is the tube (channel) inner perimeter, and A_{wr} is the surface area of the tube cross-section that water flows through. The hydraulic diameter on the water side is $d_{h,w} = 0.00706$ m, for the modelled radiator presented in Figs. 17.1 and 17.2.

Correlation (17.13) is proposed based on the results of experimental testing performed for the radiator under analysis [313]. In the case of other heat exchangers, the heat transfer coefficient on the tube inner surface can be found using formula (17.16) [310, 325] presented in Chap. 5:

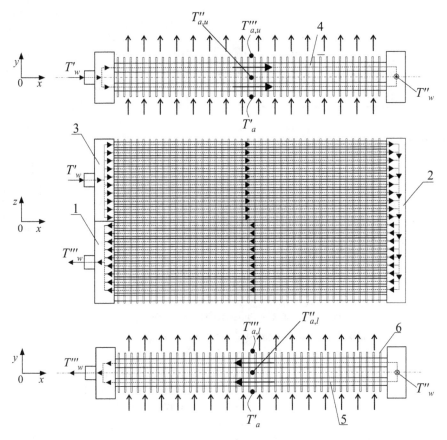

Fig. 17.1 Flow system of the two-pass two-row radiator under analysis: **a** first pass horizontal cross-section, **b** front view, **c** second pass horizontal cross-section, 1—inlet header, 2—intermediate header, 3—outlet header, 4—second tube row, 5—first tube row, 6—plate fin (lamella)

$$\mathrm{Nu}_w = \mathrm{Nu}_{m,q}(\mathrm{Re}_w = 2300) + \frac{\frac{\xi}{8}(\mathrm{Re}_w - 2300)\,\mathrm{Pr}_w^{1.008}}{1.08 + 12.39\sqrt{\frac{\xi}{8}}\left(\mathrm{Pr}_w^{2/3} - 1\right)}\left[1 + \left(\frac{d_{h,w}}{L_{ch}}\right)^{2/3}\right],$$

$$2300 \leq \mathrm{Re}_w \leq 10^6, \quad 0.1 \leq \mathrm{Pr}_w \leq 1\,000, \quad \frac{d_{h,w}}{L_{ch}} \leq 1. \tag{17.16}$$

Symbol $\mathrm{Nu}_{m,q}$ denotes the mean Nusselt number on the entire length of the tube (L_{ch}) at a constant heat flux set on the tube inner surface. The advantage of correlation (17.16) is the Nu_w continuity for $\mathrm{Re}_w = 2300$. If the tube wall temperature is constant, e.g. in some cocurrent heat exchangers, or if steam condensation occurs or liquid boils on the tube outer surface, $\mathrm{Nu}_{m,q}$ ($\mathrm{Re}_w = 2300$) in correlation (17.16) should be replaced with $\mathrm{Nu}_{m,T}$ ($\mathrm{Re}_w = 2300$)—the mean Nusselt number for the tube wall constant temperature.

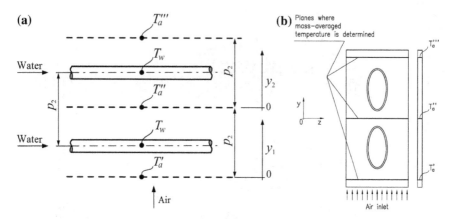

Fig. 17.2 Two-row cross-flow heat exchanger: **a** the exchanger geometrical model assumed in the analytical model, **b** diagram illustrating determination of mass-averaged air temperature T_a'' downstream the first tube row and temperature T_a''' downstream the second tube row at the air constant temperature T_a' set at the exchanger inlet (CFD simulations)

17.3 Determination of Correlations for the Air-Side Nusselt Number Using CFD Modelling

The air-side Nusselt number correlation will be determined for the radiator of a 1600 cm^3 combustion engine discussed in detail in [298, 306, 307]. The flow system of the radiator, which is a two-pass, two-row lamella heat exchanger, is shown in Fig. 17.1.

The symbols in Fig. 17.1 denote as follows:

- T_a'—air temperature upstream the radiator,
- $T_{a,u}''$—air temperature downstream the first tube row in the first pass,
- $T_{a,u}'''$—air temperature downstream the second tube row in the first pass,
- $T_{a,l}''$—air temperature downstream the first tube row in the second pass,
- $T_{a,l}'''$—air temperature downstream the second tube row in the second pass,
- T_w'—water temperature at the radiator inlet,
- T_w''—water temperature at the first pass outlet and the second pass inlet,
- T_w'''—water temperature at the radiator outlet.

The process of the heat transfer coefficient determination using CFD simulations can be simplified significantly by assuming that the water temperature in the first and the second row is the same and remains constant along the tube length. Considering that the water temperature at the first tube row outlet is only by a few degrees lower compared to the second tube row outlet, it may be assumed that in both rows the temperature is the same—especially in view of the fact that changes in the air temperature have only a small impact on the heat transfer coefficient. Temperature variations do not affect the air-side Prandtl number Pr_a significantly.

At the temperature of 0 °C and the pressure of 1 MPa, the Prandtl number for air is 0.7, whereas, in the temperature of 200 °C, it is almost the same—0.68. The heat transfer coefficients on the inner surface of tubes (α_w) are equal to each other, and their value is found from correlation (17.13) or correlation (17.16), or from other correlations available in the literature, such as the Gnielinski formula.

The tube pitches are as follows: longitudinal to the air flow $p_2 = 17$ mm, transverse to the air flow $p_1 = 18.5$ mm (cf. Figs. 17.2 and 17.3). The mean heat transfer coefficient on the surface of fins and tubes for the two-row heat exchanger under analysis will be found using the following equation:

$$\Delta T_t^a(\alpha_a) - \Delta T_t^{CFD} = 0, \qquad (17.17)$$

where ΔT_t^a denotes the increment in the mass-averaged air temperature on both tube rows, which is defined as:

$$\Delta T_t^a = T_a''' - T_a'. \qquad (17.18)$$

$\Delta T_t^a(\alpha_a)$ stands for the increment in the air temperature calculated using the analytical formula, and ΔT_t^{CFD} is the increment in the mass-averaged air temperature on both tube rows determined by CFD simulations using the ANSYS-CFX program. ΔT_t^{CFD} is treated as the air temperature increment found experimentally. The CFD simulation is performed omitting thermal resistance on the fin base/tube interface, i.e. it is assumed that $R_{tc} = 0$.

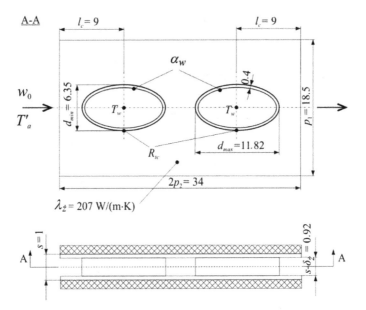

Fig. 17.3 Air flow through a channel created by two adjacent fins

The increment in temperature ΔT_t^a defined by formula (17.18) can be determined solving the differential equations that describe the air temperature distribution taking account of the boundary condition on the exchanger inlet.

Assuming that the water temperature in both tubes is constant and equal to T_w and the heat transfer coefficients in the first and the second tube row are α_a^I and α_a^{II}, respectively, changes in the air temperature in the first and the second tube row along the air flow path (cf. Fig. 17.2a) are described by the following differential equations:

$$\frac{d\,T_a^I(y_1^+)}{d\,y_1^+} = N_a^I\left[T_w - T_a^I(y_1^+)\right], \quad 0 \le y_1^+ \le 1, \tag{17.19}$$

$$\frac{d\,T_a^{II}(y_2^+)}{d\,y_2^+} = N_a^{II}\left[T_w - T_a^{II}(y_2^+)\right], \quad 0 \le y_2^+ \le 1, \tag{17.20}$$

where the numbers of the heat transfer units N_a^I and N_a^{II} are expressed as:

$$N_a^I = k^I A_{zrg}/(\dot{m}_a\, c_{pa}), \quad N_a^{II} = k^{II} A_{zrg}/(\dot{m}_a\, c_{pa}). \tag{17.21}$$

In Eqs. (17.21) A_{zrg} denotes the bare tube outer surface area and $\dot{m}_a$ is the air mass flow rate per tube. The air specific heat at constant pressure c_{pa} in the first and the second tube row is assumed as constant because it does not change much with changes in temperature. Overall heat transfer coefficients k^I and k^{II} are calculated from the following formulae:

$$\frac{1}{k^I} = \frac{A_{zrg}}{A_w}\frac{1}{\alpha_w} + \frac{A_{zrg}}{A_m}\frac{\delta_w}{\lambda_w} + \frac{1}{\alpha_{zr}^I}, \tag{17.22}$$

$$\frac{1}{k^{II}} = \frac{A_{zrg}}{A_w}\frac{1}{\alpha_w} + \frac{A_{zrg}}{A_m}\frac{\delta_w}{\lambda_w} + \frac{1}{\alpha_{zr}^{II}}. \tag{17.23}$$

The following symbols are used in formulae (17.22)–(17.23): A_w—tube inner surface area, α_w—heat transfer coefficient on the tube inner surface in the first and in the second tube row, $A_m = (A_w + A_{zrg})/2$—mean surface area of the tube, δ_w—tube wall thickness, λ_w—wall material thermal conductivity.

Weighted heat transfer coefficients α_{zr}^I and α_{zr}^{II} for the first and the second tube row are defined by the following expressions, respectively:

$$\alpha_{zr}^I = \alpha_a^I\left[\frac{A_{m\dot{z}}}{A_{zrg}} + \frac{A_{\dot{z}}}{A_{zrg}}\eta_{\dot{z}}(\alpha_a^I)\right], \tag{17.24}$$

$$\alpha_{zr}^{II} = \alpha_a^{II}\left[\frac{A_{m\dot{z}}}{A_{zrg}} + \frac{A_{\dot{z}}}{A_{zrg}}\eta_{\dot{z}}(\alpha_a^{II})\right], \tag{17.25}$$

where: α_a^I and α_a^{II}—mean heat transfer coefficient on the surface of the tubes and fins in the first and the second tube row, respectively, A_{mz}—tube surface area in between the fins, A_z—surface area of fins, η_z—efficiency of fins.

The boundary conditions for Eqs. (17.19) and (17.20), respectively, are as follows (cf. Fig. 17.2a):

$$T_a^I\big|_{y_1^+=0} = T_a', \tag{17.26}$$

$$T_a^{II}\big|_{y_2^+=0} = T_a''. \tag{17.27}$$

The solution of Eq. (17.19) obtained using the variables separation method using boundary condition (17.26) has the following form:

$$T_a^I(y_1^+) = T_w + (T_a' - T_w)\,e^{-N_a^I y_1^+}, \quad 0 \le y_1^+ \le 1. \tag{17.28}$$

Substituting $y_1^+ = 1$, expression (17.28) gives the air temperature at the second tube row inlet T_a'':

$$T_a'' = T_a^I(y_1^+ = 1) = T_w + (T_a' - T_w)\,e^{-N_a^I}. \tag{17.29}$$

The solution of Eq. (17.20) using boundary condition (17.27) has the following form:

$$T_a^{II}(y_2^+) = T_w + (T_a' - T_w)\,e^{-\left(N_a^I + N_a^{II} y_2^+\right)}, \quad 0 \le y_2^+ \le 1. \tag{17.30}$$

The air temperature downstream the second tube row T_a''' is:

$$T_a''' = T_a^{II}(y_2^+ = 1) = T_w + (T_a' - T_w)\,e^{-\left(N_a^I + N_a^{II}\right)}. \tag{17.31}$$

The increments in the air temperature downstream the first and the second tube row are defined by the following formulae:

$$\Delta T_I = T_a\big|_{y_1^+=1} - T_a\big|_{y_1^+=0} = T_a'' - T_a' = (T_w - T_a')\left(1 - e^{-N_a^I}\right), \tag{17.32}$$

$$\Delta T_{II} = T_a\big|_{y_2^+=1} - T_a\big|_{y_2^+=0} = T_a''' - T_a'' = (T_w - T_a')\,e^{-N_a^I}\left(1 - e^{-N_a^{II}}\right). \tag{17.33}$$

The increment in the air temperature on both tube rows is expressed as:

$$\Delta T_t = \Delta T_I + \Delta T_{II} = (T_w - T_a')\left[1 - e^{-\left(N_a^I + N_a^{II}\right)}\right]. \tag{17.34}$$

The heat transfer coefficients α_a^I and α_a^{II} are assumed to be equal to each other and total α_a, formula (17.34) is written as:

$$\Delta T_t = \Delta T_I + \Delta T_{II} = \left(T_w - T_a'\right)\left(1 - e^{-2N_a}\right), \qquad (17.35)$$

where the number of the heat transfer units N_a is expressed by $N_a = kA_g/(\dot{m}_a\,c_{pa})$.

Substitution of (17.35) in (17.17) gives a nonlinear algebraic equation with respect to α_a. CFD simulations performed for n different air velocities upstream the radiator: $w_{0,i}$, $i = 1,\dots, n$, give n values of the air temperature increment $\Delta T_{t,i}^{CFD}$. Equation (17.17) is then used to find coefficients $\alpha_{a,i}$, $i = 1,\dots, n$.

17.3.1 Fin Efficiency

ΔT_t cannot be calculated without function $\eta_z(\alpha_a)$, needed to determine the fin efficiency depending on the air-side heat transfer coefficient.

The impact of thermal resistance between the tube surface and the fin base on the temperature distribution in the fin and on the fin efficiency is modelled at selected values of thermal resistance R_{tc} (cf. Fig. 17.3).

The thermal conductivity of the fin material (aluminium) is $\lambda_z = 207$ W/(m^2 K). The fin efficiency η_z depending on α_z is defined as the ratio between the heat flow rate $\dot{Q}$ transferred by the fin and the maximum heat flow rate $\dot{Q}_{max}$ that would be transferred from the fin to the environment if the fin temperature was equal to the tube outer surface temperature $T_r|_{A=A_z}$:

$$\eta_z = \frac{\dot{Q}}{\dot{Q}_{max}} = \frac{\int_{A_z}\alpha_a(T_z - T_a)dA}{\alpha_a\left(T_r|_{A=A_z} - T_a\right)A_z} = \frac{\int_{A_z}(T_z - T_a)dA}{A_z\left(T_r|_{A=A_z} - T_a\right)}. \qquad (17.36)$$

The calculations of the fin efficiency η_z are based on the assumption that the heat transfer coefficient α_a and the tube outer surface temperature $T_r|_{A=A_z}$ are constant. Selected results of the fin efficiency calculations are listed in Table 17.1. The temperature distribution in the fin is found using the ANSYS-CFX program (Figs. 17.4 and 17.5).

Analysing the results presented in Table 17.1 and Fig. 17.6, it can be seen that the fin efficiency gets smaller as the heat transfer coefficient and the thermal resistance values get higher. This is the effect of the decrease in the mean temperature of the fin what causes the reduction in the heat flow rate from the fin to surrounding air with a rise in α_a and R_{tc}.

The results of the fin efficiency calculations performed using the ANSYS-CFX program are approximated using the following function:

Table 17.1 Fin efficiency calculation results for selected values of the heat transfer coefficient α_a and thermal resistance R_{tc}

Case	R_{tc} [(m² K)/W]	α_a [W/(m² K)]	$\dot{Q}$ (W)	$\dot{Q}_{max}$ (W)	$\eta_{\dot{z}}$
I	0	300	1.951	3.056	0.64
II	7.32×10^{-6}	30	0.276	0.295	0.94
III	3.83×10^{-6}	150	1.074	1.539	0.70
IV	1.40×10^{-5}	130	0.798	1.333	0.60
V	6.43×10^{-6}	300	1.533	3.056	0.50
VI	1.40×10^{-5}	300	1.226	3.056	0.40

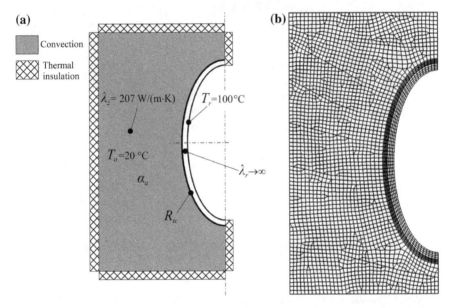

(a)

Convection

Thermal insulation

$\lambda_z = 207$ W/(m·K) $T_r = 100$°C

$T_a = 20$ °C

$\lambda_r \to \infty$

α_a

R_{tc}

(b)

Fig. 17.4 Boundary conditions and the fin division into finite elements for the calculation of the fin temperature and efficiency: **a** the fin geometrical model with set boundary conditions, **b** fin model division into finite elements

$$\eta_{\dot{z}}(R_{tc}, \alpha_a) = \frac{(c_1 + c_3 R_{tc} + c_5 \alpha_a + c_7 R_{tc}^2 + c_9 \alpha_a^2 + c_{11} R_{tc} \alpha_a)}{(1 + c_2 R_{tc} + c_4 \alpha_a + c_6 R_{tc}^2 + c_8 \alpha_a^2 + c_{10} R_{tc} \alpha_a)}. \qquad (17.37)$$

The unknown coefficients in function (17.37) are found using the least squares method where the function minimum is determined using the Levenberg-Marquardt method. The calculations are performed using the Table Curve 3D program, version 4.0 [293], and the results are presented in Table 17.2.

Function $\eta_{\dot{z}}(R_{tc}, \alpha_a)$ defined by formula (17.37) is used to calculate the weighted heat transfer coefficient α_{zr}. Determining α_a from formula (17.17) based on the increments in the air temperature obtained from CFD simulations ΔT_t^{CFD}, thermal

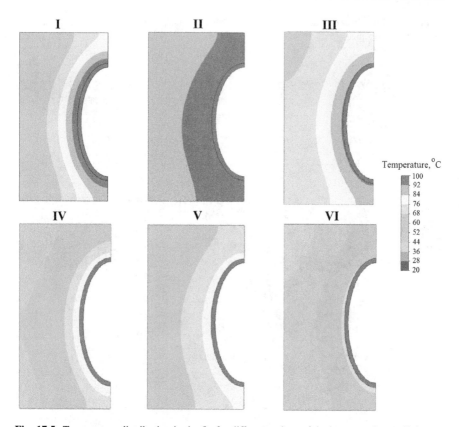

Fig. 17.5 Temperature distribution in the fin for different values of the heat transfer coefficient α_a and thermal resistance R_{tc} listed in Table 17.1

resistance on the interface between the tube outer surface and the fin base is omitted, i.e. it is assumed that $R_{tc} = 0$.

17.3.2 Correlation for the Air-Side Nusselt Number

CFD simulations using the ANSYS-CFX program were performed for different air velocities upstream the radiator, and the results obtained for the following data: $w_0 = 0.8$ m/s, $T_a' = 14.98$ °C, $\alpha_w = 1512$ W/(m² K), $T_w = 73.85$ °C, $R_{tc} = 0$ (m² K)/W are presented in Fig. 17.7.

Analysing the results presented in Fig. 17.7, it can be seen that dead zones are created downstream the first-row tube and upstream the tube in the second row, where the air velocity is very low (cf. Fig. 17.7a). Due to that, the air in the dead zones is heated to a value close to the fin surface temperature, which means that the difference between the temperatures of the fin surface and the air is tiny. As a result,

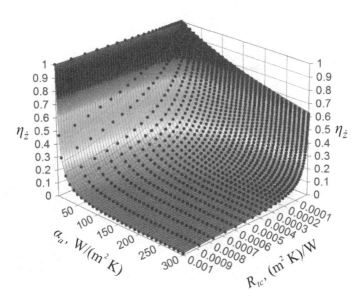

Fig. 17.6 The efficiency of the fin presented in Fig. 17.4 ($\eta_{\dot{z}}$) as a function of the heat transfer coefficient α_a and thermal resistance R_{tc}

Table 17.2 Values of coefficients in function (17.37) used to calculate the fin efficiency $\eta_{\dot{z}}(R_{tc}, \alpha_a)$

Coefficient	Value
c_1	0.999
c_2	3.100
c_3	4.850
c_4	2.100×10^{-3}
c_5	9.626
c_6	-1625.550
c_7	-3192.846
c_8	-6.763
c_9	-2.013
c_{10}	221.620
c_{11}	3.260×10^{-3}

the heat flux in this region is slight, too. The zones are dead regarding the heat transfer because their share in the total heat flow rate transferred from the fin to passing air is almost none.

In the next step, the air-side thermal and flow phenomena were modelled using the following data:

- air velocity upstream the radiator: $1 \text{ m/s} \leq w_0 \leq 2.5 \text{ m/s}$,
- air temperature upstream the radiator $T'_a = 14.98 \, °\text{C}$,
- water temperature $T_w = 68.3 \, °\text{C}$,
- heat transfer coefficient on the tube inner surface $\alpha_w = 4793.95 \text{ W/(m}^2 \text{ K)}$.

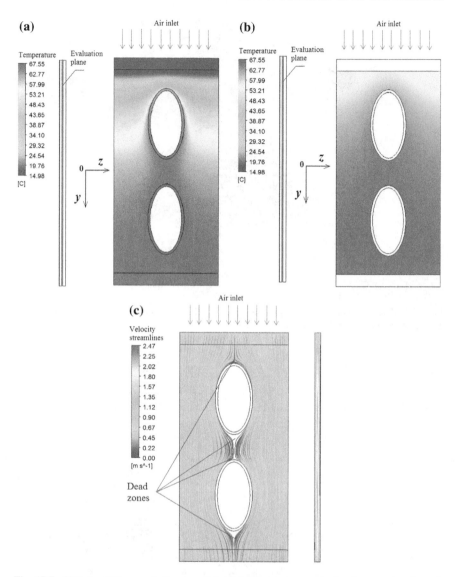

Fig. 17.7 CFD simulation results for $w_0 = 0.8$ m/s; **a** air temperature in the plane passing through the middle of the gap in between the fins, **b** fin surface temperature, **c** streamlines

The calculation results of the increment in the mass-averaged air temperature on the first and on the second tube row, ΔT_I^{CFD} and ΔT_{II}^{CFD}, respectively, and the total difference in temperatures on both tube rows ΔT_t^{CFD} are listed in Table 17.3.

The calculation results of the heat transfer coefficient α_a determined based on the air temperature increment obtained from CFD simulations are presented in Table 17.4. They are compared to the coefficient values calculated using the

Table 17.3 Air temperature increments on the first and the second tube row ($\Delta T_{I,i}^{CFD}$, $\Delta T_{II,i}^{CFD}$) and the total increment in the air temperature in the exchanger ($\Delta T_{t,i}^{CFD}$) determined using CFD simulations performed for different air velocities $w_{0,i}$, $i = 1,..., n$

i	$w_{0,i}$ (m/s)	$\Delta T_{I,i}^{CFD}$ (°C)	$\Delta T_{II,i}^{CFD}$ (°C)	$\Delta T_{t,i}^{CFD}$ (°C)
1	1.0	41.26	6.37	47.63
2	1.2	39.03	7.84	46.87
3	1.4	36.80	9.00	45.80
4	1.6	34.71	9.84	44.55
5	1.8	32.79	10.44	43.23
6	2.0	31.04	10.82	41.86
7	2.2	29.47	11.05	40.52
8	2.5	27.39	11.19	38.58

Table 17.4 Heat transfer coefficient $\alpha_{a,i}$ determined based on temperature difference $\Delta T_{t,i}^{CFD}$ found by means of the CFD simulation

$w_{0,i}$ (m/s)	$Re_{a,i}$	$Pr_{a,i}$	$j_{a,i}^{CFD}$	$\alpha_{a,i}^{CFD}$ [W/(m² K)]	$\alpha_{a,i}^{exp}$ [W/(m² K)]
1.0	149.87	0.694	0.026233	67.54	52.31
1.2	180.01		0.026226	81.02	59.19
1.4	210.29		0.025386	91.49	65.68
1.6	240.70		0.024134	99.39	71.87
1.8	271.22		0.022781	105.53	77.81
2.0	301.86		0.021425	110.26	83.53
2.2	332.60		0.020175	114.20	89.06
2.5	378.86		0.018529	118.94	97.04

correlation for the Nusselt number Nu_a established experimentally (17.39). The nonlinear algebraic Eq. (17.17), which in this case has the following form:

$$\left(T_w - T_a'\right)\left(1 - e^{-2N_a}\right) - \Delta T_t^{CFD} = 0 \tag{17.38}$$

is solved with respect to α_a using the secant method.

The experimental correlation $Nu_a = Nu_a(Re_a, Pr_a)$ for the radiator under consideration is derived in [321]. It is expressed as follows:

$$Nu_a = 0.1012\, Re_a^{0.6704} Pr_a^{1/3}, \quad 150 \leq Re_a \leq 350. \tag{17.39}$$

The following formulae define the Nusselt, the Reynolds and the Prandtl numbers (Nu_a, Re_a and Pr_a, respectively):

$$Nu_a = \frac{\alpha_a d_{h,a}}{\lambda_a}, \quad Re_a = \frac{w_{max} d_{h,a}}{\nu_a}, \quad Pr_a = \frac{c_{pa}\mu_a}{\lambda_a}. \tag{17.40}$$

Symbol w_{max} denotes the maximum velocity of air in the exchanger that occurs in the narrowest cross-section (with the surface area A_{min}) in between two adjacent tubes.

$$w_{max} = \frac{s p_1}{A_{min}} \frac{\bar{T}_a + 273.15}{T'_a + 273.15} w_0. \tag{17.41}$$

The minimum cross-section surface area A_{min} per tube pitch p_1 and tube pitch s is defined by the following formula:

$$A_{min} = (s - \delta_{\dot{z}})(p_1 - d_{min}). \tag{17.42}$$

The equivalent hydraulic diameter d_h is calculated using the formula proposed by Kays and London for compact heat exchangers [149]:

$$d_{h,a} = \frac{4 A_{min}(2 p_2)}{A_{\dot{z}} + A_{m\dot{z}}}, \tag{17.43}$$

where the surface area of a single fin $A_{\dot{z}}$ is defined as:

$$A_{\dot{z}} = 2 \cdot 2(p_1 p_2 - A_{oval}) = (4 p_1 p_2 - \pi d_{min} d_{max}), \tag{17.44}$$

and the tube surface area in between the fins is calculated from the following formula:

$$A_{m\dot{z}} = 2 \cdot U_z(s - \delta_{\dot{z}}). \tag{17.45}$$

Symbol U_z denotes the bare tube outer surface perimeter. For the radiator under analysis, the mean hydraulic diameter on the air side is $d_{h,a} = 0.00141$ mm.

Physical properties of air are determined in mean temperature $\bar{T}_a$, which is defined as:

$$\bar{T}_a = (T'_a + T'''_a)/2. \tag{17.46}$$

The CFD simulation was performed at constant temperatures T'_a and T_w at a constant heat transfer coefficient α_w, whereas velocity w_0 varied from $w_0 = 1.0$ m/s to $w_0 = 2.5$ m/s (cf. Table 17.4). It is assumed that thermal resistance on the fin/tube surface interface $R_{tc} = 0$.

Knowing the values of the heat transfer coefficient $\alpha_{a, CFD}$ (cf. Table 17.4), values of the Colburn parameter $j_{a,i}^{CFD} = Nu_{a,i}^{CFD}/(Re_{a,i} Pr_{a,i}^{1/3})$, $i = 1,...,$ 8, are calculated, where $Nu_{a,i}^{CFD} = \alpha_{a,i}^{CFD} d_{h,a}/\lambda_a$ is the Nusselt number for the i-th measuring series. The Colburn parameter j_a is approximated using the following power function:

$$j_a = x_1 Re_a^{x_2} \tag{17.47}$$

where the Colburn parameter is defined as:

$$j_a = \mathrm{Nu_a}/(\mathrm{Re}_a \mathrm{Pr}_a^{1/3})$$

The unknown coefficients x_1 and x_2 in formula (17.47) are determined using the least squares method so that the sum of squares of the differences between the parameter found using CFD modelling and parameter j_a calculated using function (17.47) should be minimal:

$$S = \sum_{i=1}^{n=8} \left(j_{a,i}^{CFD} - x_1\, \mathrm{Re}_{a,i}^{x_2} \right)^2. \tag{17.48}$$

Symbol n denotes the number of the Colburn parameters found using the CFD simulation. In the case under consideration $n = 8$ (cf. Table 17.4). The minimum of a function (17.48) is found using the Levenberg-Marquardt method using the MATLAB program [195]. It is reached for $x_1 = 0.188$ and $x_2 = -0.382$.

Figure 17.8 presents curves illustrating changes in the values of a function $j_a(\mathrm{Re}_a)$ defined by formula (17.47) and the 95% confidence intervals. Transforming correlation (17.47), the following is obtained:

$$\mathrm{Nu}_a = x_1 \mathrm{Re}_a^{(1+x_2)} \mathrm{Pr}_a^{1/3} \tag{17.49}$$

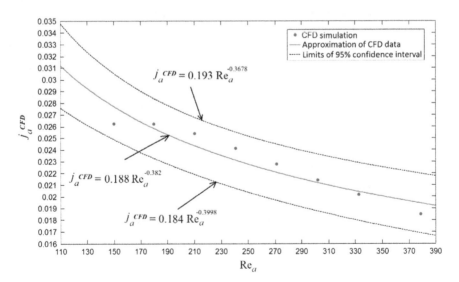

Fig. 17.8 Correlation $j_a^{CFD}(\mathrm{Re}_a) = 0.188 \mathrm{Re}_a^{-0.382}$ and 95% confidence intervals determined based on data obtained from the CFD simulation using the ANSYS-CFX program

After the found coefficients x_1 and x_2 are substituted in formula (17.49), the result is as follows [319]:

$$\alpha_a^{CFD} = \frac{\lambda_a}{d_{h,a}} \mathrm{Nu}_a = \frac{\lambda_a}{d_{h,a}} 0.188 \mathrm{Re}_a^{0.618} \mathrm{Pr}_a^{1/3}, \quad 150 \leq \mathrm{Re}_a \leq 380. \tag{17.50}$$

The obtained correlation is similar to the Kröger formula [161, 162] determined for round tubes with high fins

$$\alpha_a = \frac{\lambda_a}{d_{h,a}} 0.174 \mathrm{Re}_a^{0.613} \mathrm{Pr}_a^{1/3}. \tag{17.51}$$

Formula (17.39) established for the same radiator experimentally gives slightly lower values of the heat transfer coefficient α_a. This is the effect of thermal resistance on the interface between the fin root and the tube surface. A method for thermal resistance determination is presented in [304, 319].

It should be emphasized that CFD simulations are a useful tool for modelling thermal and flow processes in finned-tube heat exchangers. However, CFD simulation results should be verified through experimental testing taking account of the conditions created by the exchanger location in the entire installation, the liquid different mass flow rates in individual tubes and the thermal resistance between the tube and the roots of the fins fixed on the tube surface.

The uncertainty of the CFD simulation results is also caused by the fact that there are no methods that would be accurate enough to simulate the fluid flow, especially in the transitional and in the turbulent region.

Chapter 18
Automatic Control of the Liquid Temperature at the Car Radiator Outlet

The control system regulating the car engine coolant temperature has a considerable impact on the engine performance, reliability, and life. One of the system tasks is to ensure the optimum temperature, possibly constant in time, to make sure that the engine is properly lubricated and high thermal stresses are avoided in the engine head and cylinder block [42, 71]. Industrial automatic control systems, including control systems of heat exchangers, usually make use of PID controllers [77, 133, 156, 251, 262, 280, 290, 306, 313, 320, 322, 350, 372]. However, in automatic control systems of such heat exchangers as steam superheaters or finned-tube exchangers, the operation of PID controllers may be unstable. The controller needs re-adjusting if the mass/volume flow rate of the fluid inside the tubes changes over time. The exchanger characteristics are then altered, and the regulation process will not be stable unless new settings of the PID controller are selected. Despite the advancements in various methods enabling automatic selection of the PID controller settings [156, 280, 290, 372], the need to re-adjust the settings during the exchanger operation is the PID controller disadvantage. Heat exchangers can also use digital control systems based on their mathematical models.

This chapter presents a regulation system which is based on the exchanger mathematical model and a PID digital controller. The system regulates the temperature of water at the exchanger outlet [306, 320, 322] by changing the number of revolutions of the fan forcing the air flow. Appropriate computer programs were developed in the C++ language. The system regulating the water temperature at the car radiator outlet [313, 320, 322] was developed based on the car radiator analytical model presented in [296, 299, 310].

The computer takes measuring data from the ALMEMO 5590 data acquisition system using an RS232 serial port. Another RS232 serial port ensures direct communication between the computer and the inverter. The setting calculated using the computer program is transmitted to the frequency converter controlling the fan revolutions. The program uses the USS communication protocol to communicate with the Siemens MicroMaster Vector inverter.

© Springer International Publishing AG, part of Springer Nature 2019
D. Taler, *Numerical Modelling and Experimental Testing of Heat Exchangers*,
Studies in Systems, Decision and Control 161,
https://doi.org/10.1007/978-3-319-91128-1_18

18.1 System Based on the Heat Exchanger Mathematical Model

The diagram of the control system based on the exchanger nonlinear mathematical model is shown in Fig. 18.1 [313, 320].

The aim of the control system is to select the fan revolution number n for a given time instant t_i so that the water temperature measured at the radiator outlet $T''_{w,meas}$ should be equal to set temperature $T''_{w,set}$.

$$T''_{w,meas}(t_i, n) - T''_{w,set}(t_i) = 0, \tag{18.1}$$

where time instants t_i are defined as:

$$t_i = i\,\Delta t_c, i = 1, \ldots \tag{18.2}$$

Time step Δt_c should be at least five times bigger than the exchanger time constant. Considering that the radiator time constant is about 10 s for high velocities of the water and air flows, and about 20 s [158, 306, 307] if the water and air velocities are low, the assumed time step is $\Delta t_c = 120$ s. The time step Δt_c is long enough for the exchanger operating in a steady state in time instant t_i to reach a new steady state after time Δt_c, i.e., to reach a steady state also in time instant $t_{i+1} = t_i + \Delta t_c$. The water temperature at the radiator outlet $T''_{w,meas}$ is a nonlinear function of the fan revolution number n. The nonlinear algebraic Eq. (18.1) is

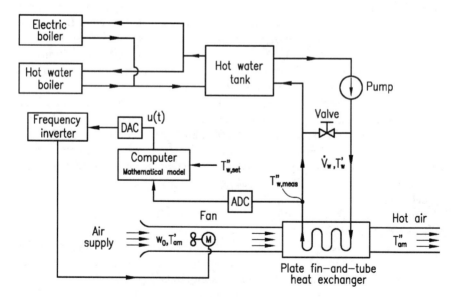

Fig. 18.1 Diagram of the digital system regulating the water temperature at the car radiator outlet using changes in the number of the fan revolutions (system based on the exchanger nonlinear mathematical model); ADC—analog-to-digital converter, DAC—digital-to-analog converter

solved using the secant method. Temperature $T''_{w,meas}[t_i, w_0(n)]$ is calculated from the analytical mathematical model of the radiator. Based on the measurements, the following relation between the air velocity upstream the radiator w_0 and the fan revolution number n is found using the least squares method:

$$w_0 = -4.1336 \times 10^{-7} n^2 + 2.5569 \times 10^{-3} n - 0.6268, \quad 250 \le n \le 1440, \quad (18.3)$$

where air velocity w_0 is expressed in m/s, and n is the number of the fan revolutions per minute.

Equation (18.1) was solved iteratively using the secant method, assuming that the difference between $n^{(k)}$ and $n^{(k-1)}$ should be less than adopted tolerance: ε 0.001 rpm. The regulation system input data are as follows: water and air temperature at the radiator inlet, T'_w and T'_{am}, respectively, water volume flow rate $\dot{V}'_w$ and water temperature at the radiator outlet T''_w. The regulation system takes the data with the time step $\Delta t = 10$ s. All the measured quantities mentioned above include random measuring errors that may cause oscillations in the determined history of the fan rotational speed depending on time. For this reason, the input measuring signals were filtered before they were used in the car radiator calculations using the mathematical model. The measured histories were smoothed using a seven-point moving filter based on a third-degree polynomial. The third-degree polynomial value of y in time t_i is found by means of the least squares method using seven measuring points (t_j, f_j), $j = i - 3, \ldots, i + 3$.

$$y_i = y(t_i) = \frac{(-2f_{i-3} + 3f_{i-2} + 6f_{i-1} + 7f_i + 6f_{i+1} + 3f_{i+2} - 2f_{i+3})}{21}, \quad (18.4)$$

where symbols $f_{i-3}, \ldots, f_{i+3}$ denote the quantity values measured in seven subsequent time points.

The moving averaging filter (18.4) is a low-pass filter eliminating from the measured signal a great part of the random high-frequency disturbances. Based on 57 measuring series, the following air- and water-side Nusselt number correlations are determined [309]:

$$\mathrm{Nu}_a = x_1 \mathrm{Re}_a^{x_2} \mathrm{Pr}_a^{1/3}, \quad 150 \le \mathrm{Re}_a \le 350, \quad (18.5)$$

$$\mathrm{Nu}_w = \frac{\frac{\xi}{8}(\mathrm{Re}_w - x_3)\,\mathrm{Pr}_w}{1 + x_4\sqrt{\frac{\xi}{8}}(\mathrm{Pr}_w^{2/3} - 1)}\left[1 + \left(\frac{d_{h,w}}{L_{ch}}\right)^{2/3}\right], \quad 4000 \le \mathrm{Re}_w \le 12,000, \quad (18.6)$$

where friction factor ξ is defined as:

$$\xi = \frac{1}{(1.82 \log \mathrm{Re}_w - 1.64)^2} = \frac{1}{(0.79 \ln \mathrm{Re}_w - 1.64)^2}. \quad (18.7)$$

Two sets of coefficients x_1, x_2, x_3, and x_4 are used in the temperature regulation system. The first ($x_1 = 0.08993$, $x_2 = 0.699$, $x_3 = 1079$, $x_4 = 16.38$) was determined in Chap. 16. In the second set ($x_1 = 0.1117$, $x_2 = 0.6469$, $x_3 = 1000$, $x_4 = 12.7$), only two coefficients x_1 and x_2 that appear in the air-side correlation (18.5) were determined. Coefficients x_1 and x_2 were found in Chap. 15 based on experimental data. Coefficients $x_3 = 1000$ and $x_4 = 12.7$ in the water-side correlation (18.6) are assumed the same as in the Gnielinski formula.

The results of regulation of the water temperature at the radiator outlet using the control system based on the exchanger analytical mathematical model are presented in Fig. 18.2a, b.

Analysing the results presented in Fig. 18.2a, b, it can be seen that the measured histories of the water temperature at the radiator outlet are very close to the values set for the two sets of coefficients x_1, x_2, x_3, and x_4. Slight differences between measured and set temperatures $T''_{w,meas}$ and $T''_{w,set}$, respectively, can be observed after sudden changes in the value of set temperature $T''_{w,set}$ or after a change in the volume flow rate $\dot{V}_w$. This is the effect of the big time step $\Delta t_c = 120$ s applied to determine the number of the fan revolutions n. After a change in the set temperature or the volume flow rate value, the fan rotational speed will be adjusted only in the next time instant t_i. The differences in the values of temperatures $T''_{w,set}$ and $T''_{w,meas}$ are also an effect of the radiator mathematical model inaccuracy.

However, it has to be emphasized that the system regulating the water temperature at the radiator outlet based on the mathematical model works very well. The number of the fan revolutions found in the process ensures fast adjustment of the water temperature at the radiator outlet to the set value.

The advantage of the developed method is the capacity for stable control of the water temperature in a wide range of changes in the exchanger operating conditions. There is no danger that periods will occur of the regulation system unstable operation or that the system will operate too slowly, needing a very long time to reach the set value of the water temperature at the exchanger outlet.

18.2   Digital PID Controller

A control system was built to regulate the water temperature at the exchanger outlet using a digital PID controller. PID controllers are the most common control devices applied in industrial practice. The PID controller equation has the following form [156, 262]:

$$u(t) = \bar{u} + K_p \left[e(t) + K_i \int_0^t e(t)dt + K_d \frac{de(t)}{dt} \right], \tag{18.8}$$

Fig. 18.2 Results of
regulation of the water
temperature at the radiator
outlet obtained by means of
the system based on the
exchanger mathematical
model; **a** $x_1 = 0.08993$,
$x_2 = 0.699$, $x_3 = 1079$,
$x_4 = 16.38$; **b** $x_1 = 0.1117$,
$x_2 = 0.6469$, $x_3 = 1000$,
$x_4 = 12.7$

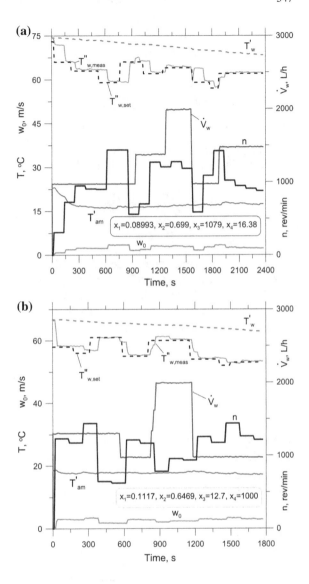

where: $u(t)$—controller output, t—time, $\bar{u}$—mean output value, K_p—PID controller
proportional gain, $e(t)$—difference between set and measured values, K_i—PID
controller integral gain, K_d—PID controller derivative gain, τ_i—integration time,
τ_d—differentiation time.

The last relationship can be written in a discrete form. Finding the integral in the
controller Eq. (18.8) using the rectangle method, and approximating the derivative
using the backward difference quotient, it is possible to obtain the expression used in
the digital PID controller. The values of output signal u_k for time t_k and signal u_{k-1}
for time t_{k-1} will be determined first.

$$u_k = \bar{u} + K_p\left(e_k + K_i \Delta t \sum_{i=1}^{k} e_i + K_d \frac{e_k - e_{k-1}}{\Delta t}\right), \quad k = 1, 2, \ldots, \qquad (18.9)$$

$$u_{k-1} = \bar{u} + K_p\left(e_{k-1} + K_i \Delta t \sum_{i=1}^{k-1} e_i + K_d \frac{e_{k-1} - e_{k-2}}{\Delta t}\right), \quad k = 1, 2, \ldots, \qquad (18.10)$$

where: $u_k = u(t_k)$, $e_k = e(t_k)$, $t_k = k\Delta t$, $k = 1, 2, \ldots$
Subtracting Eq. (18.10) from Eq. (18.9) by sides, the following is obtained:

$$u_k = u_{k-1} + K_p\left(e_k - e_{k-1} + K_i \Delta t e_k + K_d \frac{e_k - 2e_{k-1} + e_{k-2}}{\Delta t}\right), \quad k = 1, 2, \ldots. $$

$$(18.11)$$

Equation (18.11) constitutes the basis of the digital PID controller operation.
A diagram of the digital PID controller is presented in Fig. 18.3. The controller is based on Eq. (18.11), which was implemented in a PC computer software.
Figure 18.4 presents examples of the PID controller operation at different K_p, K_i and K_d settings. Analysing the results presented in Fig. 18.4a, it can be seen that the time needed to reach the set temperature of water at the radiator outlet after a change in the previously set value is rather long.
This is caused by too small a value of the controller proportional gain K_p compared to the integral and derivative gains K_i and K_d. After K_p is raised to

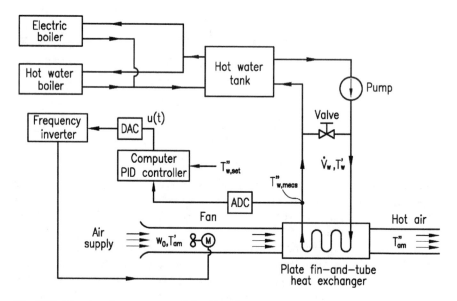

Fig. 18.3 Diagram of a system regulating the water temperature at the car radiator outlet by means of a digital PID controller

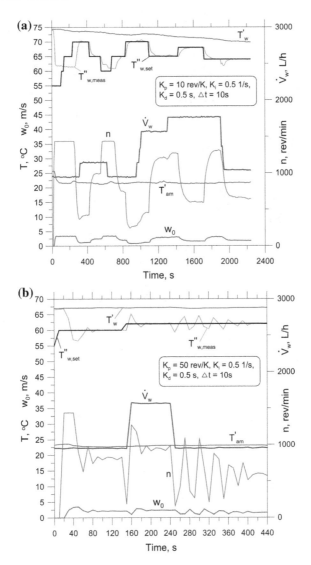

Fig. 18.4 Results of regulation of the water temperature at the car radiator outlet using a digital PID controller; **a** $\Delta t = 10$ s, $K_p = 10$ rpm/K, $K_i = 0.5$ 1/s, $K_d = 0.5$ s; **b** $\Delta t = 10$ s, $K_p = 50$ rpm/K, $K_i = 0.5$ 1/s, $K_d = 0.5$ s; **c** $\Delta t = 1$ s, $K_p = 10$ rpm/K, $K_i = 0.2$ 1/s, $K_d = 0.2$ s; **d** $\Delta t = 10$ s, $K_p = 70$ rpm/K, $K_i = 0.5$ 1/s, $K_d = 0.5$ s

50 rpm/K, the time needed to reach the set temperature value is reduced, but after a change in the water volume flow rate $\dot{V}_w$ in a time instant about 240 s, oscillations appear in the number of the fan revolutions and in the water temperature at the radiator outlet. The oscillations fade away only after about 160 s (cf. Fig. 18.4b). Comparing the results presented in Fig. 18.4a and c, it can be seen that after a

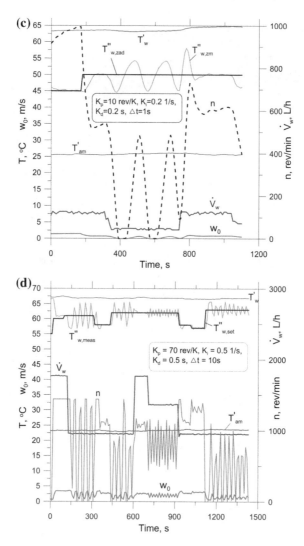

Fig. 18.4 (continued)

reduction in the integral and in the derivative gain K_i and K_d, respectively, the PID controller operation after a change in the water volume flow rate becomes oscillatory (cf. Fig. 18.4c). An increase in the controller proportional gain K_p has a similar effect. A rise in proportional gain K_p from $K_p = 50$ rpm/K (cf. Fig. 18.4b) to $K_p = 70$ rpm/K (cf. Fig. 18.4d) results in the PID controller oscillatory operation after a change in the water volume flow rate $\dot{V}_w$. Oscillations in the PID controller operation have a negative impact on the regulation quality because the values of the water temperature at the radiator outlet oscillate around set temperature $T''_{w,set}$. They also reduce the life of the fan and the fan drive.

The experimental testing results indicate that tuning a PID controller is a painstaking process which does not guarantee good quality of regulation at different values of the water volume flow rate. The computational results show that the PID controller settings should be selected automatically during the device operation to ensure stable performance of the controller at more significant variations in the exchanger operating parameters, especially in the case of bigger changes in the water volume flow rate $\dot{V}_w$.

The systems regulating temperature or heat flow rates in cross-flow tube heat exchangers can also make use of the mathematical models presented in [158, 306, 307, 308, 313], which make it possible to simulate the heat exchanger operation in transient conditions.

Chapter 19
Concluding Remarks

This monograph is devoted to issues related to calculations and testing of heat exchangers. It presents a detailed discussion of the mass, momentum and energy conservation equations, which—after appropriate simplifications—are used in the heat exchanger mathematical modelling. Considerable attention is given to cross-flow tube heat exchangers. The first part presents the laminar and the turbulent heat transfer at the fluid forced flow inside a tube, taking account of the tube entrance length. Formulae are proposed for the Nusselt number calculation as a function of the Reynolds number for the laminar, transitional and turbulent flows in tubes with finite lengths. The developed correlations can be used to calculate heat exchangers. The obtained formulae are validated by comparing the Nusselt numbers calculated using the proposed correlations with the experimental testing results. Analytical formulae are derived that define temperature distributions of fluids circulating in cocurrent, countercurrent, cross-flow, as well as in cross-flow single-row tube heat exchangers. New methods of numerical modelling of tube heat exchangers are presented. They can be used to perform design and operating calculations of exchangers characterized by complex geometries of the flow system. The proposed methods make it possible to conveniently take account of the variations in the heat flux and specific heat that arise with changes in temperature.

A large part of the book is devoted to experimental testing of heat exchangers. Methods are presented that enable assessment of the indirect measurement uncertainty. The calculation of confidence intervals of unknown coefficients appearing in the exchanger nonlinear mathematical models is discussed in detail. The coefficients are determined using the least squares method. Techniques are described for the determination of the local and the mean heat transfer coefficient in tubes and heat exchangers. Different methods of finding dimensionless numbers are discussed, including the dimensional analysis. The methods make it possible to express experimental data in the form of the Nusselt number relationship as a function of Reynolds and the Prandtl number. The Wilson method, commonly used to determine the mean heat transfer coefficient in exchangers, is presented. A novel method for parallel determination of the air- and the water-side Nusselt number correlations

© Springer International Publishing AG, part of Springer Nature 2019
D. Taler, *Numerical Modelling and Experimental Testing of Heat Exchangers*,
Studies in Systems, Decision and Control 161,
https://doi.org/10.1007/978-3-319-91128-1_19

based on the weighted least squares method is described. A comparison is made between correlations for the water- and the air-side Nusselt numbers obtained for different forms of the $\text{Nu} = f(\text{Re}, \text{Pr})$ function and different numbers of unknown coefficients appearing in the functions.

The monograph also presents the potential of the application of Computational Fluid Dynamics (CFD) modelling tools to determine the air-side Nusselt number correlations. Determination of the air-side Nusselt number correlation using the ANSYS-CFX program used to simulate the thermal and flow processes taking place in a car radiator is discussed. The obtained correlation does not differ much from the formula established based on experimental data.

Chapter 18 presents the application of the heat exchanger mathematical model for automatic control of the liquid temperature at the car radiator outlet by changing the number of revolutions of the fan forcing the air flow through the radiator. The developed control system is compared to a digital PID controller.

The results of the calculations and the experimental testing make it possible to set future research directions, the most important of which are as follows:

- calculation and testing of tube heat exchangers in the range of the laminar and the transitional flow of fluids in tubes,
- improving CFD simulation methods to optimize the design of finned-tube and lamella heat exchangers,
- application of CFD simulations to determine the gas-side Nusselt number correlations so that expensive experimental testing can be limited or eliminated in future,
- determination of experimental and CFD simulation-based formulae needed to find correlations for the Nusselt number for individual tube rows, as the air-side heat transfer coefficients in different tube rows differ from each other considerably,
- testing and modelling of heat exchangers in transient-state conditions.

The monograph is addressed to students, university scholars and research institute employees, as well as to designers and manufacturers of heat exchangers.

Appendix A
Selected Elements of the Vector and Tensor Calculus

A.1 Basic Vector Operations in a Cartesian System of Coordinates

If the scalar is denoted as s and vectors as $\mathbf{v}$ and $\mathbf{w}$, then the basic vector operations are as follows [112, 250]:

- gradient

$$\text{grad } s = \nabla s = \mathbf{i}\frac{\partial s}{\partial x} + \mathbf{j}\frac{\partial s}{\partial y} + \mathbf{k}\frac{\partial s}{\partial z}, \tag{A.1}$$

- divergence

$$\text{div } \mathbf{w} = \nabla \cdot \mathbf{w} = \frac{\partial w_x}{\partial x} + \frac{\partial w_y}{\partial y} + \frac{\partial w_z}{\partial z}, \tag{A.2}$$

- derivative of scalar field s with respect to field $\mathbf{w}$ (derivative of scalar s in the direction of vector $\mathbf{w}$)

$$\frac{ds}{dw} = \mathbf{w} \cdot \nabla s = w_x\frac{\partial s}{\partial x} + w_y\frac{\partial s}{\partial y} + w_z\frac{\partial s}{\partial z}, \tag{A.3}$$

- scalar product

$$\mathbf{v} \cdot \mathbf{w} = \mathbf{w} \cdot \mathbf{v} = v_x w_x + v_y w_y + v_z w_z, \tag{A.4}$$

- vector product

$$\mathbf{v} \times \mathbf{w} = -\mathbf{w} \times \mathbf{v} = \begin{vmatrix} \mathbf{i} & \mathbf{j} & \mathbf{k} \\ v_x & v_y & v_z \\ w_x & w_y & w_z \end{vmatrix}$$

$$= \mathbf{i}(v_y w_z - v_z w_y) + \mathbf{j}(v_z w_x - v_x w_z) + \mathbf{k}(v_x w_y - v_y w_x), \tag{A.5}$$

© Springer International Publishing AG, part of Springer Nature 2019
D. Taler, *Numerical Modelling and Experimental Testing of Heat Exchangers*,
Studies in Systems, Decision and Control 161,
https://doi.org/10.1007/978-3-319-91128-1

– substantial derivative in a medium moving with velocity $\mathbf{w}$

$$\frac{Ds}{Dt} = \frac{\partial s}{\partial t} + \mathbf{w} \cdot \nabla s, \qquad (A.6)$$

– Laplace operator (Laplacian)

$$\nabla^2 s = \operatorname{div} \operatorname{grad} s = \nabla \cdot \nabla s = \frac{\partial^2 s}{\partial x^2} + \frac{\partial^2 s}{\partial y^2} + \frac{\partial^2 s}{\partial z^2}, \qquad (A.7)$$

A.2 Basic Tensor Operations in a Cartesian System of Coordinates

The following basic tensor operations [28, 112, 250, 278] will be discussed: the vector gradient, the vector divergence, the vector substantial derivative, the vector field Laplacian, the dyadic product of two vectors and the scalar product of two vectors:

– gradient of vector $\mathbf{w}$

Vector gradient $\nabla \mathbf{w}$ is second-order tensor $\mathbf{T}$ with the following components:

$$T_{ij} = \frac{\partial w_i}{\partial x_j}; \quad i = 1, 2, 3; \ j = 1, 2, 3, \qquad (A.8)$$

where i and j are the row and the column number, respectively.
In the Cartesian coordinates tensor $\mathbf{T}$ is defined as follows:

$$\mathbf{T} = \nabla \mathbf{w} = \begin{bmatrix} \frac{\partial w_x}{\partial x} & \frac{\partial w_x}{\partial y} & \frac{\partial w_x}{\partial z} \\ \frac{\partial w_y}{\partial x} & \frac{\partial w_y}{\partial y} & \frac{\partial w_y}{\partial z} \\ \frac{\partial w_z}{\partial x} & \frac{\partial w_z}{\partial y} & \frac{\partial w_z}{\partial z} \end{bmatrix}. \qquad (A.9)$$

– divergence of tensor $\nabla \cdot \boldsymbol{\sigma}$

The components of vector $\nabla \cdot \boldsymbol{\sigma}$ are defined by the following formulae:

$$(\nabla \cdot \boldsymbol{\sigma})_x = \frac{\partial \sigma_{xx}}{\partial x} + \frac{\partial \tau_{yx}}{\partial y} + \frac{\partial \tau_{zx}}{\partial z},$$

$$(\nabla \cdot \boldsymbol{\sigma})_y = \frac{\partial \tau_{xy}}{\partial x} + \frac{\partial \sigma_{yy}}{\partial y} + \frac{\partial \tau_{zy}}{\partial z}, \qquad (A.10)$$

$$(\nabla \cdot \boldsymbol{\sigma})_z = \frac{\partial \tau_{xz}}{\partial x} + \frac{\partial \tau_{yz}}{\partial y} + \frac{\partial \sigma_{zz}}{\partial z}.$$

Product $\mathbf{v} \cdot \nabla \mathbf{w}$ can be calculated in a similar manner:

$$(\mathbf{v} \cdot \nabla \mathbf{w})_x = v_x \frac{\partial w_x}{\partial x} + v_y \frac{\partial w_x}{\partial y} + v_z \frac{\partial w_x}{\partial z},$$

$$(\mathbf{v} \cdot \nabla \mathbf{w})_y = v_x \frac{\partial w_y}{\partial x} + v_y \frac{\partial w_y}{\partial y} + v_z \frac{\partial w_y}{\partial z}, \qquad (A.11)$$

$$(\mathbf{v} \cdot \nabla \mathbf{w})_z = v_z \frac{\partial w_z}{\partial x} + v_y \frac{\partial w_z}{\partial y} + v_z \frac{\partial w_z}{\partial z}$$

It should be noted that product $\mathbf{w} \cdot \nabla \mathbf{w}$ can be determined as follows:

$$(\mathbf{w} \cdot \nabla)\mathbf{w} = \nabla(w^2/2) - \mathbf{w} \times (\nabla \times \mathbf{w}). \qquad (A.12)$$

Expression $(\mathbf{w} \cdot \nabla)\mathbf{w}$ denotes the directional derivative of vector field $\mathbf{w}$ with respect to field $\mathbf{w}$. In a Cartesian system of coordinates, the directional derivative vector components can be calculated as follows:

$$\begin{aligned}
(\mathbf{w} \cdot \nabla)\mathbf{w} &= (\mathbf{w} \cdot \nabla)(w_x\mathbf{i} + w_y\mathbf{j} + w_z\mathbf{k}) \\
&= \left(w_x \frac{\partial}{\partial x} + w_y \frac{\partial}{\partial y} + w_z \frac{\partial}{\partial z} \right)(w_x\mathbf{i} + w_y\mathbf{j} + w_z\mathbf{k}) \\
&= \left(w_x \frac{\partial w_x}{\partial x} + w_y \frac{\partial w_x}{\partial y} + w_z \frac{\partial w_x}{\partial z} \right)\mathbf{i} \qquad (A.13) \\
&+ \left(w_x \frac{\partial w_y}{\partial x} + w_y \frac{\partial w_y}{\partial y} + w_z \frac{\partial w_y}{\partial z} \right)\mathbf{j} \\
&+ \left(w_x \frac{\partial w_z}{\partial x} + w_y \frac{\partial w_z}{\partial y} + w_z \frac{\partial w_z}{\partial z} \right)\mathbf{k}
\end{aligned}$$

The components of product $\mathbf{w} \cdot \nabla \mathbf{w}$ in other systems of coordinates (e.g. cylindrical or spherical) can be found more easily using the last term in formula (A.12).

– substantial derivative of vector $\mathbf{w}$

$$\frac{D\mathbf{w}}{Dt} = \frac{\partial \mathbf{w}}{\partial t} + (\mathbf{w} \cdot \nabla \mathbf{w}) = \frac{\partial \mathbf{w}}{\partial t} + (\mathbf{w} \cdot \nabla)\mathbf{w} = \frac{\partial \mathbf{w}}{\partial t} + \nabla(w^2/2) - \mathbf{w} \times (\nabla \times \mathbf{w}),$$

$$(A.14)$$

- Laplacian of vector field $\mathbf{w}$

$$\nabla^2\mathbf{w} = \nabla \cdot \nabla\mathbf{w} = (\nabla \cdot \nabla)\mathbf{w} = \mathbf{i}\nabla^2 w_x + \mathbf{j}\nabla^2 w_y + \mathbf{k}\nabla^2 w_z, \qquad (A.15)$$

where the components are defined as:

$$\left(\nabla^2\mathbf{w}\right)_x = \nabla^2 w_x = \frac{\partial^2 w_x}{\partial x^2} + \frac{\partial^2 w_x}{\partial y^2} + \frac{\partial^2 w_x}{\partial z^2},$$

$$\left(\nabla^2\mathbf{w}\right)_y = \nabla^2 w_y = \frac{\partial^2 w_y}{\partial x^2} + \frac{\partial^2 w_y}{\partial y^2} + \frac{\partial^2 w_y}{\partial z^2}, \qquad (A.16)$$

$$\left(\nabla^2\mathbf{w}\right)_z = \nabla^2 w_z = \frac{\partial^2 w_z}{\partial x^2} + \frac{\partial^2 w_z}{\partial y^2} + \frac{\partial^2 w_z}{\partial z^2}.$$

- dyadic product of two vectors: $\mathbf{v}\,\mathbf{w}$

  The dyadic product of two vectors

$$\mathbf{T} = \mathbf{u}\mathbf{w} = \mathbf{u} \otimes \mathbf{w} \qquad (A.17)$$

is a second-order tensor, the components of which are defined by the following formula:

$$T_{ij} = u_i w_j, \quad i = 1, 2, 3. \qquad (A.18)$$

Numbers 1, 2 and 3 correspond to coordinates x, y and z.

The dyadic product of two vectors, also referred to as the exterior product of two vectors

$$\mathbf{u} = \begin{bmatrix} u_x \\ u_y \\ u_z \end{bmatrix} \text{ and } \mathbf{w} = \begin{bmatrix} w_x \\ w_y \\ w_z \end{bmatrix} \qquad (A.19)$$

is defined by the following tensor:

$$\mathbf{u} \otimes \mathbf{w} = \begin{bmatrix} u_x w_x & u_x w_y & u_x w_z \\ u_y w_x & u_y w_y & u_y w_z \\ u_z w_x & u_z w_z & u_z w_z \end{bmatrix} \qquad (A.20)$$

- the scalar product of two tensors is defined as:

$$\boldsymbol{\sigma} : \nabla \mathbf{w} = \begin{bmatrix} \sigma_{xx} & \tau_{xy} & \tau_{xz} \\ \tau_{yx} & \sigma_{yy} & \tau_{yz} \\ \tau_{zx} & \tau_{zy} & \sigma_{zz} \end{bmatrix} : \begin{bmatrix} \frac{\partial w_x}{\partial x} & \frac{\partial w_x}{\partial y} & \frac{\partial w_x}{\partial z} \\ \frac{\partial w_y}{\partial x} & \frac{\partial w_y}{\partial y} & \frac{\partial w_y}{\partial z} \\ \frac{\partial w_z}{\partial x} & \frac{\partial w_z}{\partial y} & \frac{\partial w_z}{\partial z} \end{bmatrix}$$

$$= \sigma_{xx}\frac{\partial w_x}{\partial x} + \tau_{xy}\frac{\partial w_y}{\partial x} + \tau_{xz}\frac{\partial w_z}{\partial x}$$

$$+ \tau_{yx}\frac{\partial w_x}{\partial y} + \sigma_{yy}\frac{\partial w_y}{\partial y} + \tau_{yz}\frac{\partial w_z}{\partial y}$$

$$+ \tau_{zx}\frac{\partial w_x}{\partial z} + \tau_{zy}\frac{\partial w_y}{\partial z} + \sigma_{zz}\frac{\partial w_z}{\partial z}.$$

(A.21)

Appendix B
The Navier-Stokes Equation in a Cylindrical and a Spherical System of Coordinates

The Navier-Stokes equation (2.81) (for an incompressible fluid and constant dynamic viscosity)

$$\rho \frac{D\mathbf{w}}{Dt} = \rho \mathbf{g} - \nabla p + \mu \nabla^2 \mathbf{w} \tag{B.1}$$

will be written in the cylindrical and the spherical system of coordinates [28, 40, 278] presented in Fig. 2.9b, c, respectively.

Calculating substantial derivative $D\mathbf{w}/Dt$ and Laplacian $\nabla^2 \mathbf{w}$ in different systems of coordinates, the following relations can be used [28, 278]:

$$\frac{D\mathbf{w}}{Dt} = \frac{\partial \mathbf{w}}{\partial t} + (\mathbf{w} \cdot \nabla)\mathbf{w}, \tag{B.2}$$

$$\nabla^2 \mathbf{w} = \nabla(\nabla \cdot \mathbf{w}) - \nabla \times (\nabla \times \mathbf{w}). \tag{B.3}$$

– cylindrical system of coordinates (cf. Fig. 2.9b)

$$\rho \left(\frac{\partial w_r}{\partial t} + w_r \frac{\partial w_r}{\partial r} + \frac{w_\theta}{r} \frac{\partial w_r}{\partial \theta} + w_z \frac{\partial w_r}{\partial z} - \frac{w_\theta^2}{r} \right)$$
$$= \rho g_r - \frac{\partial p}{\partial r} + \mu \left(\frac{\partial w_r}{\partial r^2} + \frac{1}{r} \frac{\partial w_r}{\partial r} + \frac{1}{r^2} \frac{\partial^2 w_r}{\partial \theta^2} + \frac{\partial^2 w_r}{\partial z^2} - \frac{w_r}{r^2} - \frac{2}{r^2} \frac{\partial w_\theta}{\partial \theta} \right), \tag{B.4}$$

$$\rho \left(\frac{\partial w_\theta}{\partial t} + w_r \frac{\partial w_\theta}{\partial r} + \frac{w_\theta}{r} \frac{\partial w_\theta}{\partial \theta} + w_z \frac{\partial w_\theta}{\partial z} + \frac{w_r w_\theta}{r} \right)$$
$$= \rho g_\theta - \frac{1}{r} \frac{\partial p}{\partial \theta} + \mu \left(\frac{\partial^2 w_\theta}{\partial r^2} + \frac{1}{r} \frac{\partial w_\theta}{\partial r} + \frac{1}{r^2} \frac{\partial^2 w_\theta}{\partial \theta^2} + \frac{\partial^2 w_\theta}{\partial z^2} + \frac{2}{r^2} \frac{\partial w_r}{\partial \theta} - \frac{w_\theta}{r^2} \right), \tag{B.5}$$

© Springer International Publishing AG, part of Springer Nature 2019
D. Taler, *Numerical Modelling and Experimental Testing of Heat Exchangers*,
Studies in Systems, Decision and Control 161,
https://doi.org/10.1007/978-3-319-91128-1

$$\rho \left(\frac{\partial w_z}{\partial t} + w_r \frac{\partial w_z}{\partial r} + \frac{w_\theta}{r} \frac{\partial w_z}{\partial \theta} + w_z \frac{\partial w_z}{\partial z} \right)$$

$$= \rho g_z - \frac{\partial p}{\partial z} + \mu \left(\frac{\partial^2 w_z}{\partial r^2} + \frac{1}{r} \frac{\partial w_z}{\partial r} + \frac{1}{r^2} \frac{\partial w_z}{\partial \theta^2} + \frac{\partial^2 w_z}{\partial z^2} \right), \tag{B.6}$$

- spherical system of coordinates (cf. Fig. 2.9c)

$$\rho \left(\frac{\partial w_r}{\partial t} + w_r \frac{\partial w_r}{\partial r} + \frac{w_\theta}{r} \frac{\partial w_r}{\partial \theta} + \frac{w_\varphi}{r \sin \varphi} \frac{\partial w_r}{\partial \varphi} - \frac{w_\theta^2 + w_\varphi^2}{r} \right)$$

$$= \rho g_r - \frac{\partial p}{\partial r} + \mu \left[\frac{1}{r^2} \frac{\partial}{\partial r} \left(r^2 \frac{\partial w_r}{\partial r} \right) + \frac{1}{r^2 \sin \theta} \frac{\partial}{\partial \theta} \left(\sin \theta \frac{\partial w_r}{\partial \theta} \right) \right. \tag{B.7}$$

$$\left. + \frac{1}{r^2 \sin \theta} \frac{\partial w_r}{\partial \varphi^2} - \frac{2 w_r}{r^2} - \frac{2}{r^2} \frac{\partial w_\theta}{\partial \theta} - \frac{2 w_\theta \cot \theta}{r^2} - \frac{2}{r^2 \sin \theta} \frac{\partial w_\varphi}{\partial \varphi} \right],$$

$$\rho \left(\frac{\partial w_\theta}{\partial t} + w_r \frac{\partial w_\theta}{\partial r} + \frac{w_\theta}{r} \frac{\partial w_\theta}{\partial \theta} + \frac{w_\phi}{r \sin \theta} \frac{\partial w_\theta}{\partial \varphi} + \frac{w_r w_\theta}{r} - \frac{w_\varphi^2 \cot \theta}{r} \right)$$

$$= \rho g_\theta - \frac{1}{r} \frac{\partial p}{\partial \theta} + \mu \left[\frac{1}{r^2} \frac{\partial}{\partial r} \left(r^2 \frac{\partial w_\theta}{\partial r} \right) + \frac{1}{r^2 \sin \theta} \frac{\partial}{\partial \theta} \left(\sin \theta \frac{\partial w_\theta}{\partial \theta} \right) \right. \tag{B.8}$$

$$\left. + \frac{1}{r^2 \sin^2 \theta} \frac{\partial^2 w_\theta}{\partial \varphi^2} + \frac{2}{r^2} \frac{\partial w_r}{\partial \theta} - \frac{w_\theta}{r^2 \sin \theta} - \frac{2 \cos \theta}{r^2 \sin^2 \theta} \frac{\partial w_\varphi}{\partial \varphi} \right],$$

$$\rho \left(\frac{\partial w_\varphi}{\partial t} + w_r \frac{\partial w_\varphi}{\partial r} + \frac{w_\theta}{r} \frac{\partial w_\varphi}{\partial \theta} + \frac{w_\varphi}{r \sin \theta} \frac{\partial w_\varphi}{\partial \varphi} - \frac{w_\varphi w_r}{r} + \frac{w_\theta w_\varphi \cot \theta}{r} \right)$$

$$= \rho g_\varphi - \frac{1}{r \sin \theta} \frac{\partial p}{\partial \varphi} + \mu \left[\frac{1}{r^2} \frac{\partial}{\partial r} \left(r^2 \frac{\partial w_\varphi}{\partial r} \right) + \frac{1}{r^2} \frac{1}{\sin \theta} \frac{\partial}{\partial \theta} \left(\sin \theta \frac{\partial w_\varphi}{\partial \theta} \right) \right. \tag{B.9}$$

$$\left. + \frac{1}{r^2 \sin^2 \theta} \frac{\partial^2 w_\varphi}{\partial \varphi^2} - \frac{w_\varphi}{r^2 \sin^2 \theta} + \frac{2}{r^2 \sin^2 \theta} \frac{\partial w_r}{\partial \varphi} + \frac{2 \cos \theta}{r^2 \sin^2 \theta} \frac{\partial w_\theta}{\partial \varphi} \right],$$

Appendix C
The Energy Conservation Equation in a Cartesian, a Cylindrical and a Spherical System of Coordinates

The energy balance equation (2.183) for Newtonian fluids

$$\rho \frac{Di}{Dt} = \nabla \cdot (\lambda \nabla T) - \nabla \cdot \dot{\mathbf{q}}_{\mathbf{r}} + \dot{q}_v + \frac{Dp}{Dt} + \dot{\phi} \qquad (C.1)$$

will be presented in three basic systems of coordinates: Cartesian, cylindrical and spherical:

- Cartesian system of coordinates (x, y, z) (cf. Fig. 2.9a):

$$\frac{D}{Dt} = \frac{\partial}{\partial t} + w_x \frac{\partial}{\partial x} + w_y \frac{\partial}{\partial y} + w_z \frac{\partial}{\partial z}, \qquad (C.2)$$

where:

$$\nabla \cdot (\lambda \nabla T) = \frac{\partial}{\partial x}\left(\lambda \frac{\partial T}{\partial x}\right) + \frac{\partial}{\partial y}\left(\lambda \frac{\partial T}{\partial y}\right) + \frac{\partial}{\partial z}\left(\lambda \frac{\partial T}{\partial z}\right), \qquad (C.3)$$

$$\nabla = \mathbf{i}\frac{\partial}{\partial x} + \mathbf{j}\frac{\partial}{\partial y} + \mathbf{k}\frac{\partial}{\partial z}, \qquad (C.4)$$

$$\dot{\phi} = \mu \left\{ 2\left[\left(\frac{\partial w_x}{\partial x}\right)^2 + \left(\frac{\partial w_y}{\partial y}\right)^2 + \left(\frac{\partial w_z}{\partial z}\right)^2\right] + \left(\frac{\partial w_y}{\partial x} + \frac{\partial w_x}{\partial y}\right)^2 + \left(\frac{\partial w_z}{\partial y} + \frac{\partial w_y}{\partial z}\right)^2 \right.$$
$$\left. + \left(\frac{\partial w_x}{\partial z} + \frac{\partial w_z}{\partial x}\right)^2 - \frac{2}{3}\left(\frac{\partial w_x}{\partial x} + \frac{\partial w_y}{\partial y} + \frac{\partial w_z}{\partial z}\right)^2 \right\};$$

$$(C.5)$$

© Springer International Publishing AG, part of Springer Nature 2019
D. Taler, *Numerical Modelling and Experimental Testing of Heat Exchangers*,
Studies in Systems, Decision and Control 161,
https://doi.org/10.1007/978-3-319-91128-1

– cylindrical system of coordinates (r, θ, z) (cf. Fig. 2.9b):

$$\frac{D}{Dt} = \frac{\partial}{\partial t} + w_r \frac{\partial}{\partial r} + \frac{w_\theta}{r} \frac{\partial}{\partial \theta} + w_z \frac{\partial}{\partial z}, \tag{C.6}$$

$$\nabla \cdot (\lambda \nabla T) = \frac{1}{r} \frac{\partial}{\partial r}\left(r\lambda \frac{\partial T}{\partial r}\right) + \frac{1}{r^2} \frac{\partial}{\partial \theta}\left(\lambda \frac{\partial T}{\partial \theta}\right) + \frac{\partial}{\partial z}\left(\lambda \frac{\partial T}{\partial z}\right), \tag{C.7}$$

$$\nabla = \mathbf{i}_r \frac{\partial}{\partial r} + \mathbf{i}_\theta \frac{1}{r} \frac{\partial}{\partial \theta} + \mathbf{i}_z \frac{\partial}{\partial z} \tag{C.8}$$

$$\dot\phi = \mu\left\{ 2\left[\left(\frac{\partial w_r}{\partial r}\right)^2 + \left(\frac{1}{r}\frac{\partial w_\theta}{\partial \theta} + \frac{w_r}{r}\right)^2 + \left(\frac{\partial w_z}{\partial z}\right)^2 \right] + \left[r\frac{\partial}{\partial r}\left(\frac{w_\theta}{r}\right) + \frac{1}{r}\frac{\partial w_r}{\partial \theta} \right]^2 \right.$$
$$\left. + \left(\frac{1}{r}\frac{\partial w_z}{\partial \theta} + \frac{\partial w_\theta}{\partial z}\right)^2 + \left(\frac{\partial w_r}{\partial z} + \frac{\partial w_z}{\partial r}\right)^2 - \frac{2}{3}\left[\frac{1}{r}\frac{\partial}{\partial r}(rw_r) + \frac{1}{r}\frac{\partial w_\theta}{\partial \theta} + \frac{\partial w_z}{\partial z}\right]^2 \right\};$$
$$\tag{C.9}$$

– spherical system of coordinates (r, θ, φ) (cf. Fig. 2.9c):

$$\frac{D}{Dt} = \frac{\partial}{\partial t} + w_r \frac{\partial}{\partial r} + \frac{w_\theta}{r} \frac{\partial}{\partial \theta} + \frac{w_\varphi}{r \sin\theta} \frac{\partial}{\partial \varphi}, \tag{C.10}$$

$$\nabla \cdot (\lambda \nabla T) = \frac{1}{r^2} \frac{\partial}{\partial r}\left(r^2\lambda \frac{\partial T}{\partial r}\right) + \frac{1}{r^2 \sin\theta} \frac{\partial}{\partial \theta}\left(\lambda \sin\theta \frac{\partial T}{\partial \theta}\right) + \frac{1}{r^2 \sin\theta} \frac{\partial}{\partial \varphi}\left(\lambda \frac{\partial T}{\partial \varphi}\right),$$
$$\tag{C.11}$$

$$\nabla = \mathbf{i}_r \frac{\partial}{\partial r} + \mathbf{i}_\theta \frac{1}{r} \frac{\partial}{\partial \theta} + \mathbf{i}_\varphi \frac{1}{r \sin\theta} \frac{\partial}{\partial \varphi}, \tag{C.12}$$

$$\dot\phi = \mu\left\{ 2\left[\left(\frac{\partial w_r}{\partial r}\right)^2 + \left(\frac{1}{r}\frac{\partial w_\theta}{\partial \theta} + \frac{w_r}{r}\right)^2 + \left(\frac{1}{r \sin\theta}\frac{\partial w_\varphi}{\partial \varphi} + \frac{w_r + w_\theta \cot\theta}{r}\right)^2 \right] \right.$$
$$+ \left[r\frac{\partial}{\partial r}\left(\frac{w_\theta}{r}\right) + \frac{1}{r}\frac{\partial w_r}{\partial \theta} \right]^2 + \left[\frac{\sin\theta}{r}\frac{\partial}{\partial \theta}\left(\frac{w_\varphi}{\sin\theta}\right) + \frac{1}{r \sin\theta}\frac{\partial w_\theta}{\partial \varphi} \right]^2$$
$$\left. + \left[\frac{1}{r \sin\theta}\frac{\partial w_r}{\partial \varphi} + r\frac{\partial}{\partial r}\left(\frac{w_\varphi}{r}\right) \right]^2 - \frac{2}{3}\left[\frac{1}{r^2}\frac{\partial}{\partial r}(r^2 w_r) + \frac{1}{r \sin\theta}\frac{\partial}{\partial \theta}(w_\theta \sin\theta) + \frac{1}{r \sin\theta}\frac{\partial w_\varphi}{\partial \varphi} \right]^2 \right\}.$$

Appendix D
Principles of Determination of the Uncertainty of Experimental Measurements and Calculation Results According to the ASME [232]

Experimental testing results should include the following information:

Z1. Precision limit (limit of accuracy) P with respect to the nominal (single or averaged) result, which is the 95% confidence interval which should comprise the mean values obtained from many measurements if the experiment is repeated many times in identical conditions and using the same equipment. Consequently, the precision limit is an estimator of the lack of repeatability due to random errors and unsteadiness of parameters during the experiment.

Z2. The bias limit B, which is an estimator of the systematic error. It is assumed here that the experimenter is 95% sure that the systematic error real value, if known, should be less than $|B|$.

Z3. Uncertainty U, where the $\pm U$ interval with respect to the nominal result is a range within which real results lie with the 95% confidence. Assuming the 95% confidence interval, the result uncertainty is calculated using the following formula:

$$U = \left(B^2 + P^2\right)^{\frac{1}{2}}. \tag{D.1}$$

Z4. An uncertainty analysis conducted according either to the brief description presented above or to literature guidelines.

The precision and systematic error limits should be estimated in a time representative of the experiment. The uncertainty assessment should include additional information, presented preferably in a table format. The required items are as follows:

(a) precision and systematic error limits for each variable and parameter used during the testing,

(b) the equation needed to develop the results (indirect measurement),

(c) presentation of results, including a comparison between the spread of results obtained from the experiment repeated many times and anticipated spread ($\pm P$) determined from the analysis of the measurement uncertainty.

© Springer International Publishing AG, part of Springer Nature 2019   565
D. Taler, *Numerical Modelling and Experimental Testing of Heat Exchangers*,
Studies in Systems, Decision and Control 161,
https://doi.org/10.1007/978-3-319-91128-1

A discussion of the experimental error sources without the uncertainty assessment described above does not satisfy the ASME requirements. All presented results must include the uncertainty assessment. All figures illustrating new experimental data should show assessment of the data uncertainty, either in the figure itself or in the figure caption. The publications devoted to uncertainty assessment are listed at the end of this appendix.

Example

Heat flow rate $\dot{Q}$ collected in the exchanger by cooling air is calculated from the following formula:

$$\dot{Q} = \dot{m}c_p(T_o - T_i),\tag{D.2}$$

where: $\dot{m}$—air mass flow rate, T_i and T_o—air inlet and outlet temperature, respectively, c_p—air specific heat at constant pressure.

Measurement uncertainty $U_{\dot{Q}}$ of the calculated value of $\dot{Q}$ at the assumed confidence level of 95% is the effect of random uncertainty $P_{\dot{Q}}$ and systematic error $B_{\dot{Q}}$:

$$U_{\dot{Q}} = \sqrt{P_{\dot{Q}}^2 + B_{\dot{Q}}^2}\tag{D.3}$$

The two components of uncertainty $U_{\dot{Q}}$ can be calculated separately as a function of sensitivity coefficients of the calculated value of $\dot{Q}$ with respect to measured quantities (e.g. $\frac{\partial \dot{Q}}{\partial \dot{m}}$) according to the error propagation rule [62, 63, 205–207]:

$$P_{\dot{Q}}^2 = \left(\frac{\partial \dot{Q}}{\partial \dot{m}}\right)^2 P_{\dot{m}}^2 + \left(\frac{\partial \dot{Q}}{\partial c_p}\right)^2 P_{c_p}^2 + \left(\frac{\partial \dot{Q}}{\partial T_o}\right)^2 P_{T_o}^2 + \left(\frac{\partial \dot{Q}}{\partial T_i}\right)^2 P_{T_i}^2\tag{D.4}$$

and

$$B_{\dot{Q}}^2 = \left(\frac{\partial \dot{Q}}{\partial \dot{m}}\right)^2 B_{\dot{m}}^2 + \left(\frac{\partial \dot{Q}}{\partial c_p}\right)^2 B_{c_p}^2 + \left(\frac{\partial \dot{Q}}{\partial T_o}\right)^2 B_{T_o}^2 + \left(\frac{\partial \dot{Q}}{\partial T_i}\right)^2 B_{T_i}^2 + 2\left(\frac{\partial \dot{Q}}{\partial T_o}\right)$$
$$\cdot \left(\frac{\partial \dot{Q}}{\partial T_i}\right) B'_{T_o} B'_{T_i}\tag{D.5}$$

where B'_{T_o} and B'_{T_i} are parts of B_{T_o} and B_{T_i} caused by identical sources of errors (such as the calibration error of thermocouples calibrated using the same standards, equipment and procedures); it is therefore assumed that they are perfectly correlated. Using equation (D.2) to find derivatives and using $\Delta T = T_o - T_i$, the following is obtained after transformations:

$$\frac{P_{\dot{Q}}}{\dot{Q}} = \left(\frac{P_{\dot{m}}}{\dot{m}}\right)^2 + \left(\frac{P_{c_p}}{c_p}\right)^2 + \left(\frac{P_{T_o}}{\Delta T}\right)^2 + \left(\frac{P_{T_i}}{\Delta T}\right)^2\tag{D.6}$$

and

$$\frac{B_{\dot{Q}}}{\dot{Q}} = \left(\frac{B_{\dot{m}}}{\dot{m}}\right)^2 + \left(\frac{B_{c_p}}{c_p}\right)^2 + \left(\frac{B_{T_o}}{\Delta T}\right)^2 + \left(\frac{B_{T_i}}{\Delta T}\right)^2 - 2\left(\frac{B'_{T_o}}{\Delta T}\right) \cdot \left(\frac{B'_{T_i}}{\Delta T}\right). \qquad (D.7)$$

The derivatives in formulae (D.4) and (D.5) can be calculated numerically, using a data development program, or analytically.

Precision (accuracy) limits $P_{\dot{m}}, P_{T_o} i\, P_{T_i}$ can be assumed as two times standard deviations of the set of observations (measurements) of $\dot{m}, T_o$ and T_i, respectively, obtained using appropriate meters in normal operating conditions. The expressions must also take account of the unsteadiness (randomness) of the process; the inaccuracy of the measuring instrument only is insufficient.

A sufficient number of samples (>30) should be considered during a sufficiently long sampling time with respect to the biggest unsteadiness period so that the unsteadiness parameters should be representative of the process. The specific heat determination precision (accuracy) limit may be affected by changes in the mean temperature in which the heat value is determined using tables or the $c_p(T)$ function derived by means of the least squares method. The c_p value can be found if the dependence of c_p on T is defined. In most practical cases, unlike the other terms in formula (z.4), this term can be omitted in the precision (accuracy) limit $P_{\dot{Q}}$ calculation.

The limits of systematic errors $B_{\dot{m}}, B_{T_o}$ and B_{T_i} are determined either by means of calibration performed prior to and after the experiment or estimation of partial systematic errors in relevant variables; they are calculated as the root of the sum of squares of partial errors. Partial systematic errors include estimated errors in calibration standards, the calibration procedure and errors arising due to imperfect approximation of calibration results. One of the components of the limit of systematic error B_c is the "organic" error representing the systematic error included in the specific heat value read from tables. The error is an effect of errors arising when these properties are measured and when the results are put into tables. Such shares are usually no less than $0.25 \div 0.5\%$, and quite often they are much higher [63, 207]. While estimating the precision (accuracy) and the systematic error limits, the real variable must be identified. For example, in equation (D.2) temperatures T_o are T_i are the fluid mean temperatures in the outlet and in the inlet cross-section, respectively. If temperatures T_o and T_i in equation (D.2) are measured in specific points, the systematic error (which Moffat [205–207] called the concept systematic error) is equal to the difference between point-measured temperatures T_o and T_i and the fluid mean temperatures corresponding to them in the outlet and in the inlet cross-section, respectively.

In order to take this difference into consideration, the indications should be corrected, and while calculating the systematic error, the correction residual uncertainty should be taken into account by adding it to systematic errors in calibration, etc.

Consider a situation where the systematic error limits are non-correlated and total 0.5K, with the systematic error limit for specific heat of 0.5%. The estimated systematic error in the mass flow rate measurement totals 0.25% in the range of indications from 10 to 90% of the entire range. A discussion with the manufacturer shows that this is assessment of a constant error (which cannot be reduced by averaging many measurements and which is in fact a systematic error). For $\Delta T = 20$ K, the following is obtained from equation (D.7):

$$\frac{B_{\dot{Q}}}{\dot{Q}} = \sqrt{(0.0025)^2 + (0.005)^2 + \left(\frac{0.5K}{20K}\right)^2 + \left(\frac{0.5K}{20K}\right)^2} = 0.036 \ (= 3.6\%) \quad (D.8)$$

It can be seen that the temperature measurement systematic error is dominant and has the greatest impact on the pooled systematic error totalling 3.6%.

If random errors and the process unsteadiness were such that precision limit $P_{\dot{Q}}$ calculated from equation (D.6) was 2.7%, total uncertainty of $\dot{Q}$ determination, denoted as $U_{\dot{Q}}$, would total:

$$\frac{U_{\dot{Q}}}{\dot{Q}} = \sqrt{\left(\frac{B_{\dot{Q}}}{\dot{Q}}\right)^2 + \left(\frac{P_{\dot{Q}}}{\dot{Q}}\right)^2} = \sqrt{(0.36)^2 + (0.027)^2} = 0.045 \ (= 4.5\%). \quad (D.9)$$

Appendix E
Prediction Interval Determination

This appendix presents a more detailed discussion of determination of the prediction (forecast) interval in a linear least squares problem [25, 261, 375].

The indirectly measured quantity η is defined as:

$$\eta = a_0 + a_1 x_1 + \ldots + a_N x_N, \tag{E.1}$$

where coefficients $a_0, a_1 \ldots a_N$ are determined by means of the least squares method.

The predicted value of η in point $\mathbf{x}^0$:

$$\mathbf{x}^0 = \left\{ \begin{array}{c} 1 \\ x_1^0 \\ \vdots \\ x_N^0 \end{array} \right\} \tag{E.2}$$

is calculated from the following expression:

$$\tilde{\eta}_0 = \tilde{a}_0 + \tilde{a}_1 x_1 + \ldots + \tilde{a}_N x_N, \tag{E.3}$$

which can be written in the following vector form:

$$\tilde{\eta}_0 = \left(\mathbf{x}^0 \right)^T \mathbf{a}. \tag{E.4}$$

In order to find the mean prediction error, the value of the measured quantity y in point $\mathbf{x}^0$ will be determined first:

$$y_0 = \eta_0 + \varepsilon_0 = a_0 + a_1 x_1^0 + \ldots + a_N x_N^0 + \varepsilon_0, \tag{E.5}$$

where ε_0 is the measurement random error.

© Springer International Publishing AG, part of Springer Nature 2019

D. Taler, *Numerical Modelling and Experimental Testing of Heat Exchangers*,
Studies in Systems, Decision and Control 161,
https://doi.org/10.1007/978-3-319-91128-1

The expected value of y_0 is defined as:

$$E(y_0) = E\left(a_0 + a_1 x_1^0 + \ldots + a_N x_N^0 + \varepsilon_0\right) = \eta_0, \qquad (E.6)$$

where:

$$\eta_0 = a_0 + a_1 x_1^0 + \ldots + a_N x_N^0. \qquad (E.7)$$

It can be noticed that

$$E(\tilde{\eta}_0) = E\left[\left(\mathbf{x}^0\right)^T \tilde{\mathbf{a}}\right] = \left(\mathbf{x}^0\right)^T E(\tilde{\mathbf{a}}) = \left(\mathbf{x}^0\right)^T \mathbf{a} = E(y_0) \qquad (E.8)$$

because estimators $\tilde{a}_0, \tilde{a}_1, \ldots, \tilde{a}_N$ of the model parameters $a_0, a_1, \ldots, a_N$ are unbiased, i.e. $E(\tilde{a}_i) = a_i$.

Variance var $(y_0 - \tilde{\eta}_0)$ is found in the following way:

$$
\begin{aligned}
\text{var}(y_0 - \tilde{\eta}_0) &= E\left[(y_0 - \tilde{\eta}_0)^2\right] = E\left\{\left[\left(\mathbf{x}^0\right)^T \mathbf{a} + \varepsilon_0 - \left(\mathbf{x}^0\right)^T \tilde{\mathbf{a}}\right]^2\right\} \\
&= E\left\{\left[\left(\mathbf{x}^0\right)^T (\mathbf{a} - \tilde{\mathbf{a}}) + \varepsilon_0\right]^2\right\} = E\left[\left(\mathbf{x}^0\right)^T (\mathbf{a} - \tilde{\mathbf{a}})(\mathbf{a} - \tilde{\mathbf{a}})^T \mathbf{x}^0 \right. \\
&\quad \left. + 2\left(\mathbf{x}^0\right)^T (\mathbf{a} - \tilde{\mathbf{a}})\varepsilon_0 + \varepsilon_0^2\right] \\
&= E\left[\left(\mathbf{x}^0\right)^T (\mathbf{a} - \tilde{\mathbf{a}})(\mathbf{a} - \tilde{\mathbf{a}})^T \mathbf{x}^0\right] + E\left[2\left(\mathbf{x}^0\right)^T (\mathbf{a} - \tilde{\mathbf{a}})\varepsilon_0\right] + E\left(\varepsilon_0^2\right) \\
&= \left(\mathbf{x}^0\right)^T \left[(\mathbf{a} - \tilde{\mathbf{a}})(\mathbf{a} - \tilde{\mathbf{a}})^T\right]\mathbf{x}^0 + 0 + \sigma_0^2 = \left(\mathbf{x}^0\right)^T \mathbf{C}_{\tilde{\mathbf{a}}}\mathbf{x}^0 + \sigma_0^2 \\
&= \sigma_0^2 \left(\mathbf{x}^0\right)^T \left(\mathbf{X}^T \mathbf{G_y} \mathbf{X}\right)^{-1} \mathbf{x}^0 + \sigma_0^2.
\end{aligned}
$$

$$\qquad (E.9)$$

Assuming that $\sigma^2 = \sigma_0^2$, formula (E.9) gives (11.149).

Approximating standard deviation of random measuring errors by means of s_y, the $(1 - \alpha) \times 100\%$ prediction interval for the measured quantity y at any point $\mathbf{x}^0$, which does not need to be a measuring point, is obtained from the following expression:

$$P\left\{-t_{n-N-1,\alpha} < \frac{y_0 - \tilde{\eta}_0}{s_y \sqrt{1 + \left(\mathbf{x}^0\right)^T \left(\mathbf{X}^T \mathbf{G_y} \mathbf{X}\right)^{-1} \mathbf{x}^0}} < t_{n-N-1,\alpha}\right\} = 1 - \alpha. \qquad (E.10)$$

For the 95% confidence interval $\alpha = 0.05$. Transforming the term in the curly bracket in expression (E.10), formula (11.151) is obtained.

Bibliography

1. Abell, M. L., Braselton, J. P., & Rafter, J. A. (1999). *Statistics with Mathematica*. San Diego: Academic Press.
2. Abernathy, R. B., Benedict, R. P., & Diwdell, R. B. (1985). ASME measurement uncertainty. *ASME Journal of Fluids Engineering, 107,* 161–164.
3. Abraham, J. P., Sparrow, E. M., & Minkowycz, W. J. (2011). Internal-flow Nusselt numbers for the low-Reynolds-number end of the laminar-to-turbulent transition regime. *International Journal of Heat and Mass Transfer, 54,* 584–588.
4. Abraham, J. P., Sparrow, E. M., & Tong, J. C. K. (2009). Heat transfer in all pipe flow regimes: laminar, transitional/intermittent and turbulent. *International Journal of Heat and Mass Transfer, 52,* 557–563.
5. Acharya, S., Baliga, B., Karki, K., Murthy, J. Y., Prakash, C., & Vanka, S. P. (2007). Pressure-based finite-volume methods in computational fluid dynamics. *ASME Journal of Heat Transfer, 129,* 407–424.
6. Agrawal, H. C. (1960). Heat transfer in laminar flow between parallel plates at small Péclét numbers. *Applied Scientific Research, 9*(1), 177–189.
7. Allen, J. J., Shockling, M. A., Kunkel, G. J., & Smits, A. J. (2007). Turbulent flow in smooth and rough pipes. *Philosophical Transactions of Royal Society A, 365*(1852), 699–714.
8. Allen, R. W., & Eckert, E. R. G. (1964). Friction and heat-transfer measurements to turbulent pipe flow of water (Pr = 7 and 8) at uniform wall heat flux. *Transactions of the ASME, Journal of Heat Transfer, 86,* 301–310.
9. Anderson, J. D. (1995). *Computational fluid dynamics: The basics with applications.* Boston: McGraw-Hill.
10. Anderson, J. D. (2000). *Hypersonic and high temperature gas dynamics.* Reston, Virginia: American Institute of Aeronautics and Astronautics.
11. ANSYS CFX, Release 13.0, Documentation, 2011.
12. Aoki, S. (1963). A consideration on the heat transfer in liquid metal. *Bulletin of the Tokyo Institute of Technology, 54,* 63–73.
13. Aravinth, S. (2000). Prediction of heat and mass transfer for fully developed turbulent fluid flow through tubes. *International Journal of Heat and Mass Transfer, 43,* 1399–1408.
14. *ASHRAE Handbook, Fundamentals volume*, American Society of Heating, Refrigerating and Air-Conditioning Engineers, Inc. Atlanta, GA. 1997.
15. Baclic, B. S. (1978). A simplified formula for cross-flow heat exchanger. *Transactions of the ASME, Journal of Heat transfer, 100,* 746–747.
16. Baehr, H. D., & Stephan, K. (1994). *Wärme-und Stoffübertragung.* Berlin: Springer.
17. Baker, R. C. (2000). *Flow measurement handbook.* Cambridge: Cambridge University Press.
18. Baliga, B. R., & Patankar, S. V. (1983). A control-volume finite element method for two-dimensional fluid flow and heat transfer. *Numerical Heat Transfer, 6,* 245–261.

© Springer International Publishing AG, part of Springer Nature 2019   571
D. Taler, *Numerical Modelling and Experimental Testing of Heat Exchangers*,
Studies in Systems, Decision and Control 161,
https://doi.org/10.1007/978-3-319-91128-1

19. Baliga, B. R., & Patankar, S. V. (1988). *Elliptic systems: Finite element II.* Wiley, New York: Handbook of numerical heat transfer.
20. Beck, J. V., Balckwell, B., Clair, Ch. R. St. Jr. (1985). *Inverse heat conduction. Ill-posed problems.* New York: A Wiley Interscience.
21. Bejan, A. (1993). *Heat transfer.* New York: Wiley.
22. Bejan, A. (2004). *Convection heat transfer* (3rd ed.). Hoboken: Wiley.
23. Bejan, A., & Kraus, A. D. (2004). *Heat transfer handbook.* Hoboken: Wiley.
24. Bertin, J. J. (2002). *Aerodynamics for engineers* (4th ed.). Upper Saddle River, New Jersey: Prentice Hall.
25. Bevington, P. R. (1969). *Data reduction and error analysis for the physical sciences.* New York: McGraw-Hill.
26. Bhatti, M. S., & Shah, R. K. (1987). Turbulent and transition flow convective heat transfer in ducts (Chapter 4). In S. Kakaç, R. K. Shah, & W. Aung (Eds.), *Handbook of single-phase convective heat transfer.* New York: Wiley Interscience.
27. Binnie, A. M., & Poole, E. G. C. (1937). The theory of the single-pass cross-flow heat exchange. *Proceedings of the Cambridge Philosophical Society, 33,* 403–411.
28. Bird, R. B., Stewart, W. E., & Lightfoot, E. N. (2007). *Transport phenomena* (2nd ed.). New York: Wiley.
29. Björck, Å., & Dahlguist, G. (1987). *Metody numeryczne.* Warszawa: Wydawnictwo Naukowe PWN [Björck, Å., Dahlguist, G. (1987). *Numerical methods.* Warsaw: Scientific Publishing House PWN].
30. Black, A. W. (1966). *The effect of circumferentially-varying boundary conditions on turbulent heat transfer in a tube,* Ph.D. Thesis. University of Minnesota, Minneapolis.
31. Blasius, P. R. H. (1913). Das Ähnlichkeitsgesetz bei Reibungsvorgängen in Flüssigkeiten. *VDI Forschungsheft, 131,* 1–41.
32. Boam, D., & Sattary, J. (2001). Uncertainty in flow measurement. In rozdział w Spitzer D. W. (Ed.), *Flow measurement,* ISA—The Instrumentation, Systems and Automation Society, Research Triangle Park, USA 2001, pp. 713–730.
33. Borkowski, B., Dudek H., & Szczesny W. (2004). *Ekonometria. Wybrane zagadnienia.* Warszawa: Wydawnictwo Naukowe PWN [Borkowski, B., Dudek, H., & Szczesny, W. (2004). *Econometrics. Selected problems.* Warsaw: Scientific Publishing House PWN].
34. Brandt, F. (1985). *Wärmeübertragung in Dampferzeugern und Wärmeaustauschern. Essen,* FDBR Fachverband Dampfkessel-, Behälter - und Rohrleitungsbau e.V.. Essen: Vulkan Verlag.
35. Brandt, S. (1998). *Analiza danych. Metody statystyczne i obliczeniowe.* Warszawa: Wydawnictwo Naukowe PWN [Brandt, S. (1998). *Data analysis. Statistical and computational methods.* Warszawa: Scientific Publishing House PWN].
36. Brower, W. B. (1999). *A Primer in fluid mechanics. Dynamics of flows in one space dimension.* Boca Raton: CRC Press.
37. Brown, G. M. (1960). Heat or mass transfer in a fluid in laminar flow in a circular or flat conduit. *AIChE Journal, 6,* 179–183.
38. Burden, R. L., & Faires, J. D. (1985). *Instructor's manual to accompany numerical analysis* (3rd ed.). Boston: PWS Publishers.
39. Burden, R. L., & Faires, J. D. (1985). *Numerical analysis* (3rd ed.). Boston: PWS Publishers.
40. Burmeister, L. C. (1993). *Convective heat transfer* (2nd ed.). New York: Wiley Interscience.
41. Buyukalaca, O., Ozceyhan, V., & Gunes, S. (2012). Experimental investigation of thermal performance in a tube with detached circular ring turbulators. *Heat Transfer Engineering, 33*(8), 682–692.
42. Carrigan, R. C., & Eichelberger, J. (2006). *Automotive heating and air conditioning.* Clifton Park, NY, USA: Thomson Delmar Learning.

43. Carslaw, H. S., & Jaeger, J. G. (2008). *Conduction of heat in solids* (2nd ed.). Oxford: Clarendon Press.

44. Cebeci, T. (1973). A model for eddy conductivity and turbulent Prandtl number. *ASME Journal of Heat Transfer, 95*, 227.

45. Cebeci, T. (2002). *Convective heat transfer* (2nd rev. ed.). Long Beach: Horizons Publishing and Berlin: Springer.

46. Cebeci, T. (2004). *Turbulence models and their application.* Long Beach: Horizons Publishing and Berlin: Springer.

47. Çengel, Y. A., & Turner, R. H. (2001). *Fundamentals of thermal-fluid sciences.* Boston: McGraw-Hill International Edition.

48. Cess, R. D., & Shaffer, E. C. (1959). Heat transfer to laminar flow between parallel plates with a prescribed wall heat flux. *Applied Scientific Research, Section A, 8*, 339–344.

49. Chen, C. J., & Chiou, J. S. (1981). Laminar and turbulent heat transfer in the pipe entrance region for liquid metals. *International Journal of Heat and Mass Transfer, 24*(7), 1179–1189.

50. Chen, N. H. (1979). An explicit equation for friction factor in pipe. *Industrial and Engineering Chemistry Fundamentals, 18*(3), 296–297.

51. Cheng, N. S. (2008). Formulas for friction factor in transitional regions. *Journal of Hydraulic Engineering, ASCE, 134*(9), 1357–1362.

52. Cheng, X., & Tak, N.-I. (2006). Investigation on turbulent heat transfer to lead-bismuth eutectic flows in circular tubes for nuclear applications. *Nuclear Engineering and Design, 236*, 385–393.

53. Ching-Jen, C., & Jenq, S. C. (1981). Laminar and turbulent heat transfer in the pipe entrance region for liquid metals. *International Journal of Heat and Mass Transfer, 24*(7), 1179–1189.

54. Chow, C. C. (1983). *Econometric methods.* New York: McGraw-Hill.

55. Churchill, S. W. (1973). Empirical expressions for the shear stress in turbulent flow in commercial pipe. *American Institute of Chemical Engineering Journal, 19*(2), 375–376.

56. Churchill, S. W. (1977). Friction-factor equation spans all fluid-flow regimes. *Chemical Engineering, 84*, 91–92.

57. Churchill, S. W., & Ozoe, H. (1973). Correlations for laminar forced convection in flow over an isothermal flat plate and in developing and fully developed flow in an isothermal tube. *ASME Journal of Heat Transfer, 95*, 416–419.

58. Churchill, S. W., & Usagi, R. (1972). A general expression for the correlation of rates of transfer and other phenomena. *AIChE Journal, 18*, 1121–1128.

59. Colburn, A. P. (1933). A method of correlating forced convection heat transfer data and a comparison with fluid friction. *Transactions of the American Institute of Chemical Engineers, 29*, 174–210.

60. Colebrook, C.F. (1938–1939). Turbulent flow in pipes, with particular reference to the transition region between the smooth and rough pipe laws. *Journal of the Institution of Civil Engineers, 11*, 133–156.

61. Colebrook, C. F., & White, C. M. (1937). Experiments with fluid friction in roughened pipes. *Proceedings of the Royal Society of London, 161*, 367–381.

62. Coleman, H. W., & Steele, W. G. (1998). Uncertainty analysis (Chapter 39). In R. W. Johnson (Ed.), *The handbook of fluid dynamics.* Boca Raton, Heidelberg: CRC Press i Springer.

63. Coleman, H. W., & Steele, W. G. (2009). *Experimentation, validation, and uncertainty analysis for engineers* (3rd ed.). New York: Wiley.

64. Coles, D. (1955). The law of the wall in turbulent shear flow (Chapter 50). In H. Görtler & W. Tollmien (Eds.), *Jahre Grenzschichtforschung* (pp. 153–163). Braunschweig: Vieweg.

65. Cornwell, K. (1977). *The flow of heat.* Wokingham, Berkshire, England: Van Nostrand Reinhold Company Ltd.

66. Coulson, J. M., Richardson, J. F., Backhurst, J. R., & Harker, J. H. (2000). Fluid flow, heat transfer and mass transfer. In *Coulson and Richardson's chemical engineering* (6th ed., vol. 1). Oxford: Butterworth-Heinemann.

67. Crowe, C. T., Elger, D. F., & Roberson, J. A. (2005). *Engineering fluid mechanics* (8th ed.). Hoboken: Wiley.

68. Cuevas, C., Makaire, D., Dardenne, L., & Ngendakumana, P. (2011). Thermo-hydraulic characterization of a louvered fin and flat tube heat exchanger. *Experimental Thermal and Fluid Science, 35*, 154–164.

69. Cutlip, M. B., & Shacham, M. (1999). *Problem solving in chemical engineering with numerical methods.* Upper Saddle River: Prentice Hall PTR.

70. Czermiński, J. B., Iwasiewicz, A., Paszek, Z., Sikorski, A. (1992). *Metody statystyczne dla chemików,* Wydanie drugie zmienione. Warszawa: Wydawnictwo Naukowe PWN [Czermiński, J. B., Iwasiewicz, A., Paszek, Z., & Sikorski, A. (1992). *Statistical methods in chemistry* (2nd ed.). Warsaw: Scientific Publishing House PWN].

71. Daly, S. (2006). *Automotive air-conditioning and climate control systems.* Amsterdam: Elsevier.

72. Date, A. W. (2005). *Introduction to computational fluid dynamics.* New York: Cambridge University Press.

73. Davidson, J. (2000). *Econometric theory.* Oxford: Blackwell.

74. Davidson, R., & MacKinnon, J. G. (1993). *Estimation and inference in econometrics.* Oxford: Oxford University Press.

75. Deissler, R. G. (1950). Analytical and experimental investigation of adiabatic turbulent flow in smooth tubes. *NACA Technical Notes, No. 2138.*

76. Deissler, R. G. (1955). *Analysis of turbulent heat transfer, mass transfer, and friction in smooth tubes at high Prandtl and Schmidt numbers.* NACA Report 1210.

77. Diaz-Mendez, S. E., Patiño-Carachure, C., & Herrera-Castillo, J. A. (2014). Reducing the energy consumption of an earth-air heat exchanger with a PID control system. *Energy Conversion and Management, 77*, 1–6.

78. Dieterich Company Catalogue, Standard annubar-flow measurement systems. An ISO 9001 Certified Manufacturer, March 1995.

79. Dittus, F. W., Boelter, L. M. K. (1985). *Heat transfer in automobile radiators of the tubular type,* University of California Publications on Engineering 2, 1930, s. 443–461. (Reprinted in: *International Communications on Heat and Mass Transfer, 12*, pp. 3–22, 1985).

80. Dixon, S. L. (1998). *Fluid mechanics and thermodynamics of turbomachinery* (4th ed.). Boston: Butterworth-Heinemann.

81. Dormand, J. R. (1996). *Numerical methods for differential equations. A computational approach.* Boca Raton: CRC Press.

82. Draper, N. R., & Smith, H. (1998). *Applied regression analysis.* New York: Wiley.

83. Duda, P., Taler, J., & Roos, E. (2004). Inverse method for temperature and stress monitoring in complex-shaped bodies. *Nuclear Engineering and Design, 227*, 331–347.

84. Dukelow, S. G. (1991). *The control of boilers.* Instrument Society of America 67 Alexander Drive, P.O. Box 12277, Research Triangle Park, NC 27709, 1991.

85. Eckert, E. R. G., & Drake, R. M. Jr. (1972). *Analysis of heat and mass transfer.* Routledge: Taylor & Francis.

86. Eiamsa-Ard, S., Kongkaitpaiboon, V., & Promvonge, P. (2011). Thermal performance assessment of turbulent tube flow through wire coil turbulators. *Heat Transfer Engineering, 32*(11–12), 957–967.

87. *Ekonometria. Zbiór zadań,* praca zbiorowa pod redakcją Welfe A., Wydanie trzecie zmienione, Polskie Wydawnictwo Ekonomiczne, Warszawa 2003 [Welfe A. (ed.), *Econometrics. Excercises,* 3rd ed., Warsaw: Polish Publishing House for Economy 2003].

88. ESDU (Engineering Sciences Data Units) International plc, *Forced convection heat transfer in circular tubes. Part 1: correlation for fully developed flow,* ESDU Item 67016.

89. Fang, X., Xu, Y., & Zhou, Z. (2011). New correlations of single-phase friction factor for turbulent pipe flow and evaluation of existing single-phase friction factor correlations. *Nuclear Engineering and Design, 241,* 897–902.

90. Favre-Marinet, M., & Tardu, S. (2008). *Convective heat transfer.* Hoboken: Wiley.

91. Figliola, R. S., & Beasley, D. E. (2000). *Theory and design for mechanical measurements* (3rd ed.). New York: Wiley.

92. Filonienko, G. K. (1954). Friction factor for turbulent pipe flow. *Teploenergetika, 1*(4), 40–44.

93. Spitzer, D. W. (ed.), (2001). *Flow measurement. Practical guides for measurement and control.* Reseach Triangle Park: ISA.

94. (1996). *Fluid mechanics. REA's problem solvers.* Piscataway, New Jersey: Research and Education Association.

95. Forsythe, G. E., Malcolm, M. A., & Moler, C. B. (1977). *Computer methods for mathematical computations.* Englewood Cliffs, New Jersey: Prentice-Hall.

96. Gajda, J. B. (2004). *Ekonometria.* Warszawa: Wydawnictwo C.H.Beck [Gajda, J. B. (2004). *Econometrics.* Warsaw: Publishing House C.H.Beck].

97. Ganapathy, V. (2003). *Industrial boilers and heat recovery steam generators. Design, applications, and calculations.* Boca Raton: CRC Press.

98. Genić, S., Arandjelović, I., Kolendić, P., Jarić, M., Budimir, N., & Genić, V. (2011). A review of explicit approximations of Colebrook's equation. *FME Transactions, 39,* 67–71.

99. Gerald, C. F., & Wheatley, P. O. (1994). *Applied numerical analysis* (5th ed.). Reading, Massachusetts: Addison-Wesley.

100. Ghajar, A. J., & Madon, K. F. (1992). Pressure drop measurements in the transition region for a circular tube with three different inlet configurations. *Experimental Thermal and Fluid Science, 5,* 129–135.

101. Ghajar, A. J., Tam, L. M., & Tam, S. C. (2004). Improved heat transfer correlation in the transition region for a circular tube with three inlet configurations using artificial neural networks. *Heat Transfer Engineering, 25*(2), 30–40.

102. Ghiaasiaan, S. M. (2014). *Convective heat and mass transfer* (2nd ed.). Cambridge: Cambridge University Press.

103. Glockner, P. S., & Naterer, G. F. (2005). Near-wall velocity profile with adaptive shape functions for turbulent forced convection. *International Communications in Heat and Mass Transfer, 32,* 72–79.

104. Gnielinski, V. (1975). Neue Gleichungen für den Wärme- und den Stoffübergang in turbulent durchströmten Rohren und Kanälen. *Forschung im Ingenieurwesen, 41*(1), 8–16.

105. Gnielinski, V. (1995). Ein neues Berechnungsverfahren für die Wärmeübertragung im Übergangsbereich zwischen laminarer und turbulenter Rohrströmung. *Forschung im Ingenieurwesen-Engineering Research, 61*(9), 240–248.

106. Gnielinski, V. (2010). Heat transfer in concentric annular and parallel plate ducts. In: *Rozdział G2 w VDI Heat Atlas* (2nd ed.). Berlin: Springer, pp. 701–708.

107. Gnielinski, V. (2013). Durchströmte Rohre (Chapter G1). In *VDI-Wärmeatlas 11., bearbeitete und erweiterte Auflage.* Berlin: Springer, pp. 785–791.

108. Gnielinski, V. (2013). On heat transfer in tubes. *International Journal of Heat and Mass Transfer, 63,* 134–140.

109. Gnielinski, V., Žukauskas, A., & Skrinska, A. (1992). Banks of plain and finned tubes (Chapter 2.5.3). In G. F. Hewitt (ed.), *Handbook of heat exchanger design.* New York: Begell House, pp. 2.5.3–1–2.5.3–16.

110. Graebel, W. P. (2007). *Advanced fluid mechanics.* Amsterdam: Elsevier-Academic Press.

111. Graetz, L. *Über die Wärmeleitungsfähigkeit von Flüssigkeiten,* Part 1, Annals of Physical Chemistry (pp. 79–94). 18, 1883; Part 2, Annals of Physical Chemistry (pp. 337–357). 25, 1885.

112. Greenberg, M. D. (1988). *Advanced engineering mathematics* (2nd ed.). Upper Saddle River, New Jersey: Prentice Hall.
113. Greene, W. H. (2000). *Econometric analysis.* Upper Saddle River, New Jersey: Prentice Hall.
114. Grote, K., Meyer, J. P., & McKrell, T. (2013) *The influence of multi-walled carbon nanotubes on the pressure drop characteristics in the transitional flow regime of smooth tubes.* Poster Session 1, Paper 489 w *Proceedings of the 8th World Conference on Experimental Heat Transfer, Fluid Mechanics and Thermodynamics (ExHFT-8), June 16–20, 2013.* Lisbon, Portugal.
115. Grötzbach, G. (2013). Challenges in low-Prandtl number heat transfer simulation and modelling. *Nuclear Engineering and Design, 264,* 41–55.
116. *Guide to the expression of uncertainty in measurement* (1995). Geneva: International Organization for Standardization.
117. Haaland, S. E. (1983). Simple and explicit formulas for the friction factor in turbulent pipe-flow. *Transactions of the ASME, Journal of Fluids Engineering, 105*(1), 89–90.
118. Han, X., & Liu, G. R. (2003). *Computational inverse technique in nondestructive evaluation.* Boca Raton: CRC Press.
119. Hausen, H. (1959). Neue Gleichungen für die Wärmeübertragung bei freier oder erzwungener Strömung. *Allgemeine Wärmetechnik, 9*(4/5), 75–79.
120. Hausen, H. (1974). Erweiterte Gleichungen für den Wärmeübertragung in Rohren bei turbulenter Strömung. *Wärme- und Stoffübertragung, 4.*
121. Hausen, H. (1983). *Heat transfer in counter flow, parallel flow, and cross-flow.* New York: McGraw-Hill.
122. Hewitt, G. F. (Ed.). (1992). *Handbook of heat exchanger design.* New York: Begell House.
123. Hewitt, G. F., Shires, G. L., & Bott, T. R. (1994). *Process heat transfer.* Boca Raton: CRC Press i Begell House.
124. Hickman, W. H. (1975). Annubar properties investigation. In *Proceedings of ISA's Industry Conference and Exhibit, Milwaukee, October 6–9, 1975,* pp. 1–14. Instrument Society of America 30, Part 3.
125. Hobler, T. (1986). *Ruch ciepła i wymienniki.* Warszawa: Wydanie szóste, WNT [Hobler, T. (1986). *Heat transfer and heat exchangers* (6th ed.). Warsaw: WNT].
126. Hodge, B. K., & Taylor, R. P. (1999). *Analysis and design of energy systems* (3rd ed.). Upper Saddle River: Prentice Hall.
127. Holman, J. P. (2001). *Experimental methods for engineers* (7th ed.). Boston: McGraw-Hill.
128. Holman, J. P. (2010). *Heat transfer* (10th ed.). Boston: McGraw Hill.
129. Hsu, C. J. (1965). Heat transfer in a round tube with sinusoidal wall heat flux distribution. *AIChE Journal, 11,* 690–695.
130. Huber, D., & Walter, H. (2010). Forced convection heat transfer in the transition region between laminar and turbulent flow for a vertical tube. In Mastorakis, N., Mladenov, V., & Bojkovic, Z. (eds.), *Latest trends on theoretical and applied mechanics, fluid mechanics and heat transfer* (pp. 132–136). Wisconsin: WSEAS Press.
131. Hughes, W. F., & Brighton, J. A. (1999). *Fluid dynamics* (3rd ed., Schaum`s Outline Series). New York: McGraw-Hill.
132. Incropera, F. P., & DeWitt, D. P. (1996). *Fundamentals of heat and mass transfer* (4th ed.). New York: Wiley.
133. Ingham, J., Dunn, I. J., Heinzle, E., Přenosil, J. E., & Snape, J. B. (2007). *Chemical engineering dynamics. An introduction to modelling and computer simulation.* Weinheim: Wiley VCH.
134. International Mathematical and Scientific Library IMSL Math/Library, IMSL Inc., 2500 CityWest Boulevard, Houston TX 77042.
135. *International vocabulary of basic and general terms in metrology* (1993). Geneva: International Organization for Standarization.
136. Inverse engineering handbook (2003). In Woodbury K. A. (ed.). Boca Raton: CRC Press.

137. Jaluria, Y. (1980). *Natural convection. Heat and mass transfer.* Oxford: Pergamon Press.
138. Janke, E., Emde, F., & Lösch, F. (1960). *Tafeln höherer Funktionen.* Stuttgart: Teubner.
139. Ji, W.-T., Zhang, D.-C., He, Y.-L., & Tao, W.-Q. (2012). Prediction of fully developed turbulent heat transfer of internal helically ribbed tubes—An extension of Gnielinski equation. *International Journal of Heat and Mass Transfer, 55,* 1375–1384.
140. Jischa, M. (1982). *Konvektiver Impuls-, Wärme- und Stoffaustausch.* Braunschweig/Wiesbaden: Friedr. Vieweg & Sohn.
141. Jischa, M., & Rieke, H. B. (1979). About the prediction of turbulent Prandtl and Schmidt numbers. *International Journal of Heat and Mass Transfer, 22,* 1547–1555.
142. Kakaç, S., & Yener, Y. (1995). *Convective heat transfer* (2nd ed.). Boca Raton: CRC Press.
143. Kakaç, S., & Liu, H. (2002). *Heat exchangers, selection, rating, and thermal design* (2nd ed.). Boca Raton: CRC Press.
144. Kandlikar, S. G., Garimella, S., Li, D., Colin, S., & King, M. R. (2013). *Heat transfer and fluid flow in minichannels and microchannels* (2nd ed.). Amsterdam: Elsevier.
145. von Kármán, Th. (1939). The analogy between fluid friction and heat transfer. *Transactions of ASME, 61,* s. 705–710 [von Kármán, Th. (1930). *Mechanische Ähnlichkeit und Turbulenz,* Ges. der Wiss. zu Gött., Nachrichten, Math.-Phys. Kl., pp. 58–76].
146. Kast, W., & Nirschl, H. (2013). *Druckverlust in durchströmten Rohren,* Rozdział L1.2 w VDI-Wärmeatlas, 11., bearbeitete und erweiterte Auflage. Berlin: Springer.
147. Kays, W. (1994). Turbulent Prandtl Number—Where are we? *Transactions of the ASME, Journal of Heat Transfer, 116,* 284–295.
148. Kays, W., Crawford, M., & Weigand, B. (2005). *Convective heat and mass transfer* (4th ed.). Boston: McGraw-Hill.
149. Kays, W. M., & London, A. L. (1998). *Compact heat exchangers* (3rd ed.). Malabar, Florida, USA: Krieger.
150. Kemink, R. (1997). *Heat transfer in a downstream tube of a fluid withdrawal branch (Ph. D. Thesis).* Minneapolis: University of Minnesota.
151. Kirillov, P. L., Yuriev, Y. S., & Bobkov, W. P. (1984). *Handbook of hydraulic calculations (Nuclear reactors, heat exchangers, steam generators).* Moscow: Energoatomizdat.
152. Kiusalaas, J. (2005). *Numerical methods in engineering with MATLAB®.* New York: Cambridge University Press.
153. Kline, S. J., & McClintock, F. A. (1955). Describing uncertainties in single-sample experiments. *Mechanical Engineering, 75,* 3–8.
154. Kline, S. J. (1985). Symposium on uncertainty analysis closure. *ASME Journal of Fluids Engineering, 107*(1983), 181–182.
155. Kline, S. J. (1985). The purposes of uncertainty analysis. *ASME Journal of Fluids Engineering, 107*(1983), 153–160.
156. Koenig, D. M. (2009). *Practical control engineering.* New York: McGraw-Hill.
157. Konakov, P. K. (1946). A new correlation for the friction coefficient in smooth tubes. *Izvestija AN SSSR, 51*(7), 503–506.
158. Korzeń, A., & Taler, D. (2015). Modeling of transient response of a plate fin and tube heat exchanger. *International Journal of Thermal Sciences, 92,* 188–198.
159. Kraus, A. D., Aziz, A., & Welty, J. (2001). *Extended surface heat transfer.* New York: Wiley.
160. Kreith, F., Manglik, R. M., & Bohn, M. S. (2011). *Principles of heat transfer* (7th ed.). Stamford, USA: Cengage Learning.
161. Kröger, D. G. (1985). *Radiator characterization and optimization.* SAE Paper 840380, pp. 2.984–2.990.
162. Kröger, D. G. (1998). *Air-cooled heat exchangers and cooling towers.* Matieland, South Africa: University of Stellenbosch.
163. Kruger, W. (1972). Theorie der Volumenstrommessung in Rohren mit Rechteckquerschnitt. *HLH, 23*(4), 121–123.

164. Kundu, B., & Das, P. K. (2000). Performance of symmetric polygonal fins with and without tip loss—a comparison of different methods of prediction. *The Canadian Journal of Chemical Engineering, 78,* 395–401.

165. Kundu, B., & Das, P. K. (2009). Performance and optimum dimensions of flat fins for tube-and-fin heat exchangers: A generalized analysis. *International Journal of Heat and Fluid Flow, 30,* 658–668.

166. Kundu, B., Maiti, B., & Das, P. K. (2006). Performance analysis of plate fins circumscribing elliptic tubes. *Heat Transfer Engineering, 27,* 86–94.

167. Kundu, P. K., & Cohen, I. M. (2008). *Fluid mechanics* (4th ed.). Amsterdam: Elsevier.

168. Kunze, H. J. (1986). *Physikalische Messmethoden. Eine Einführung in Prinzipien klassischer und moderner Verfahren.* Stuttgart: B. G. Teubner.

169. Kuppan, T. (2013). *Heat exchanger design handbook* (2nd ed.). Boca Raton: CRC Press Taylor & Francis Group.

170. Langtry, R. B., & Menter, F. R. (2005). *Transition modeling for general CFD applications in aeronautics.* AIAA 2005, paper 552.

171. Larkin, B. K. (1961). High-order eigenfunctions of the Graetz problem. *AIChE Journal, 7* (191), 530.

172. Lassahn, G. D. (1985). Uncertainty definition. *ASME Journal of Fluids Engineering, 107,* 179.

173. Lau, S. (1981). *Effect of plenum length and diameter of turbulent heat transfer in a downstream tube and on plenum-related pressure loss (Ph.D. Thesis).* Minneapolis: University of Minnesota.

174. Lawson, Ch L, & Hanson, R. J. (1974). *Solving least squares problems.* Englewood Cliffs: Prentice-Hall.

175. Lee, T.-W. (2008). *Thermal and flow measurements.* Boca Raton: CRC Press Taylor & Francis Group.

176. Lévêque, M. A. (1928). *Les lois de la transmission de chaleur par convection.* Annales des Mines, Memoires, Series 12, 13, pp. 201–299, 305–362, 381–415.

177. Lewis, P. E., & Ward, J. P. (1991). *The finite element method. Principles and applications.* Wokingham, England: Addison-Wesley Publishing Company.

178. Li, C. H. (1987). A new simplified formula for cross-flow heat exchange effectiveness. *Transactions of the ASME, Journal of Heat transfer, 109,* 521–522.

179. Li, Q., & Xuan, Y. (2002). Convective heat transfer and flow characteristics of Cu-water nanofluid. *Science in China Series E: Technolgical Science, 45*(4), 408–416.

180. Li, X.-W., Meng, J.-A., & Guo, Z.-Y. (2009). Turbulent flow and heat transfer in discrete double inclined ribs tube. *International Journal of Heat and Mass Transfer, 52,* 962–970.

181. Li, X.-W., Meng, J.-A., & Li, Z.-X. (2007). Experimental study of single-phase pressure drop and heat transfer in a micro-fin tube. *Experimental Thermal and Fluid Science, 32,* 641–648.

182. Lienhard, J. H. V., & Lienhard, J. H., IV. (2011). *A Heat transfer textbook* (4th ed.). Mineola, New York: Dover Publications Inc.

183. Lingfield, G., & Penny, J. (1995). *Numerical methods using Matlab.* New York: Ellis Horwood.

184. Liu, Z. H., & Liao, L. (2010). Forced convective flow and heat transfer characteristics of aqueous drag-reducing fluid with carbon nanotubes added. *International Journal of Thermal Sciences, 49,* 2331–2338.

185. Ludowski, P., Taler, D., & Taler, J. (2013). Identification of thermal boundary conditions in heat exchangers of fluidized bed boilers. *Applied Thermal Engineering, 58,* 194–204.

186. Lyon, R. N. (1951). Liquid metal heat transfer coefficients. *Chemical engineering progress, 47*(2), 75–79.

187. Madejski, J. (1996). *Teoria wymiany ciepła*. Wydawnictwo Uczelniane, Szczecin: Zachodniopomorski Uniwersytet Technologiczny w Szczecinie [Madejski, J. (1996). *Theory of heat transfer*. Szczecin: Publishing House of West Pomerania University in Szczecin].

188. *Mały poradnik mechanika, Tom II*, Wydanie 18 poprawione i uaktualnione, WNT, Warszawa 1994 [*Small handbook for mechanical engineers* (volume II, 18th ed.). Warsaw: Publishing House WNT].

189. Manadilli, G. (1997). Replace implicit equations with sigmoidal functions. *Chemical Engineering, 104*(8), 187.

190. Martin, H. (1990). *Vorlesung Wärmeübertragung I*. Karlsruhe: Universität Karlsruhe (TH).

191. Martin, H. (1996). A theoretical approach to predict the performance of chevron-type plate heat exchangers. *Chemical Engineering and Processing, 35*, 301–310.

192. Martin, H. (2002). *The generalized Lévêque equation and its use to predict heat or mass transfer from fluid friction*. In Invited Lecture, *Proceedings 20th National Heat Conference, UIT, Maratea, Italy, 27–29 June 2002*, pp. 21–29.

193. Martin, H. (2002). The generalized Lévêque equation and its practical use for the prediction of heat and mass transfer rates from pressure drop. *Chemical Engineering Science, 57*, 3217–3223.

194. Mathpati, C. S., & Joshi, J. B. (2007). Insight into theories of heat and mass transfer at the solid-fluid interface using direct numerical simulation and large eddy simulation. *Industrial & Engineering Chemistry Research, 46*, 8525–8557.

195. MATLAB online documentation. http://www.mathworks.com/help/matlab.

196. McAdams, W. H. (1954). *Heat transmission* (3rd ed.). New York: McGraw-Hill.

197. McQuiston, F. C., Parker, J. D., & Spitler, J. D. (2005). *Heating, ventilating, and air conditioning. Analysis and design*. Hoboken: Wiley.

198. Menter, F. R. (1994). Two-equation eddy-viscosity turbulence models for engineering applications. *AIAA Journal, 32*(8), 1598–1605.

199. Mercer, A. M. (1960). The growth of the thermal boundary layer at the inlet to a circular tube. *Applied Scientific Research, Section A, 9*, 450–456.

200. *Method of testing air-to-air heat exchangers*, ASHRAE STANDARD, ANSI/ASHRAE 84-1991, American Society of Heating, Refrigerating and Air-Conditioning Engineers, 1791 Tullie Circle, Atlanta 1992.

201. Mikielewicz, J. (1994). *Modelowanie procesów cieplno-przepływowych*, Tom 17, Wydawnictwo Instytutu Maszyn Przepływowych PAN, Gdańsk [Mikielewicz, J. (1994). *Modeling of flow-thermal processes* (vol. 17). Gdańsk: Publishing House of the Institute of Fluid Flow Machinery, Polish Academy of Sciences].

202. Mills, A. F. (1999). *Basic heat & mass transfer*. Upper Saddle River, New Jersey: Prentice Hall.

203. Mirth, D. R., Ramadhyani, S., & Hittle D. C. (1993). Thermal performance of chilled-water cooling coils at low water velocities. *ASHRAE Transactions, 99*(Part 1), 43–53.

204. Mirth, D. R., & Ramadhyani, S. (1994). Correlations for predicting the air-side Nusselt numbers and friction factors in chilled-water cooling coils. *Experimental Heat Transfer, 7*, 143–162.

205. Moffat, R. J. (1982). Contributions to the theory of single-sample uncertainty analysis. *ASME Journal of Fluids Engineering, 104*, 250–260.

206. Moffat, R. J. (1985). Using Uncertainty analysis in the planning of an experiment. *ASME Journal of Fluids Engineering, 107*, 173–178.

207. Moffat, R. J. (1988). Describing the uncertainties in experimental results. *Experimental Thermal and Fluid Science, 1*, 3–17.

208. Moody, L. F. (1944). Friction factors for pipe flow. *Transactions of the American Society of Mechanical Engineers, 66*, 671–684.

209. Muzychka, Y. S., & Yovanovich, M. M. (2004). Laminar forced convection heat transfer in the combined of non-circular ducts. *Transactions of the ASME, Journal of Heat Transfer, 126*, 54–60.

210. Nellis, G., & Klein, S. (2009). *Heat transfer*. New York: Cambridge University Press.

211. Newman, J. (1969). Extension of the Lévêque solution. *ASME Journal of Heat Transfer, 91*, 177–178.

212. Nielsen, A. A. (2013). *Least squares adjustment: Linear and nonlinear weighted regression analysis* (pp. 1–52). http://www.imm.dtu.dk/alan. September 19, 2013.

213. Nikuradse, J. (1932). Gesetzmäßigkeit der turbulenten Strömung in glatten Rohren. *Forschung im Ingenieurwesen, 356* (NASA TT F-IO 359, 1966, Trans.).

214. Nikuradse, J. *Strömungsgesetze in rauhen Rohren*, VDI Forschungsheft 361, Beilage zu Forsch. Arb. Ing. Wes., Ausgabe B, Band 4, Juli/August 1933 VDI Verlag, Berlin (*Laws of flow in rough pipes*, NACA TM 1292, 1950, Trans.).

215. Nowak, E. (2002). *Zarys metod ekonometrii*. Wydanie trzecie poprawione, Warszawa: Wydawnictwo Naukowe PWN [Nowak, E. (2002). *Introduction to econometrics methods* (3rd ed.). Warsaw: Scientific Publishing House PWN].

216. Nunge, R. J., Porta, E. W., & Bentley, R. (1970). A correlation of local Nusselt numbers for laminar flow heat transfer in annuli. *International Journal of Heat and Mass Transfer, 13*, 927–931.

217. Nusselt, W. (1930). Eine neue Formel für den Wärmedurchgang im Kreuzstrom. *Technische Mechanik und Thermodynamik, 1*, 417–42.

218. Olander, D. R. (2008). *General thermodynamics*. Boca Raton: CRC Press Taylor&Francis Group.

219. Oleśkowicz-Popiel, C., & Wojtkowiak, J. (2007). *Eksperymenty w wymianie ciepła* (Wydanie drugie rozszerzone). Poznań: Wydawnictwo Politechniki Poznańskiej [Oleśkowicz-Popiel, C., & Wojtkowiak, J. (2007). *Experiments in heat transfer* (2nd ed.). Poznań: Publishing House of Poznań University of Technology].

220. Olivier, J. A., & Meyer, J. P. (2010). Single-phase heat transfer and pressure drop of the cooling water inside smooth tubes for transitional flow with different inlet geometries. *HVAC&R Research, 16*(4), 471–496.

221. Oosthuizen, P. H., & Naylor, D. (1999). *Introduction to convective heat transfer analysis*. New York: WCB-McGraw-Hill.

222. Orłowski, P., Dobrzański, W., Szwarc, E. (1979). *Kotły parowe, konstrukcja i obliczenia* (Wydanie trzecie). Warszawa: WNT [Orłowski, P., Dobrzański, W., & Szwarc E. (1979). *Steam boilers, design and calculations* (3rd ed.). Warsaw: Publishing house WNT].

223. Özisik, M. N., & Orlande, H. R. B. (2000). *Inverse heat transfer*. Fundamentals and Applications: Taylor & Francis, New York.

224. Peng, W. W. (2008). *Fundamentals of turbomachinery*. Hoboken: Wiley.

225. Petchers, N. (2012). *Combined heating, cooling & power handbook. Technologies & applications* (2nd ed.). Lilburn: The Fairmont Press.

226. Pietukhov, B. S. (1970). Heat transfer and friction in turbulent pipe flow with variable physical properties. In praca w Hartnett J. P., Irvine T.F. (eds.), *Advances in heat transfer* (vol. 6, pp. 503–564). New York: Academic Press.

227. Петухов, Б. С., Генин, Л. Г., & Ковалев, С. А. (1974). *Теплообмен в ядерных энергетических установках*. Москва: Атомиздат [Pietukhov, B. S., Genin, L. G., & Kovalev, S.A. (1974). *Heat transfer in nuclear energy installations*. Moskwa: Atomizdat].

228. Петухов, Б. С., Кириллов, В. В. (1958). *К вопросу о теплообмене при турбулентном течении жидкости в трубах*, Теплоэнергетика 5(4), s. 63–68 [Petukhov, B. S., & Kirillov, V. V. (1958). The problem of heat exchange in the turbulent flow of liquids in tubes. *Teploenergetika, 5*(4), 63–68].

229. Петухов, Б. С., & Курганов, В. А., Гладунцов, А. И. (1972). Теплообмен в трубах при турбулентном течении газов с переменными свойствами, в сб. *Тепло-и массоперенос, 1*, ч. 2, 117 [Pietukhov, B. S., Kurganov, V. A., Gladuncov, A. I. (1972). Heat transfer in turbulent flow in tubes of gases with variable properties. *Heat and mass transfer, 1*, Part 2, 1972, 117).

230. Петухов, Б. С., & Попов, В. Н. (1963). Теоретический расчет теплообмена и сопротивления трения при турбулентном течении в трубах несжимаемой жидкости с переменными физическими свойствами. *Теплофизика высоких температур, 1*(1), 69–83 [Pietukhov, B. S., Popov V. N., Theoretical calculation of heat transfer in turbulent flow in tubes of an inompressible fluid with variable properties. *High Temperature Physics, 1*(1), 69–83].

231. Pohlhausen, E. (1921). Wärmeaustausch zwischen festen Körpern und Flüssigkeiten mit kleiner Wärmeleitung. *Zeitschrift für Angewandte Mathematik und Mechanik, 1*(2), 115–121.

232. (2000). Policy on reporting uncertainties in experimental measurements and results. *ASME Journal of Heat Transfer, 122*, 411–413.

233. (1973). *Poradnik inżyniera. Matematyka.* Warszawa: WNT [(1973). *Hanbook for Engineers. Mathematics.* Warsaw: Publishing House WNT].

234. Prandtl, L. (1910). Eine Beziehung zwischen Wärmeaustausch und Strömungswiderstand der Flüssigkeit. *Z. Physik, 11*, 1072–1078.

235. Prandtl, L. (1925). Über die ausgebildete Turbulenz. *Zeitschrift für angewandte Mathematik und Mechanik, 5*, 136–139.

236. Prandtl, L. (1949). *Führer durch die Strömungslehre*, Vieweg und Sohn, Braunschweig [Prandtl, L. (1949). *Essentials of fluid dynamics.* London: Blackie & Son, 117, Trans.].

237. Prandtl, L. (1961). *Gesammelte Abhandlungen.* In W. Tollmien, H. Schlichting, & H. Görtler (eds.), *Band 2.* Berlin: Springer.

238. Prandtl, L. (1905). *Über Flüssigkeitsbewegung bei sehr kleiner Reibung.* Lepzig: Verhandlungen des III. Internationalen Mathematiker- Kongresses.

239. Press, W. H., Teukolsky, S. A., Vetterling, W. T., & Flannery, B. P. (2006). *Numerical recipes in Fortran 77, The art of scientific computing.* Cambridge: Cambridge University Press.

240. Purtell, L. P., Klebanoff, P. S., & Buckley, F. T. (1981). Turbulent boundary layer at low Reynolds number. *The Physics of Fluids, 24*, 802–811.

241. Rabas, T. J., & Taborek, J. (1987). Survey of turbulent forced convection heat transfer and pressure drop characteristics of low finned tube banks in crossflow. *Heat Transfer Engineering, 8*, 49–61.

242. *Rachunek wyrównawczy w technice cieplnej*, Praca zbiorowa pod redakcją J. Szarguta, Wydawnictwo PAN, Zakład Narodowy imienia Ossolińskich, Wrocław 1984 (*Data reconciliation in thermal engineering*, Szargut J., Publishing House of Polish Academy of Sciences PAN, Ossoliński Publishing House, Wrocław 1984).

243. Reichardt, H. (1940). *Die Wärmeübertragung in turbulenten Reibungsschichten*, Zeitschrift für angewandte Mathematik und Mechanik 20, pp. 297 (tłumaczenie na język angielski: *Heat transfer through turbulent friction layers*, NACA TM1047, 1943).

244. Reichardt, H. (1944). Impuls-und Wärmeaustausch in freier Turbulenz. *Zeitschrift für angewandte Mathematik und Mechanik, Bd, 24*, 268.

245. Reichardt, H. (1951). Vollständige Darstellung der turbulenten Geschwindigkeitsverteilung in glatten Leitungen. *Zeitschrift für angewandte Mathematik und Mechanik, Bd, 31*(7), 208–219.

246. Rennels, D. C., & Hudson, H. M. (2012). *Pipe flow. A practical and comprehensive guide.* Hoboken: AIChE, Wiley.

247. Rennie, T. J., & Raghavan, V. G. S. (2005). Experimental studies of a double-pipe helical heat exchanger. *Experimental Thermal and Fluid Science, 29*, 919–924.

248. Reynolds, O. (1874). On the extent and action of the heating surface for steam boilers. *Proceedings of the Literary and Philosophical Society, 14,* 7–12.
249. Rich, D. G. (1966). The efficiency of thermal resistance of annular and rectangular fins. In Proceedings of the Third International Heat Transfer Conference (vol. III, pp. 281–289).
250. Riley, K. F., Hobson, M. P., & Bence, S. J. (1998). *Mathematical methods for physics and engineering.* Cambridge: Cambridge University Press.
251. Roffel, B., & Betlem, B. (2006). *Process dynamics and control, modeling for control and prediction.* Chichester, England: Wiley.
252. Rose, J. W. (2004). Heat-transfer coefficients, Wilson plots and accuracy of thermal measurements. *Experimental Thermal and Fluid Science, 28,* 77–86.
253. Sacharczuk, J., & Taler, D. (2013). Wykorzystanie kształtek betonowych lub ceramicznych do magazynowania ciepła pozyskiwanego w instalacji solarnej. *Rynek Energii, 107,* 37–42 [Sacharczuk, J., & Taler, D. (2013). Applying of concrete or ceramic elements for heat storage in sollar installation. *Rynek Energii, 107,* 37–42].
254. Sacharczuk, J., & Taler, D. (2014). A concrete heat accumulator for use in solar heating systems - a mathematical model and experimental verification. *Archives of Thermodynamics, 35*(3), 281–295.
255. Schenk, J., & Beckers, H. L. (1954). Heat transfer in laminar flow between parallel plates. *Applied Scientific Research, Section A, 4*(5–6), 405–413.
256. Schlichting, H. (1982). *Grenzschicht-Theorie.* In G. Braun (ed), Karlsruhe [Schlichting, H. (1968). *Boundary layer theory* (6th ed.). New York: McGraw-Hill].
257. Schlünder, E. U., & Martin, H. (1995). *Einführung in die Wärmeübertragung,* 8., neubearbeitete Auflage, Vieweg, Braunschweig-Wiesbaden.
258. Schmidt, Th. E. (April, 1949). Heat transfer calculations for extended surfaces. *Journal of ASHRAE, Refrigerating Engineering.* 351–357.
259. Schmidt, Th. E. (1950). *Die Wärmeleistung von berippten Oberflächen,* Abh. Deutsch. Kältetechn. Verein., Nr. 4, C.F. Müller, Karlsruhe.
260. Schroeder, D. W. (2001). *A tutorial on pipe flow equations.* Carlisle, Pennsylvania: Stoner Associates Inc.
261. Seber, G. A. F., & Wild, C. J. (1989). *Nonlinear regression.* New York: Wiley.
262. Seborg, D. F., Edgar, T. F., Mellichamp, D. A., & Doyle, F. J., III. (2011). *Process dynamics and control* (3rd ed.). Hoboken: Wiley.
263. Segerlind, L. J. (1984). *Applied finite element analysis* (2nd ed.). New York: Wiley.
264. Sellars, J., Tribus, M., Klein, J. (1954). *Heat transfer to laminar flow in a round tube or flat conduit. The Graetz problem extended, Wright Air Development Center.* Technical report 54-255.
265. Sellars, R. J., Tribus, M., & Klein, J. S. (1956). Heat transfer to laminar flow in a round tube or flat conduit—Graetz problem extended. *Transactions of the ASME, 78,* 441–448.
266. Серов, Е. П., Корольков, Б. П. (1981). *Динамика парогенераторов,* Энергоиздат, Москва [Serov. E. P., & Korolkov, B. P. (1981). *Dynamics of steam generators.* Moscow: Energoizdat].
267. Serth, R. W. (2007). *Process heat transfer, Principles and applications.* Amsterdam: Academic Press.
268. Shah, R. K., & Bhatti, M. S. (1987). Laminar convective heat transfer in ducts. In Rozdział 3 w Kakaç S., & Shah, R.K., Aung, W. (eds.), *Handbook of single-phase convective heat transfer.* New York: Wiley.
269. Shah, R. K., & London, A. L. (1978). *Laminar flow forced convection in ducts, Suplement 1 w advances in heat transfer.* New York: Academic Press.
270. Shah, R. K., & Sekulić, D. P. (2003). *Fundamentals of heat exchanger design.* Hoboken: Wiley.
271. Sheriff, N., & O'Kane, D. J. (1981). Sodium eddy diffusivity of heat measurements in a circular duct. *International Journal of Heat Mass Transfer, 24,* 205–211.

272. Sieder, E. N., & Tate, G. E. (1936). Heat transfer and pressure drop of liquids in tubes. *Industrial and Engineering Chemistry, 28*, 1429–1436.

273. Siegel, R., Sparrow, E. M., & Hallman T. M. (1958). Steady laminar heat transfer in a circular tube with prescribed wall heat flux. *Applied Scientific Research, Section A, 7*(5), 386–392.

274. Sinalski, E. G. (2011). *Hydromechanics, theory and fundamentals.* Weinheim: Wiley VCH.

275. Skelland, A. H. P. (1974). *Diffusional mass transfer.* London: Krieger.

276. Składzień, J. (1974). *Zależność bezwymiarowej temperatury podgrzania od bezwymiarowej powierzchni grzejnej dla najczęściej stosowanych typów rekuperatorów konwekcyjnych,* Hutnik, Nr 10, pp. 507–512. [Składzień, J. (1974). *Dimensionless fluid temperature increase as a function of heating surface area for the most used types of recuperators.* Hutnik (vol. 10, pp. 507–512).

277. Skupinski, E., Tortel, J., & Vautrey, L. (1965). Détermination de coefficients de convection d'un alliage sodium-potassium dans une tube circulaire. *International Journal of Heat and Mass Transfer, 8*, 937–951.

278. Slattery, J. C. (1999). *Advanced transport phenomena.* Cambridge: Cambridge University Press.

279. Sleicher, C. A., & Rouse, M. W. (1975). A convenient correlation for heat transfer to constant and variable property fluids in turbulent pipe flow. *International Journal Heat Mass Transfer, 18*, 677–683.

280. Smith, C. L. (2009). *Practical process control.* Hoboken: Wiley.

281. Smith, R. E., Jr., & Wehofer, S. (1985). From measurement uncertainty to measurement communications, Credibility and cost control in propulsion ground test facilities. *ASME Journal of Fluids Engineering, 107*, 165–172.

282. Sobota, T. (2011). *Eksperymentalne wyznaczanie średnich współczynników wnikania ciepła,* Rozdział 3.7 w Taler J. (ed.), *Procesy cieplne i przepływowe w dużych kotłach energetycznych,* Wydawnictwo Naukowe. Warszawa: PWN, 351–366 [Sobota T. (2011). Experimental determining mean heat transfer coefficients (Chapter 3.7). In Taler J. (ed.), *Thermal and flow processes in large steam boilers.* Warsaw: Publishing House PWN, 351–366].

283. Sapali, S. N., & Patil, P. A. (2010). Heat transfer during condensation of HFC-134a and R-404A inside of a horizontal smooth and micro-fin tube. *Experimental Thermal and Fluid Science, 34*, 1133–1141.

284. Spang, B. (1996). Einfluss der thermischen Randbedingungen auf den laminaren Wärmeübergang im Kreisrohr bei hydrodynamischer Einlaufströmung. *Heat and Mass Transfer, 31*, 199–204.

285. Sparrow, E. M., & Lin, S. H. (1964). Heat-transfer characteristics of polygonal and plate fins. *International Journal of Heat and Mass Transfer, 7*, 951–953.

286. Sparrow, E. M., & Ohadi, M. M. (1977). Numerical and experimental studies of turbulent heat transfer in a tube. *Numerical Heat Transfer, 11*, 461–476.

287. Speziale, C. G. (1991). Analytical methods for the development of Reynolds-stress closures in turbulence. *Annual Review of Fluid Mechanics, 23*, 107–157.

288. Stoer, J., & Bulirsch, R. (1987). *Wstęp do analizy numerycznej,* Wydawnictwo Naukowe PWN, Warszawa [Stoer, J., & Bulirsch, R. (1987). *Introduction to numerical analysis.* Warsaw: Publishing House PWN].

289. Stultz, S. C., & Kitto, J. B. (Eds.). (1992). *Steam/Its generation and use.* Barberton, USA: The Babcock & Wilcox Company.

290. Sung, S. W., Lee, J., & Lee, I. (2009). *Process identification and PID control.* Singapore: Wiley.

291. Swamee, P. K., & Jain, A. K. (1976). Explicit equations for pipe-flow problems. *Journal of the Hydraulics Division, 102*(HY5), 657–664.

292. Szydłowski, H. (1994). *Pracownia fizyczna.* Warszawa: Wydawnictwo Naukowe PWN [Szydłowski, H. (1994). *Physics laboratory.* Warsaw: Publishing House PWN].

293. *Table Curve. Automated curve fitting software*, Jandel Scientific, San Rafael 2014, CA94901.
294. Taborek, J. (1994). Design method for tube-side laminar and transition flow regime with effects of natural convection. Paper OPF-11-21, 9th International Heat Transfer Conference. Jerusalem.
295. Taitel, Y., & Dukler, A. E. (1976). A model for predicting flow regime transition in horizontal and near horizontal gas-liquid flow. *AIChE J, 22,* 47–55.
296. Taler, D. (2002). *Theoretical and experimental analysis of heat exchangers with extended surfaces*, Polska Akademia Nauk Oddział w Krakowie, Teka Komisji Naukowo-Problemowej Motoryzacji, Kraków, Volume 25, Monograph 3.
297. Taler, D. (2004). Determination of heat transfer correlations for plate-fin-and-tube heat exchangers. *Heat and Mass Transfer, 40,* 809–822.
298. Taler, D. (2004). Experimental determination of heat transfer and friction correlations for plate fin-and-tube heat exchangers. *Journal of Enhanced Heat Transfer, 11*(3), 183–204.
299. Taler, D. (2005). Prediction of heat transfer correlations for compact heat exchangers. *Forschung im Ingenieurwesen (Engineering Research)* 69, 137–150.
300. Taler, D. (2006). *Pomiar ciśnienia, prędkości i strumienia przepływu płynu*, Uczelniane Wydawnictwa Naukowo-Dydaktyczne, Kraków (Taler, D. (2006). *Measurements of pressure, velocity, and mass flow rate.* Cracow: Publishing House of AGH.
301. Taler, D. (2006). Dynamic response of a cross-flow tube heat exchanger. *Chemical and Process Engineering, 27,* 1053–1067.
302. Taler, D. (2007). Experimental and numerical predictions of the heat transfer correlations in the cross-flow plate fin and tube heat exchangers. *Archives of Thermodynamics, 28*(2), 3–18.
303. Taler, D. (2007). Analytical and numerical model of transient heat transfer in a single tube row heat exchanger. *Archives of Thermodynamics, 28*(1), 51–64.
304. Taler, D. (2008). Effect of thermal contact resistance on the heat transfer in plate finned tube heat exchangers. In praca w Müller-Steinhagen H., Reza Malayeri M., & Watkinson P. (eds.), *ECI Symposium Series, Volume RP5: Proceedings of 7th International Conference on Heat Exchanger Fouling and Cleaning—Challenges and Opportunities*, Engineering Conferences International, Tomar, Portugal, July 1–6, 2007 (pp. 362–371). Berkeley: Berkeley Electronic Press. Available in electronic form. http://services.bepress.com/eci/heatexchanger2007/47.
305. Taler, D. (2009). Control of the rate of heat flow in a compact heat exchanger by changing the speed of fan rotation. *Archives of Thermodynamics, 30*(4), 67–80.
306. Taler, D. (2009). *Dynamika rurowych wymienników ciepła*, Rozprawa habilitacyjna, Rozprawy, habilitacje – nr 193, Uczelniane Wydawnictwa Naukowo-Dydaktyczne, Kraków [Taler, D. *Dynamics of tubular heat exchangers*. Habilitation Thesis, Monograph no. 193. Cracow: Publishing House of AGH].
307. Taler, D. (2011). Direct and inverse heat transfer problems in dynamics of plate fin and tube heat exchangers (Chapter 3). In A. Belmiloudi (Ed.), *Heat transfer, mathematical modelling, numerical methods and information technology* (pp. 77–100). Rijeka: InTech.
308. Taler, D. (2011). *Kotłowe wymienniki ciepła.* In Rozdział 3.5 w książce Taler J. (ed.), *Procesy cieplne i przepływowe w dużych kotłach energetycznych. Modelowanie i monitoring.* Warszawa: Wydawnictwo Naukowe PWN, pp. 273–308 [Taler, D. (2011). Boiler heat exchangers (Chapter 3.5). In Taler J. (ed.), *Thermal and flow processes in large steam boilers. Modeling and monitoring* (pp. 273–308). Warsaw: Publishing House PWN].
309. Taler, D. (2013). Experimental determination of correlations for average heat transfer coefficients in heat exchangers on both fluid sides. *Heat and Mass Transfer, 49*(8), 1125–1139.
310. Taler, D. (2018). Mathematical modeling and experimental study of heat transfer in a low-duty air-cooled heat exchanger. *Energy Conversion and Management, 159,* 232–243.

311. Taler, D. (2014). Fins of rectangular and hexagonal geometry. In R. B. Hetnarski (Ed.), *Encyclopedia of thermal stresses* (pp. 1658–1670). Dordrecht Heidelberg, New York, London: Springer.

312. Taler, D. (2014). Fins of straight and circular geometry. In R. B. Hetnarski (Ed.), *Encyclopedia of thermal stresses* (pp. 1670–1683). Dordrecht Heidelberg, New York, London: Springer.

313. Taler, D. (2015). Mathematical modeling and control of plate fin and tube heat exchangers. *Energy Conversion and Management, 96*, 452–462.

314. Taler, D., & Cebula, A. (2009). Modelling of air flow and heat transfer in compact heat exchangers. *Archives of Thermodynamics, 30*(4), 45–66.

315. Taler, D., & Cebula, A. (2010). A new method for determination of thermal contact resistance of a fin-to-tube attachment in plate fin-and-tube heat exchangers. *Chemical and Process Engineering, 31*, 839–855.

316. Taler, D., & Korzeń, A. (2011*)*. Modelowanie wymiany ciepła w żebrach o złożonych kształtach. *Rynek Energii, 97,* 61–65 [Taler, D., & Korzeń, A. (2011). Modeling of heat transfer in fins of complicated shapes. *Rynek Energii, 97,* 61–65].

317. Taler, D., & Korzeń, A., & Madejski, P. (). Wyznaczanie temperatury rur w grodziowym przegrzewaczu pary w kotle fluidalnym. *Rynek Energii, 93,* 56–60 [Taler, D., Korzeń, A., & Madejski P. (). *Determination of tube temperature in platen steam superheaters in a fluidized bed boiler. Rynek Energii, 93,* 56–60].

318. Taler, D., & Ocłoń, P. (2014). Determination of heat transfer formulas for gas flow in fin-and-tube heat exchanger with oval tubes using CFD simulations. *Chemical Engineering and Processing, 83,* 1–11.

319. Taler, D., & Ocłoń, P. (2014). Thermal contact resistance in plate fin-and-tube heat exchangers determined by experimental data and CFD simulations. *International Journal of Thermal Sciences, 84,* 309–322.

320. Taler, D., & Sury, A. (2011). Inverse heat transfer problem in digital temperature control in plate fin and tube heat exchangers. *Archives of Thermodynamics, 32*(4), 17–32.

321. Taler, D. (2013). Experimental determination of correlations for average heat transfer coefficients in heat exchangers on both fluid sides. *Heat and Mass Transfer, 49*(8), 1125–1139.

322. Taler, D., & Sury, A. (2013). *Modele matematyczne rurowych krzyżowo-prądowych wymienników ciepła i ich zastosowanie do regulacji temperatury wylotowej czynnika,* Wydawnictwo Politechniki Krakowskiej, seria Mechanika, Monografia 437, Kraków [Taler, D., & Sury, A. (2013). *Mathematical models of tubular cross-flow heat exchangers and their application for control of fluid temperature at the heat exchanger outlet.* Cracow: Publishing House of Cracow University of Technology, Mechanical Engineering Series, Monograph no. 437].

323. Taler, D., & Taler, J. (2014). Steady-state and transient heat transfer through fins of complex geometry. *Archives of Thermodynamics, 35*(2), 117–133.

324. Taler, D. (2016). Determining velocity and friction factor for turbulent flow in smooth tubes. *International Journal of Thermal Sciences, 105,* 109–122.

325. Taler, D. (2016). A new heat transfer correlation for transition and turbulent fluid flow in tubes. *International Journal of Thermal Sciences, 108,* 108–122.

326. Taler, D. (2017). Simple power-type heat transfer correlations for turbulent pipe flow in tubes. *Journal of Thermal Science, 26*(4), 339–348.

327. Taler, D. (2018). Semi-empirical heat transfer correlations for turbulent tube flow of liquid metals. *International Journal of Numerical Methods for Heat and Fluid Flow, 28*(1), 151–172.

328. Taler, D., & Korzeń, A. (2018). Numerical modeling transient response of tubular cross flow heat exchanger. *International Journal of Numerical Methods for Heat and Fluid Flow, 28*(1), 81–91.

329. Taler, J. (1995). *Teoria i praktyka identyfikacji procesów przepływu ciepła*. Wrocław: Ossolineum [Taler, J. (1995). *Theory and practice of identification of heat flow processes*. Wrocław: Publishing House Ossolineum].

330. Taler, J., & Duda, P. (1999). A space marching method for multidimensional transient inverse heat conduction problems. *Heat and Mass Transfer, 34*, 349–356.

331. Taler, J., & Duda, P. (2006). *Solving direct and inverse heat conduction problems*. Berlin: Springer.

332. Taler, J., Taler, D., Sobota, T., & Cebula, A. (2012). Theoretical and experimental study of flow and heat transfer in a tube bank. In Rozdział 1 w książce Petrova V. M. (ed.), *Advances in engineering research* (vol. 1, pp. 1–56). New York: Nova Science Publishers.

333. Tam, L. M., & Ghajar, A. J. (1997). Effect of inlet geometry and heating on the fully developed Friction factor in the transition region of a horizontal tube. *Experimental Thermal and Fluid Science, 15*, 52–64.

334. Tam, L. M., & Ghajar, A. J. (2006). Transitional heat transfer in plain horizontal tubes. *Heat Transfer Engineering, 27*(5), 23–38.

335. Tam, L. M., & Ghajar, A. J. (2008). Contribution analysis of dimensionless variables for laminar and turbulent flow convection heat transfer in a horizontal tube using artificial neural network. *Heat Transfer Engineering, 29*(9), 793–804.

336. Tavoularis, S. (2009). *Measurement in fluid mechanics*. New York: Cambridge University Press.

337. Taylor, G. I. (1960). *Scientific papers*. In Batchelor G. K. (Ed.), *Volume II: Meteorology, oceanography and turbulent flow*. Cambridge: Cambridge University Press.

338. *Teoria pomiarów*, praca pod redakcją Szydłowski H., Wydawnictwo Naukowe PWN, Warszawa 1981 [*Theory of measurements*, in Szydłowski H., Editor, Publishing House Wydawnictwo Naukowe PWN, Warsaw 1981].

339. Thakre, S. S., & Joshi, J. B. (2002). Momentum, mass and heat transfer in single-phase turbulent flow. *Reviews in Chemical Engineering, 18*, 283–293.

340. The handbook of fluid dynamics. (1998). In praca pod redakcją Johnson R. W (ed.). Boca Raton, Heidelberg: CRC Press, Springer.

341. *Thermally developing laminar flow*, z Thermal-FluidsPedia. https://www.thermalfluidscentral.org/encyclopedia/index.php.

342. Thomson, W. J. (1997). *Atlas for computing mathematical functions*. New York: Wiley Interscience.

343. TORBAR. Flowmeters Ltd, Unit 21, Star Road, Partridge Green West Sussex RH 138RA, United Kingdom.

344. Tucker, A. S. (1996). The LMTD correction factor for single-pass cross-flow heat exchangers with both fluids unmixed. *Transactions of the ASME, Journal of Heat transfer, 118*, 488–490.

345. Tucker, P. G. (2001). *Computation of unsteady internal flows*. Dordrecht: Kluwer Academic Publishers.

346. Van Wylen, G., Sonntag, R., & Borgnakke, C. (2008). *Fundamentals of classical thermodynamics* (4th ed.). Hoboken: Wiley.

347. *VDI-Wärmeatlas, 11.*, bearbeitete und erweiterte Auflage, Springer Vieweg, Berlin, Heidelberg 2013.

348. Venkateshan, S. P. (2008). *Mechanical measurements*. Boca Raton: CRC Press Taylor & Francis Group.

349. Versteeg, H. K., & Malalasekera, W. (1995). *An introduction to computational fluid dynamics. The finite volume method*. England, Harlow: Addison Wesley Longman Limited.

350. Visioli, A. (2006). *Practical PID control*. London: Springer.

351. Wang, C. C., Webb, R. L., & Chi, K. Y. (2000). Data reduction for air-side performance of fin-and-tube heat exchangers. *Experimental Thermal and Fluid Science, 21*, 218–226.

352. Webb, R. L. (1971). A critical evaluation of analytical solutions and Reynolds analogy equations for turbulent heat and mass transfer in smooth tubes. *Wärme und Stoffübertragung, 4*, 197–204.

353. Webb, R. L. (1994). *Principles of enhanced heat transfer.* New York: Wiley.

354. Weigand, B. (2004). *Analytical methods for heat transfer and fluid flow problems.* Berlin: Springer.

355. Weigand, B., Ferguson, J. R., & Crawford, M. E. (1997). An extended Kays and Crawford turbulent Prandtl number model. *International Journal of Heat and Mass Transfer, 40*(17), 4191–4196.

356. Welfe, A. (2003). *Ekonometria,* Wydanie trzecie zmienione, Polskie Wydawnictwo Ekonomiczne, Warszawa [Welfe, A. (2003). *Econometrics* (3rd ed.). Warsaw: Polish Publishing House for Economics].

357. Welty, J. R., Wicks, C. E., Wilson, R. E., & Rorrer, G. L. (2008). *Fundamentals of momentum, heat, and mass transfer* (5th ed.). Hoboken: Wiley.

358. Welty, R. W. (1974). *Engineering heat transfer.* New York: Wiley.

359. Wendt, J. F. (Ed.). (2009). *Computational fluid dynamics, An introduction* (3rd ed.). Berlin, Heidelberg: Springer.

360. Wesley, D. (1976). *Heat transfer in pipe downstream of a tee, (Ph.D. Thesis).* Minneapolis: University of Minnesota.

361. White, F. M. (1991). *Heat and mass transfer.* Reading: Addison Wesley.

362. Wilcox, D. C. (1988). Reassessment of scale—determining equation for advanced turbulence models. *AIAA Journal, 26*, 1299–1310.

363. Wilcox, D. C. (1993). Comparison of two-equation turbulence models for boundary layers with pressure gradient. *AIAA Journal, 31*, 1414–1421.

364. Wilcox, D. C. (1994). Simulation of transition with a two-equation turbulence model. *AIAA Journal, 26*, 247–255.

365. Wilson, E. E. (1915). A basis for rational design of heat transfer apparatus. *Transactions of the ASME, 37*, 47–82.

366. Wilson, N. W., & Azad, R. S. (1975). A continuous prediction method for fully developed laminar transitional and turbulent flows in pipes. *Journal of Applied Mechanics, 42*, 51–54.

367. Wiśniewski, S., & Wiśniewski, T. S. (1994). *Wymiana ciepła.* Wydanie trzecie zmienione, WNT, Warszawa [Wiśniewski, S., & Wiśniewski, T. S. (1994). *Heat transfer* (3rd ed.). Warsaw: Publishing House WNT].

368. Worsøe-Schmidt, P. M. (1967). Heat transfer in the thermal entrance region of circular tubes and annular passages with fully developed laminar flow. *International Journal of Heat and Mass Transfer, 10*, 541–551.

369. Wright, T. (1999). *Fluid machinery performance, analysis, and design.* Boca Raton: CRC Press.

370. Xiao-wei, L., Ji-an, M., & Zhi-xin, L. (2011). Roughness enhanced mechanism for turbulent convective heat transfer. *International Journal of Heat and Mass Transfer, 54*, 1775–1781.

371. Yildrim, G. (2009). Computer-based analysis of explicit approximations to the implicit Colebrook-White equation in turbulent flow friction factor calculation. *Advances in Engineering Software, 40*, 1183–1190.

372. Yu, C. C. (2006). *Autotuning of PID controllers.* London: Springer.

373. Zamzamian, A., Oskouie, S. N., Doosthoseini, A., Joneidi, A., & Pazouki, M. (2011). Experimental investigation of forced convective heat transfer coefficient in nanofluids of Al2O3/EG and CuO/EG in a double pipe and plate heat exchangers under turbulent flow. *Experimental Thermal and Fluid Science, 35*, 495–502.

374. Zdaniuk, G. J., Chamra, L. M., & Mago, P. J. (2008). Experimental determination of heat transfer and friction in helically-finned tubes. *Experimental Thermal and Fluid Science, 32*, 761–775.

375. Zeliaś, A., & Pawełek, B., & Wanat, S. (2004). *Prognozowanie ekonomiczne. Teoria, przykłady, zadania*, Wydawnictwo Naukowe PWN, Warszawa [Zeliaś, A., Pawełek, B., & Wanat, S. (2004). *Prediction in economics. Theory, examples, and excercises*. Warsaw: Publishing House PWN].

376. Zeller, M., & Grewe, M. (1994). Verallgemeinerte Näherungsgleichung für den Wirkungsgrad von Rippen auf kreisförmigen und elliptischen Kernrohren. *Wärme-und Stoffübertragung, 29,* 379–382.

377. Zhang, Z., Yang, W., Guan, C., Ding, Y., Li, F., & Yan, H. (2012). Heat transfer and friction characteristics of turbulent flow through plain tube inserted with rotor-assembled strands. *Experimental Thermal and Fluid Science, 38,* 33–39.

CPSIA information can be obtained
at www.ICGtesting.com
Printed in the USA
LVHW05*2155210618
581517LV00001B/14/P